Deep Learning for the Earth Sciences

Deep Learning for the Earth Sciences

A Comprehensive Approach to Remote Sensing, Climate Science, and Geosciences

Edited by

Gustau Camps-Valls
Universitat de València, Spain

Devis Tuia
EPFL, Switzerland

Xiao Xiang Zhu
German Aerospace Center and Technical University of Munich, Germany

Markus Reichstein
Max Planck Institute, Germany

WILEY

This edition first published 2021

Registered Offices
John Wiley & Sons, Inc., 111 River Street, Hoboken, NJ 07030, USA
John Wiley & Sons Ltd, The Atrium, Southern Gate, Chichester, West Sussex, PO19 8SQ, UK

Editorial Office
The Atrium, Southern Gate, Chichester, West Sussex, PO19 8SQ, UK

For details of our global editorial offices, customer services, and more information about Wiley products visit us at www.wiley.com.

Wiley also publishes its books in a variety of electronic formats and by print-on-demand. Some content that appears in standard print versions of this book may not be available in other formats.

Library of Congress Cataloging-in-Publication Data

Name: Camps-Valls, Gustau, editor.
Title: Deep learning for the earth sciences : a comprehensive approach to remote sensing, climate science and geosciences / edited by Gustau Camps-Valls [and three others].
Description: Hoboken, NJ : Wiley, 2021. | Includes bibliographical references and index.
Identifiers: LCCN 2021012965 (print) | LCCN 2021012966 (ebook) | ISBN 9781119646143 (cloth) | ISBN 9781119646150 (adobe pdf) | ISBN 9781119646167 (epub)
Subjects: LCSH: Earth sciences–Study and teaching. | Algorithms–Study and teaching.
Classification: LCC QE26.3 .D44 2021 (print) | LCC QE26.3 (ebook) | DDC 550.71–dc23
LC record available at https://lccn.loc.gov/2021012965
LC ebook record available at https://lccn.loc.gov/2021012966

Cover Design: Wiley
Cover Image: © iStock.com/monsitj, Emilia Szymanek/Getty Images

MIX
Paper from responsible sources
FSC® C013604

To Adrian Albert, in memoriam

Contents

Foreword

Earth science, like many other scientific disciplines, is undergoing a data revolution. In particular, a massive amount of data about Earth and its environment is now continuously being generated by Earth observing satellites as well as physics-based earth system models running on large-scale computational platforms. These information-rich datasets offer huge potential for understanding how the Earth's climate and ecosystem have been changing, and for addressing societal grand challenges relating to food/water/energy security and climate change.

Deep learning, which has already revolutionized many disciplines (e.g., computer vision, natural language processing) holds tremendous promise to revolutionize earth and environmental sciences. In fact, recent years have seen an exponential growth in the use of deep learning in Earth Science, with many amazing results. Deep learning also faces challenges that are unique to earth science data: multimodality; high degree of heterogeneity in space and time; and the fact that earth science data can only provide an incomplete and noisy view of the underlying eco-geo-physical processes that are interacting and unfolding at different spatial and temporal scales. Addressing these challenges requires development of entirely new approaches that can effectively incorporate existing earth science knowledge inside the deep learning learning framework. Success in addressing these challenges stands to revolutionize deep learning itself and accelerate discovery across many other scientific domains.

The book does a fantastic job of capturing the state of the art in this fast evolving area. It is logically organized in to 3 coherent parts, each containing chapters written by experts in the field. Each chapter provides an easily to understand introductory material followed by an in-depth treatment of the applications of deep learning to specific earth science applications as well as ideas for future research. This book is a must read for the students and researchers alike who would like to harness the data revolution in earth sciences to address pressing societal challenges.

Vipin Kumar
Regents Professor, Department of Computer Science &
Engineering, University of Minnesota, USA

Acknowledgments

We would like to acknowledge the help of all involved in the collation and review process of the book, without whose support the project could not have been satisfactorily completed. A further special note of thanks goes also to all the staff at Wiley, whose contributions throughout the whole process, from inception of the initial idea to final publication, have been valuable. Special thanks also go to the publishing team at Wiley, who continuously prodded via e-mail, keeping the project on schedule.

We wish to thank all of the authors for their insights and excellent contributions to this book. Most of the authors of chapters included in this book also served as referees for chapters written by other authors. Thanks go to all those who provided constructive and comprehensive reviews.

This book was possible without any dedicated funding, but editors' and authors' research was partially supported by research projects that made it possible. We want to thank all agencies and organizations for supporting our research in general, and this book indirectly. Gustau Camps-Valls acknowledges support by the European Research Council (ERC) under the ERC-CoG-2014 project 647423.

Thanks all!

Gustau Camps-Valls, Devis Tuia, Xiao Xiang Zhu, Markus Reichstein
València+Sion+Munich+Jena, August, 2021

List of Contributors

Adriana Romero
School of Computer Science,
McGill University, Canada

Ankur Mahesh
Lawrence Berkeley National Lab
UC Berkeley
USA

Basil Kraft
Max Planck Institute for Biogeochemistry
Jena & Technical University of Munich,
Germany

Begüm Demir
Faculty of Electrical Engineering and
Computer Science
Technische Universität Berlin
Germany

Benjamin Kellenberger
Wageningen University and Research
The Netherlands

Bertrand Le Saux
ESA / ESRIN Φ-lab
Italy

Bharath Bhushan Damodaran
IRISA-OBELIX Team
France

Burlen Loring
Lawrence Berkeley National Lab
UC Berkeley
USA

Carlo Gatta
Vintra Inc.
Barcelona
Spain

Chaopeng Shen
Civil and Environmental Engineering
Pennsylvania State University
University Park
USA

Christian Reimers
German Aerospace Center (DLR) &
Friedrich-Schiller-Universität
Jena
Germany

Christian Requena-Mesa
German Aerospace Center (DLR) &
Max-Planck Institute for Biogeochemistry
& Friedrich-Schiller-Universität
Jena
Germany

Christopher Beckham
MILA / Polytechnique Montreal
Montreal, Canada

Christopher Pal
Polytechnique Montreal
Canada

Danfeng Hong
German Aerospace Center
Germany

David Malmgren-Hansen
Department of Applied Mathematics and Computer Science
Technical University of Denmark
Kgs. Lyngby
Denmark

Devis Tuia
EPFL
Switzerland

Diego Marcos
Wageningen University and Research
The Netherlands

Dit-Yan Yeung
CSE Department
HKUST
Hong Kong

Evan Racah
Lawrence Berkeley National Lab
UC Berkeley
USA

Gencer Sumbul
Faculty of Electrical Engineering and Computer Science
Technische Universität Berlin
Germany

Giuseppe Scarpa
University of Naples Federico II
Italy

Gonzalo Mateo-García
Image Processing Laboratory
Universitat de València
Spain

Gui-Song Xia
State Key Lab. LIESMARS
and School of Computer Science
Wuhan University
China

Gustau Camps-Valls
Image Processing Laboratory
Universitat de València
Spain

Hao Wang
Department of Computer Science
Rutgers University
USA

Jakob Runge
German Aerospace Center (DLR)
Jena
Germany

Javier García-Haro
Environmental Remote Sensing group (UV-ERS)
Universitat de València
Spain

Jian Ding
State Key Lab. LIESMARS
Wuhan University
China

Jian Kang
Faculty of Electrical Engineering and Computer Science
Technische Universität Berlin
Germany

Jim Biard
Lawrence Berkeley National Lab
UC Berkeley
USA

Jinwang Wang
School of Electronic Information
Wuhan University
China

Jose E. Adsuara
Image Processing Laboratory
Universitat de València
Spain

Karthik Kashinath
Lawrence Berkeley National Lab
UC Berkeley
USA

Kathryn Lawson
Civil and Environmental Engineering
Pennsylvania State University
USA

Kenneth E. Kunkel
North Carolina State University (NCSU)
US

Konrad Schindler
ETH Zurich
Switzerland

Laure Zanna
New York University
USA

Lin Liu
Earth System Science Programme
Faculty of Science
The Chinese University of Hong Kong
Hong Kong SAR
China

Luis Gómez-Chova
Image Processing Laboratory
Universitat de València
Spain

Manuel Campos-Taberner
Environmental Remote Sensing Group (UV-ERS)
Universitat de València
Spain

Marc Russwurm
Technical University of Munich
Germany

Marco Körner
Technical University of Munich
Germany

Maria Vakalopoulou
CentraleSupelec
University Paris Saclay
Inria Saclay
France

Markus Reichstein
Max-Planck Institute for Biogeochemistry
Jena
Germany

Mayur Mudigonda
Lawrence Berkeley National Lab
UC Berkeley
USA

Michael F. Wehner
Lawrence Berkeley National Lab
UC Berkeley
USA

Mihir Sahasrabudhe
CentraleSupelec
Universite Paris Saclay
Inria Saclay
France

Naoto Yokoya
The University of Tokyo and RIKEN Center for Advanced Intelligence Project
Japan

Nicolas Courty
Université de Bretagne Sud
Laboratoire IRISA
France

Nikos Paragios
CentraleSupelec
Universite Paris Saclay
Inria Saclay
France

Onur Tasar
Inria Sophia Antipolis
France

Peter A. G. Watson
School of Geographical Sciences
University of Bristol
UK

Peter Bauer
European Centre for Medium Range Weather Forecasts (ECMWF)
Reading
UK

Peter D. Dueben
European Centre for Medium Range Weather Forecasts (ECMWF)
Reading
UK

Pierre Gentine
Columbia University
USA

Prabhat Ram
Lawrence Berkeley National Lab
UC Berkeley
USA

Ribana Roscher
Institute of Geodesy and Geoinformation
University of Bonn
Germany

Samantha Adams
Met Office Informatics Lab
Exeter
UK

Samira Kahou
École de technologie supérieure
Montreal
Quebec
Canada

Simon Besnard
Max Planck Institute for Biogeochemistry
Jena Germany
Laboratory of Geo-Information Science and Remote Sensing
Wageningen University & Research
The Netherlands

Sookyung Kim
Lawrence Berkeley National Lab
UC Berkeley
USA

Stergios Christodoulidis
Institut Gustave Roussy
Paris
France

Sujan Koirala
Max Planck Institute for Biogeochemistry
Jena
Germany

Tatsumi Uezato
RIKEN Center for Advanced Intelligence Project
Japan

Tegan Maharaj
Montreal Polytechnique & Mila,
Montreal, Canada

Thomas Bolton
University of Oxford
UK

Thorsten Kurth
Lawrence Berkeley National Lab
UC Berkeley
USA

Tom Beucler
Columbia University & University of California
Irvine
USA

Travis O'Brien
Lawrence Berkeley National Lab
UC Berkeley
USA

Valero Laparra
Image Processing Laboratory
Universitat de València
Spain

Veronika Eyring
German Aerospace Center (DLR) and University of Bremen
Germany

Wai-Kin Wong
Hong Kong Observatory

Wang-chun Woo
Hong Kong Observatory

Wei He
RIKEN Center for Advanced Intelligence Project
Japan

Wen Yang
School of Electronic Information
Wuhan University
China

William D. Collins
Lawrence Berkeley National Lab
UC Berkeley
USA

Xavier-Andoni Tibau
German Aerospace Center (DLR)
Jena
Germany

Xiao Xiang Zhu
Technical University of Munich and
German Aerospace Center (DLR)
Munich
Germany

Xingjian Shi
Amazon
USA

Yunjie Liu
(Formerly) Lawrence Berkeley National Lab
UC Berkeley
US

Zhihan Gao
Hong Kong University of Science and
Technology
Hong Kong

List of Acronyms

AE	Autoencoder
AI	Artificial Intelligence
AIC	Akaike's Information Criterion
AP	Access Point
AR	Autoregressive
ARMA	Autoregressive and Moving Average
ARX	Autoregressive eXogenous
AWGN	Additive white Gaussian noise
BCE	Binary Cross-Entropy
BER	Bit Error Rate
BP	Back-propagation
BPTT	Back-propagation through Time
BRT	Bootstrap Resampling Techniques
BSS	Blind Source Separation
CAE	Contractive Autoencoder
CBIR	Content-based Image Retrieval
CCA	Canonical Correlation Analysis
CCE	Categorical Cross-Entropy
CGAN	Conditional Generative Adversarial Network
CNN	Convolutional Neural Network
CONUS	Conterminous United States
CPC	Contrastive Predicting Coding
CSVM	Complex Support Vector Machine
CV	Cross Validation
CWT	Continuous Wavelet Transform
DAE	Denoising Autoencoder
DCT	Discrete Cosine Transform
DFT	Discrete Fourier Transform
DL	Deep Learning
DNN	Deep Neural Network
DSM	Dual Signal Model
DSP	Digital Signal Processing
DSTL	Deep Self-taught Learning

DWT	Discrete Wavelet transform
ELBO	Evidence Lower Bound
EM	Expectation–Maximization
EO	Earth Observation
EPLS	Enforcing Population and Lifetime Sparsity
ERM	Empirical Risk Minimization
ET	Evapotranspiration
EUMETSAT	European Organisation for the Exploitation of Meteorological Satellites
FC	Fully Connected
FFT	Fast Fourier Transform
FIR	Finite Impulse Response
FT	Fourier Transform
GAE	Generalized Autoencoder
GAN	Generative Adversarial Network
GCM	General Circulation Model
GM	Gaussian Mixture
GP	Gaussian Process
GPR	Gaussian Process Regression
GRNN	Generalized Regression Neural Network
GRU	Gated Recurrent Unit
HMM	Hidden Markov Model
HP	Hyper-parameter
HRCN	High Reliability Communications Networks
HSIC	Hilbert-Schmidt Independence Criterion
i.i.d.	Independent and Identically Distributed
IASI	Infrared Atmospheric Sounding Interferometer
ICA	Independent Component Analysis
IIR	Infinite Impulse Response
KF	Kalman Filter
KKT	Karush–Kuhn–Tucker
KM	Kernel Method
KPCA	Kernel Principal Component Analysis
KRR	Kernel Ridge Regression
LAI	Leaf Area Index
LASSO	Least Absolute Shrinkage and Selection Operator
LCC	Leaf-Chlorophyll-Content
LE	Laplacian eigenmaps
LiDAR	Light Detection and Ranging of Laser Imaging Detection and Ranging
LLE	Locally Linear Embedding
LMS	Least Mean Squares
LS	Least Squares
LSTM	Long-Short-Term-Memory
LTSA	Local Tangent Space Alignment
LUT	Look-up Tables
MAE	Mean Absolute Error

MDN	Mixture Density Network
ME	Mean Error
MGU	Minimal Gated Unit (MGU)
ML	Maximum Likelihood
MLP	Multilayer Perceptron
MNF	Minimum Noise Fractions
MSE	Mean Square Error
NDVI	Normalized-Vegetation-Difference-Index
NMR	Nuclear Magnetic Resonance
NN	Neural Networks
NOAA	National Oceanic and Atmospheric Administration
NSE	Nash-Sutcliffe model efficiency coefficient
NWP	Numerical Weather Prediction
OAA	One Against All
OAO	One Against One
OLS	Ordinary Least Square
OMP-k	Orthogonal Matching Pursuit
PAML	Physics-aware Machine Learning
PCA	Principal Component Analysis
PINN	Physics-informed Neural Network
PSD	Predictive Sparse Decomposition
RAE	Relational Autoencoder
RBF	Radial Basis Function
RBM	Restricted Boltzmann Machine
RKHS	Reproducing Kernel in Hilbert Space
RMSE	Root Mean Square Error
RNN	Recurrent Neural Network
ROC	Receiver Operating Characteristic
RS	Remote Sensing
RTRL	Real-Time Recurrent Learning
SAE	Sparse Auto-Encoders
SAE	Sparse Autoencoder
SAR	Synthetic Aperture Radar
SC	Sparse Coding
SNR	Signal-to-Noise Ratio
SRM	Structural Risk Minimization
SSL	Semi-Supervised Learning
STL	Self-taught Learning
SV	Support Vector
SVAE	Sparse Variational Autoencoder
SVM	Support Vector Machine
tBPTT	truncated Back-propagation through Time
VAE	Variational Autoencoder
XAI	Explainable Artificial Intelligence

1

Introduction

Gustau Camps-Valls, Xiao Xiang Zhu, Devis Tuia, and Markus Reichstein

Machine learning methods are widely used to extract patterns and insights from the ever-increasing data streams from sensory systems. Recently, deep learning, a particular type of machine learning algorithm (Goodfellow et al. 2016), has excelled in tackling data science problems, mainly in the fields of computer vision, natural language processing, and speech recognition. Since a few years ago, it has become impossible to ignore deep learning. Started as a curiosity in the 1990s, deep learning has imposed itself as the prime machine learning paradigm in the last ten years, especially thanks to the availability of large datasets and of the advances in hardware and parallelization allowing them to be learned from. Nowadays, most machine learning research is somehow deep learning-based and new heights in performance have been reached in virtually all fields of data science, both applied and theoretical. Adding to this the community efforts in sharing code and the availability of computational resources, deep learning seems to be the winner to unlock data science research.

In recent years, deep learning has shown increased evidence of the potential to address problems in Earth and climate sciences as well (Reichstein et al. 2019). As for many applied fields of science, Earth observation and climate science are more and more strongly data-driven. Deep learning strategies are currently explored by more and more researchers and neural networks are used in many operational systems. The advances in the field are impressive, but there is still much ground to cover to understand the complex systems that are our Earth and its climate. Why deep learning is working in Earth data problems is also a challenging question, for which one could argue a statistical reason. As in computer vision or language processing, Earth Sciences also consider spatial and temporal data that exhibit high autocorrelation functions which deep learning methods treat very well. But what is the physical reason, if any? Is deep learning discovering guiding or first principles in the data automatically? Why do convolutions in space or time lead to appropriate feature representations? Are those representations sparse, physically consistent, or even causal? Explaining what the deep learning model actually learned is a challenge itself. Even though AI has promised to change the way we often do science, with DL the first step in this endeavor, this will not be the case unless we resolve these questions.

Deep Learning for the Earth Sciences: A Comprehensive Approach to Remote Sensing, Climate Science, and Geosciences, First Edition. Edited by Gustau Camps-Valls, Devis Tuia, Xiao Xiang Zhu, and Markus Reichstein.

The field of deep learning for Earth and climate sciences is so wide and fast-evolving that we could not cover all different methodological approaches and geoscientific problems. A selected representative subset of methods, problems, and promising approaches were selected for the book. With this introduction (and more in general with this book), we want to take a picture of the state of the art of the efforts in machine learning (section 1.1), in the remote sensing (section 1.2) and geosciences and climate (section 1.3) communities to integrate, use, and improve deep learning methods. We also want to provide resources for researchers that want to start including neural networks-based solutions in their data problems.

1.1 A Taxonomy of Deep Learning Approaches

Given the current pace of the advances in deep learning, providing a taxonomy of approaches is not an easy task. The field is full of creativity and new inventive approaches can be found on a regular basis. Without the pretension of being exhaustive, most deep learning approaches can be placed along the lines of the following dimensions:

- *Supervised vs. unsupervised*. This is probably the most traditional distinction in machine learning and also applies in deep learning methodologies. Basically, it boils down to knowing whether the method uses labeled information to train or not. The best known examples of supervised deep methods are the Convolutional Neural Network (CNN, Fukushima (1980); LeCun et al. (1998a); Krizhevsky (1992)) and the recurrent neural network (RNN, Hochreiter and Schmidhuber (1997)), both using labels to evaluate the loss function and backpropagate errors to update weights, the former for image data and the latter for data sequences. As for unsupervised methods, they do not use ground truth information and therefore rely on unsupervised criteria to train. Among unsupervised methods, autoencoders (Kramer 1991; Hinton and Zemel 1994) are the most well known. They use the error in reconstructing the original image to train and are often used to learn low-dimensional representations (Hinton and Salakhutdinov 2006a) or for denoising images (Vincent and Larochelle 2010).
  In between these two endpoints, one can find a number of approaches tuning the level and the nature of supervision: *weakly supervised* models (Zhou 2018), for instance, use image-level supervision to predict phenomena at a finer resolution (e.g. localize objects by only knowing whether they are present in the image), while *self-supervised* models use the content of the image itself as a supervisory signal; proceeding this way, the labels to train the model come for free. For example, self-supervised tasks include predicting the color values from a greyscale version of the image (Zhang et al. 2016c), predicting relative position of patches to learn part to object relations (Doersch et al. 2015), or predicting the rotation that has been applied to an image (Gidaris et al. 2018).
- *Generative vs. discriminative*. Most methods described above are discriminative, in the sense that they minimize an error function comparing the prediction with the true output (a label or the image itself when reconstructing). They model the conditional probability of the target Y given an observation x, i.e., $P(Y|X = x)$. A generative model generates possible inputs that respect the joint input/outputs distribution. In other words it models

the conditional probability of the data X given an output y, i.e. $P(X|Y = y)$. Generative models can therefore sample instances (e.g. patches, objects, images) from a distribution, rather than only choosing the most likely one, which is a great advantage when data are complex and show multimodalities. For instance, when generating images of birds, they could generate different instances of birds of the same species with subtle shape or color differences. Examples of generative deep models are the variational autoencoders (VAE, Kingma and Welling (2014); Doersch (2016)) and the generative adversarial networks (GAN, Goodfellow et al. (2014a)), where a generative model is trained to generate images that are so realistic that a model trained to recognize real from fake ones fails.

- *Forward vs. recurrent*. The third dimension concerns the functioning of the network. Most models described above are forward models, meaning that the information flows once from the input to prediction before errors are backpropagated. However, when dealing with data structured as sequences (e.g. temporal data) one could make information flow across the sequence dimension. Recurrent models (RNNs, firstly introduced in Williams et al. (1986)) exploit this structure to inform the next step in the sequence of the hidden representations learned by the previous. Backpropagating information along the sequence also has its drawbacks, especially in terms of vanishing gradients, i.e. gradients that, after few recursion steps become zero and do not update the model any more: to cope with this, network including skip connections called memory gates have been proposed: the Long-Short Memory Network (LSTM, Hochreiter and Schmidhuber (1997)) and the Gated Recurrent Unit (GRU, Cho et al. (2014)) are the most known.

1.2 Deep Learning in Remote Sensing

Taking off in 2014, deep learning in remote sensing has become a blooming research field, almost a hype. To give an example, to date there are more than 1000 published papers related to the topic (Zhu et al. 2017; Ma et al. 2019). Such massive and dynamic developments are triggered by, on the one hand, methodological advancements in deep learning and the open science culture in the machine learning and computer vision communities which resulted in open access to codes, benchmark datasets, and even pre-trained models. On the other hand, it is due to the fact that Earth observation (EO) has become an operational source of open big data. Fostered by the European Copernicus program with its high-performance satellite fleet and open access policy, the user community has increased and widened considerably during the last years. This raises high expectations for valuable thematic products and intelligent knowledge retrieval. In the private sector, NewSpace companies launch(ed) hundreds of small satellites which have become a complementary and affordable source of EO data. This requires new data-intensive – or even data-driven – analysis methods from data science and artificial intelligence, among others – deep learning.

To summarize the development in the past six years, deep learning in remote sensing has been through three main phases with temporal overlapping: exploration, benchmarking, and EO-driven methodological developments. In the following, we overview these three phases. Given the huge number of existing literature, it is unavoidable to give just a selection of examples subject to bias.

- *Phase 1: Exploration (2014 to date)*: The exploration phase is characterized by quick wins, often achieved by the transfer and tailoring of network architectures from other fields, most notably from computer vision. To name a few early examples, stacked autoencoders are applied to extract high-level features from hyperspectral data for classification purposes in Chen et al. (2014). Bentes et al. have exploited deep neural networks for the detection and classification of objects, such as ships and windparks, in oceanographic SAR images (Bentes et al. 2015). In 2015, Marmanis et al. (2015) have fine-tuned ImageNet pre-trained networks to boost the performance of land use classification with aerial images. Since then, researchers explore the power of deep learning for a wide range of classic tasks and applications in remote sensing, such as classification, detection, semantic segmentation, instance segmentation, 3D reconstruction, data fusion, and many more. Whether using pre-trained models or training models from scratch, it is always about addressing new and intrinsic characteristics of remote sensing data (Zhu et al. 2017):
  - Remote sensing data are often *multi-modal*. Tailored architectures must be developed for, e.g. optical (multi- and hyperspectral) (Audebert et al. 2019) and synthetic aperture radar (SAR) data (Chen et al. 2016; Zhang et al. 2017; Marmanis et al. 2017; Shahzad et al. 2019), where both the imaging geometries and the content are completely different. Data and information fusion use these complementary data sources in a synergistic way (Schmitt and Zhu 2016). Already prior to a joint information extraction, a crucial step is to develop novel architectures for the matching of images taken from different perspectives and even different imaging modality, preferably without requiring an existing 3D model (Marcos et al. 2016; Merkle et al. 2017; Hughes et al. 2018). Also, besides conventional decision fusion, an alternative is to investigate transfer learning from deep features of different imaging modalities (Xie et al. 2016).
  - Remote sensing data are *geo-located*, i.e., each pixel in a remote sensing imagery corresponds to a geospatial coordinate. This facilitates the fusion of pixel information with other sources of data, such as GIS layers (Chen and Zipf 2017; Vargas et al. 2019; Zhang et al. 2019b), streetview images (Lefèvre et al. 2017; Srivastava et al. 2019; Kang et al. 2018; Hoffmann et al. 2019a), geo-tagged images from social media (Hoffmann et al. 2019b; Huang et al. 2018c), or simply other sensors as above.
  - Remote sensing *time series* data is becoming standard, enabled by Landsat, ESA's Copernicus program, and the blooming NewSpace industry. This capability is triggering a shift from individual image analysis to time-series processing. Novel network architectures must be developed for optimally exploiting the temporal information jointly with the spatial and spectral information of these data. For example, convolutional recurrent neural networks are becoming baselines in multitemporal remote sensing data analysis applied to change detection (Mou et al. 2018), crop monitoring (Rußwurm and Körner 2018b; Wolanin et al. 2020), as well as land use and land cover classification (Qiu et al. 2019). An important research direction is unsupervised or weakly supervised learning for change detection (Saha et al. 2019b) or anomaly detection (Munir et al. 2018) from time series data.
  - Remote sensing has irreversibly entered the *big data era*. We are dealing with very large and ever-growing data volumes, and often on a global scale. On the one hand this allows large-scale or even global applications, such as monitoring global urbanization (Qiu et al. 2020), large-scale mapping of land use/cover (Li et al. 2016a), large-scale

cloud detection (Mateo-García et al. 2018) or cloud removal (Grohnfeldt et al. 2018), and retrieval of global greenhouse gas concentrations (Buchwitz et al. 2017) and a multitude of trace gases resolved in space, time, and vertical domains (Malmgren-Hansen et al. 2019). On the other hand, algorithms must be fast enough and sufficiently transferable to be applied for the whole Earth surface/atmosphere, which in turn calls for large and representative training datasets, which is the main topic of phase 2.

In addition, it is important to mention that – unlike in computer vision – classification and detection are only small fractions of remote sensing and Earth observation problems. Actually, most of the problems are related to the retrieval of *bio-geo-physical or bio-chemical variables*. This will be discussed in section 1.3.

- *Phase 2: Benchmarking (2016 to date)*: To train deep learning methods with good generalization abilities and to compare different deep learning models, large-scale benchmark datasets are of great importance. In the computer vision community, there are many high-quality datasets available which are dedicated to, for example, image classification, semantic segmentation, object detection, and pose estimation tasks. To give an example, the well-known ImageNet image classification database consists of more than 14 million hand-annotated images cataloged into more than 20,000 categories (Deng et al. 2009). It is debatable whether the computer vision community is too much driven by the benchmark culture, instead of caring about real-world challenges. In remote sensing it is, however, the other extreme – we are lacking sufficient training data. For example, most classic methodological developments in hyperspectral remote sensing have been based on only a few benchmark images of limited sizes, let alone the annotation demanding deep learning methods. To push deep learning related research in remote sensing, community efforts in generating large-scale real-world scenario benchmarks are due. Motivated by this, since 2016 an increasing number of large-scale remote sensing datasets have become available covering a variety of problems, such as instance segmentation (Chiu et al. 2020; Weir et al. 2019; Gupta et al. 2019), object detection (Xia et al. 2018; Lam et al. 2018), semantic segmentation (Azimi et al. 2019; Schmitt et al. 2019; Mohajerani and Saeedi 2020), (multi-label) scene classification (Sumbul et al. 2019; Zhu et al. 2020), and data fusion (Demir et al. 2018; Le Saux et al. 2019). To name a few examples:
  - DOTA (Xia et al. 2018): This is a large-scale dataset for object detection in aerial images, which collect 2806 aerial images from different sensors and platforms containing objects exhibiting a wide variety of scales, orientations, and shapes. In total, it contains 188,282 object instances in 15 common object categories and serves as a very important benchmark for development of advanced object detection algorithms in very high resolution remote sensing.
  - So2Sat LCZ42 (Zhu et al. 2020): This is a benchmark dataset for global local climate zones classification. It is a rigorously labeled reference dataset in EO. Over one month 15 domain experts carefully designed the labeling workflow, the error mitigation strategy, the validation methods, and conducted the data labeling. It consists of manually assigned local climate zone labels of 400,673 Sentinel-1 and Sentinel-2 image patch pairs globally distributed in 42 urban agglomerations covering all the inhabited continents and 10 cultural zones. In particular, it is the first EO dataset that provides a quantitative measure of the label uncertainty, achieved by letting a group of domain experts cast 10 independent votes on 19 cities in the dataset.

An exhaustive list of remote sensing benchmark datasets is summarized by Rieke et al. (2020). There is no doubt that these high-quality benchmarks are essential for the next phase – EO-driven methodological research.

- *Phase 3: EO-driven methodological research (2019 to date)*: Going beyond these successful yet still EO-driven but application-oriented researches mentioned in phase 1, EO-driven fundamental yet rarely addressed methodological challenges are attracting attention in the remote sensing community.
  - Reasoning: the capability to link meaningful transformation of entities over space or time is a fundamental property of intelligent species and also the way people understand visual data. Recently, in computer vision several efforts have been made to enable such capability of deep networks. For instance, Santoro et al. (2017) proposed a relational reasoning network for the problem of visual question answering. This network achieves a so-called super-human performance. Zhou et al. (2018) presented a temporal relation network to enable multiscale temporal relational reasoning in networks for video classification tasks. Reasoning is particularly relevant for Earth observation, as every measurement in remote sensing data is associated with a spatial-temporal coordinate and characterized by spatial and temporal contextual relations, in particular when it comes to geo-physical processes. As to reasoning networks in remote sensing, a first attempt can be found in Mou et al. (2019), where the authors propose reasoning modules in a fully convolutional network for semantic segmentation in aerial scenes. Further extending the relational reasoning to semantics, Hua et al. (2020) proposed an attention-aware label relational reasoning network for multilabel aerial image classification. Another pioneering work of reasoning in remote sensing is visual question answering, the so-called let remote sensing imagery speaks for itself (Lobry et al. 2019). More remote sensing tasks benefiting from reasoning networks are yet open for discovery.
  - Uncertainty: EO applications target at retrieving physical or bio-chemical variables in a large scale. These predicted physical quantities are often used in data assimilation and in decision making, for example in support of and for monitoring of the UN Sustainable Development Goals (SDGs). Therefore, besides high accuracy, traceability, and reproducibility of results, quantifying the uncertainty of these predictions from a deep learning algorithm is indispensable towards a quality and reliable Artificial Intelligence in Earth observation. Although quantifying uncertainty of parameter estimates in EO is common practice in traditional model-driven approaches, this has not caught up with the rapid development of deep learning, where the model can also be learned. Only a handful of literature addressed it in the past (Zhu et al. 2017). But the EO community is realizing its indispensability for a responsible AI. For example, the "Towards a European AI4EO R&I Agenda" (ESA, 2018) mentioned uncertainty estimation as one of the future challenges of AI4EO. To give one encouraging example, one active research direction in uncertainty quantification focuses on using Bayesian neural networks (BNNs), which are a type of network which not only gives point estimates of model parameters and output predictions, but also provides the whole distribution over these values. For example, Kendall and Gal (2017) proposed a BNN that uses a technique called Learned Loss Attenuation to learn the noise distribution in input data, which can be used to find uncertainty in the final output. More recent studies,

Ilg et al. (2018); Kohl et al. (2018) proposed BNNs that output a number of plausible hypotheses enabling creation of distribution over outputs and measuring uncertainties. Actually, Bayesian deep learning (BDL) offers a probabilistic interpretation of deep learning models by inferring distributions over the models' weights (Wang and Yeung 2016; Kendall and Gal 2017). These models, however, have not been applied extensively in the Earth Sciences, where, given the relevance of uncertainty propagation and quantification, they could find wide adoption. Only some pilot applications of deep Gaussian processes (Svendsen et al. 2018) for parameter retrieval and BNNs for time series data analysis (Rußwurm et al. 2020) are worth mentioning. In summary, the Bayesian deep learning community has developed model-agnostic and easy-to-implement methodology to estimate both data and model uncertainty within deep learning models, which has great potential when applied to remote sensing problems (Rußwurm et al. 2020).

Other open issues that recently caught the attention in the remote sensing community include but are not limited to: hybrid models integrating physics-based modeling into deep neural networks, efficient deep nets, unsupervised and weakly supervised learning, network architecture search, and robustness in deep nets.

1.3 Deep Learning in Geosciences and Climate

A vast number of algorithms and network architectures have been developed and applied in the geosciences too. Here, the great majority of applications have to do with estimation of key biogeophysical parameters of interest or forecasting essential climate variables (ECV). The (ab)use of the standard multilayer perceptron in many studies has given rise to the use of more powerful techniques like convolutional networks that can exploit spatial fields of view while providing vertical estimations of parameters of interest in the atmosphere (Malmgren-Hansen et al. 2019) and recurrent neural nets, and the long short-term memory (LSTM) unit in particular, which has demonstrated good potential to deal with time series of biogeophysical parameters estimation, forecasting, and memory characterization of processes (Besnard et al. 2019).

While deep learning approaches have classically been divided into spatial learning (for example, convolutional neural networks for object detection and classification) and sequence learning (for example, forecasting and prediction), there is a growing interest in blending these two perspectives. After all, Earth data can be cast as spatial structures evolving through time: weather forecasting or hurricane tracking are clear examples, but also is the case of the solid Earth (Bergen et al. 2019). We often face time-evolving multi-dimensional structures, such as organized precipitating convection which dominates patterns of tropical rainfall, vegetation states that influence the flow of carbon, and volcanic ash particles whose shape describe different physical eruption mechanisms, just to name a few (Reichstein et al. 2019; Bergen et al. 2019). Studies are starting to apply combined convolutional-recurrent deep networks for precipitation nowcasting (Xingjian et al. 2015) or extreme weather forecasting (Racah et al. 2017); for example. modeling atmospheric and ocean transport, fire spread, soil movements, or vegetation dynamics are other examples of problems where spatio-temporal dynamics are important. This is the natural scenario where DL excels; exploiting spatial and/or temporal regularities in huge amounts of data.

Physical modeling and machine learning have often been treated as completely different and irreconcilable fields; scientists should adhere to either a theory-driven or a data-driven approach. Yet these approaches are indeed complementary: physical approaches are interpretable and allow extrapolation beyond the observation space by construction, and data-driven approaches are highly flexible and adaptive to data. Their synergy has gained attention lately in the geosciences (Karpatne et al. 2017a; Camps-Valls et al. 2018b). Interactions can be diverse (Reichstein et al. 2019). There are several ways in which Physics and DL can interact within Earth Sciences:

- *Improving parameterizations.* Physical models require parameters that can be seldom derived from first principles. Deep learning can learn such parameterizations to optimally describe the ground truth that can be observed or generated from detailed and high-resolution models of clouds (Gentine et al. 2018). In the land domain, for example, instead of assigning parameters of the vegetation in an Earth system model to plant functional types, one can allow these parameterizations to be learned from proxy covariates with machine learning, allowing them to be more dynamic, interdependent, and contextual (Moreno-Martínez et al. 2018).
- *Surrogate modeling and emulation.* Surrogate modeling, also known as *emulation*, is gaining popularity in remote sensing (Camps-Valls et al. 2016; Reichstein et al. 2019; Camps-Valls et al. 2019). Emulators are essentially statistical models that learn to mimic the energy transfer code using a small yet representative dataset of simulations. Emulators allow one to readily perform fast forward simulations, which in turn allow improved inversion. However, replacing an simulator (e.g. RTM or climate model (sub)component) with a deep model requires running expensive evaluations off first. Recent more efficient alternatives construct an approximation to the forward model starting with a set of optimal RTM simulations selected iteratively (Camps-Valls et al. 2018a; Svendsen et al. 2020). This topic is related to active learning and Bayesian optimization, which might push results further in accuracy and sparsity, especially when modeling complex codes such as climate model components.
- *Blending networks and process-based models.* Including knowledge through extra regularization that forces DL models to respect some physics laws can be seen as a form of inductive bias for which ML is prepared with many optimization techniques (Kashinath et al. 2019; Wu et al. 2018). A fully coupled net can be devised: here, layers describing complicated and uncertain processes feed physics-layers that encode known relations of intermediate feature representation with the target variables (Reichstein et al. 2019). The integration of physics into DL models not only achieves improved generalization but, more importantly, endorses DL models with consistency and faithfulness. As a by-product, the hybridization process has an interesting regularization effect, as physics discards implausible models and promotes simpler, sparser structures.

An important point and active research field is that of *explainability* of the derived DL models. Interpreting what the DL model learned is important to understand how the system works to debug it or improve it, to anticipate unforeseen circumstances, to build up trust in the technology, to understand the strengths and limitations, to audit a prediction/decision, to facilitate monitoring and testing, and to guide users into actions or behaviors. A plethora of techniques have been developed to gain insight from a model (Molnar 2019; Samek et al. 2017): (1) feature visualization to characterize the network architecture (e.g. how redundant, outlier-prone, or adversarial-sensitive the network is); (2) feature attribution to analyze how each input contributed to a particular prediction; and (3) model distillation that explains a neural network with a surrogate simpler (often linear or tree-based) model. Several works in remote sensing and geosciences have studied interpretability of deep nets. For example, Kraft et al. (2019) introduced an agnostic-based method through time-series permutation which allows memory effects of climate and vegetation affecting net ecosystem CO_2 fluxes in forests to be studied. In Wolanin et al. (2020), activation maps of hidden units in convolutional nets were studied for crop yield estimation from remote sensing data; analysis suggested that networks mainly focus on growing seasons and can provide a ranking of more important covariates. Recently, in Toms et al. (2019a), the method of layer-wise relevance propagation (LRP) (Montavon et al. 2018) was used to study patterns in Earth System variability.

1.4 Book Structure and Roadmap

This book is not conceived as a mere compilation of works about Deep Learning in Earth Sciences but rather aims to put a carefully selected set of puzzle pieces together for highlighting the scope of relevant milestones in the intersection of both fields. We start the book with an introductory chapter *Ch.* that treats the main challenges and opportunities in Earth Sciences. After this, the book is split into four main Parts:

Part I. Deep learning to extract information from remote sensing images. The first part is devoted to extract information from remote sensing images. We depart from novel developments in unsupervised learning, move to weakly supervised models and then follow reviewing the main applications that involve supervised learning.
The field of unsupervised learning in Earth observation problems is of paramount relevance, given the high cost of obtaining labeled data in resources, human and economic terms. The concepts of sparse representations, compactness and expressive features has emerged in canonical – yet unsupervised – convolutional neural networks (*Ch. 2*). Simulating processes with neural networks have also found wide application in geoscientific problems, in particular with Generative Adversarial Networks (GANs) (*Ch. 3*). When a few labeled data are also available, the field of semisupervised and self-taught learning emerge as potentially useful fields (*Ch. 4*). Supervised learning is, however, the most active one in the field, and we included dedicated chapters to the most relevant aspects: image classification and segmentation (*Ch. 5*) and object recognition in remote sensing images (*Ch. 6*). However, often all problems implying data classification fail because domains of training and test differ in their statistics. Here either adapting the classifier or the feature representation becomes crucial. The key problem of adapting domains for data classification is surveyed in *Ch. 7*. When time is involved in detection and classification of land use and land covers, recurrent neural networks excel; an extensive review of these deep learning techniques is provided in *Ch. 8*. Yet deep learning has been also used in contexts where information extraction does not mean straightforward goals like classification or detection. For instance, whenever one needs to perform particular operations

like image matching and co-registration (*Ch. 9*), multisource image fusion *Ch. 10*, and search and retrieval of images in huge data archives (*Ch. 11*)

Part II. Making a difference in the geosciences with deep learning. The second part of the book deals with a selected collection of problems where deep learning has made a big difference compared to previous approaches: problems that involve real target variables, particular data structures (spatio-temporal, strongly time autocorrelated, extremely high dimensional volumetric atmospheric data), challenging problems like the detection of climate extremes, weather forecasting and nowcasting, and the study of the cryosphere.
The part starts with a chapter dedicated to the detection of extreme weather patterns in *Ch. 12*). Spatio-temporal data and teleconnections is present in many weather and climate studies; here spatio-temporal autoencoders can discover the underlying modes of variability in data cubes to represent the underlying processes (*Ch. 13*). Deep learning to improve weather predictions is treated thoroughly in *Ch. 14*. *Ch. 15* reviews the problem of weather forecasting and in particular that of precipitation nowcasting architectures. Deep learning has found enormous challenges when working with high-dimensional data for parameter retrieval; *Ch. 16* shows developments to approach the problem in the atmosphere and the cryosphere. An extensive review of DL for cryospheric studies is provided in *Ch. 17*. The part ends with the application of recurrent networks to emulate and learn ecological memory (*Ch. 18*).

Part III. Linking physics and deep learning models. The field has grown enormously in methods and applications. Yet a wide adoption in the broad field of Earth Sciences is still missing, mainly due to the fact that (a) models are often overparameterized black boxes, hence interpretability is often compromised, (b) deep learning models often do not respect the most elementary laws of physics (like advection, convection, or the conservation of mass and energy), and (c) they are after all costly models to train from scratch needing a huge amount of labeled data, hence democratization of machine learning is not actually a reality. These issues have been recently tackled by pioneer works in the interface between machine learning and physics, and will be reviewed as well in this last part of the book, where physics-aware deep learning hybrid models will be treated.
We start the part with a chapter dedicated to the impact of deep learning in hydrology, another field where DL has impacted recently; applications and new architectures suited to the problems are treated in detail in *Ch. 19*. A field where DL has found enormous interest and adoption recently is that of parametrization of subgrid processes for unresolved turbulent ocean processes (*Ch. 20*) and climate models in general (*Ch. 21*). Using deep learning to correct theoretically-derived models opens a new path of interesting applications where machine learning and physics interact (*Ch. 22*).

We end up the book with some final words and perspectives in *Ch. 23*. We review where we are now, and the challenges ahead. We treat issues such as adapting architectures to data characteristics (to deal with e.g. large-range relations), interpretability and explainability, hybrid modeling as evolved forms of data assimilation techniques, learning plausible physics models, and the more ambitious goal of learning expressive causal representations from data, DL models, and assumptions.

Supporting material is also included in two forms. On the one hand, examples real and advanced application examples are provided in each chapters. But we also provide scripts, code, and pointers to toolboxes and applications of deep learning in the geosciences in a dedicated GitHub site:

https://github.com/DL4ES

In this dedicated repository, many links are maintained to other widely used software toolboxes for deep learning and applications in the Earth Sciences. This repository is periodically updated with the latest contributions, and can be helpful for the Earth and climate data scientist.

We sincerely hope you enjoy the material and that it serves your purposes!

Part I

Deep Learning to Extract Information from Remote Sensing Images

2

Learning Unsupervised Feature Representations of Remote Sensing Data with Sparse Convolutional Networks

Jose E. Adsuara, Manuel Campos-Taberner, Javier García-Haro, Carlo Gatta, Adriana Romero, and Gustau Camps-Valls

2.1 Introduction

Fields like remote sensing, computer vision, or natural language processing typically work in the so-called structured domains, in which the original data representation has temporal and/or spatial dimensions defined in uniform grids. From a geometrical viewpoint, data can be represented in their original coordinates, but visualizing, understanding, and designing algorithms therein is challenging, mainly due to the high dimensionality and correlation between covariates. This is why learning alternative, typically simpler and compact, feature representations of the data has captured a lot of interest in the scientific community. This is the field of dimensionality reduction or feature extraction, for which one has both supervised and unsupervised algorithms (Rojo-Álvarez et al. 2018; Bengio et al. 2013; Hinton and Salakhutdinov 2006a).

Unsupervised learning is the preferred approach in cases of label sparsity. Different algorithms implementing different criteria are available. Principal Component Analysis (PCA) (Jolliffe 1986) is one of the most popular methods for dimensionality reduction due to its easy implementation and interpretability. Two relevant, and often unrealistic, assumptions are done though: linearity and orthogonality. In the last decade, a profusion of non-linear dimensionality reduction methods including both manifold (Lee and Verleysen 2007) and dictionary learning (Kreutz-Delgado et al. 2003) have sprung up in the literature. Non-linear manifold learning methods can be mainly categorized as either local or global approaches (Silva and Tenenbaum 2003). Local methods retain local geometry of data, and are computationally efficient since only sparse matrix computations are needed. Global manifold methods, on the other hand, keep the entire topology of the dataset, yet are computationally expensive for large datasets. They have higher generalization power, but local ones can perform well on datasets with different sub-manifolds. Local manifold learning methods include, inter alia, locally linear embedding (LLE) (Roweis and Saul 2000), local tangent space alignment (LTSA) (Zhang and Zha 2004), and Laplacian eigenmaps (LE) (Belkin and Niyogi 2003). Basically, these approaches build local structures to obtain a global manifold. Among the most widely used global manifold methods the ISOMAP (Tenenbaum et al. 2000) and the kernel version of PCA (kPCA) (Schölkopf

Deep Learning for the Earth Sciences: A Comprehensive Approach to Remote Sensing, Climate Science, and Geosciences, First Edition. Edited by Gustau Camps-Valls, Devis Tuia, Xiao Xiang Zhu, and Markus Reichstein.

et al. 1998) stand out, as well as kernel-based and spectral decompositions that learn mappings optimizing for maximum variance, correlation, entropy, or minimum noise fraction (Arenas-García et al. 2013), and their sparse versions (Tipping 2001). In addition there exist neural networks that generalize PCA to encode non-linear data structures via autoencoding networks (Hinton and Salakhutdinov 2006a), as well as projection pursuit approaches leading to convenient Gaussian domains (Laparra et al. 2011).

In the last years, the use of deep learning techniques has become a trending topic in remote sensing and geosciences due to the increasing availability of many and large datasets. Some excellent reviews on this topic have been published by Zhang et al. (2016b), Zhu et al. (2017), and Ma et al. (2019). Deep learning methods can deal with the intrinsic problems related with the analysis of non-linear spatial-spectral datasets.

Several unsupervised neural nets are available too; for example, an autoencoder is a deep learning architecture in which the input signal is reconstructed/decoded at the output layer through an intermediate layer with reduced number of hidden nodes. Basically, the autoencoder aims at reproducing the inputs at the output layer by using the high-abstraction features learned in the intermediate layer. The use of autoencoders implies, however, the tuning of several free parameters, addressing the regularization issue mainly by limiting the structure of the network heuristically. The use of autoencoders in remote sensing is widespread for a wide range of applications, including feature extraction and image classification (Zabalza et al. 2016; Othman et al. 2016), spectral unmixing (Guo et al. 2015; Su et al. 2019), image fusion (Azarang et al. 2019), and change detection (Lv et al. 2018). However, autoencoders require tuning several critical hyperparameters.

While the issue of *non-linear* feature extraction can be resolved with deep networks efficiently, there is still the issue of what a sensible criterion should be employed. Natural data, namely, data generated by a physical process or mechanism, often show strong autocorrelation functions (in space and time), heavy tails, strong feature correlations, not necessarily linear, and (from a manifold learning perspective) data lives in subspaces of (much) lower dimensionality. This can be interpreted as that the observations were generated by a much simpler mechanisms, and that one should seek for a sparse representation.

In this sense, sparse coding and dictionary learning has an efficient way to learn sparse image features in unsupervised settings, which are eventually used for image classification and object recognition. A dictionary can be understood as a set of bases that sparsely represent the original data. The main goal in dictionary learning is to find a dataset to best represent the inputs by only using a small subset (also known as atoms) of the dictionary. Many applications and approaches have been developed using dictionary learning: image denoising via sparse and redundant representations (Elad and Aharon 2006), spatial-spectral sparse-representation and image classification using discriminative dictionaries (Wang et al. 2013b), change detection based on joint dictionary learning (Lu et al. 2016), image classification with sparse kernel networks (Yang et al. 2014), image pansharpening by means of sparse representations over learned dictionaries (Li et al. 2013), large-scale remote sensing image representation based on including particle swarm optimization into online dictionary learning (Wang et al. 2015), image segmentation with saliency-based codes (Dai and Yang 2010; Rigas et al. 2013), image super-resolution based on patch-wise sparse recovery (Yang et al. 2012), automatic target detection employing sparse bag-of-words codes (Sun et al. 2012), unsupervised learning of sparse features for aerial image classification (Cheriyadat 2014), and cloud removal based on sparse representation using multitemporal dictionary learning (Xu et al. 2016). These methods describe the input images in sparse representation spaces but do not take advantage of the high non-linear nature of deep architectures.

However, attaining sparse non-linear deep networks is still unresolved in the literature, especially in unsupervised learning. In the next section, we introduce a methodology to learn sparse spatial-spectral feature representations in deep convolutional neural network architectures in an unsupervised way.

2.2 Sparse Unsupervised Convolutional Networks

The proposed methodology relies on the use of the standard convolutional neural network (CNN) for which efficient implementations, architectures, and training algorithms are available. Training of (deep) convnets is typically done in a supervised fashion, e.g. by means of standard back-propagation (LeCun et al. 1998b; Krizhevsky et al. 2012; Sermanet et al. 2014). Alternatively, one can train convnets by means of greedy layer-wise pre-training (Hinton et al. 2006; Lee et al. 2009; Masci et al. 2011) in an unsupervised way by using greedy layer-wise pre-training (Hinton et al. 2006; Ngiam et al. 2011; Kavukcuoglu et al. 2010; Lee et al. 2009; Masci et al. 2011). We propose to couple this standard architecture with an efficient algorithm called "Enforcing Population and Lifetime Sparsity" (EPLS) (Romero et al. 2014) that generates so-called "pseudo-labels" by imposing sparsity. The methodology is very modular, and allows using different architectures or alternative methods for generating proxy codes of the labels.

2.2.1 Sparsity as the Guiding Criterion

Sparsity is among the properties of a good feature representation (Field 1994; Olshausen and Field 1997; Ranzato et al. 2006; Ngiam et al. 2011; Bengio et al. 2013). Sparsity can be defined in terms of population sparsity and lifetime sparsity. On one hand, population sparsity ensures simple representations of the data by allowing only a small subsets of outputs to be active at the same time (Willmore and Tolhurst 2001). On the other hand, lifetime sparsity controls the frequency of activation of each output throughout the dataset, ensuring rare but high activation of each output (Willmore and Tolhurst 2001). State-of-the-art unsupervised learning methods such as sparse Restricted Boltzmann Machines (RBM) (Hinton et al. 2006), Sparse Auto-Encoders (SAE) (Ranzato et al. 2006), Sparse Coding (SC) (Olshausen and Field 1997), Predictive Sparse Decomposition (PSD) (Kavukcuoglu et al. 2010), Sparse Filtering (Ngiam et al. 2011), and Orthogonal Matching Pursuit (OMP-k) (Coates and Ng 2011) have been successfully used in the literature to extract sparse feature representations. However, the great majority of these methods optimize for either lifetime of population sparsity, not both altogether, and have numerous meta-parameters to tune. See the related concept of sparse representation treated in Chapter 4.

2.2.2 The EPLS Algorithm

In Romero et al. (2014), we introduced EPLS, a novel, meta-parameter free, off-the-shelf and simple algorithm for unsupervised sparse feature learning. The method provides discriminative features that can be very useful for classification as they capture relevant spatial and spectral image features jointly. The method iteratively builds a sparse target from the output of a layer and optimizes for that specific target to learn the filters. The sparse target is defined such that it ensures both population and lifetime sparsity. Figure 2.1 summarizes the steps of the method in Romero et al. (2014). Essentially, given a matrix of input patches to train layer l, $\mathbf{H}^{l-1}$, we (1) compute the output of the patches $\mathbf{H}^l$ by applying the learned weights and biases to the input, and subsequently the non-linearity; (2) call the EPLS algorithm to generate a sparse target $\mathbf{T}^\mathbf{l}$ from the output of the layer; and (3) optimize the parameters of the layer by minimizing the L_2-norm of the difference between the layer's output and the EPLS sparse target:

$$\theta^{l*} = \arg\min_{\theta^l} \|\mathbf{H}^l - \mathbf{T}^l\|_2^2. \tag{2.1}$$

The optimization is performed by means of Stochastic Gradient Descent (SGD) with adaptive learning rates (Schaul et al. 2013).

2.2.3 Remarks

The learned hierarchical representations of the input data (in our case, remote sensing images) are used for classification, where lower layers extract low-level features and higher layers exhibit more abstract and complex representations. The methodology is fully unsupervised, which is a different (and more challenging) setting to the common supervised use of convolutional nets.

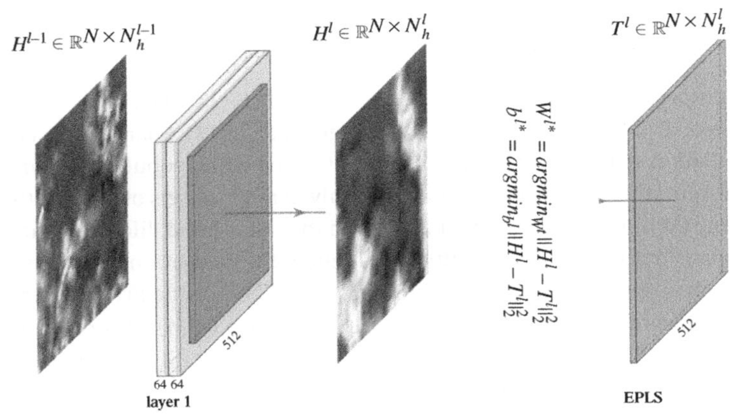

Figure 2.1 Scheme of the proposed method for unsupervised and sparse learning of image feature representations, where a convolutional neural network is trained iteratively driven by the EPLS algorithm that generates a sparse output (pseudo-labels) target matrix. The EPLS algorithm selects the output that has the maximal activation value, thus ensuring population sparsity, and the element that most frequently activates, ensuring population sparsity. More details on the EPLS algorithm can be found in Romero et al. (2014).

After training the parameters of a network, we can proceed to extract feature representations. To do so, we must choose an encoder to map the input feature map of each layer to its representation, i.e. we must choose the non-linearity to be used after applying the learned filters to all input locations. A straightforward choice is the use of a natural encoding, i.e. the non-linearity used to compute the output of each layer. However, different training and encoding strategies might be combined together. Summarizing, we train deep architectures by means of greedy layer-wise unsupervised pre-training in conjunction with EPLS and choose a feature encoding strategy for each specific problem. The interested reader may find an implementation of the EPLS algorithm in https://sites.google.com/site/adriromsor/epls.

2.3 Applications

To make the potentiality of the introduced methodology clear, we will use it for classification of hyperspectral images, and for multisensor multispectral and LiDAR image fusion.

2.3.1 Hyperspectral Image Classification

We start exemplifying the behavior of the method applying it on a hyperspectral image (HS) classification problem. Specifically, we use the whole image AVIRIS Indiana's Indian Pinest test site, which is a very challenging land-cover classification scenario, consisting of 614×2166 pixels and 220 spectral bands. In particular, we only use the 38 classes with more than 1000 samples and the 200 least noisy bands of the image.

On one side, and to start with, we extract different number of features using both PCA/kPCA, denoted as $N_f = \{5, 10, 20, 50, 100, 200\}$. In the case of the kPCA, the kernel used is the Radial Basis Function (RBF) with lengthscale set to the average distance among all training samples. On the other side, we devise CNN with different depths, but the same number of outputs per layer $N_h^l = \{5, 10, 20, 50, 100, 200\}$ which coincides with the number of features extracted. We compare the expression power of these two sets of features by feeding a 1-NN classifier with Euclidean distance, and measuring the performance through the estimated Cohen's kappa statistic (κ) for an independent test set.

Figure 2.2(a) shows the κ statistic for several numbers of extracted features (N_f and N_h^1, respectively) using PCA, kPCA, and single-layer networks. Both kPCA and the networks yield poor results when a low number of features are extracted, and drastically improve their performance for more than 50 features. Single-layer networks stick around $\kappa = 0.3$, even with an increased number of features. Nevertheless, there is a relevant gain when spatial information is considered. The best results are obtained for $N_h^1 = 200$ features and 5×5 receptive fields. With these encouraging results, we decided to train deeper CNNs using 30% of the available training samples per class and $N_h^l = 200$ output features per layer.

Another question to be addressed is the robustness of the features in terms of number of training examples. So the feature extraction, for different rates of training samples per class in all scenarios, has also been analyzed. The set of rates used is $\{1\%, 5\%, 10\%, 20\%, 30\%, 50\%\}$. We chose different receptive field sizes (1×1, 3×3, or 5×5) for each deep architecture, and using it for all layers. A CNN has been trained in a

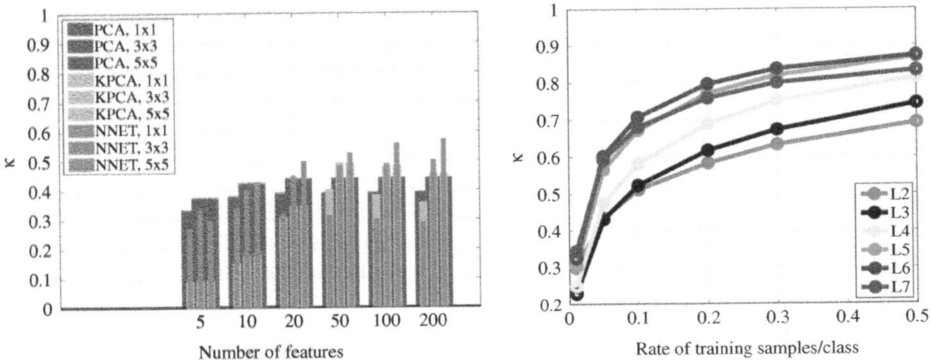

Figure 2.2 Kappa statistic (classification accuracy estimated) for several numbers of features (left), spatial extent of the receptive fields (for the single-layer network) or the included Gaussian filtered features (for PCA and kPCA) using 30% of data for training; and for different rates of training samples (right), $\{1\%, 5\%, 10\%, 20\%, 30\%, 50\%\}$, with pooling.

layer-wise fashion by means of EPLS with logistic non-linearity. Then, natural encoding has been used without polarity split to extract the network's features. Figure 2.2(b) highlights that using a few supervised samples to train a deep CNN can provide better results than using far more supervised samples to train a single-layer one. Note, for instance, that the 6-layer network using 5% samples/class outperforms the best single-layer network using 30% of the samples/class.

An important aspect of the proposed deep architectures lies in the fact that they typically give rise to compact hierarchical representations. In Figure 2.3, and for a subset of the whole image, the best three features extracted by the networks according to the mutual information between features and labels are depicted. The deeper we go, the more complicated and abstract features we retrieve, except for the seventh layer that provides spatially over-regularized features due to the downscaling impact of the max-pooling stages. Interestingly, it is also observed that, the deeper structures we use, the higher spatial decorrelation of the best features we obtain.

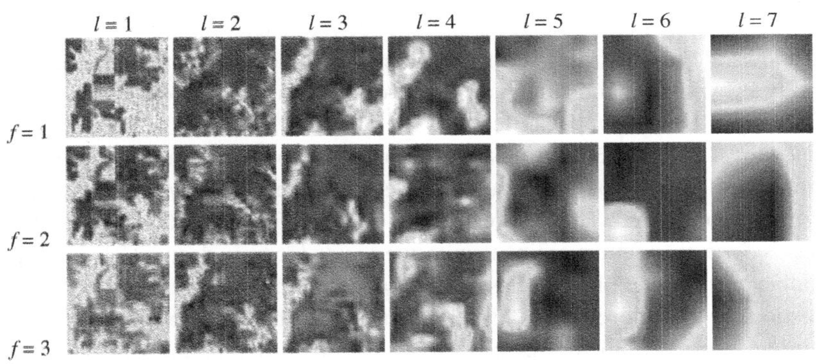

Figure 2.3 For the outputs of the different layers 1st to 7th, in columns, most informative features three features, in rows, according to the mutual information with the labels for a subregion of the whole image.

2.3.2 Multisensor Image Fusion

In this second example, we blend optical and LiDAR images and focus on the learned representation. This dataset corresponds to the finest spatial resolution addressed so far by the 2015 IEEE GRSS Data Fusion Contest, and made for a very challenging image analysis and fusion competition (Campos-Taberner et al. 2016).

Optical and LiDAR data represent objects in different ways: color or passive radiance versus altitude or active return intensity. The question is, how does this different physical acquisition principles translate into the neural network feature representation? With this in mind, we analyze spatial-spectral feature representations with convolutional neural networks using RGB, LiDAR, and the combined RGB+LiDAR representation. Such feature representations were studied in terms of sparsity and topological visualization.

For illustration, 100,000 image patches of size 10×10 were extracted from the original image, and 30,000 images patches were used for training the networks. We always use a maximum of $N_H = 1000$ hidden nodes, for all the architectures, and several symmetric receptive fields. EPLS with logistic non-linearity was used for training the networks on contrast-normalized image patches. We retrieved the sparse features applying the network parameters with natural encoding (i.e. with the logistic non-linearity) and polarity split which takes into account the positive and negative components of a code (weights) and hence doubles the number of outputs and is usually applied to the output layer of the network.

In Figure 2.4 we can see both the learned basis for RGB, LiDAR, RGB+LiDAR, and the corresponding topological representations using the first two ISOMAP components. For the sake of simplicity in the visualization, the neighborhood was fixed to $c = 1$ in ISOMAP's epsilon distance. Figure 2.4(a) displays that the EPLS algorithm applied to very high resolution color images learns not only common basis like oriented edges/ridges in many directions and colors, but also uncommon ones such as corner detectors, tri-banded colored filters, center surrounds or Laplacian of Gaussians among others (Romero et al. 2014). We might think, therefore, that imposing lifetime sparsity benefits the system to learn a set of complex and richer basis. With respect to the learned LiDAR basis, Figure 2.4(b), these are edge detectors related to "height variations" of the objects, e.g. containers-vs-ground, roof-vs-ground, ground-vs-sea, roofs, train rails vs. ground in the image. When combining RGB+LiDAR, Figure 2.4(c) shows that the learned basis inherits properties of both modalities, resembling altitude-colored detectors.

We also see in Figure 2.4d that RGB bases scatter twofold for the projections onto the first two ISOMAP components: the first a color-predominant diagonal on top of a typical edge, and the second tri-band grayscale textures. Also higher frequency, both grayscale and colored, are far from the subspace center. Figure 2.4e shows the LiDAR case, where the scatter is simpler: low frequencies in the center and height-edges surrounding this center. Finally, Figure 2.4f shows RGB+LiDAR, where color and texture clusters are unraveled, but height-edges become slightly more colored, and again high-frequency patterns are far from the mean.

Our experiments suggest that RGB and LiDAR carry complementary pieces of information and then its feature representation is no longer sparse. Further, the induced topological spaces through ISOMAP embeddings also have revealed this complementarity so that RGB+LiDAR leads to a representation in which color and altitude are combined.

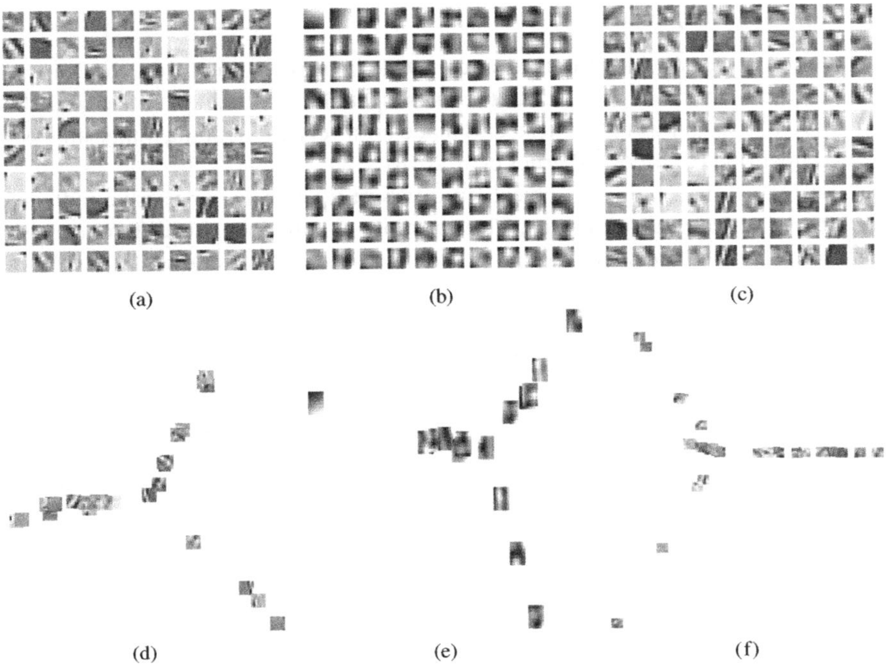

Figure 2.4 Top: for RGB (a), LiDAR (b) and RGB+LiDAR (c), learned bases by the convolutional network using EPLS. Bottom: the corresponding topological representations using the first two ISOMAP components.

2.4 Conclusions

Unsupervised learning remains an important and challenging endeavor for artificial intelligence in general, and for Earth Sciences in particular. We treated the topic of learning feature representations from multivariate remote sensing data when no labels are available. All the chapters of this book deal with deep generative models like Generative Adversarial Networks, Variational AutoEncoders, and self-taught learning, and rely on the idea of finding a good latent embedding space autonomously.

Extracting meaningful representations in this scenario is challenging because an objective criterion is missing. We want to highlight that whatever criterion will be absolutely arbitrary. Here we focused on including sparsity in standard convnets, even if the framework around the EPLS algorithm could be in principle applied to any other deep learning architecture. Sparsity is a very sensible criterion, motivated by the neuroscience literature, and very useful in practice: sparse representations enforce faster, more compact and interpretable models. We reviewed the field of sparse coding in deep learning, illustrated the framework in several remote sensing applications, and paid special attention to accuracy, robustness and interpretability of the extracted features. We have confirmed that the deeper the neural network, the more advantageous the sparse representation, thus confirming the results widely observed in the supervised literature. We have also shown that the network results in less compact representation when it fuses data with (physically orthogonal and complementary) information content, such as multispectral and LiDAR data. Visualization of the learned features from data was done explicitly and in a geometric space, which shed light on the way the network interprets information and fuses it. The methodology allows exploitation and exploration of very high-dimensional spaces, not necessarily attached to remote sensing images, or even natural images in general, but to Earth global products or climate model simulations.

The current literature is offering new ways: (van den Oord et al. 2018) propose Contrastive Predicting Coding (CPC), which tries to be a universal unsupervised learning approach pretending to learn representations that encode the underlying stable structure and shared information between different parts of highly dimensional data. This has inspired the Contrastive Learning Framework (Chen et al. 2020), which, in the particular area of visual representations, tries to find a representations by maximizing agreement between differently augmented views of the same data example via a contrastive loss in the latent space. We anticipate that sparse and deep convolutional representations learned from data will open new venues to find expressive and sparse feature representations of the Earth system.

3

Generative Adversarial Networks in the Geosciences

Gonzalo Mateo-García, Valero Laparra, Christian Requena-Mesa, and Luis Gómez-Chova

3.1 Introduction

In the last years, deep learning has been applied to develop generative models using mainly three different approaches.

Variational autoencoders (VAEs) (Doersch 2016) are a class of autoencoders that use deep neural networks architectures for the encoder and the decoder, and the parameters are optimized to enforce the statistical properties of the data in the latent space. VAEs allow new synthetic data to be generated following the distribution of the training data. In order to do so, one only has to generate data following the distribution defined in the intermediate stage and apply the decoder network. While VAE is an interesting technique, it has not been widely adopted in remote sensing yet. Further insight into VAEs and their use for Earth System Science can be found in Chapter 13.

Another popular approach is based on normalizing flows (Jimenez Rezende and Mohamed 2015), which rely on architectures that respect some properties of the data, like the dimensionality. Similarly to VAEs, they enforce a particular distribution in the transformed domain, being the selected distribution a multivariate Gaussian in most of the cases. However, this technique has not been extensively used in remote sensing problems.

Probably the most used generative methods based on deep learning are generative adversarial networks (GANs) (Goodfellow et al. 2014b). Of the three mentioned methods, GANs have excelled in several problems, and in the Earth Sciences in particular. Application of GANs has had an enormous impact on fields like image and language processing. Some of the applications of GANs have become the state of the art in these fields where there is a clear spatial and/or temporal data structure (see for instance Gonog and Zhou (2019)).

In the last decade, a plethora of models and architectures based on the fundamentals of GANs have been developed theoretically and widely used in real-world applications[1]. While presenting a taxonomy of this huge number of methods is far from the scope of this chapter, most of the approaches can be divided in three main families. The first family corresponds to the regular GANs, where the architecture is similar to the VAE or the normalizing flow methods. The second family is the conditional GANs, where an extra input, on which the

1 See for instance: https://github.com/hindupuravinash/the-gan-zoo.

Deep Learning for the Earth Sciences: A Comprehensive Approach to Remote Sensing, Climate Science, and Geosciences, First Edition. Edited by Gustau Camps-Valls, Devis Tuia, Xiao Xiang Zhu, and Markus Reichstein.

model conditions some properties of the generated data, is added to the model. Finally, the third family is devoted to merge different GANs structures, which helps to deal with unaligned datasets, as we will see in detail in this chapter.

In the field of Earth Sciences, GANs have been used for multiple problems, as we will review in section 3.3. Firstly, GANs have been used for generating synthetic samples to be used in supervised, semi-supervised, and unsupervised problems. However, their use is not restricted to these tasks. For instance, an interesting problem in remote sensing is domain adaptation (DA), since some GANs architectures can be used to adapt existing datasets or algorithms to different satellite features.

3.2 Generative Adversarial Networks

In this section, we review the three main different families of GANs architectures. The original GANs were proposed as an unsupervised method, in which the only requirement is to have data from the distribution we want to synthesize new samples. However, in practice, when we generate new samples, we might want to control the characteristics of the generated data; for instance, the class which the generated samples belongs to. Conditional GANs were proposed to address this issue. In conditional GANs, the generator synthesizes data taking into account particular auxiliary information given as input. Conditional GANs thus need to be provided with this extra information. Note that this approach loses the main advantage of GANs, which is being an unsupervised method. The idea of the third family is a little bit different from the previous ones but probably is the most interesting one. This family is based on an autoencoder architecture where the encoder and decoder networks are the generator part of two different GANs. Therefore, it is devoted to finding transformations to convert from one class of data to another class. The extra terms of the discriminators to the autoencoder architecture help the learning process when using unpaired datasets for training, since having paired data is a strong requirement, e.g. for domain adaptation problems.

3.2.1 Unsupervised GANs

The scheme for a GAN is shown in Figure 3.1. Both the generator (G) and the discriminator (D) are implemented using either shallow or deep neural networks (Radford et al. 2016).

During the training procedure, the generator converts a random vector ($\mathbf{r} \in R$) into a synthetic sample ($\hat{\mathbf{x}}$), which is passed to the discriminator. The discriminator could receive data either from the generator ($\hat{\mathbf{x}}$) or from the real dataset ($\mathbf{x} \in X$), and its goal is to discern whether the sample is real or is a *fake* coming from the generator. The discriminator produces as output a real number between 0 and 1 that is interpreted as the probability of

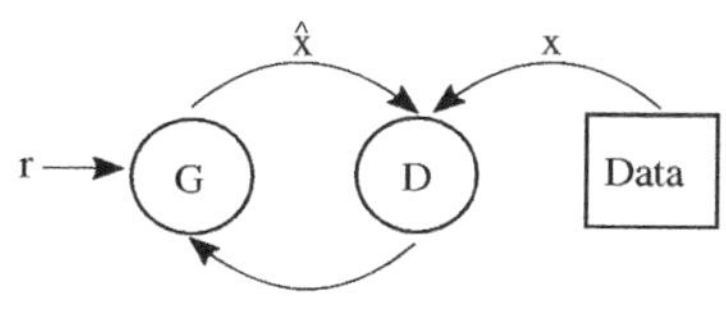

Figure 3.1 Generative adversarial network scheme. It shows the flow for the original data (**x**), synthesized data ($\hat{\mathbf{x}}$), and the random data (**r**) from the different parts of the method: generator (G), discriminator (D), and the training dataset (Data).

the sample being real. When the provided data comes from the generator, this probability is used as the error metric in order to improve the generator.

In particular, the loss functions of the discriminator (D) and the generator (G) are given by:

$$\mathcal{L}_{GAN}(D) = -\mathbb{E}_X[\log(D(\mathbf{x}))] - \mathbb{E}_R[\log(1 - \overbrace{D(G(\mathbf{r}))}^{\hat{\mathbf{x}}}].\tag{3.1}$$

$$\mathcal{L}_{GAN}(G) = -\mathbb{E}_R[\log(\overbrace{D(G(\mathbf{r}))}^{\hat{\mathbf{x}}}]\tag{3.2}$$

Note that Equations 3.1 and 3.2 are *adversarial* since minimizing Equation 3.1 will push $D(\hat{\mathbf{x}})$ towards zero, whereas minimizing 3.2 will push $D(\hat{\mathbf{x}})$ to 1. GANs' proposal consists of minimizing these equations iteratively using a gradient descent based method w.r.t. the weights of the generator and discriminator networks, respectively.

The main novelty of GANs is proposing a novel non-parametric and adaptive cost function. This cost function can be seen as a way to evaluate the likelihood of the generated data but without using a fixed likelihood function (Gutmann et al. 2018). Instead, we use a non-parametric method (e.g. a neural network) in order to learn the likelihood function while training the generator. This allows the likelihood function to change from less to more restrictive during the training. This helps the generator learning procedure since, at the beginning of the training, it is complicated for the generator to produce reliable samples. Therefore, having a very restrictive likelihood function at this stage would stall the generator learning process, i.e., the likelihood of the generated samples would be always zero and the generator would have no useful gradients towards improving the generated samples.

While GANs generated samples are *state-of-the-art*, several extensions have been proposed in order to restrict the generated data to have particular characteristics. For instance, the Info-GAN (Chen et al. 2016) maximizes the mutual information between some latent space features and the target space. By doing so, we can have control over imposing different features when generating the data while, at the same time, we can explore the feature space.

3.2.2 Conditional GANs

The scheme for the conditional GANs (Mirza and Osindero 2014) is shown in Figure 3.2. The main difference with the classical GANs is that the input to the generator is a combination of a random vector ($\mathbf{r}$) and a feature vector ($\mathbf{y}$). Actually, in most of the implementations, the random vector is neglected (Isola et al. 2016).

The cost functions of the discriminator (D) and the generator (G) of the conditional GANs, are given by:

$$\mathcal{L}_{CGAN}(D) = -\mathbb{E}_{(X,Y)}[\log(D(\mathbf{x},\mathbf{y}))] - \mathbb{E}_{(R,Y)}[\log(1 - D(\overbrace{G(\mathbf{r},\mathbf{y})}^{\hat{\mathbf{x}}},\mathbf{y}))].\tag{3.3}$$

$$\mathcal{L}_{CGAN}(G) = -\mathbb{E}_{(R,Y)}[\log(D(\overbrace{G(\mathbf{r},\mathbf{y})}^{\hat{\mathbf{x}}},\mathbf{y}))]\tag{3.4}$$

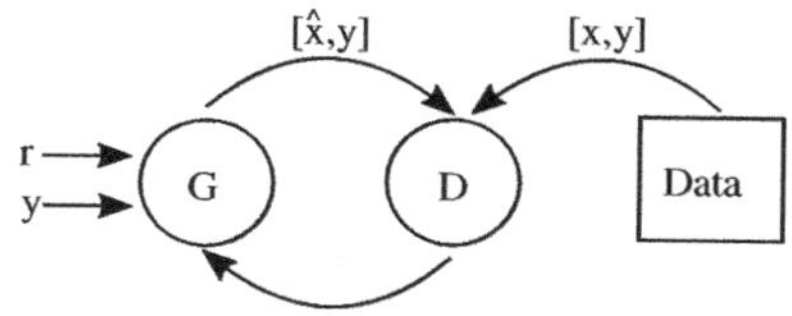

Figure 3.2 Conditional generative adversarial network scheme. It shows the flow for the original data (**x**), synthesized data (**x̂**), the random data (**r**), and the auxiliary information (**y**) among the different parts of the method: generator (*G*), discriminator (*D*), and the training dataset (Data).

We see that, in this approach, the discriminator also has two inputs which are either real (**x**) or fake data (**x̂**) and auxiliary information (**y**). Hence, the discriminator tries to distinguish samples from the joint distribution of X and Y. This in turn forces the generator to be consistent with its input (the auxiliary information **y**), avoiding the *mode collapse* problem (the generator memorizes one sample which is always produced as output). But, on the other hand, this makes the method dependent on paired samples.

3.2.3 Cycle-consistent GANs

The scheme for the cycle-consistent GANs is shown in Figure 3.3. Different authors proposed this scheme simultaneously in Zhu et al. (2017); Kim et al. (2017b); Yi et al. (2017). In general, having paired samples is costly or impossible in some cases. This family of GANs tries to solve this problem. While these works have been proposed for the particular problem of image-to-image translation, the scheme is generic and it can be used to find a DA procedure using unpaired samples, i.e. we have inputs and outputs but we do not know which particular input corresponds to which output.

This scheme consists of two coupled conditional GANs, where the output of one generator is the auxiliary input information for the other generator. This makes the method unsupervised since we only need to have samples from both domains but not necessarily paired (the discriminators in this case require only one input). The cost function of the coupled GANs consists of three parts:

$$\mathcal{L}_{CYCGAN} = \mathcal{L}_{GAN1} + \mathcal{L}_{GAN2} + \lambda \mathcal{L}_{CYC}. \tag{3.5}$$

The first ($\mathcal{L}_{GAN1}$) and second ($\mathcal{L}_{GAN2}$) parts of the cost functions correspond to the cost function of the conditional GANs, where the auxiliary input information corresponds to the output of the other generator. The third part ($\mathcal{L}_{CYC}$) is the classical autoencoder cost

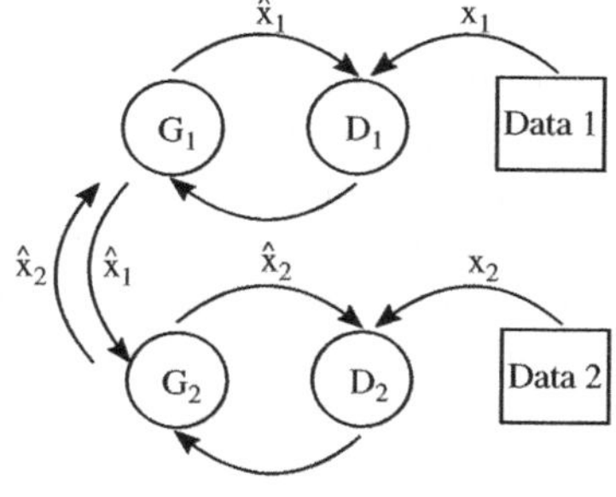

Figure 3.3 Cycle-consistent generative adversarial network scheme. It shows the flow for the original data from the first ($\mathbf{x}_1$) and the second ($\mathbf{x}_2$) datasets. The synthesized data from one generator (**x̂**) are used as the desired signal (**y**) for the other generator.

function. It enforces that the sample that passes through both generators has to be similar to itself:

$$\mathcal{L}_{CYC} = ||\mathbf{x}_1 - G_2(\overbrace{G_1(\mathbf{x}_1)}^{\hat{\mathbf{x}}_2})||. \tag{3.6}$$

As in classical autoencoders, different norm functions can be used in this part depending on the final goal.

3.3 GANs in Remote Sensing and Geosciences

Since 2017 the use of generative adversarial networks in remote sensing and geosciences has increased exponentially. GANs were proposed originally in the field of remote sensing for synthesizing Earth observation images. In particular, in Guo et al. (2017) they used GANs to improve the performance of synthetic aperture radar (SAR) image simulators from a given database in an unsupervised way. However, the use of GANs was rapidly extended to other supervised, semi-supervised, and unsupervised problems, and the generative adversarial methods were adapted to the particularities of the application field. The unsupervised use of GANs expanded to domain adaptation (DA) problems and to the extraction of deep features from remote sensing data. DA approaches have a great applicability in remote sensing due to the large number of available data sources. The adaptation of data distributions can be applied for instance to different datasets but is also of paramount importance when considering multimodal and multisensor approaches (Gómez-Chova et al. 2015). Details of an illustrative DA application in Earth observation will be presented in section 3.4.1. GANs have also been used as feature extractors that enable the computation of deep features. These feature extraction approaches are usually unsupervised or semi-supervised, and the extracted features can be used for further classification or change detection studies. It is also relevant for mitigating the small-datasets problem, which affects most remote sensing applications due to the high cost and the difficulties of creating supervised labeled sets to be used as ground truth. In the following sections, we will present an overview of a range of remote sensing and geosciences applications in which the use of different generative adversarial network architectures has been exploited.

3.3.1 GANs in Earth Observation

In Earth Sciences applications, data sources are diverse, heterogeneous, and highly correlated. In some cases, one is interested in generating synthetic data from a particular sensor following a particular distribution, i.e., simulating data of a single source and for given conditions (Guo et al. 2017). However, in most cases, we have a problem where data from several sensors have to be combined and further processed. In these cases, when the sensors present different characteristics and provide data with slightly different distributions, results might be improved by adapting the different source domains to a common domain where the problem is more easily solved.

In the context of DA, a lot of different applications can be found in the literature. On the one hand, in classification problems, when considering a single sensor, one can assume that the training and test datasets come from different domains, and GANs can be useful to find invariant representations for both the training and test data (Elshamli et al. 2017). This happens, for example, in the classification of aerial images when one wants to transfer previously labeled data to the new images (Zhu et al. 2019b), and DA with GANs helps to reduce the bias between the source and target distributions increasing the discriminative ability of generated features (Yan et al. 2019 ; Liu et al. 2019). On the other hand, a similar but more difficult situation arises when dealing with two or more different satellite platforms. The idea is still the same and GANs are used to obtain a better adaptation between the two satellites' data during the test phase without carrying out a separate training for each platform. For example, in Ye et al. (2019), an unsupervised DA model was presented to learn invariant features between SAR and optical images for image retrieval. Even when working with multiple optical remote sensing platforms, DA can increase robustness of the models allowing to transfer the knowledge gained from one trained domain to the target domain, in terms of both transfer learning and data augmentation (Segal-Rozenhaimer et al. 2020).

Following a similar reasoning as for DA, different adversarial architectures can be also used to extract the intrinsic data features in a deep latent space. The idea is to exploit the powerful ability of deep learning models to extract spatial and spectral features in an unsupervised manner, providing a feature representation that can be eventually useful for: anomaly detection in hyperspectral imagery (Xie et al. 2019), hyperspectral image processing (Zhang et al. 2019), or aerial scene classification (Yu et al. 2020).

The list of specific remote sensing and geosciences applications that have benefited from the advent of generative adversarial methods is large. However, two of these applications deserve a special attention due to the number of generative adversarial approaches that have been proposed to deal with change detection and super-resolution. On the one hand, the first change detection approach based on GANs was proposed by Gong et al. (2017), where change detection was handled as a generative learning procedure that modeled the relation between bitemporal images and the desired change map. In Zhao et al. (2019), a seasonal invariant term was introduced to avoid undesired changes in the final maps due to seasonality trends. Finally, in Hou et al. (2019), GANs were used to reformulate change detection as an image translation problem, differencing bitemporal images in the feature domain rather than in the traditional image domain. On the other hand, in the last years, a lot of attention has been also put on super-resolution techniques based on generative adversarial methods. GAN-based methods have shown to provide high-resolution images with higher perceptual quality than mean-square-error-based methods, which tend to generate smoothed images. In Jiang et al. (2019), a GAN-based edge-enhancement network was proposed for the super-resolution reconstruction along with a noise insensitive adversarial learning. In Li et al. (2020), a multiband super-resolution method based on GANs was proposed to exploit the correlation of spectral bands and avoid the spectral distortion in hyperspectral images. Finally, in Zhang et al. (2020a), a visual saliency GAN was proposed to enhance the perceptual quality of generated super-resolution images avoiding undesired pseudo textures.

3.3.2 Conditional GANs in Earth Observation

Despite the excellent performance of GANs synthesizing realistic images, in remote sensing problems we usually intend to use the generated data for a particular purpose that also depends on additional input variables. In these cases, the data generation has to be conditioned on these related features. In fact, in most conditional GANs (CGANs) approaches (see section 3.2.2 for details), in addition to replacing the random inputs of the generator by these features, the cost function of the conditional GANs is also modified to adapt the training of the Generator weights to the specific problem.

Conditional GANs were firstly applied to remote sensing to generate or simulate Earth observation data given a different observation from other data source. It is worth to note that, for the training of CGAN the input features have to be paired with the synthesized samples. For example, in Ghamisi and Yokoya (2018), the network is trained on images where both a digital surface model (DSM) and optical data are available in order to define an image-to-DSM transformation. In Merkle et al. (2018), a CGAN generates a set of ground control points (GCPs) from a SAR satellite image by matching optical and SAR images. In Niu et al. (2019), a CGAN translates an optical image to the corresponding SAR image in order to enable change detection in heterogeneous images by reducing their pixelwise differences and making the direct comparison feasible.

This later translation approach has been also followed to synthesize images from one source domain that are then used to restore or enhance images from the target domain. In Bermudez et al. (2019), missing optical data is generated by exploiting the corresponding SAR image from the same area but at a different acquisition date. In Dong et al. (2019), they proposed a loss function for an inpainting network, which adds a supervision term to remove cloud occlusions in historical records of sea surface temperature images. Finally, in Dong et al. (2020), a CGAN is trained using in-situ elevation measurements for filling voids due to mountain shadows in incomplete radar data.

In a similar way, CGANs have been used for spatial interpolation and image pansharpening. In Oliveira et al. (2019), CGANs are used as an interpolation tool for improving seismic data resolution. In Shao et al. (2019), a CGAN is used to preserve the spatial and spectral information of the panchromatic and multispectral bands, simultaneously. And, lately, multiconditional GANs have been also used to generate contours for surface reconstruction from large point clouds (Zhang et al. 2020b).

3.3.3 CycleGANs in Earth Observation

One of the last adversarial approaches applied to Earth Sciences problems has been the Cycle-Consistent Adversarial Networks (sec 3.2.3). This coupled GAN architecture was firstly proposed in remote sensing to handle the difficult problem of image-to-image translation between synthetic aperture radar (SAR) and optical images (Liu and Lei 2018). In Wang et al. (2019), the authors presented a supervised version of the CycleGAN in order to control the generation process of optical images from the SAR images. The objective was to preserve both the land cover and structure information during the SAR-to-optical image translation at the cost of training the models in a supervised manner with paired images. One of the consequences of the SAR-to-optical image translation is that the generated optical images are free of clouds, which is critical for land studies. In this context, CycleGANs have also been applied for cloud removal in optical images by learning the mapping between unpaired cloudy images and cloud-free images (Singh and Komodakis 2018). On the other hand, CycleGANs have been also used in remote sensing to exploit their DA capabilities. In Liu et al. (2018b), a CycleGAN was used to adapt simulated samples to be more similar to real samples, which allows an improved data augmentation with simulation approaches. In a similar way, in Saha et al. (2019a), CycleGANs are used to mitigate multisensor differences and to adapt the different source domains before applying an unsupervised change detection.

3.4 Applications of GANs in Earth Observation

In this section, we are going to present two illustrative applications of GANs in real remote sensing problems. The goal is to pass from theory to practice in two relevant case studies. These applications are domain adaptation of images coming from two different satellites and landscape emulation using climate, geological, and anthropogenic variables as input.

3.4.1 Domain Adaptation Across Satellites

In this section, we will focus on using GANs to align images coming from different sensors. Notice that, in a multi-sensor scenario, the DA problem (see Chapter 7) arises constantly since there is a constantly growing number of different satellites whose images share some properties (since they are all imaging the Earth) and are different in others (spatial resolution, spectral response, quality of the sensors, or active vs. passive instruments). Those differences mean that derived models using those images as input must be tailored for each input data source (both traditional remote sensing models and machine learning based ones have the same problem). Hence, finding a DA transformation from one satellite to another could potentially make models independent of the source data used to train them, i.e. of the source domain that they were originally designed to work in. In addition, a DA transformation between satellite imagery can be used to exploit existing databases of images and ground truth measurements (such as in-situ data or manually annotated images) for new satellites. These databases, which are very costly to produce, are collected once more for each new satellite that is launched, since the distribution of satellite imagery changes.

As explained also in Chapter 7, GANs can be applied for DA to align *source* and *target* data distributions. In our case the *source* and *target* distributions will be represented by images coming from two different sensors; hence, following the notation of (see Chapter 7), we will focus on image to image translation approaches (I2I) where the goal is to align the input distributions (see section).

In order to apply I2I translation with GANs, we can rely on CGANs (see Chapter 7) or in CycleGANs (section 3.2.3). CGANs require paired samples for training; in our application it would mean that *simultaneous* images of the same location and same acquisition time from both sensors are needed. In some cases, the time constraint can be relaxed to *close in time* images, however, in other cases this might not be enough; for instance, for applications that look for sudden changes in images such as clouds, floods, and wildfire detection. In those cases, the CycleGANs formulation is more appealing since it does not require paired images for training. See the work of Hoffman et al. (2018) for a comprehensive reference of the use of Conditional GANs and CycleGANs to DA problems in computer vision.

One illustrative example of DA using conditional GANs applied to the multi-sensor scenario is the work in Mateo-García et al. (2019). In this work, the authors propose a slight modification of the conditional GANs formulation to build a DA transformation between Landsat-8 and Proba-V satellites that does not require paired samples. The goal is to exploit Landsat-8 manually annotated cloud masks to build a cloud detection algorithm for Proba-V. In order to build the DA transformation, firstly the overlapping bands of Landsat-8 and Proba-V are selected and then Landsat-8 images are upscaled from the 30m Landsat-8 spatial resolution to the 333m Proba-V resolution using the physical characteristics of both sensors: the point spread function of Landsat-8 and Proba-V and their spectral response. After this physically based transformation, the spatio-spectral properties of the images are the same (similar spectral bands and same spatial resolution), however, statistical differences between upscaled Landsat-8 and Proba-V images still remain as shown in Figure 3.4 (see e.g. the blueish color of clouds in Proba-V). These differences are probably due to differences in the Proba-V and Landsat-8 instruments. Since Proba-V is a smaller and somehow less accurate satellite, authors build a DA model from Proba-V to Landsat-8 upscaled images. This model can be seen as a noise removal method for Proba-V images. Figure 3.5 summarizes the procedure to train the DA model (the generator in the CGAN scheme). The conditional GAN model is trained using unpaired Landsat-8 and Proba-V images adding a consistency loss for the generator between the real and the generated image. This is required since only the generated data is used as input to the discriminator, but not the input and the labels as in equations 3.3 and 3.4. Results show a better quality of adapted Proba-V images with a lower amount of saturated values as seen in Figure 3.4. In addition, the cloud detection model trained in Landsat-8 upscaled images of Mateo-García et al. (2020) performs better in the Proba-V denoised images than in the raw Proba-V imagery.

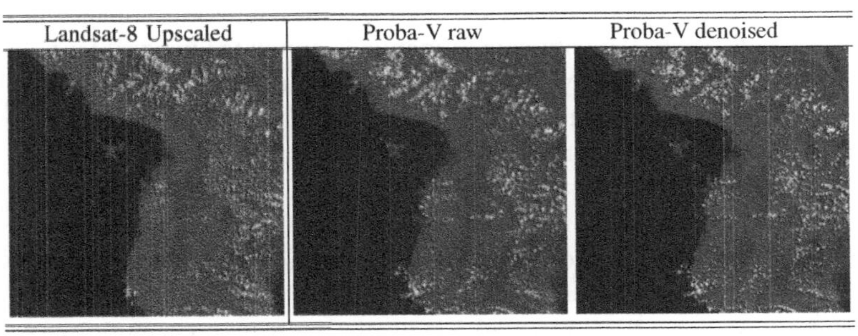

Figure 3.4 Close in time upscaled (333m resolution) Landsat-8 and Proba-V images before and after the domain adaptation (Generator).

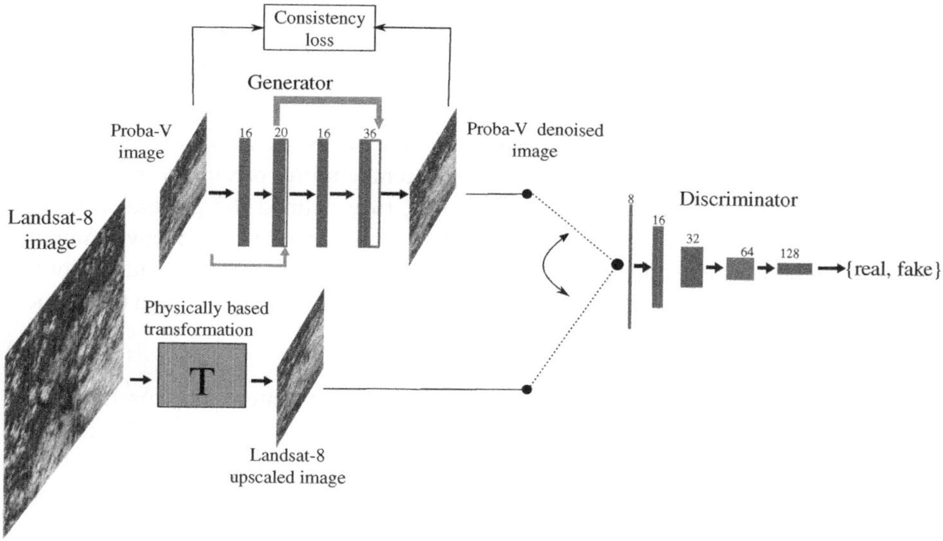

Figure 3.5 Example of architecture for Domain Adaptation between two satellites proposed in Mateo-García et al. (2019).

3.4.2 Learning to Emulate Earth Systems from Observations

Earth system modeling is mostly based on the understanding of the underlying processes. Such understanding is formalized in a set of equations and rules forming up the programs that allow us to virtually recreate our system and make valuable predictions.

With the advent of learned generative models such as Variational Autoenconders and Generative Adversarial Networks, a new approach to modeling Earth becomes viable: learning the behavior of the complex spatio-temporal dynamics from raw data. This becomes very interesting, especially on systems for which it is notoriously difficult to build classical simulations but we have plenty of observational data. One example of a learned predictive model on such system is landscape prediction based on other climatic variables (Requena-Mesa et al. 2019). Such learned models allow us to make predictions on systems for which we do not have good knowledge-based models. On the other hand, it is also possible to emulate existing classical simulators in order to speed up the data generation process. The ability of generative models to link the given input to the distribution of possible outcomes makes them behave alike an ensemble run of classical simulations. It produces not only a single prediction, but an array of them, allowing for uncertainty measurement. In such sense, variational and adversarial spatio-temporal generators can be trained to emulate real Earth systems. These generators can be used for nowcasting by feeding them some context frames of an observed Earth system, and are able to forecast the next few time steps.

Further elaborating on the ability to build predictive models for hard to simulate systems, some works have deployed generative networks in order to build successfully.

In Requena-Mesa et al. (2019), a model capable of predicting landscapes as seen from space is introduced for the first time. Raw remote sensing data is used as a proxy of the landscapes of Earth. Landscapes are complex systems, and their evolution is linked to the interplay of climatic, geological and anthropogenic factors. They built a predictive model based on conditional GANs capable of generating new landscapes, as seen from space, given a set of landscape forming variables (e.g. average temperature, precipitation, geological substrate, etc.). To model the problem uncertainty, they defined the ground truth as a probability distribution over the remote sensing data conditioned on a set of environmental conditions C. They then trained a generative neural model G as an approximation of the unknown function that relates environmental conditions to landscapes as seen from space:

$$G(C, r; \theta) \approx f(Clim, Geo, AI), \tag{3.7}$$

where θ denotes the network parameters and r is a probabilistic latent space useful to sample multiple plausible landscapes for each set of environmental variables (Climatic variables, Geological variables, and Anthropogenic Intervention indicators).

The study deployed a spatial to spatial generative model (see Figure 3.6). Such a model convolutionally encodes into the latent code spatial information describing higher-order relationships across the environmental variables, and deconvolutionally decodes satellite imagery features from the latent code. The model also makes use of skip connections, as these are needed to keep the landscape features on the right spatial locations, e.g., if there is a high slope on the top-right corner of the environmental predictors, there probably should be a mountain on the generated imagery. The study shows that the conditional GANs can generate landscapes that closely resemble the real ones as measured based on patch level metrics, while simpler models cannot replicate these high level metrics to an usable degree. In addition, they show that both the use of a convolutional-deconvolutional network architecture and a discriminator-based training are key to achieve a good landscape prediction. Both the convolutions and the adversarial training are some of the greatest recent advances in deep learning and it is only now that their applications to relevant Earth system problems is being displayed.

While there are just very few works deploying generative models for systems that are hard to predict numerically, it is not hard to imagine it could be used towards many other challenging tasks. For example, wildfire nowcasting, long term fluvial and coastal sediment dynamics, landscape evolution over time, or patterns of urban growths among others.

The use of spatio-temporal generator networks to emulate existing numerical simulations is in its infancy. However, the dimensionality of the data, both large in space and time, make the current deep learning architectures used for video prediction of special interest for Earth System science. Specially, those architectures that model explicitly temporal dependencies (with LSTM-like structures), spatial dependencies (with convolutional steps), and the inherent stochastic and ambiguity of the ground truth (with probabilistic latent space), such as Babaeizadeh et al. (2017) or Lee et al. (2018a), seem to have all the key ingredients to be able to emulate complex Earth System models. There are currently ongoing works examining the suitability of conditioned stochastic spatio-temporal generator to emulate some of the current classical models showing preliminary promising results. These models might trickle down into a more massive usage in the following years.

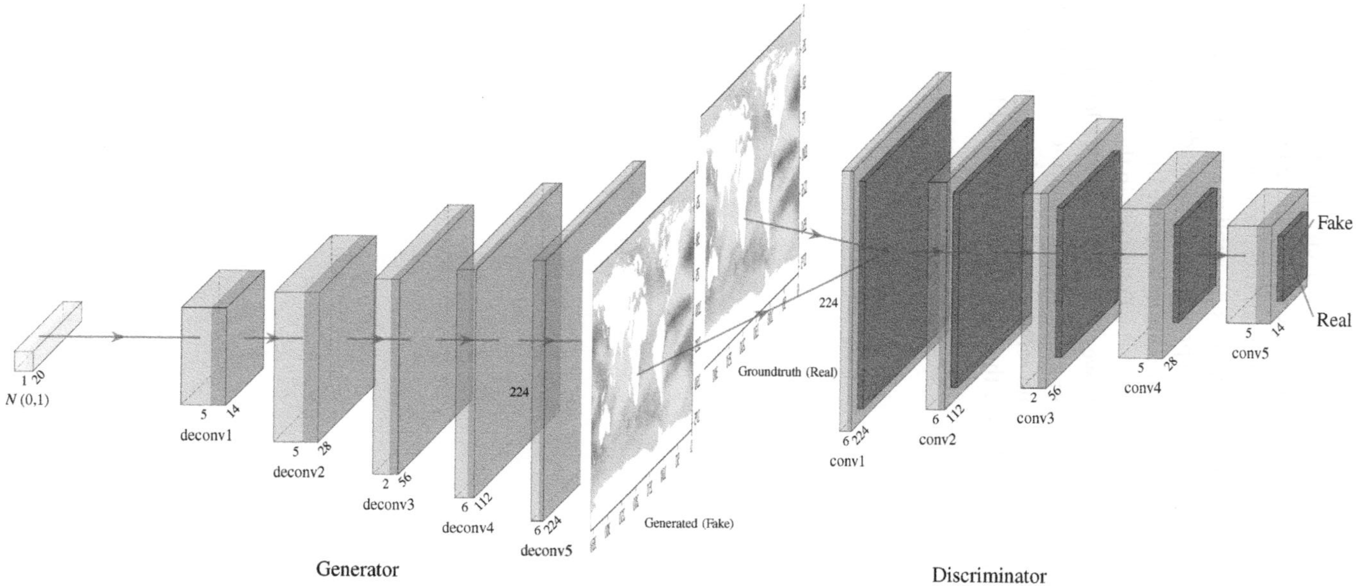

Figure 3.6 An example architecture of a convolutional generative adversarial model. We can use generative models to unsupervisedly learn the distribution of Earth system variables and expand our available datasets.

3.5 Conclusions and Perspectives

In this chapter, we have reviewed a particular type of deep learning generative models based on Generative Adversarial Networks. Main GAN families have been presented describing their architectures, motivations, and how these families have been applied to remote sensing problems. Finally, two different remote sensing applications have been presented to illustrate the use of GANs in two challenging problems: domain adaptation across different satellites and landscape emulation using climate, geological, and anthropogenic variables.

While the apparent hype for GANs seems to be fading out in the computer science community, these generative models and the newer ones to come have a lot of potential to show in the geosciences field. If further development, refinement, and understanding of them is pursued, as well as their adaptation into remote sensing and geoscientific applications, generative models can ultimately change the basis by which we build our predictive and inference models.

4

Deep Self-taught Learning in Remote Sensing

Ribana Roscher

4.1 Introduction

STL, originally proposed by Raina et al. (2007), has become a promising paradigm to exploit large amounts of unlabeled data for analysis and interpretation tasks. Its main strength is to exploit unlabeled data samples, which do not have to belong to the same classes as the labeled data samples nor do the unlabeled samples have to follow the same data distribution as the labeled data samples (see Figure 4.1). This makes the approach advantageous over common approaches such as semi-supervised learning or active learning. The common procedure to STL is sparse representation (SR), which learns features in an unsupervised way and uses them for supervised classification.

DSTL extends this approach by combining STL and deep learning. It has the same goal as deep neural networks, namely to learn a representation that is better suited for the chosen task than the original representation. However, the basis of self-taught learning is sparse representation instead of a neural network, and thus other possibilities exist such as designing and learning of an interpretable model. In the literature, several variants of deep sparse representations have been proposed, which show considerable improvements over shallow sparse representations. Recent approaches learn multiple layers of sparse representations: He et al. (2014) propose a fully unsupervised feature learning procedure and combines it with hand-crafted feature extractions and pooling layers. Combining the representations from multiple layers, they achieve state-of-the-art performance on object recognition. Lin et al. (2010) use deep belief networks with local coordinate coding, which represents all data samples as a sparse linear combination of anchor points. All these approaches use labeled information only as a final step for classification purposes, and thus the representation is not optimized given labeled information. In contrast, Gwon et al. (2016) applied the concept of backpropagation which is commonly used for optimizing the parameters in neural networks.

The above approaches determine the presentation in such a way that the data is represented as well as possible and in some cases additionally the classification task is solved as accurately as possible. Since the representation does not consist of real data samples and no restrictions are imposed, the representation cannot be interpreted and explained. However, this is often necessary, for example, for unmixing tasks (Bioucas-Dias et al. 2012). One

Deep Learning for the Earth Sciences: A Comprehensive Approach to Remote Sensing, Climate Science, and Geosciences, First Edition. Edited by Gustau Camps-Valls, Devis Tuia, Xiao Xiang Zhu, and Markus Reichstein.

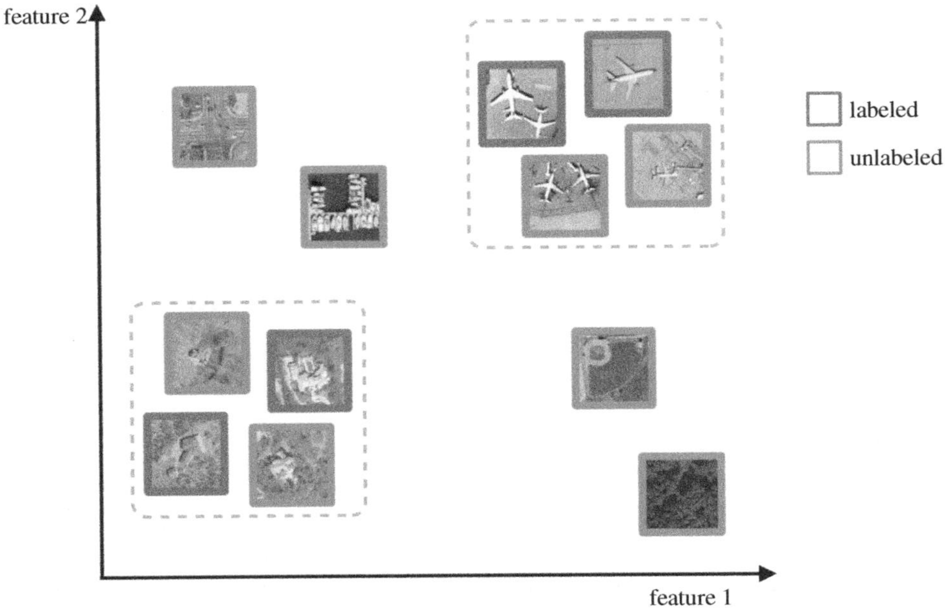

Figure 4.1 Schematic illustration of different learning paradigms and their use of labeled (dark gray) and unlabeled (light grey) data samples. In contrast to semi-supervised learning (data samples used shown in dotted boxes), self-taught learning also uses unlabeled data, which need not belong to the same classes as the labeled data. Images are from the UC Merced dataset (Yand and Newsam 2010).

approach in this direction is presented by Bettge et al. (2017), where the representation can be directly related to real data samples.

The first section of the chapter describes the basic principle of STL and introduces an interpretable version if it. Next, the motivation behind a deep version of STL is given, and the deep framework and the single steps are explained in detail. Moreover, the main components sparse representation and dictionary learning are introduced. We refer the reader to Goodfellow et al. (2016) for details about the basic concept of deep learning. The goal of this chapter is to describe an interpretable and explainable deep learning framework which is able to learn deep features with the help of large amounts of unlabeled data. The chapter shows how the set of unlabeled data is optimized to be used for classification and how interpretability can be enforced, the latter being important for various tasks in remote sensing (Roscher et al. 2020; Reichstein et al. 2019).

4.2 Sparse Representation

The following section explains sparse representation, which is the most commonly used approach to STL. The basic idea of sparse representation is that each sample can be represented by a weighted linear combination of a few elements from a dictionary. For STL, the dictionary contains unlabeled, yet powerful basis elements which are relevant to a given task (e.g., classification, anomaly, or change detection). For classification tasks, for example, the estimated weights of the sparse linear combination are used as input into a classifier and thus, the dictionary elements need to be chosen accordingly such that the weights are highly discriminative. Earth observation data such as remote sensing images are particularly suitable for an efficient representation as a sparse linear combination of dictionary elements due to its statistical structure, especially its high spatial redundancy among neighbored observations (Thiagarajan et al. 2015).

In terms of basic sparse representation a $(V \times 1)$-dimensional sample $\boldsymbol{x}$ is represented by a weighted linear combination of a few elements taken from a $(V \times T)$-dimensional dictionary $\boldsymbol{D}$ such that

$$\boldsymbol{x} = \boldsymbol{D}\boldsymbol{\alpha} + \epsilon, \tag{4.1}$$

with $\|\epsilon\|_2$ being the reconstruction error, V the dimension of the sample, and T the number of dictionary elements. The coefficient vector comprising the weights is given by $\boldsymbol{\alpha}$. The sample $\boldsymbol{x}$ can be, for example, an $(M \times 1)$-dimensional pixel from an image, so that $V = M$, or an $((M \cdot Z) \times 1)$-dimensional vectorized image patch $\boldsymbol{x} = \text{vec}(\boldsymbol{X})$ with Z being the number of pixels in patch $\boldsymbol{X}$ and $V = M \cdot Z$.

The sparse coding optimization problem for the determination of optimal $\hat{\boldsymbol{\alpha}}$ is given by

$$\hat{\boldsymbol{\alpha}} = \text{argmin } \|\boldsymbol{D}\boldsymbol{\alpha} - \boldsymbol{x}\|_2 \quad \text{subject to} \quad \|\boldsymbol{\alpha}\|_f < \rho \,. \tag{4.2}$$

The norm $\|\cdot\|_f$ needs to be chosen accordingly to induce sparsity. In case L_0-norm is used, that means $f = 0$, ρ is the number of non-zero elements in the coefficient vector. Less intuitive, L_1-norm with $f = 1$ is defined as the sum of absolute values and demands a suitable choice of a threshold ρ. A commonly used optimization procedure is orthogonal matching pursuit for the L_0-optimization task (Zhang et al. 2015; Elad 2010). Further constraints can also be applied, for example non-negativity constraints or a summation of the coefficients to 1, which will enhance the interpretability of the result.

4.2.1 Dictionary Learning

Well established approaches to learn dictionaries are, for example, K-SVD and its variants such as discriminative K-SVD. Here, the learning of the dictionary is integrated in a two-step approach comprising alternating a sparse coding step with fixed dictionary and a dictionary update step. These approaches generally result in high approximation ability, which is necessary for tasks such as image denoising or inpainting (Aharon et al. 2006; Mairal et al. 2008). However, the learned dictionaries do not contain real data samples, so that the presentation cannot be explained from the point of view of certain applications like unmixing.

In the context of unsupervised learning, alternative approaches aim at the construction of a suitable dictionary by selecting representative samples such as cluster centers (Raina et al. 2007), extremal points (Roscher et al. 2015; Chen et al. 2014), or those identified by auxiliary approaches such as autoencoders (Feng et al. 2020b). In a similar way, for classification tasks, dictionaries can be built from labeled data samples (Wang et al. 2013a; Aharon et al. 2006). This is accomplished by using class-wise dictionaries, i.e., using a structured dictionary based on the class assignment of the dictionary elements. An additional enforcement of samples to be reconstructed by one class-specific dictionary only can be introduced, showing improved classification results (Chen and Zhang 2011). Approaches which construct dictionaries from given samples or representatives of them have the advantage to provide interpretable and explainable results, which is of great interest in a wide variety of applications in the Earth Sciences. Typical applications based on satellite-based and close-range Earth observation data comprise unmixing tasks (Bioucas-Dias et al. 2012) or plant phenotyping (Wahabzada et al. 2016; Römer et al. 2012). However, for Earth observation data, the number of labeled samples is limited and thus, class-wise dictionaries may not be representative enough.

4.2.2 Self-taught Learning

The STL procedure uses labeled training data following an underlying distribution and unlabeled data, which can belong to arbitrary classes and need not to follow the same distribution as the labeled data. Although both datasets do not have to follow the same distribution, a relation between them can be helpful and may lead to an increase in accuracy (Feng et al. 2020b). In detail, STL uses unlabeled data ${}^{u}\boldsymbol{X} = [{}^{u}\boldsymbol{x}_q], q = 1, \ldots, Q$, training data ${}^{\text{tr}}\boldsymbol{X} = [{}^{\text{tr}}\boldsymbol{x}_n], n = 1, \ldots, N$ with labels ${}^{\text{tr}}\boldsymbol{y} = \left[{}^{\text{tr}}y_n\right]$ given, and test data ${}^{\text{t}}\boldsymbol{X} = [{}^{\text{t}}\boldsymbol{x}_u], u = 1, \ldots, U$, also with labels ${}^{\text{t}}\boldsymbol{y}$. All data samples consist of V-dimensional feature vectors $\boldsymbol{x} \in \mathrm{I\,R}^V$, and labels are given by $y \in C = \{C_1, \ldots, C_K\}$, where K is the number of classes. The labels are also represented by target vectors $\boldsymbol{t} = \left[t_k\right]$ of length k coding the label with $t_k = 1$ for $y = C_k$ and $t_k = 0$ otherwise. The dictionary $\boldsymbol{D} = \left[{}^{u}\boldsymbol{x}_1, \ldots, {}^{u}\boldsymbol{x}_T\right]$ is embodied by unlabeled data samples, whereas generally $T \leq U$. In case of supervised classification, a classifier model is trained with ${}^{\text{tr}}\hat{\boldsymbol{\alpha}}_n$ being the new higher-level feature representations of ${}^{\text{tr}}\boldsymbol{x}_n$ with respect to the dictionary $\boldsymbol{D}$. In the same way, higher-level features are extracted for the test data ${}^{\text{t}}\boldsymbol{x}_u$, which are classified by the learned model.

4.3 Deep Self-taught Learning

The shallow STL approach can be extended to DSTL by stacking STL modules yielding deep representations (Gwon et al. 2016; Bettge et al. 2017). In the following, a simple DSTL framework is explained in detail which is used for a classification task. The baseline approach, being a stack of sparse representations, is defined as follows:

$$\boldsymbol{X} = \boldsymbol{D}^{(1)}\boldsymbol{A}^{(1)} + \boldsymbol{E}^{(1)}, \tag{4.3}$$

$$\boldsymbol{A}^{(1)} = \boldsymbol{D}^{(2)}\boldsymbol{A}^{(2)} + \boldsymbol{E}^{(2)}, \tag{4.4}$$

$$\vdots$$

$$\boldsymbol{A}^{(L-1)} = \boldsymbol{D}^{(L)}\boldsymbol{A}^{(L)} + \boldsymbol{E}^{(L)}, \tag{4.5}$$

with $\boldsymbol{X}$ being the original data samples (pixels or image patches), $D^{(\cdot)}$ being the dictionaries embodied by unlabeled data samples, $\boldsymbol{A}^{(\cdot)} = [\boldsymbol{\alpha}]^{(\cdot)}$ being the sparse coefficients, and $||\boldsymbol{E}^{(\cdot)}||$ being the reconstruction error calculated by Frobenius norm.

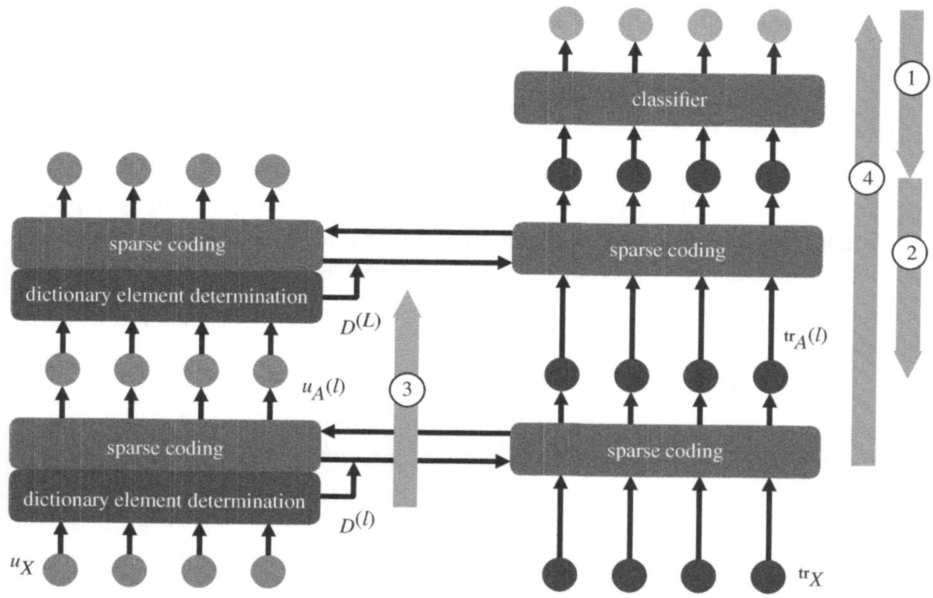

Figure 4.2 Schematic illustration of the deep self-taught learning framework: Deep feature representations are learned layer-wise by an iterative procedure of updating the sparse activations (steps 1, 2, and 4) and learning updated dictionaries (step 3).

The schematic structure and the workflow of DSTL is illustrated in Figure 4.2. The left side illustrates the part of the architecture which uses unlabeled samples and the right side illustrates the part of the architecture which uses labeled samples, where the samples and their representations are depicted with blue and red circles. The network consists of L layers and is trained layer-wise. The following steps describe the procedure.

Initialization The initialization is performed layer-wise, so that in each layer $l = 1, \ldots, L$ the dictionary elements $\boldsymbol{D}^{(l)}$ are determined. The first layer $l = 1$ is the first one to be initialized by extracting relevant dictionary elements from unlabeled data ${}^{u}\boldsymbol{X}$. After one of the procedures specified in section 4.2.1, the dictionary elements either represent real data so that the dictionary is interpretable, or the dictionary elements are calculated with respect to an optimization criterion. Given the dictionaries, ${}^{u}\boldsymbol{A}^{(1)}$ is estimated using sparse coding. Likewise, in all subsequent layers $l > 1$, representative samples are extracted from ${}^{u}\boldsymbol{A}^{(l-1)}$ to build $\boldsymbol{D}^{(l)}$, and sparse coding is performed yielding the representations for the labeled data ${}^{\mathrm{tr}}\boldsymbol{A}^{(l)} = \boldsymbol{D}^{(l+1)\mathrm{tr}}\boldsymbol{A}^{(l+1)} + \boldsymbol{E}^{(l+1)}$.

Classifier Training The last layer of the DSTL framework is intended to perform the classification given the learned representation. Regarding neural networks, the most common method is the use of a softmax layer. Transferred to DSTL, the layer performs a logistic regression for classification given the learned representation in the last layer, see (Bishop 2006, chapter 4). The posterior probabilities derived by logistic regression are given by

$$P\left(C_k \mid {}^{\mathrm{tr}}\boldsymbol{\alpha}_n^{(L)}\right) = h_k = \frac{\exp\left(\boldsymbol{w}_k^{\mathrm{T}}\,{}^{\mathrm{tr}}\boldsymbol{\alpha}_n^{(L)}\right)}{\sum_k \exp\left(\boldsymbol{w}_k^{\mathrm{T}}\,{}^{\mathrm{tr}}\boldsymbol{\alpha}_n^{(L)}\right)} \tag{4.6}$$

where the weight matrix $\boldsymbol{W} = [\boldsymbol{w}_k]$ contains the parameters of the separating hyperplanes in feature space. Given sparse representations ${}^{\mathrm{tr}}\boldsymbol{A}^{(L)} = [{}^{\mathrm{tr}}\boldsymbol{\alpha}_n^{(L)}]$, the goal is to learn a classifier $h_k({}^{\mathrm{tr}}\boldsymbol{A}^{(L)})$ for all classes $C = \{C_1, \dots, C_K\}$. For this, the posterior probabilities

$${}^{\mathrm{tr}}\tilde{\boldsymbol{T}} = \begin{bmatrix} P(C_1 \mid {}^{\mathrm{tr}}\boldsymbol{\alpha}_n^{(L)}) \\ \vdots \\ P(C_K \mid {}^{\mathrm{tr}}\boldsymbol{\alpha}_n^{(L)}) \end{bmatrix} \tag{4.7}$$

are compared to reference targets $\boldsymbol{T} = [\boldsymbol{t}_n]$. The error is minimized by updating the classifier parameters, the learned dictionaries and sparse representations using the update steps explained in the following.

Update Procedure In order to learn these parameters, we perform the following steps illustrated by (1)–(4) in Figure 4.2.

Step 1: The output from the last layer L is fed into a classifier and the classification loss function value is computed. Therefore, in the first update step, the dictionaries are fixed and only the training representations are updated. Given the reference targets ${}^{\mathrm{tr}}\boldsymbol{t}_n$ and estimated posterior probabilities ${}^{\mathrm{tr}}\tilde{\boldsymbol{t}}_n$, the loss function is given by

$$\mathcal{J}\left({}^{\mathrm{tr}}\boldsymbol{\alpha}_n^{(L)}\right) = \frac{1}{2}\|{}^{\mathrm{tr}}\boldsymbol{t}_n - {}^{\mathrm{tr}}\tilde{\boldsymbol{t}}_n\|^2. \tag{4.8}$$

With the gradient of the loss function with respect to the training representations ${}^{\mathrm{tr}}\boldsymbol{\alpha}_n^{(L)}$, the representations in the last layer are updated with

$${}^{\mathrm{tr}}\alpha_{rn}^{*(L)} = {}^{\mathrm{tr}}\alpha_{rn}^{(L)} + \rho\,\frac{\partial\,\mathcal{J}\left({}^{\mathrm{tr}}\boldsymbol{\alpha}_n^{(L)}\right)}{\partial\,{}^{\mathrm{tr}}\alpha_{rn}^{(L)}}. \tag{4.9}$$

with the upper index $(\cdot)^*$ indicating the updated representations, ρ being the learning rate, and the index r denoting the r-th row.

Step 2: In order to update the representations in layers $l = 2, \dots, L-1$, the updated representation in the last layer is used subsequently with

$${}^{\mathrm{tr}}\boldsymbol{A}^{*(l)} = \boldsymbol{D}^{(l+1)}\,{}^{\mathrm{tr}}\boldsymbol{A}^{*(l+1)}, \tag{4.10}$$

starting with $L-1$.

Step 3: Given the updated training representations, the dictionaries $\boldsymbol{D}^{(l)}$ are updated using the gradient descent method as proposed by Gwon et al. (2016). The following loss function is used to compute the dictionary update

$$J_D\left(\boldsymbol{D}^{(l)}\right) = \frac{1}{2}\|\boldsymbol{D}^{(l)}\,{}^{\mathrm{tr}}\boldsymbol{A}^{*(l)} - {}^{\mathrm{tr}}\boldsymbol{A}^{*(l-1)}\|^2, \tag{4.11}$$

which will be minimized. In the zeroth layer (input layer), ${}^{\mathrm{tr}}\boldsymbol{A}^{*(0)}$ is set to be the original data ${}^{\mathrm{tr}}\boldsymbol{X}$. The gradient descent updating rule for the rth dictionary element of the lth layer is

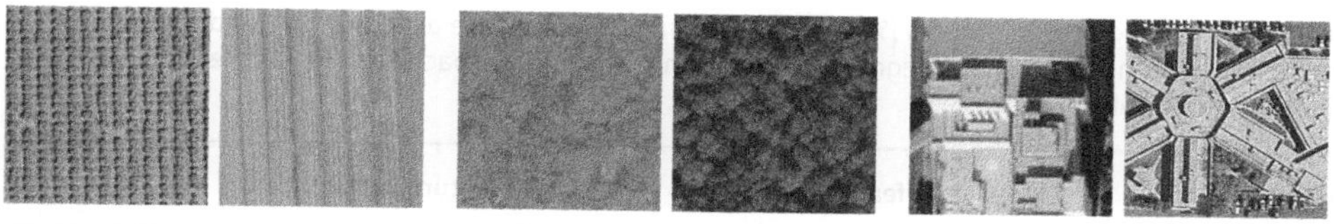

Figure 4.3 Example images from UC Merced dataset for the classes `agriculture`, `forest`, and `buildings`.

given by

$$d_r^{(l)} = d_r^{(l)} - \gamma \left(D^{(l)\text{tr}} A^{(l)} - {}^{\text{tr}}A^{*(l-1)} \right) {}^{\text{tr}}\alpha_r^{(l)}, \tag{4.12}$$

where γ is the learning rate. In the same way, step 2 is repeated for the unlabeled data and dictionary updates are repeated using the unlabeled data representation ${}^{u}A^{(l)}$.

Due to the dictionary updates, their entries no longer represent real samples of data, so they can no longer be interpreted and explained in the context of a specific application. As an optional step and to keep them interpretable, the dictionary elements in the first layer are limited to real data samples. This can be achieved, for example, by moving a sufficiently changed dictionary element to the nearest neighbor in the set of unlabeled data points in the feature space. A sufficiently large change is necessary to ensure that the data samples are shifted in such a way that the updated dictionary elements do not match the original ones and the optimization gets stuck.

Step 4: This step readjusts the labeled representations with the updated dictionaries $D^{(l)}$ by minimizing the reconstruction error of ${}^{\text{tr}}X$ using the sparse coding procedure and updates the classifier. We iterate steps (1)–(4) until convergence of the dictionaries or by applying early stopping.

Extensions: The presented workflow illustrates the basic variant of the DSTL framework. However, it only represents a linear representation, which may not be flexible enough to solve an intended task. To learn a more complex, yet interpretable, representation, operations like pooling can be introduced. An application of further operations as they are common in neural networks is also possible.

4.3.1 Application Example

Dataset The example dataset is a subset of the well-known UC Merced dataset (Yang and Newsam 2010) for image categorization with 21 classes. For our application, we use the classes `agriculture`, `forest`, and `buildings` as labeled samples and the rest of the classes are treated as unlabeled samples. Overall, the dataset contains 300 labeled samples and 1800 unlabeled samples, where one vectorized image is treated as one data sample.

Experimental Setup The used RGB images are resized to $32 \times 32 \times 3$ pixels, leading to 3072-dimensional input feature vectors. All samples are normalized to a range $[0, 1]$. We randomly extract 150 training samples (${}^{\text{tr}}X$) and use the remaining 150 samples as test samples (${}^{\text{t}}X$). We use archetypal analysis on the unlabeled dataset to extract dictionary elements, where the size of the dictionary is limited by how many archetypes can be extracted (Cutler and Breiman 1994).

Table 4.1 Class-wise accuracies [%], overall accuracy [%], average accuracy [%], and Kappa coefficient obtained by logistic regression (LR) using various approaches. The best results are highlighted in bold-print.

class	original features + LR	STL features + LR	DSTL + LR
agriculture	74.0	84.0	**86.0**
forest	82.0	98.0	**100.0**
buildings	42.0	78.0	**88.0**
overall	66.0	86.7	**92.0**
average	66.0	86.7	**92.0**
Kappa	0.49	0.80	**0.88**

The experiment compares logistic regression using the original samples, STL with logistic regression, and DSTL with two layers and logistic regression to investigate whether the accuracy benefits from the DSTL approach over basic logistic regression and STL. The learning rate to update the sparse representation of the training data in the last layer is set to $\rho = 0.1$ and the dictionary learning rate is set to $\gamma = 0.1$. Logistic regression is performed with gradient descent with a learning rate of 0.1 and 1000 iterations in each DSTL iteration. The weights are initialized by the solution from the last iteration. The DSTL update is iterated 100 times to find the best dictionary, judged by application to the validation data.

Results Table 4.1 shows the class-wise, overall, and average test accuracy. In all our experiments, the STL and DSTL with logistic regression achieve an improvement over the original representations with logistic regression. For DSTL initialization, 10 archetypes were extracted in the first layer and 16 archetypes could be extracted in the second layer.

The approach was also implemented as an interpretable framework, resulting in a slight drop in accuracy of about 2%, by moving the dictionary elements to the nearest neighbors from the set of unlabeled data at every fifth iteration after the dictionary update using the unlabeled data. The used unlabeled dictionary elements show beach, storage tanks, harbor, tennis courts, runways, and free ways. Overall, similarities to known, though unlabeled, scenes can be helpful for a classification. In the context of other applications where these relations may be important, this approach can help to gain further insight into the results.

4.3.2 Relation to Deep Neural Networks

DSTL shows some parallels to other deep learning methods, especially deep neural networks. In neural networks, an application-specific representation is also learned, though the optimization strategy may differ (Mairal et al. 2011). The main focus of the DSTL framework is the interpretability and explainability of the learned model. In general, deep sparse representations are highly related to CNNs: The convolutional filters in the CNN have a similar task as dictionary elements. While the filters in CNNs are optimized using back-propagation, DSTL uses an optimization procedure also based on gradient descent, which, however, can be adjusted to yield interpretable dictionary elements. The filter responses in the CNN are derived independently in each layer by summed multiplications in contrast to the jointly derived activations in the sparse linear combination in the DSTL framework. Nevertheless, the filters and the responses in the CNN are jointly optimized beyond all layers. An even closer relation represent networks using convolutional sparse coding (Bristow et al. 2013), which replace the multiplication in the sparse coding procedure by convolutions. Also Kemker and Kanan (2017) introduce a related approach, which uses stacked convolutional autoencoders for learning deep representations.

4.4 Conclusion

In this chapter deep self-taught learning was introduced to combine the advantages of self-taught learning and deep learning. In our example experiment we could show that the framework for self-taught learning benefits from unlabeled data, so that the learned deep features can be used for an improved classification compared to a classification with the original feature representation. Since the dictionaries can be restricted to unlabeled data samples, they are interpretable and explainable in the context of a specific application and can be used to derive further insights into the learned model. Deep self-taught learning shows many parallels to other methods that use deep learning, especially neural networks. Many operations that are used in neural networks can also be used for deep self-taught learning, so the presented method can benefit from the previous knowledge of neural networks. The advantage of deep self-taught learning compared to many other methods is that it has a simple influence on the interpretability of the model. However, this also increases the computing time, which requires the development of more efficient methods for learning interpretable dictionaries. Self-taught learning has not yet been used for many applications, and especially in the field of remote sensing the potential has not yet been fully analyzed. In contrast to many other deep learning methods, self-taught learning can use a lot of unlabeled data and work with few labeled data, which is a typical scenario for remote sensing. In addition, in remote sensing the interpretability and explainability of the model is often more important than in other communities, because on the one hand we can use previous knowledge and on the other hand a scientific consistency for the quality of the result must often be provided.

5

Deep Learning-based Semantic Segmentation in Remote Sensing

Devis Tuia, Diego Marcos, Konrad Schindler, and Bertrand Le Saux

5.1 Introduction

Semantic segmentation is the task of attributing each pixel in an image to a semantic class. In the case of Earth observation images, it is also called semantic labeling and is often related to some kind of mapping, for instance of land use types, vulnerability/risk, or in order to detect changes that have occurred in between acquisitions. Semantic segmentation is generally framed as a supervised task: labeled examples of each class are provided to a model which learns the input/output relations necessary to predict the same classes in as-yet unseen data.

Segmenting images from an overhead perspective has always been related to the need to integrate some kind of a-priori about spatial structures (Fauvel et al. 2013): in urban environments, the co-occurrence of classes and typical geometrical arrangements of objects are precious information that can lead to more accurate models, while in agriculture applications, the mixture of spectral signatures of crops and soil, as well as the textures observed at the leaves level, can be used to characterize stages of growth or detect diseases attacking the crops ahead of time.

This need to integrate priors about spatial arrangements, as well as the spatio-temporal correlations observed in remote sensing signals, made the transition to deep learning algorithms very natural: convolutional neural networks are spatial feature extractors by design and were rapidly adopted by the optical remote sensing community, which had been using convolutional image filters for decades. Questions of scale and rotation invariance as well as multi-sensor processing then became drivers for new developments of algorithms custom tailored to remote sensing data, which we will review in this chapter.

The chapter is organized as follows: in section 5.2 we review recent literature on semantic segmentation for remote sensing. In section 5.3, we present the most common approaches to semantic segmentation as they were introduced in computer vision. Finally, in section 5.4 we present three approaches from the recent literature where these architectures were introduced in remote sensing and modified to cope with the data specificities or the problem's own requirements.

Deep Learning for the Earth Sciences: A Comprehensive Approach to Remote Sensing, Climate Science, and Geosciences, First Edition. Edited by Gustau Camps-Valls, Devis Tuia, Xiao Xiang Zhu, and Markus Reichstein.

5.2 Literature Review

The remote sensing community was always aware of the importance of including texture information to succeed in semantic segmentation and, from the very beginning, filters based on convolution operators were used extensively. Some examples are simple moving averages (Camps-Valls et al. 2006), occurrence and co-occurrence filters (Pacifici et al. 2009), and morphological filters (Benediktsson et al. 2003). These textural features were designed and extracted in a feature engineering phase, prior to the classification step. More often, they were computed as activation (filter) maps, then stacked to the spectral bands as additional layers, and finally used in a classifier such as support vector machine or a random forest.

Later works started to explore the possibility of a less static approach, where filters would be learned, rather than imposed a-priori using expert knowledge. The first works towards *learning* the image filters were undertaken by performing a random search in the filters' parameter space: in Tokarczyk et al. (2015), the authors used filters generated with random mathematical operations in the pixel's neighborhood (for instance, selecting two pixels and taking the difference of their values) and then optimized the feature space by applying boosting principles. In Tuia et al. (2014), the authors used a random generator of parameters for a preset family of texture and morphological filters that were then learned greedily by selecting those which aligned the best with the error of the current model. This approach was further improved in Tuia et al. (2015), where a hierarchical filter generation strategy was proposed, where the output of a filter that was retained in the model became available for further re-filtering, therefore becoming a source to generate new, highly nonlinear hybrid combinations. The depth of the re-filter generation was controlled by regularization. This work has strong connections with modern convolutional networks, but still the filter space was explored randomly and filters were chosen in a greedy manner. The learning of the filters' parameters themselves was explored in Flamary et al. (2012), where authors alternately optimized an SVM classifier *and* a series of filters applied to the input image channels. This was pursued with the objective of learning the filter leading to margin maximization of the SVM and the filters were learned by gradient descent.

Given all these elements, the transition to modern convolutional neural networks for semantic segmentation was quite natural in remote sensing. The first attempts at semantic segmentation with convolutional neural nets applied the classification principle (i.e. predict a single label for the entire patch, attribute it to the central pixel, and then apply the classifier as a sliding window across the image). For example, a comparison of state of art networks as VGG or AlexNet was reported in Lagrange et al. (2015), where the authors showed the effectiveness of deep learning versus traditional feature engineering. Despite such encouraging results, the sliding window approach had two major limitations: on the one hand, it involved heavy computational burden at inference time, since a patch centered around each pixel had to go through the CNN for prediction, resulting in redundant computations; on the other hand, even though a patch-based CNN was indeed encoding spatial information in the prediction, it was doing it in a spatially unstructured way, since the prediction of two nearby pixels was obtained independently in subsequent inference passes (Volpi and Tuia 2017).

To cope with these shortcomings, fully convolutional approaches were explored (Sherrah 2016) and have nowadays become the state of the art in remote sensing semantic segmentation: these approaches are mostly based on encoder-decoder structures (Audebert et al. 2016; Kampffmeyer et al. 2016; Volpi and Tuia 2017; Maggiori et al. 2017a; Daudt et al. 2019) or on the creation of multiscale tensors by stacking activation maps at different levels of the CNN (Maggiori et al. 2017b; Fu et al. 2017; Volpi and Tuia 2018). Further works pushed the accuracy of the networks, for example by making use of Conditional Random Fields for post-processing (Paisitkriangkrai et al. 2016), class-specific edge information (Marmanis et al. 2018), multiscale context (Liu et al. 2018), co-occurrences between classes (Volpi and Tuia 2018), or by post-processing the resulting maps, for instance by deploying recurrent networks refining the maps iteratively (Maggiori et al. 2017b).

The success of these remote sensing specific approaches was enabled by new, public datasets with high spatial resolution and dense ground references. A large palette of datasets that focuses on sub-metric pixel segmentation in urban areas is nowadays available, for land use mapping (ISPRS 2D segmentation benchmark[1], IEEE Data Fusion Contest 2015 (Campos-Taberner et al. 2016)[2], Zurich summer (Volpi and Ferrari 2015)[3]), and building detection (Inria dataset (Maggiori et al. 2017c)[4], Spacenet[5]). Some other datasets tackle several of these challenges in parallel, such as DeepGlobe dataset (Demir et al. 2018)[6], which includes tasks of land cover classification as well as building and road extraction and can be used for the development of multitask and lifelong learning methods. Other data modalities than multispectral very high resolution images are also gaining momentum through competitions, such as SAR imagery (see the recent SpaceNet6 (Shermeyer et al. 2020)[7] about urban classification and Sen1Floods11 (Bonafilia et al. 2020)[8] about flood water identification) and hyperspectral images (IEEE GRSS Data Fusion Contest 2018[9]) or LiDAR point clouds (DALES dataset (Varney et al. 2020)[10]). Finally, recent datasets aiming at large scale (e.g. multi-city) classification with high-resolution data (e.g. Sentinel-2) are also more and more present, for instance for the classification of Local Climate zones (IEEE GRSS Data Fusion Contest 2017 (Yokoya et al. 2018)[11] or So2Sat LCZ42 (Zhu et al. 2020)[12]), land use (MiniFrance dataset (Castillo-Navarro et al. 2020)[13]), cloud detection (38-Cloud

1 http://www2.isprs.org/commissions/comm3/wg4/semantic-labeling.html
2 http://www.grss-ieee.org/community/technical-committees/data-fusion/2015-ieee-grss-data-fusion-contest/
3 https://sites.google.com/site/michelevolpiresearch/data/zurich-dataset
4 https://project.inria.fr/aerialimagelabeling/
5 https://spacenetchallenge.github.io
6 http://deepglobe.org/
7 https://spacenet.ai/sn6-challenge/
8 https://github.com/cloudtostreet/Sen1Floods11
9 http://www.grss-ieee.org/community/technical-committees/data-fusion/2018-ieee-grss-data-fusion-contest/
10 https://udayton.edu/engineering/research/centers/vision_lab/research/was_data_analysis_and_processing/dale.php
11 http://www.grss-ieee.org/community/technical-committees/data-fusion/2017-ieee-grss-data-fusion-contest-2/
12 https://mediatum.ub.tum.de/1454690
13 http://dx.doi.org/10.21227/b9pt-8x03

dataset (Mohajerani and Saeedi 2019)[14]) or temporal land cover changes (OCSD dataset (Daudt et al. 2018)[15]).

However, obtaining sufficient ground truth data for particular tasks and locations remains a challenge. In computer vision, the use of pretrained models is common to reduce the amount of required labels, thanks to the availability of large-scale datasets such as ImageNet. This has been shown to help in some remote sensing tasks (Cheng et al. 2017), but is not straightforward to use, for instance with images with more than three input bands or statistics that differ substantially from RGB images (e.g. SAR, thermal). As alternatives, it has been proposed to pre-train models on related remote sensing tasks (Wurm et al. 2019), render synthetic images by simulating an appropriate sensor (Kemker et al. 2018) or make use of publicly available maps as ground truth (Kaiser et al. 2017). Finally, domain adaptation (Wang and Deng 2018) is also being considered to adapt deep models trained on one geographic region to another: for the interested reader, we invite to read the dedicated chapter in the book (cf. Chapter 7).

As we show in the rest of this chapter, Earth observation data also possesses unique characteristics that can be exploited for semantic segmentation. For instance, data from different modalities (Audebert et al. 2018) and acquisition times (Zhong et al. 2019) are often available, which demands segmentation models specifically adapted to each type of data.

5.3 Basics on Deep Semantic Segmentation: Computer Vision Models

Semantic segmentation as a task is rooted in the computer vision literature and many architectures that have been proposed for handling natural images can in principle be adapted for the labeling of pixels composing aerial or satellite images. In this section, we review standard architectures as they were originally proposed and in the following section we discuss potential bottlenecks when adapting them to serve Earth observation purposes.

5.3.1 Architectures for Image Data

From Classification to Semantic Segmentation Semantic segmentation deep learning architectures inherit many of their characteristics from models originally designed for image classification, such as VGG and ResNet (He et al. 2016b), which take as input an image of a fixed size ($M \times N$) and number of bands (b_0) in the form of a tensor of size $M \times N \times b_0$ and return a vector of class scores of size C. Figure 5.1a shows a typical pipeline for image classification, i.e. a model that, from an image, predicts a single label representing the content of the center pixel. By the use of strided operations, the spatial extent of intermediate feature tensors is progressively reduced. For instance, a convolutional layer with b_1 filters and stride s results in an output tensor of size $M/s \times N/s \times b_1$. This downsampling allows each tensor location in the deeper layers to receive information from increasingly large areas of the image, i.e. to have a larger receptive field, which allows spatial context to be exploited.

14 https://github.com/SorourMo/38-Cloud-A-Cloud-Segmentation-Dataset
15 https://rcdaudt.github.io/oscd/

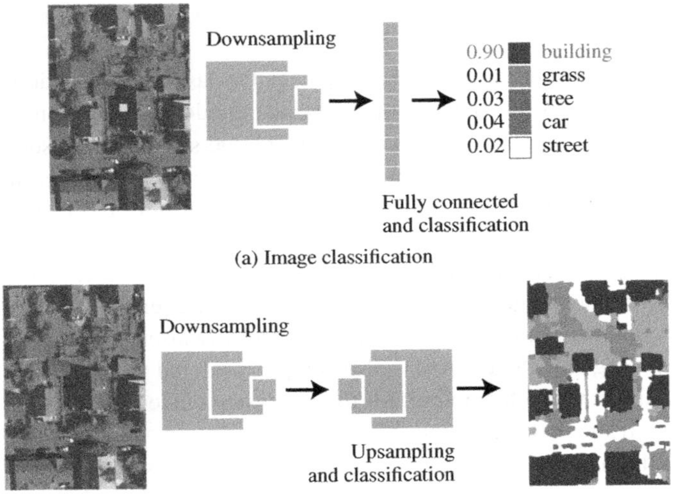

(a) Image classification

(b) Semantic segmentation

Figure 5.1 Comparison of pipelines for (a) image classification versus (b) semantic segmentation.

In image classification CNNs, the last layer is a tensor of size $1 \times 1 \times c_l$ and has the whole image as its receptive field. This tensor is treated as a vector and serves as input to the fully connected part of the model that outputs a vector of class scores of size C, summarizing the content of the image: in the case of remote sensing image semantic segmentation, this should represent the class the central pixel belongs to.

However, this aggregation of contextual information is at the price of spatial detail, which is detrimental for semantic segmentation. On the other hand, not performing any downsampling would allow to maintain the spatial detail at the cost of a smaller receptive field, which could prevent the access to enough spatial context to assign a class to each pixel.

In semantic segmentation (Figure 5.1b), the desired output is a tensor of size $M/d \times N/d \times C$, where d is the downsampling factor. Often $d = 1$, meaning that each individual pixel in the input image is assigned to a class. As pointed out in section 5.2, treating semantic segmentation as $M \times N$ classification problems, i.e. one prediction per pixel, is highly inefficient. This has driven the research of deep learning architectures for semantic segmentation that are able to simultaneously extract contextual information while maintaining spatial detail. We invite the reader to consult (Minaee et al. 2020) for a detailed survey of methods. In the following we will group and present the main architectures following an arbitrary distinction based on how they perform the upsampling from the CNN's last layer to the expected prediction support (one prediction per each original pixel).

Architectures based on Hard-coded Upsampling The upsampling operator is predefined, often bilinear interpolation. In these models, such as FCN (Long et al. 2015) and Hypercolumns (Hariharan et al. 2015), the number of learnable parameters is close to the one of the downsampling backbone. Their main limitation is that they are only able to learn about the context at the input level (e.g. the shape, size, and relative position of the features that indicate the presence of an object), and not at the output level (e.g. the typical shape that the object tends to have in the segmentation map).

The first approaches, such as Fully Convolutional Networks (FCN) (Long et al. 2015), consisted of substituting the fully connected operators with 1×1 convolutions, which are algebraically equivalent. In this way, an increase in the size on the input image results in a proportional increase in the size of the output tensor, which then becomes a map of class probabilities. This map is then upsampled to the resolution of the ground truth map in order to compute a pixel-wise loss. However, the resulting map tends to be coarse due spatial information lost through downsampling. In FCN, the authors compute multiple class probability maps, using tensors from different layers in the backbone, and average the results to minimize the loss of spatial detail.

Hypercolumns (Hariharan et al. 2015) is a similar architecture, but where feature maps obtained at different scales are upsampled and stacked, allowing each scale to specialize in different features. In the example shown in section 5.4.1, authors use variations of this method, in which the features are fused earlier by stacking them and applying several fully connected layers that allow to learn interactions between features at different scales (see Figure 5.2).

In PSPNet (Zhao et al. 2017) the authors propose to increase the receptive field to the whole image by applying average pooling with different number of bins, including a global average pooling, to a feature map. The resulting downsampled feature maps undergo an additional convolution before being upsampled back to the original resolution and stacked.

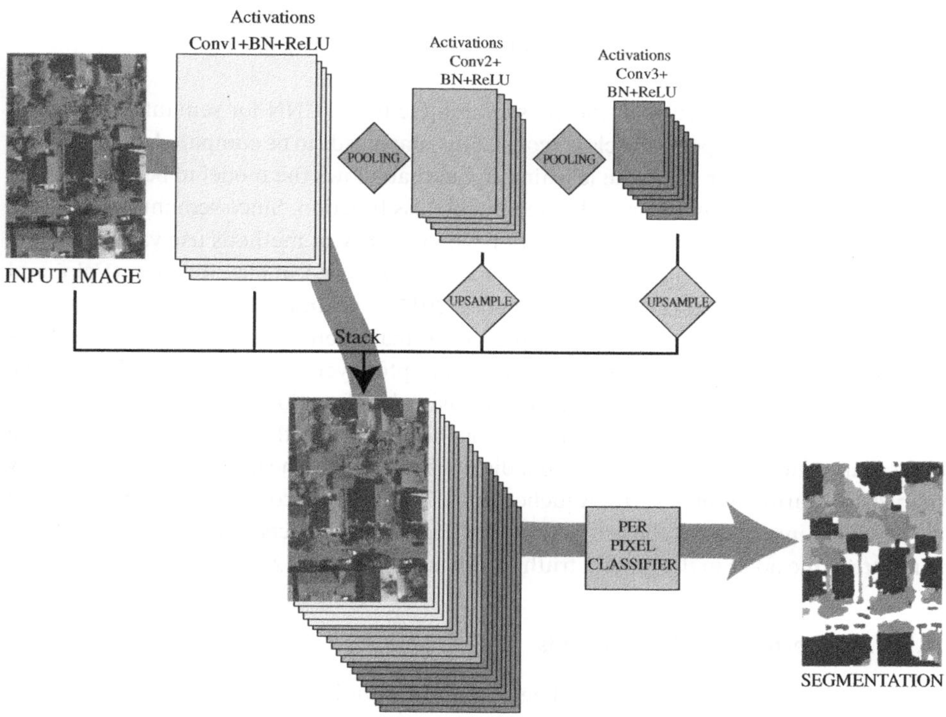

Figure 5.2 Example of architecture with a hard-coded upsampling, in which every feature map in the downsampling backbone is bilinearly interpolated to the image resolution and stacked.

Another way of improving the resolution/receptive field trade-off is to use dilated convolutions, also known as *à-trous* convolutions. Filters in dilated convolutions are sparse, resulting in larger kernels, and therefore larger receptive fields, without an increase in the number of parameters. The DeepLab (Chen et al. 2017a) pipeline uses dilated convolutions to reduce the impact of downsampling, but the authors found that a post-processing step based on Conditional Random Fields was needed to obtain a satisfactory level of spatial detail.

Architectures Learning the Upsampling A different type of approach are the *encoder-decoder* architectures (Noh et al. 2015), which consists of coupling the downsampling network, or encoder, with one that computes the upsampling in a cascade of stages, the decoder. These are often designed to be approximately symmetric, such that information from each layer in the encoder can be transmitted to the corresponding layer in the decoder.

This can be done by transferring the indices from each max-pooling layer in the encoder to un-pooling layers in the decoder, such as in SegNet (Badrinarayanan et al. 2017), see Figure 5.3a. Alternatively, each entire feature map from the encoder can be appended to the corresponding feature map in the decoder. In U-Net (Ronneberger et al. 2015a), both feature maps are stacked and upsampled with a deconvolutional layer, which is equivalent to a convolutional layer with fractional stride, see Figure 5.3b. Instead of a deconvolution, a bilinear upsampling can be applied, followed by convolutional layers to learn how to refine the result (Pinheiro et al. 2016). This learned upsampling can add a substantial amount of additional parameters and computational cost to the downsampling backbone, but allows the context both at the input and at the output levels to be learned.

Loss Functions Independently of the chosen architecture, a CNN for semantic segmentation generates a map of per-pixel class scores. This map needs to be compared to the ground truth map in order to produce the learning signal that allows the model to be trained. The exact nature of this comparison is defined by the loss function. Since semantic segmentation can be posed as pixel-wise classification, the majority of methods use variants of the cross entropy loss, also called multinomial logistic loss, used in classification (Volpi and Tuia 2017; Audebert et al. 2016; Maggiori et al. 2017a; Wurm et al. 2019). These variants often aim to compensate the imbalance present in many semantic segmentation datasets by re-weighting the importance of each class (Kampffmeyer et al. 2016). In addition, the per-pixel nature of the task can be leveraged by using loss functions that aim to exploit auxiliary information and spatial relations, such as those based on the Euclidean distance to a height map (Audebert et al. 2016) or to a distance map to the nearest semantic boundary (Yuan 2016; Marmanis et al. 2018; Audebert et al. 2019a). More complex loss functions that take explicitly into account the geometry of the segmented objects (Marcos et al. 2018b) or the nature of the noise in the ground truth (Mnih and Hinton 2012) have also been explored.

5.3.2 Architectures for Point-clouds

Not all Earth Observation data are images, and 3D data in the form of point clouds are becoming more and more common. When dealing with 3D data, semantic segmentation aims at point-wise classification of 3D points (instead of pixels). The main characteristics

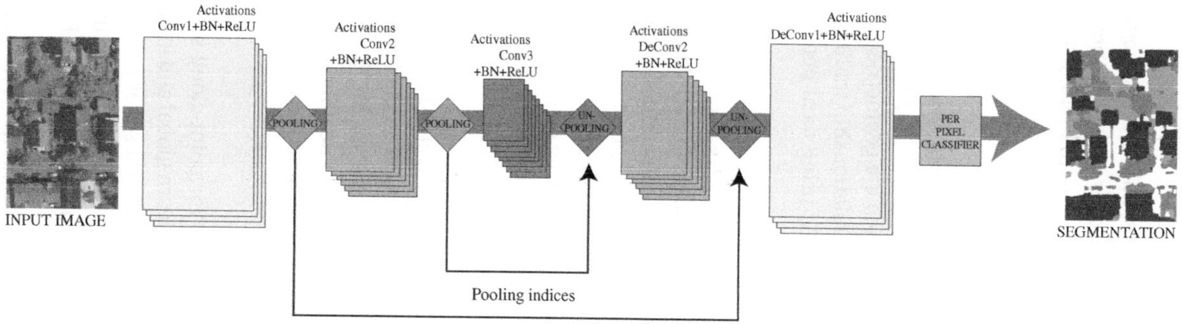

(a) SegNet (Badrinarayanan et al. 2017), propagating pooling indices.

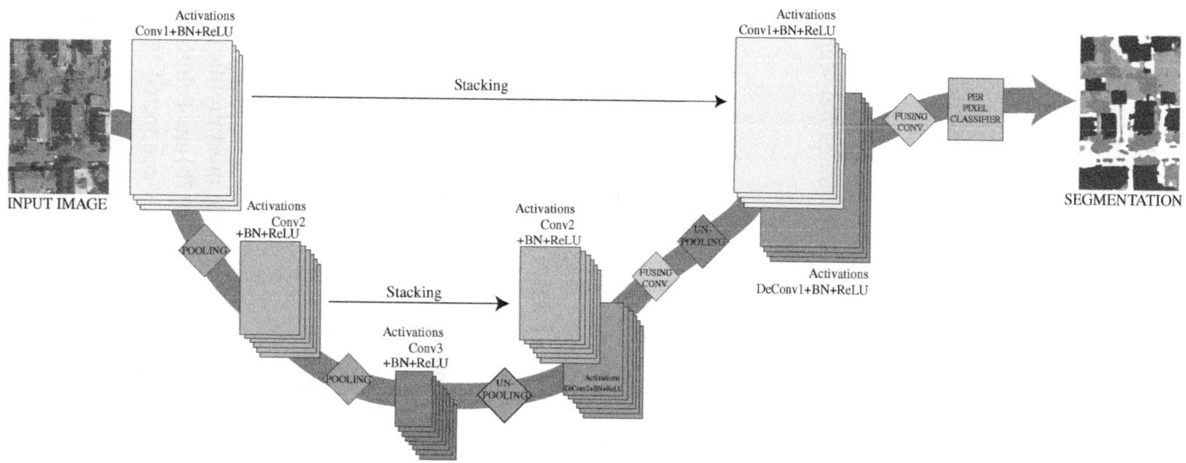

(b) U-Net (Ronneberger et al. 2015a), propagating activation maps.

Figure 5.3 Semantic segmentation architectures learning the upsampling.

emphasized previously also apply: the segmentation model is learned in a supervised manner and the spatial configuration of points matters. However, due to point cloud peculiarities, most methods described previously are not transferable directly. Therefore, research on point cloud semantic segmentation is very active and the current state of the art is blooming with several new approaches.

Statistical learning approaches for 3D aim to efficiently sample the spatial arrangement of local neighborhoods, while ensuring invariance at global scale and over various scenes. It leads to different families of approaches that we detail in the following: graph-based, 3D, 2D or 1D (i.e. approaches acting directly on every point). We also refer to the comprehensive review of Xie et al. (2019) for an overview of the state-of-the-art.

Graph-based Approaches These methods build a graph over points or locally consistent subsets of points and use graph neural networks for classification. For example, SuperPointGraph (Landrieu and Simonovsky 2018) first creates superpoints (which are geometrically simple shapes) and then builds the graph of superpoints using rich relationship features to link the superpoints. Finally, contextual segmentation is performed using local neural networks and graph learning.

3D-based Approaches These approaches are similar to the image-based CNNs described in the previous section, but consider an input space with an extra dimension. In VoxNet (Maturana and Scherer 2015), the local 3D neighborhoods are sampled with 3D convolutions over voxels (the 3D analog of pixels). Sparsity of points in 3D is a key issue here. It is handled with trilinear interpolation in SegCloud (Tchapmi et al. 2017) to refine the characterization of points with respect to their precise location. Another trick consists in using octrees as in OctNet (Riegler et al. 2017) to allocate voxels of various sizes according to the local density of points.

Sampling 2D Neighborhoods These methods consist in projecting the 3D point clouds in a 2D plane, either with multiple views as in MultiViewCNN (Su et al. 2015) or on an enclosing cylinder surface as in the Panorama representation (Shi et al. 2015a; Sfikas et al. 2017). The SnapNet approach (Boulch et al. 2018) pushes further the idea for local 2D neighborhood sampling by randomizing the view generation over the point-cloud. It is described later in section 5.4.2.

Point-based Sampling This is the most prolific category of algorithms for point-cloud segmentation. PointNet (Qi et al. 2017b) and PointNet++ (Qi et al. 2017b) learn global and local representations by applying fully-connected Multi-Layer Perceptrons over a set of points, thus encoding their geometric relationship. To offer a better characterization, point descriptors can be used as the input of PointNet rather than the simple location, as in PointSIFT (Jiang et al. 2018b).

Such approaches were surprisingly efficient and paved the way for developments of algorithms which try to emulate on point clouds the behavior of convolutions in 2D. Indeed, they define local transforms with global and local invariance properties. As for global sampling, local neighborhood characterization can be 3D, 2D, or 1D. In 3D, Flex-convolutions (Groh et al. 2018) use a local 3D voxel grid to capture the surroundings of each point. In

2D, Tangent-Conv (Tatarchenko et al. 2018) projects points on locally tangent 2D planes. Finally, local point-based approaches include PointCNN Li et al. (2018b) which introduced χ-convolutions over local subsets of points, KPConv (Thomas et al. 2019) and ConvPoint (Boulch 2019), which defined discrete convolutions weighted with respect to point distance.

5.4 Selected Examples

In the previous sections, we have presented deep learning architectures designed in other fields (mostly computer vision) and explained their functioning. As mentioned in the introduction, these architectures are very effective and can be used out of the box on aerial and satellite images, but with some points of attention. First, they only consider RGB images as inputs: to accommodate the high-dimensional input space offered, for example, by hyperspectral images, specific architectures must be designed (for a review of architectures for hyperspectral imaging, see Audebert et al. (2019b)). Second, they are not specific to remote sensing image characteristics and do not take into account priors such as the behavior of the image with respect to rotation (in optical aerial images, rotation is arbitrary and should not influence prediction) or looking geometry in SAR. Third, environmental monitoring often involves repetitive sensing, i.e. images of the same place being acquired several times: Semantic segmentation models can take advantage of this strong prior about spatial consistency, while focusing on learning temporal changes.

In this section, we discuss three case studies, each one dealing with one of the points mentioned above: first we'll see the benefits of encoding rotational invariance in a small CNN, and show that an invariant model can match performances of larger models (by orders of magnitude in terms of learnable parameters). Then, we will show a solution to process point clouds by characterizing local neighborhoods with 2D sampling. Finally, in the third case we will present a study in environmental monitoring, specifically lake ice detection, where static webcam across seasons and synthetic aperture radar images are used.

5.4.1 Encoding Invariances to Train Smaller Models: The example of Rotation

One of the main characteristics of optical overhead imagery is isotropy: there is no up-down orientation and all directions are *a priori* equivalent. However, the relative orientation of objects in the image with respect to their neighbors can still carry useful information. Semantic segmentation methods can therefore benefit from filtering out the information about the absolute orientation while keeping the information about relative orientation.

Extracting features in a convolutional manner, i.e. as a sliding window, constrains CNNs by forcing them to apply the same local operation at every location of the image, improving their capacity to generalize (LeCun et al. 1989). This is due to the translation *equivariance* that characterizes most image-based problems, meaning that translating the image and applying the function (e.g. edge detection, semantic segmentation, etc.) should produce the same result as applying the function and then translating the output.

Approach The same principle can be applied to rotation. Indeed, semantic segmentation is by nature *rotation equivariant*, since a rotation of the image would ideally result in the same rotation of the segmentation map. This can be implemented by using a sliding window that also rotates, applying the same function at all pixel locations and a discrete set of orientations, with $\alpha_r \in \{\alpha, 2\alpha, \dots, R\alpha\}, \alpha = 2\pi/R$. Given a filter $\mathbf{W} \in \mathbb{R}^{m \times n \times c_0}$, we can see it as a collection of feature vectors $\mathbf{W}_{i,j,:} \in \mathbb{R}^{c_0}$, each associated with a spatial location $[i,j]$. A rotated version $\mathbf{W}^r$ of the filter can be obtained by computing a new location for these vectors and interpolating to the nearest grid points:

$$[i',j'] = [i,j] \begin{bmatrix} \cos(\alpha_r) & \sin(\alpha_r) \\ -\sin(\alpha_r) & \cos(\alpha_r) \end{bmatrix}. \tag{5.1}$$

Applying a filter bank of c_1 filters in a sliding and rotating window fashion, which we will call RotConv filter, to an image $\mathbf{X} \in \mathbb{R}^{M \times N \times c_0}$ results in a tensor $\mathbf{Y} \in \mathbb{R}^{M \times N \times R \times c_1}$. Note how this tensor, as well as any filter we would like to apply to it, can be interpreted as a collection of feature vectors $\mathbf{Y}_{i,j,r,:} \in \mathbb{R}^{c_1}$, one per location of the roto-translational space. This increases substantially the memory and computational footprint of the model as R becomes larger. To prevent this, we could max-pool across the rotation dimension, but this would result in the loss of information related to relative orientation (e.g. we would see that a car and a road edge have been detected, but without information about their orientation with respect to each other). We propose to use max-pooling across the rotation dimension but returning *both* the maximally activating *magnitude and orientation*:

$$\mathbf{Y}^{\rho}{}_{i,j,:} = \max{}_r \mathbf{Y}_{i,j,r,:} \qquad \mathbf{Y}^{\theta}{}_{i,j,:} = \frac{360}{R} \arg\max{}_r \mathbf{Y}_{i,j,r,:}. \tag{5.2}$$

These two tensors can be interpreted as the polar representation of a 2D tensor field if $\mathbf{Y}^{\rho}{}_{i,j,:} \geq 0 \quad \forall i,j$, which can be enforced by applying a linear rectifier $\mathtt{ReLU}(\rho) = \max(\rho, 0)$. A Cartesian representation $\mathbf{Z} \in \mathbb{R}^{M \times N \times 2 \times c_1}$ is then computed as:

$$\mathbf{Z}^u = \mathtt{ReLU}(\mathbf{Y}^{\rho}) \cos(\mathbf{Y}^{\theta}) \qquad \mathbf{Z}^v = \mathtt{ReLU}(\mathbf{Y}^{\rho}) \sin(\mathbf{Y}^{\theta}). \tag{5.3}$$

Since the input tensor to the following RotConv layer is a stack of vector fields, a filter bank $\mathbf{Q} \in \mathbb{R}^{m \times n \times 2 \times c_2}$ also needs to consist of vector fields with u and v components, and the convolution operator is computed separately for each component:

$$(\mathbf{Z} * \mathbf{Q}) = (\mathbf{Z}^u * \mathbf{Q}^u) + (\mathbf{Z}^v * \mathbf{Q}^u). \tag{5.4}$$

Note that the rotation described in Equation 5.1 can be applied to $\mathbf{Q}$ but requires the additional step of rotating each 2D vector $\mathbf{Q}_{i,j,:,k}$ by α_r according to Equation 5.1.

Data and Setup We performed experiments on the ISPRS Vaihingen "2D semantic labeling contest" benchmark[16], which consists of 33 tiles of 9 cm resolution aerial imagery acquired over the town of Vaihingen (Germany), with an average tile size of 2494 × 2064, three optical bands (near infrared, red, and green), and a photogrammetric digital surface model (DSM) (see examples in Figure 5.5). Sixteen of the tiles are publicly available and contain six land-cover classes: "impervious surfaces" (roads, concrete flat surfaces), "buildings", "low vegetation", "trees", "cars", and a class of "clutter" to group uncategorized surfaces

16 http://www2.isprs.org/commissions/comm3/wg4/semantic-labeling.html

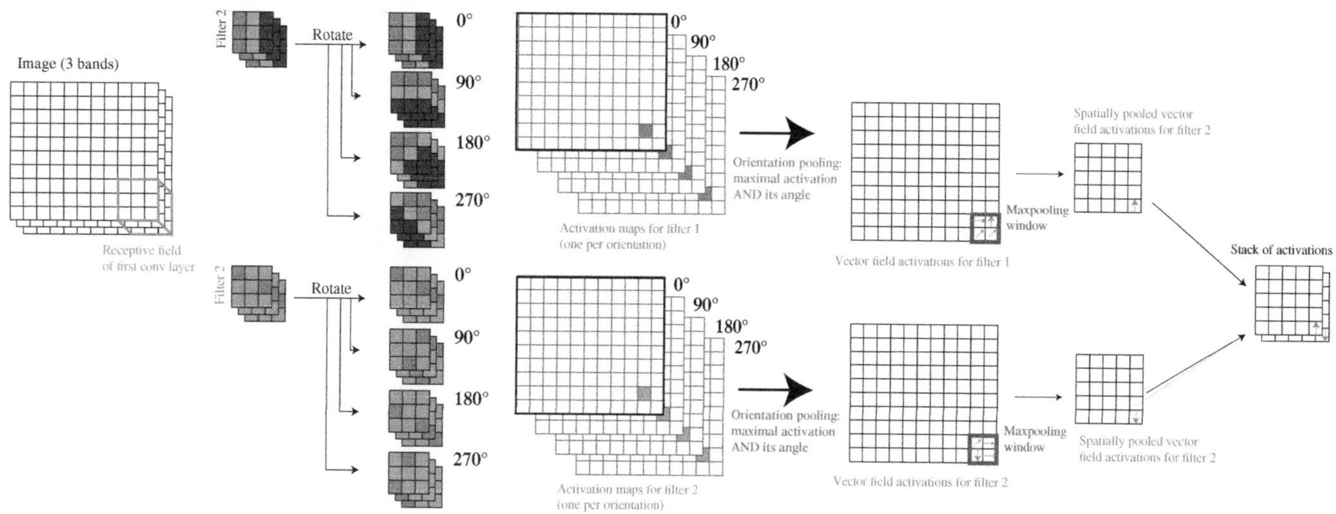

Figure 5.4 (Adapted from (Marcos et al. 2018a)) Diagram of the first RotConv layer with two filters and $R = 4$. The output is a stack of 2D vector fields that encode which rotated version of each filter is maximally activated, in terms of both magnitude and angle.

Table 5.1 Results on the Vaihingen validation set. F1 scores per class and global average (AA) and overall accuracies (OA). Best result per row is in dark gray, second in light gray.

Model	**RotEqNet**			**CNN**			**CNN-FPL**[a)]
# params.	10^5			10^6			10^7
% train set	**4%**	**12%**	**100%**	**4%**	**12%**	**100%**	**100%**
Impervious	88.0	88.7	89.5	86.9	88.8	**89.8**	-
Buildings	94.1	94.6	**94.8**	92.5	92.9	94.6	-
Low veg.	71.6	75.6	**77.5**	74.5	74.5	76.8	-
Trees	82.3	85.6	**86.5**	83.3	84.4	86.0	-
Cars	62.7	62.5	**72.6**	52.7	54.4	54.5	-
OA	84.6	86.6	87.5	84.8	85.5	87.4	**87.8**
AA	78.4	80.5	**83.9**	76.1	77.1	78.2	81.4

a) = from Volpi and Tuia (2017)

and noisy structures. Classes are highly imbalanced, with the classes "buildings" and "impervious surfaces" accounting for roughly 50% of the data, while classes such as "cars" and "clutter" account only for 2% of the total labels.

The CNN architecture we used is inspired in Hypercolumns (Hariharan et al. 2015) and follows the structure depicted in Figure 5.2. It consists of six RotConv layers with 7×7 filters, each followed by orientation pooling, factor two spatial pooling and a modified batch normalization. The magnitudes of all vector field maps are upsampled to the original resolution with bilinear interpolation and stacked before applying three layers of 1×1 convolutions, which are also equivariant to rotation because they do not capture any spatial patterns. In order to test the effect of changing the size of the model, the number of filters in each layer was parametrized by a single integer, Nf, as $[2, 2, 3, 4, 4, 4] \cdot \mathrm{Nf}$.

Results and Discussion We investigated the effect of varying the amount of training data (100%, 12% and 4% of the total) on the generalization capabilities of the model and compared against a standard CNN with an equivalent architecture. On the full training set, RotEqNet saturated in performance with $Nf = 3$ (approximately 10^5 parameters), while the standard CNN needed $Nf = 12$ (approximately 10^6 parameters). As shown in Table 5.1, RotEqNet obtains a comparable overall accuracy in all the studied settings with one order of magnitude less parameters. In addition, the average accuracy obtained by RotEqNet is higher than the one of a standard CNN, mostly because of its higher performance in the class "cars" (that can be clearly appreciated in Figure 5.5) and "buildings" (in the 4% and 12% training scenarios). These results suggest that RotEqNet offers a bigger advantage when segmenting classes with complex geometry, such as "cars" and "buildings", compared to those with simpler, texture-like characteristics, such as vegetation.

Figure 5.5 Examples of classification maps obtained in the Vaihingen validation images with the RotEqNet and the standard CNN models (from Marcos et al. (2018a)). Best viewed in color.

5.4.2 Processing 3D Point Clouds as a Bundle of Images: SnapNet

Since the world is 3D, understanding the environment often requires it to be modeled and represented in all its dimensions. Perceiving 3D is essential for planning motion, recognizing objects, and characterizing scenes but also as a first acquisition step in computer graphics, computer-aided design and virtual reality. To meet these needs, various sensing devices have emerged to capture 3D, from stereo-cameras to modern structured-light-based sensors and laser scanning devices. In particular, the latter produce 3D point clouds with unprecedented location accuracy and regularity. In the current state of sensor technology, point clouds are the *de facto* standard for representing 3D. They bear some features in common with 2D images: points, as pixels, are characterized by a location in a local system and may carry some spectral information stored in several channels. However, point clouds are sparse, that is they do not provide a dense 3D representation of a scene, and mostly unstructured, since the points are not organized according a regular grid.

In this section, we present in detail SnapNet, one of the successful approaches for semantic segmentation of point clouds. A review of alternative approaches to segment point clouds has been provided in section 5.3.2.

Approach The idea behind SnapNet (Boulch et al. 2018) comes from the *Building Rome in a Day* article (Agarwal et al. 2011) where an approach for bundle adjustment able to produce a point cloud from thousands of images (such as tourist snapshots) was proposed. Conversely, SnapNet produces thousand of views from a single point cloud and learn to classify them to retrieve the 3D semantics.

The strengths of the approach lie in making possible to leverage the power of 2D convolutional networks (including the use of pre-training) and the ability to process appearance information as well as geometric 3D features. The process is as follows (Figure 5.6):

1. The point-cloud is pre-processed to obtain a realistic mesh of the 3D scene.
2. 2D views are generated and include appearance, geometric and semantic information for training.
3. Semantic segmentation is performed in the image space.
4. At inference time 2D semantic labels are projected back in the original 3D space and vote for 3D semantic labeling.

In the following, each step is described in detail.

First step is *point-cloud preparation*. Due to sparsity, direct views of 3D point sets are very different from real images: not all the pixels have values and occlusions are not handled, so 3D points from foreground and background objects are mixed-up when projected in the 2D image plane. To avoid this effect, a simple mesh is generated using the fast reconstruction method from Marton et al. (2009). Faces of the mesh are textured with colors from the points in order to propagate the appearance information. Simultaneously, a geometric texture is computed to store local 3D structure: normal deviation to the vertical (Boulch and Marlet 2016) and local noise estimation.

View generation is processed in a 3D mesh viewer. Various camera positions and locations are chosen all around the point cloud according to the following procedure which ensures meaningful images at multiple scales: a virtual camera location and a camera axis are picked randomly, then 3 captures are taken at various origins and looking towards the point cloud. At each capture, different type of views are taken: the color image, the geometric composite (which consists of depth, i.e. the distance to the camera, normal and noise), the 2D semantic map at training, and a mesh-face identifier map, which will be useful for back projecting the semantic predictions. In practice, for urban scenes such as those of the Semantic3D dataset (Hackel et al. 2017), around 500 views per scene are picked randomly at training, and 500 views at 3 different scales for inference (number of samples chosen by cross-validation).

Image semantic labeling is performed using fully convolutional networks (Long et al. 2015), namely SegNet (Badrinarayanan et al. 2017) or U-Net (Ronneberger et al. 2015a), as described in section 5.3. The multimodal architecture with residual correction of Audebert et al. (2018) is chosen to combine efficiently the color and the depth composite images. The network is trained using the available semantic labels and used to predict dense semantic score maps at inference time.

Finally, the last step consists in *back-projection to the mesh and point-cloud semantic labeling*. By combining the semantic maps and the mesh-face identifier map, the score vectors of each pixel are transferred to the mesh and accumulated in each vertex. The final vertex class is the one with the highest score. 3D Points are then labeled as their nearest vertex (right panel of Figure 5.7).

Discussion Figure 5.7 shows results of the SnapNet approach for semantic segmentation of point clouds. The scene is one of the test point-clouds of the Semantic3D dataset (Hackel et al. 2017) and was obtained by terrestrial laser scanning in an urban environment. These point clouds have a large scale ($4 \cdot 10^8$ points on average) and require computationally efficient algorithms for processing. The obtained semantic 3D maps show precise classification of points belonging to buildings, terrain, roads, or urban hardscape.

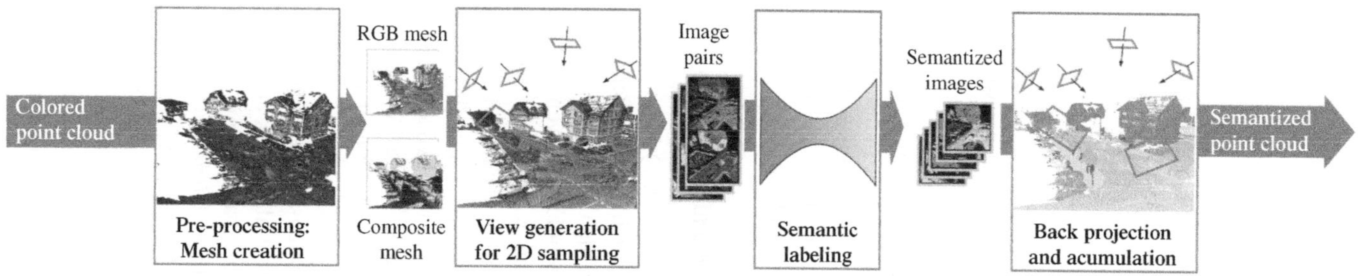

Figure 5.6 SnapNet processing: (1) The point-cloud is meshed to enable the (2) generation of random views at multiple scales, both in appearance and geometry. (3) Semantic segmentation is performed in the 2D domain, and results are (4) back-projected in 3D for voting and 3D semantic segmentation.

Figure 5.7 SnapNet results on the Semantic3D dataset (Hackel et al. 2017): colored point cloud captured in St Gall, Switzerland (left) and semantic 3D map with buildings in red, natural terrain in green, impervious surfaces in gray, etc. (right).

SnapNet integrates two features which are essential in modern point cloud segmentation approaches: learning local representations while maintaining global statistics on the scene. Local spatial patterns are encoded by 2D convolutional filters applied on both appearance and geometric features after projection in the image space. Global statistics are computed through the multiscale view generation strategy.

5.4.3 Lake Ice Detection from Earth and from Space

Lake ice is an essential climate variable (ECV) in the global climate observing system (GCOS), so there is a need to monitor it in an efficient and repeatable way. A main requirement for monitoring lake ice is high temporal resolution – ideally daily, but at least every 2 days, to precisely identify phenological events like the *ice-on* and *ice-off* dates. Among the data sources that fulfil this requirement are low-resolution optical satellite images. If ≈1 km GSD is sufficient, sensors like MODIS and VIIRS can be used to monitor lake ice (Tom et al. 2018). But their effective temporal resolution can be a lot lower than 1 day, due to cloud cover. Here, we will describe two alternatives that are largely unaffected by clouds, namely (1) high-resolution spaceborne SAR imagery from Sentinel-1 (Tom et al. 2020) and (2) close-range webcam pictures (Prabha et al. 2020).

Approach We tackle the semantic segmentation of lake ice with a state-of-the-art convolutional network, *DeepLab v3+* (Chen et al. 2018a), an encoder-decoder architecture that uses separable convolutions and Atrous Spatial Pyramid Pooling (ASPP). For SAR images, we use a variant of *DeepLab v3+* with *mobilenetv2* (Sandler et al. 2018) as encoder, and train it on 128×128 pixel patches with batch size 8, minimizing the cross-entropy loss with stochastic gradient descent. Atrous rates were set to [1, 2, 3]. For webcams we use *Xception65* (Chollet 2017) as the encoder backbone, with atrous rates [6, 12, 18], and train with 321×321 patches and batch size 8. Overall, that encoder has an output stride (spatial downsampling from input to final feature encoding) of 16, which we upsample in the decoder stage in two steps, each of factor ×4, with additional skip connections to re-inject high-frequency detail (similar to the U-Net model described in section 5.3.1). In both cases we employ models pre-trained on the PASCAL VOC 2012 close-range dataset. It turns out that pre-training on RGB amateur images greatly improves the performance not only for webcams, but, somewhat surprisingly, also for SAR amplitude images. It appears that, despite the completely different sensing principle, the local structure of SAR data after standard preprocessing (see below) is similar enough to that of optical images to benefit from the pre-trained initial weights.

Discussion – SAR Data Sentinel-1 is a constellation of two identical satellites at altitude 693 km with 180° phase shift, in a sun-synchronous, near-polar orbit with repeat cycle 6 days on the equator. Due to the large across-track area coverage at the latitude of our target area (and most other areas with regularly freezing lakes), the revisit time is reduced to < 2 days. Footprints of the four relevant orbits for *Region Sils* are shown in Figure 5.8. The satellites have the same C-band SAR system on board. In our research, we use only the amplitude information. We work with the Level-1 Ground Range Detected (GRD) product in Interferometric Wide (IW) swath mode as provided by the *Google Earth Engine*[17]. That product consists of log-scaled backscatter coefficients (no phase information), with nearly square pixel footprint of 10m×10m per pixel. It has already been corrected for thermal and border noise, radiometrically calibrated to backscatter intensities, and corrected for terrain and viewpoint effects (using the SRTM elevation model).

We collect data for two winters, 2016-2017 and 2017-2018. Labeling was only possible for the clear cases of completely frozen or non-frozen lakes, whereas the transition dates with partly frozen lakes are not used for training and quantitative evaluation, but only for qualitative assessment. For each lake, a single label (*frozen, non-frozen*) per day was assigned by a human operator after visual interpretation of the webcam data, when available supported also by optical Sentinel-2 satellite images. Some label noise likely remains due to interpretation errors, as a result of overly oblique viewing angles of webcams and compression artefacts in the images.

We employ cross-validation (CV) across different winters and different lakes. *Leave-one-winter-out CV* assesses the models' capability to generalize to the conditions of unseen years, since it would be impractical to label training data for each new year. The results are shown in Table 5.2. For both winters the intersection-over-union (IoU) scores are similar, around 90%. The results show that the model generalizes well to a new winter, without

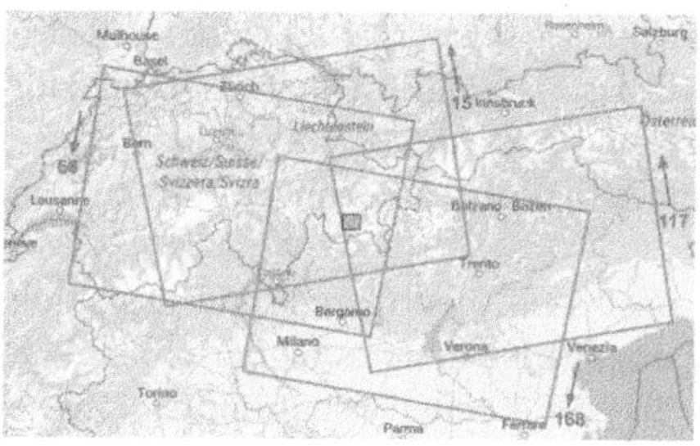

Orbit	Scan time (UTC)	Incidence angle
15	17:15	41.0°
66	05:35	32.3°
117	17:06	30.8°
168	05:26	41.7°

Figure 5.8 The four Sentinel-1 orbits (15, 66, 17, 168) that scan *Region Sils* (shown as filled rectangle).

17 https://earthengine.google.com

Table 5.2 Leave-one-winter-out results (left, over all three lakes) and Leave-one-lake-out results (right, over both winters).

	Winter		Lake		
	2016–17	**2017–18**	**Sils**	**Silvaplana**	**St. Moritz**
IoU non-frozen	91.0%	90.0%	96.7%	93.3%	85.6%
IoU frozen	90.6%	87.7%	96.4%	92.7%	82.9%
mIoU	90.8%	88.9%	96.5%	93.1%	84.3%

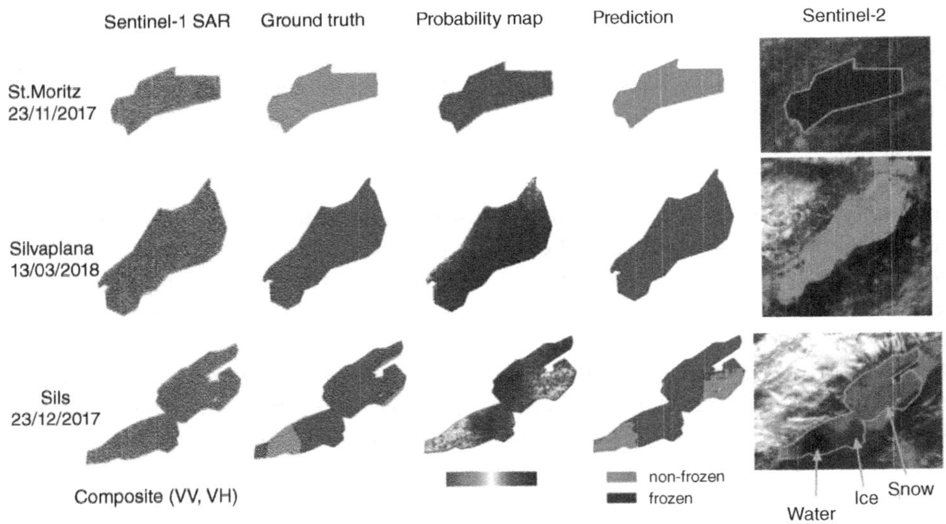

Figure 5.9 Example results for St. Moritz on a non-frozen day (row 1), Silvaplana on a frozen day (row 2), and Sils on a transition day (row 3). Best viewed in color.

having seen data from any day within the test period. *Leave-one-lake-out* CV evaluates the capability to generalize to unseen lakes. The results are shown on Table 5.2. Depending on the lake, the predictions are 84–96% correct, meaning that ice segmentation works well also for new lakes (with similar imaging conditions). In all cases, a single model was trained for images from all orbits. Fitting separate models for ascending and descending orbits (respectively, morning and afternoon) resulted in performance drops of 5–7 percent points, see Tom et al. (2018).

Figure 5.9 shows exemplary qualitative results on frozen, non-frozen, and transition dates, as well as the corresponding soft probability maps (blue denotes higher probability for frozen, red higher probability for non-frozen). To give a better visual impression we also show the corresponding image from Sentinel-2.

Discussion – Webcams As an alternative data source, we have collected webcam data from lake St. Moritz, for the same two winters 2016–17 and 2017–18, and have manually

Table 5.3 Key figures of the St. Moritz webcam data

Cam	Res	#imgs 2016-17	#imgs 2017-18
Cam0	324×1209	820	474
Cam1	324×1209	1180	443

Table 5.4 Lake ice segmentation results for webcams

Train set		Test set						
Cam	Winter	Cam	Winter	Water	Ice	Snow	Clutter	mIoU
Cam0	16-17	Cam0	16-17	0.98	0.95	0.95	0.97	0.96
Cam1	16-17	Cam1	16-17	0.99	0.96	0.95	0.79	0.92
Cam0	17-18	Cam0	17-18	0.97	0.88	0.96	0.87	0.93
Cam1	17-18	Cam1	17-18	0.93	0.84	0.92	0.84	0.89
Cam0	16-17	Cam0	17-18	0.64	0.58	0.87	0.59	0.67
Cam0	17-18	Cam0	16-17	0.98	0.91	0.94	0.58	0.87
Cam1	16-17	Cam1	17-18	0.86	0.71	0.93	0.57	0.77
Cam1	17-18	Cam1	16-17	0.93	0.76	0.86	0.65	0.80
Cam0	16-17	Cam1	16-17	0.76	0.75	0.84	0.61	0.74
Cam1	16-17	Cam0	16-17	0.94	0.75	0.92	0.48	0.77
Cam0	17-18	Cam1	17-18	0.62	0.66	0.89	0.42	0.64
Cam1	17-18	Cam0	17-18	0.59	0.67	0.91	0.51	0.67

annotated ground truth masks of the lakes, and labels *water*, *ice*, *snow*, *clutter* for the lake pixels. The numbers of images and their resolution are given in Table 5.3, for further details see (Prabha et al. 2020). There are two different, fixed webcams, which we call *Cam0* and *Cam1*, both observing lake St. Moritz at different zoom levels. Example images are shown in Figure 5.10.

We report results for different train/test settings, see Table 5.4. In the *same camera/both winters* setting, the model is trained randomly on 75% of the images from a webcam stream and tested on the remaining images from the same webcam. The *cross-winter* setting again evaluates generalization to the potentially different conditions of an unseen year. The model also generalizes quite well across winters with an average IoU scores of 78%, although not quite as well as the SAR version. In the *cross-camera* setting, we train on one camera and test on another, so as to check how well the model generalizes to an unseen viewpoint, image scale, and lighting. While there is a noticeable performance drop, the segmentation still works surprisingly well, reaching mean IoU scores around 70%.

Qualitative example results are shown in Figure 5.10. Some images are even confusing for humans to annotate correctly, e.g., row 2 shows an example of ice with smudged snow on

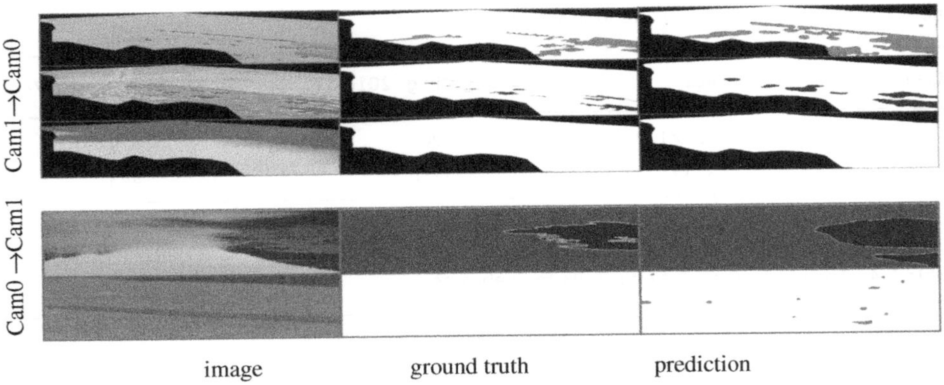

Figure 5.10 Segmentation results (cross-camera setting).

top, for which the "correct" labeling is not well-defined. One can see that the segmentation is robust against cloud/mountain shadows cast on the lake (row 3). There are also cases of label noise where the network "corrects" human errors, such as in row 5, where humans present on the frozen lake were not annotated due to their small size.

5.5 Concluding Remarks

Semantic segmentation approaches based on convolutional neural networks are increasingly used in Earth Observation to automatically generate categorical maps from spatial data. Off-the-shelf computer vision models offer a good starting point, but we show that there is much to be gained by designing semantic segmentation methods specifically tailored to the characteristics of Earth Observation data. In particular, we have demonstrated the advantages of injecting rotation equivariance into segmentation models applied to overhead optical imagery, the fusion of multiple data modalities for the monitoring of lake ice and the application of image based semantic segmentation to point cloud data.

Even if at a first glance the topic might seem saturated (if not solved), much work remains ahead, especially in applications beyond segmentation of well understood urban objects, for which several datasets exist. Tackling global challenges such as deforestation or sea level rise will need support from accurate mapping, and semantic segmentation will play an important role there, way beyond benchmarking efforts. Issues of transfer learning (also discussed in Chapter 7), data fusion (Chapter 10) and of using our expert knowledge of the world to compensate for domain shifts (e.g. physical knowledge, spatiotemporal trends, models' outputs) seems the necessary next step to prepare semantic segmentation for the challenges ahead.

6

Object Detection in Remote Sensing

Jian Ding, Jinwang Wang, Wen Yang, and Gui-Song Xia

6.1 Introduction

6.1.1 Problem Description

Object detection is a fundamental task towards an automatic understanding of remote sensing images. The aim of object detection in remote sensing images is to locate the objects of interest and identify their categories on the ground (e.g., vehicles, airplanes). To acquire remote sensing images, there are a variety of platforms, including satellites, airplanes, and drones equipped with different sensors, such as optical cameras or synthetic aperture radars (SARs). Figure 6.1 shows several images containing objects taken with optical and SAR sensors. In the past decades, extensive research has been devoted to object detection in remote sensing images (Porway et al. 2010; Lin et al. 2015; Cheng et al. 2016b; Moranduzzo and Melgani 2014; Wang et al. 2017a; Wan et al. 2017; Ok et al. 2013; Shi et al. 2013; Kembhavi et al. 2010; Proia and Pagé 2009), using hand-crafted features. For object detection in remote sensing images, traditional methods like HOG (Dalal and Triggs 2005) and SIFT (Lowe 1999) are well used for feature extraction. However, these shallow models have limited ability to detect objects in complex environments. Nowadays, with the development of deep learning and its successful application in object detection, earth vision researchers have tried methods (Liu et al. 2016d, 2017c; Liao et al. 2018b; Yang et al. 2019c; Cheng et al. 2016b) based on fine-tuning networks pre-trained on large-scale datasets of natural images such as ImageNet (Deng et al. 2009). Nevertheless, there is a huge domain shift between natural images and remote sensing ones. Thus, object detection methods developed in natural images can not be directly used in remote sensing images. We summarize the difficulties of object detection in remote sensing as follows:

- **Arbitrary orientation of objects**. Objects in remote sensing images can be arbitrarily orientation without any restrictions because the sensors observe the objects on the ground from a bird's eye view. This largely challenges conventional systems since it requires rotation-invariant features to obtain good performance.

Deep Learning for the Earth Sciences: A Comprehensive Approach to Remote Sensing, Climate Science, and Geosciences,
First Edition. Edited by Gustau Camps-Valls, Devis Tuia, Xiao Xiang Zhu, and Markus Reichstein.

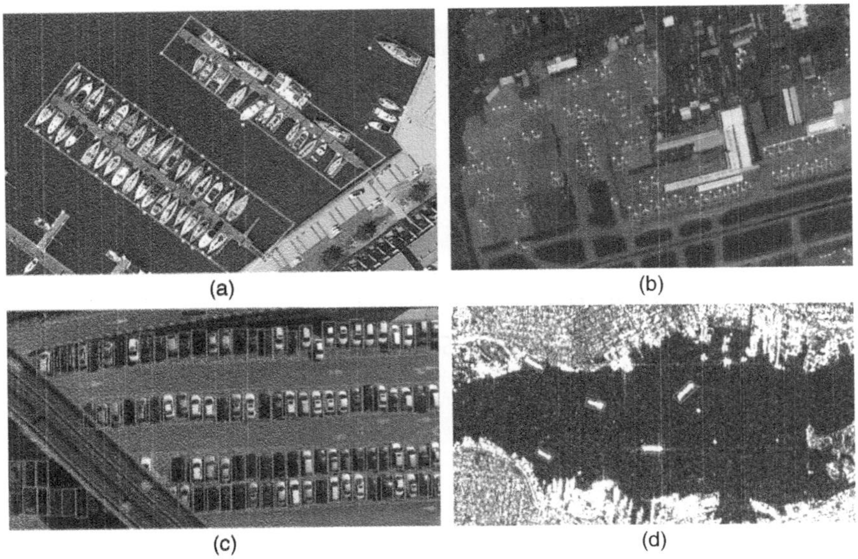

Figure 6.1 Examples of remote sensing images containing objects of interest. (a) An image from Google Earth, containing ships and harbors. (b) An image form JL-1 satellite, including planes. (c) An drone-based image containing many vehicles. (d) A SAR image, containing ships.

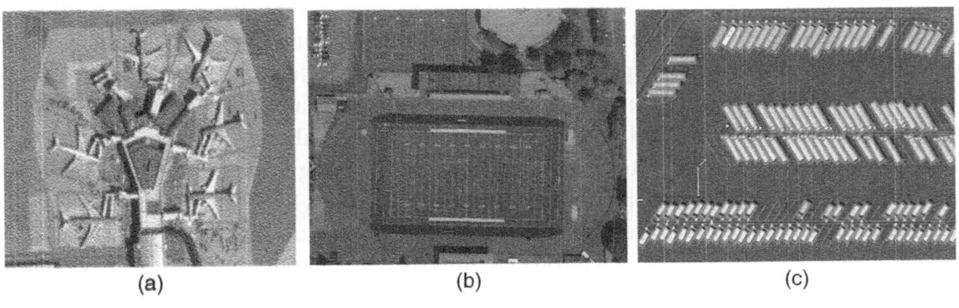

Figure 6.2 Challenges of object detection in remote sensing. (a) Arbitrary orientations. (b) Huge scale variations. (c) Densely packed instances.

- **Huge scale variations.** Objects in remote sensing vary in a wide range according to the GSD of sensors and actual physical sizes of objects, from 10 pixels (e.g., small vehicles) to above 1000 pixels (e.g., ground track field), all in one scene (Figure 6.2).
- **Large-size images.** Images in remote sensing may be very large (above 20,000 pixels), and such large-size images are always challenging for the current computational hardware. Besides, the instances are distributed non-uniformly in remote sensing images, some small-size (e.g., 1k×1k) chips can contain hundreds to thousands of instances, while some large-size images (above 20,000 pixels) sometimes only contain a few.
- **Densely packed instances.** The instances are usually densely packed in some specific scenes, such as harbor and parking lot. This makes them hard to distinguish and separate.

We choose some examples to show the difficulties in Figure 6.2.

6.1.2 Problem Settings of Object Detection

As the tasks of object detection include classifying and regression, the formal problem settings for object detection based on deep learning can be represented in the following. Assume a collection of N annotated images is given, and for image x_i there are M_i objects belonging to C classes with annotations:

$$y_i = \left\{ \left(c_1^i, b_1^i\right), \left(c_2^i, b_2^i\right), \dots, \left(c_{M_i}^i, b_{M_i}^i\right) \right\} \tag{6.1}$$

where $c_j^i (c_j^i \in C)$ and b_j^i denote categorical label and bounding box of j-th object in x_i respectively. Bounding box is the minimum rectangle that encloses the object; it tends to be represented as (c_x, c_y, w, h), which denote coordinates of the center, width, and height of the box. The model weights of detector are parameterized by θ. For each image x_i, the prediction y^i_{pred} shares an identical format with y_i:

$$y^i_{pred} = \left\{ \left(c^i_{pred_1}, b^i_{pred_1}\right), \left(c^i_{pred_2}, b^i_{pred_2}\right), \dots \right\}. \tag{6.2}$$

Finally, a loss function l is set to optimize detector as:

$$l(x, \theta) = \frac{1}{N} \sum_{i=1}^{N} l\left(y^i_{pred}, x_i, y_i, \theta\right) + \frac{\lambda}{2} \|\theta\|_2^2 \tag{6.3}$$

where the second term is a regularizer, including trade-off parameter λ.

6.1.3 Object Representation in Remote Sensing

To locate an object, the horizontal bounding box (HBB) is the most widely used formulation (Lin et al. 2014) for general object detection. It can be represented by a center, width, and height of an instance as (c_x, c_y, w, h), which we have discussed before. However, HBB is not always appropriate for object detection in remote sensing images, as HBB cannot reflect the orientation of objects for precise location or be used for rotation invariant feature extraction. The oriented bounding box (OBB) is a more suitable representation for objects with orientations. The detailed formulation is (c_x, c_y, w, h, θ), where the (c_x, c_y), w, h, θ are the center, width, height, and angle relative to the X-axis respectively. Besides remote sensing, the OBB representation is also widely used in face detection (Huang et al. 2007) and text scene detection (Liao et al. 2018a). However, the orientation variations in text scenes and faces are restricted in a relatively narrow range while the orientations of objects in remote sensing images are arbitrary values in $[0, 2\pi)$.

6.1.4 Evaluation Metrics

- **Intersection over Union.** Intersection over Union (IoU) is used to measure the overlap between two bounding boxes, i.e., a ground truth bounding box B_{gt} and a predicted bounding box B_p. By applying the IoU, we can judge a detection to be valid or not. Specifically, IoU is given by the overlapping area of two bounding boxes divided by their union area:

$$IoU = \frac{area(B_p \cap B_{gt})}{area(B_p \cup B_{gt})}. \tag{6.4}$$

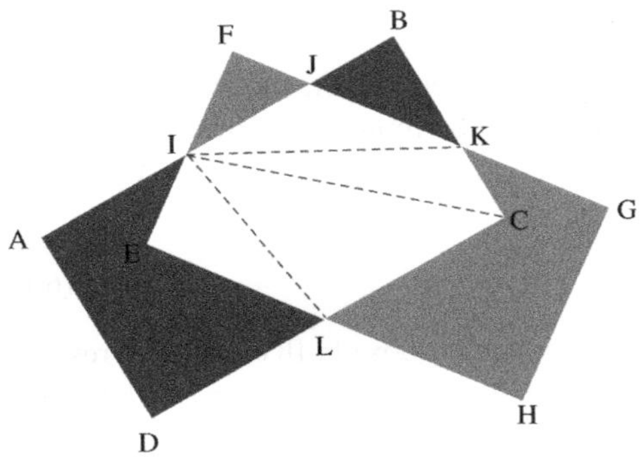

Figure 6.3 IoU calculation between two oriented bounding boxes.

Table 6.1 The rules of samples classification.

		Ground Truths	
		Positives	Negatives
Predictions	Positives	TP	FP
	Negatives	FN	TN

For oriented object detection, IoU is calculated between two OBBs using computational geometry method which is more complicated than the horizontal ones. Specifically, as shown in Figure 6.3, we need find the intersection polygon *IJKCLE* of two OBBs *ABCD* and *EFGH*, and calculate the IoU as:

$$IoU_{OBB} = \frac{S_{IJKCLE}}{S_{ABCD} + S_{EFGH}} = \frac{S_{ILE} + S_{ICL} + S_{IKC} + S_{IJK}}{S_{ABCD} + S_{EFGH}} \tag{6.5}$$

where S denotes the area of a region.

- **Precision and Recall.** For a common binary classification task, Table 6.1 illustrates the rules of samples classification. In object detection, if a detection overlaps with the nearest ground truth by an IoU pre-defined threshold (usually 0.5), it will be regarded as False Positive (*FP*) directly. For every ground truth box, at most one prediction is counted as a True Positive (*TP*). Any other prediction with IoU greater than the set threshold is discarded as a *FP* instead. False Negative (*FN*) indicates a ground truth which is not detected. Then, we can define the concept of Precision (*P*) and Recall (*R*):

$$P = \frac{TP}{TP + FP} \tag{6.6}$$

$$R = \frac{TP}{TP + FN}. \tag{6.7}$$

6.1.4.1 Precision-Recall Curve

The Precision-Recall Curve (PRC) is a common metric to evaluate the performance of a detector as the confidence is changed by plotting a curve for each class. Specifically, we classify all detections of a particular class into *TP*, *FP*, and *FN* by probabilities to get the

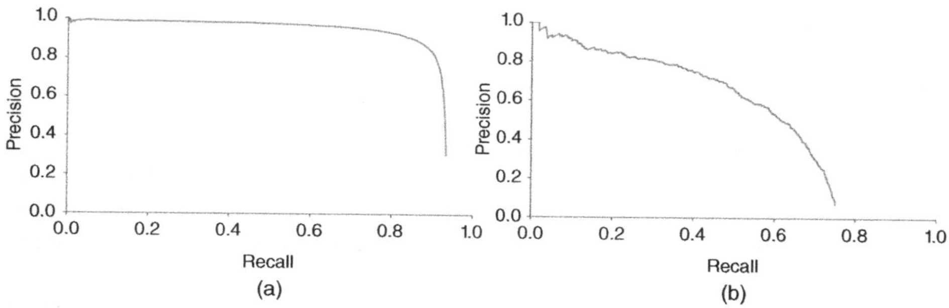

Figure 6.4 Examples of Precision-Recall Curve. As the recall increases, (a) stays high precision while the precision of (b) drops significantly.

precision and recall as the confidence threshold increases. As shown in Figure 6.4, if an object detector stays high precision as recall increases (e.g., Figure 6.4(a)), it can be considered as a good detector. This means that even you change the confidence threshold, the precision and recall can still be high. However, comparing the curves of different classes and detectors in the same plot is not an easy task, as the curves often cross with each other.

6.1.4.2 Average Precision and Mean Average Precision

Average Precision is a metric that summarizes the precision-recall curve of detectors. Many popular object detection datasets in natural images (Everingham et al. 2010; Lin et al. 2014) and aerial images (Liu et al. 2016d; Xia et al. 2018) adopt this metric. It calculates the area under the P-R curve, which can help us to compare different classes and detectors. Assume that we have obtain a P-R curve, the Average Precision (AP) of a single class can be defined as:

$$AP = \int_0^1 P(R)\, dR. \tag{6.8}$$

For multi-class object detection, we can define the mean Average Precision (mAP):

$$mAP = \frac{1}{N_{cls}} \sum_i AP_i \tag{6.9}$$

where N_{cls} is the number of classes and AP_i is the average precision of the i-th class.

6.1.5 Applications

- **Ship detection.** With the development of remote sensing technology, more and more high-resolution remote sensing images are available for ship detection and recognition. There are a wide range of applications for ship detection, such as fishery management, vessel traffic services, and maritime and docks surveillance. Both SAR images and optical images have been used for ship detection (Liu et al. 2017c; Yang et al. 2018a; Zhang et al. 2018d).
- **Vehicle detection.** Vehicle detection in aerial images is important for applications such as traffic management, parking lot utilization, urban planning. By collecting traffic and parking information from remote sensing images, we can quickly get coverage over a

large area with lower cost and fewer deployment of sensors compared with traditional approaches (e.g., road monitors). Liu and Máttyus (2015) and Deng et al. (2017) show that it is possible to detect small vehicles from large remote sensing images and still be fast enough.

- **Airport and plane detection.** Airport and plane detection in remote sensing images have attracted much attention for its importance in military and civilian fields. Xiao et al. (2017) use a multiscale fusion feature for optical remote sensing images to detect airports. Mohammadi (2018) propose a rotation and scale invariant airplane proposal generator to solve scale variances and direction of airplanes.
- **Others.** As a fundamental problem in remote sensing image analysis, object detection has many complicated applications, such as environmental monitoring, land use and land cover mapping, geographic information system update and so on.

6.2 Preliminaries on Object Detection with Deep Models

Before the further introduction of object detection in remote sensing, it is necessary to explain how the detectors work. In general, object detectors can be mainly categorized into two-stage and one-stage approaches. Two-stage approaches consist of region proposal stage and region classification stage (e.g., Girshick et al. (2014), Dai et al. (2016)). In the region proposal stage, detectors generate region proposals, which are class-agnostic and the areas that possibly contain objects. Various region proposal techniques have been proposed, such as selective search (Uijlings et al. 2013) and Region Proposal Network (Girshick 2015). In the region classification stage, region proposals are classified into object categories. As for one-stage approaches (e.g., Redmon et al. (2016), Liu et al. (2016)), they skip the first stage applied in two-stage approaches. This simplicity gains faster speed, but tends to get degradation of accuracy.

6.2.1 Two-stage Algorithms

6.2.1.1 R-CNNs

- **R-CNN.** The first two-stage deep learning detection approach is R-CNN (Girshick et al. 2014). R-CNN uses selective search (Uijlings et al. 2013) to generate region proposals, which is more efficient than the sliding window approach that scans the whole images in multiple scales. Then, R-CNN warps each proposal and feeds it into the CNN to compute features. A one-vs-all SVM is used to identify the categories. Finally, the regressors are adopted to calculate the localization offset of bounding boxes and refine the results.
- **Fast R-CNN.** R-CNN has much redundant computation and can not be trained end to end. To solve these problems, Fast R-CNN (Girshick 2015) shares computation between different region proposals. In detail, it warps the features from the feature map instead of the images by RoI pooling. The RoI pooling operation is differentiable, which makes Fast R-CNN trainable end to end. The Fast R-CNN is 213 $\times$ faster at test-time than R-CNN and more accurate.

- **Faster R-CNN.** The proposal generation in Fast R-CNN is still hand-crafted. Faster R-CNN (Ren et al. 2017) proposes an efficient fully convolutional network to generate region proposals, which is called Region Proposal Network (RPN). RPN learns the "objectness" of all instances and accumulates the proposals, which are used by the detector. The detector subsequently classifies and refines bounding boxes for those proposals. RPN and detector can be trained end-to-end. Faster R-CNN uses thousands of reference boxes, which are also called anchors. These anchors form a grid of boxes, which act as starting points for bounding boxes regression. They are then trained to regress the bounding boxes offsets and score the objectness for each anchor. The size and aspect ratio of anchors are determined by the general range of size of instances in the dataset and the receptive field of the convolutional layers. RoI Pooling then warps the proposals generated by the RPN to fixed-size features. Then the features are fed into the fully-connected layer for classification and detection.

6.2.1.2 R-FCN

In Faster R-CNN, each proposal generated by RPN is computed to offset and class individually, which is time-consuming. To minimize the repetitive computation from proposals, Dai et al. (2016) proposed R-FCN (Region-based Fully Convolutional Networks). As illustrated in Figure 6.5, R-FCN crops the last layer of features before prediction, instead of cropping features from the same layer where applied RPN. Besides, to add localization representations that respect translation variance in detection task, a position-sensitive cropping strategy is proposed in R-FCN, which is used to replace the standard ROI pooling operations used in Fast R-CNN and Faster R-CNN. R-FCN achieves comparable accuracy to Faster R-CNN while at faster running times.

6.2.2 One-stage Algorithms

6.2.2.1 YOLO

This is the first one-stage detection method presented by Redmon et al. (2016). YOLO can be optimized end-to-end directly without generating proposals. Such a design allows it to predict boxes in a single forward pass without generating proposals, thus speeding up the detector. YOLO begins by dividing an image into $S \times S$ grids and assuming B bounding boxes per grid. Every cell that contains the center of an object is responsible for detecting that instance. One bounding box prediction includes four coordinates, a class-agnostic objectness score, and class probabilities. To have a receptive field that covers the whole image, YOLO includes a fully connected layer in its design towards the end of the network.

6.2.2.2 SSD

Inspired by Faster-RCNN, SSD (Liu et al. 2016a) uses reference boxes with various aspect ratios and sizes to predict object instances, while the region proposal stage is completely gotten rid of. During training, thousands of default boxes corresponding to different anchors on different feature maps are learned to distinguish objects and background, localize, and predict class probabilities of the object instances, with the help of a multitask loss.

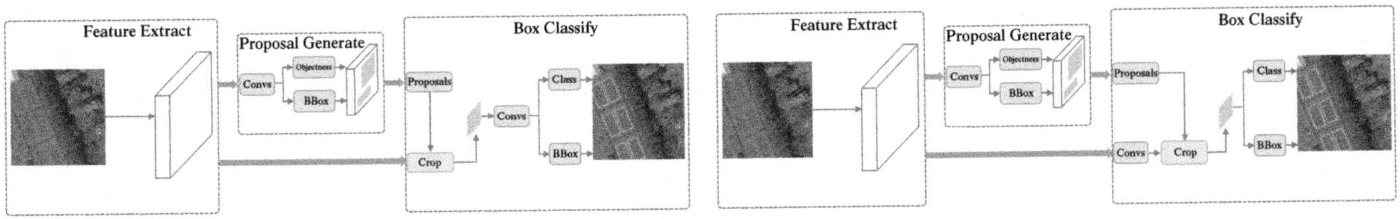

Figure 6.5 Architectures of Faster R-CNN and R-FCN.

6.3 Object Detection in Optical RS Images

6.3.1 Related Works

6.3.1.1 Scale Variance

As mentioned before, scale variation is a big challenge to object detection in remote sensing images. To address this issue, Feature Pyramid Network (FPN) and Image Pyramid are widely used in these works (Lin et al. 2017; Singh et al. 2018). FPN fuses features from multiple layers by lateral connections in a top-down way and outputs the fused features with different resolutions. The features from deep layers provide more semantic information while the features form shallow layers provide more detailed spatial information. These features are fused by element-wise summation followed by convolutions to reduce the dimensions. Image pyramid involves multi scale training and testing, which is much more time-consuming. Combining both the FPN and Image pyramid can further improve the performance, however, the computation is much complicate. The work in Azimi et al. (2018) proposes the Image Cascade Network (ICN) to compute both FPN and Image pyramid quickly. It uses an algorithm including a joint image cascade and feature pyramid network with convolutions of multiscale kernels to obtain scale robust features.

6.3.1.2 Orientation Variance

By default, CNNs are not robust to orientation variations. Some networks are designed to model the geometry transformations such as the Spatial Transformer Networks (STN) (Jaderberg et al. 2015), deformable convolution networks (DCN) (Dai et al. 2017), and oriented response network (ORN) (Zhou et al. 2017b). They are developed in natural images and widely used in object detection in remote sensing. Besides these works, Rotation-Invariant CNN (RICNN) (Cheng et al. 2016a) proposes a rotation-invariant layer and plugs it into R-CNN (Girshick et al. 2014). The learning of RICNN is based on the rotational data augmentation. These works do not need extra supervision of geometry transformation information. To extract more precise and robust rotation invariant features, Rotated RoI Pooling (RRoI Pooling) (Liu et al. 2017c) and its variations are proposed. However, the Rotated RoI Pooling needs rotated proposals as inputs. RR-CNN uses the hand-crafted ship rotated bounding box space (SRBBS) to obtain the rotated proposals, which can not be trained end to end and is time-consuming. Rotated Region Proposal Network (RRPN) design rotated anchors to generate rotated proposals. For each position, there are $(angles \times spatial\_ratio \times scales)$ anchors in total. So it is still very time-consuming.

6.3.1.3 Oriented Object Detection

Many works (Azimi et al. 2018; Ding et al. 2019; Xia et al. 2018; Liao et al. 2018b; Li et al. 2019) handle this problem as a regression task and directly regress oriented bounding boxes. We call these kinds of methods "regression-based methods". For instance, DRBox (Liu et al. 2017a) redesigns the SSD (Liu et al. 2016a) to regress oriented bounding boxes by multi-angle prior oriented bounding boxes. Xia et al. (2018) proposes the Faster R-CNN OBB (FR-O) which regresses the offsets of OBBs relative to HBBs. ICN (Image Cascade Network) (Azimi et al. 2018) joints image cascade and feature pyramid network to extract features for regressing the offsets of OBBs relative to HBBs. Ma et al. (2018) designs a

Rotation Region Proposal Network (RRPN) to generate prior proposals with the object orientation angle information, and then regress the offsets of OBBs relative to oriented proposals. R-DFPN (Yang et al. 2018a) adopts RRPN and puts forward the Dense Feature Pyramid Network to solve the narrow width problems of objects like ships. Ding et al. (2019) designs a RoI learner to transform horizontal RoIs to oriented RoIs in a supervised way. All these regression-based methods summarize the problem of regression as the offsets of OBBs relative to HBBs or OBBs, and they rely on the accurate representation of OBB. There are also some methods that intend to seek the object region at pixel-level and then utilize the post-processing methods to obtain OBBs. We call these kinds of methods segmentation-based methods. For instance, Wang et al. (2019b) proposes the Mask OBB, which uses binary segmentation map to represent oriented objects. SegmRDet (Wang et al. 2020) uses the Mask R-CNN (He et al. 2017) structure to generate box masks for detecting oriented objects.

6.3.1.4 Detecting in Large-size Images

Images with a large extent can reveal more information for precise object detection but also require more computations and memory. The most direct approach is to resize the large-size image into smaller ones to save computations and speed up inference processing. However, due to the reduction of image information and scale variance during resizing, the quality of detection will drop significantly.

Another approach crops large-size images into smaller chips. For example, we detect objects in each chip separately, and merge detections back together to the full image extent. The baselines of DOTA (Xia et al. 2018) crop 1024×1024 patches with a overlap of 500 from the original images, which have a size ranging from 800 to 4000. Many other methods also follow this pipeline of data preparation. However, to overcome the problem that an object may be cut into two patches, a big overlapping area (e.g., 512) is required between two patches, which increases the total computation time.

Though image resizing and cropping can alleviate huge memory usage in some way, they inevitably cause performance drop and introduce complicated data processing pipelines. ClusterNet, proposed by (LaLonde et al. 2018), combines both spatial and temporal information with CNN and proposes regions of objects of interest (ROOBI), which can contain from one object to clusters of objects. Then a simple network with several convolutional layers, named FoveaNet, is adopted to estimate the location of objects in a given ROOBI via heatmap estimation. This two-stage method can significantly reduce the searching space in wide area motion imagery. ClusDet (Yang et al. 2019) shares a similar idea with (LaLonde et al. 2018) that we can find object clusters firstly and then detect on those clusters. The authors observed that for grid-based uniform partition, a lot of image chips contain zero objects, and detecting on these chips is a waste of computation and memory. Based on this, it first finds object cluster regions, estimates fine object scales for each cluster region and then feeds the regions into a detection network. In this way, it can greatly reduce the number of chips for final object detection stage and achieves high inference speed.

Unlike most detectors, Uzkent et al. (2019) proposed using reinforcement learning for adaptively selecting fine resolution of each image chip provided to the detector. To reduce the dependency on high-resolution images and obtain high efficiency, a coarse level detector is applied when the image is dominated by large objects and a fine detector is applied when it is dominated by small objects. The results show that it reduces the running time by 50% while keeping similar accuracy on the challenging aerial dataset, xView (Lam et al. 2018).

6.3.2 Datasets and Benchmark

Many datasets and benchmarks for object detection in optical remote sensing have been published in the past decades. Most of the early datasets are small and contain only a limited number of categories. To advance the research for object detection in optical remote sensing images, several large-scale datasets are proposed. We describe them in detail in the following.

6.3.2.1 DOTA

The original data of DOTA (Xia et al. 2018)[1] mainly comes from China Resources Satellite Data and Application Center, Google Earth, JL-1 satellite, and GF-2 satellite remote sensing data. The dataset contains a total of 15 categories. The dataset contains 2806 remote sensing images acquired by different sensors, with resolutions ranging from 800 × 800 to 4000 × 4000. The dataset contains a total of 188,282 object instances annotated with oriented bounding boxes and is divided into a training set, a verification set, and a test set according to the ratio of 1/2, 1/6, and 1/3.

6.3.2.2 VisDrone

The VisDrone (Zhu et al. 2018)[2] dataset is collected and annotated by the AISKYEYE team of the Machine Learning and Data Mining Laboratory of Tianjin University. The dataset includes 263 drone videos and 10209 drone images. These videos contain a total of 179,264 frames with image size of 2000. The dataset contains more than 2.5 million manually annotated objects with axis-aligned bounding boxes in 10 categories, such as pedestrians, cars, bicycles, and tricycles. It also provides some additional attributes, including scene visibility, object class, and occlusion, to make better use of the data.

VisDrone contains a total of four tasks: object detection in image, object detection in video, single object tracking, and multi-object tracking.

6.3.2.3 DIOR

DIOR (Li et al. 2020) images are collected from Google Earth. It consists of 23,463 images, containing a total of 20 object categories, and each category contains about 1200 images, for a total of 192,472 instances. The objects are annotated with axis-aligned bounding boxes.

6.3.2.4 xView

xView (Lam et al. 2018)[3] contains data from WordView-3 satellites at 0.3 m ground sample distance, giving higher resolution imagery than many other satellite datasets. This dataset covers images over 1400 km^2 of the ground. It has 60 fine-grained classes and over 1 million objects. The annotation method is axis-aligned bounding boxes.

---

1 https://captain-whu.github.io/DOTA/
2 http://aiskyeye.com/
3 http://xviewdataset.org/

6.3.3 Two Representative Object Detectors in Optical RS Images

6.3.3.1 Mask OBB

Ambiguity of Regression-based OBBs As mentioned in section 6.1.3, some works (Ma et al. 2018; Ding et al. 2019) use $\{(cx, cy, h, w, \theta)\}$ to represent OBB; we call this representation θ-based OBB. Besides, (Xia et al. 2018; Liao et al. 2018b; Azimi et al. 2018) use $\{(x_i, y_i)|i = 1, 2, 3, 4\}$ (point-based OBB) to represent OBB, where (x_i, y_i) is the i-th vertex of OBB. Jiang et al. (2017) uses $\{(x_1, y_1, x_2, y_2, h)\}$ (h-based OBB) to represent OBB, where (x_1, y_1) and (x_2, y_2) are first vertex and second vertex of OBB, and h is the height of OBB. We call these formats regression-based OBB representations. Figure 6.6 (a) and (b) demonstrate these formats.

Although these representations ensure the uniqueness in OBB's definition with some rules, there still allow extreme conditions. In these conditions, a tiny change of OBB angle would result in a large change of the OBB representation. We denote angle values in these conditions as discontinuity points. For oriented object detectors, similar features extracted by the detector with close positions are supposed to generate similar position representations. However, OBB representations of these similar features would differ greatly near discontinuity points. This would force the detector to learn totally different position representations for similar features. It would impede the detector training process and deteriorate the detector's performance, obviously.

Specifically, for point-based OBB representation, to ensure the uniqueness of OBB definition, (Xia et al. 2018) chooses the vertex closest to the "top left" vertex of the corresponding horizontal bounding box as the first vertex. Then the other vertexes are fixed in clockwise order, so we get the unique representation of OBB. Nevertheless, this mode still allows discontinuity points, as illustrated in Figure 6.6 (a) and (b). When the l_1 on the horizontal bounding box is shorter than the l_2, the OBB is represented with $R_1 = (x_1, y_1, x_2, y_2, x_3, y_3, x_4, y_4)$ (point-based OBB), as Figure 6.6 (a) shows. Otherwise, the OBB is represented with $R_2 = (x_4, y_4, x_1, y_1, x_2, y_2, x_3, y_3)$ (point-based OBB), as Figure 6.6 (b) shows. When the length of l_1 increases with θ, till θ approaches and surpasses $\pi/4$, the OBB representation would jump from R_1 to R_2, and vice versa. Hence $\pi/4$ is a discontinuity point in this mode.

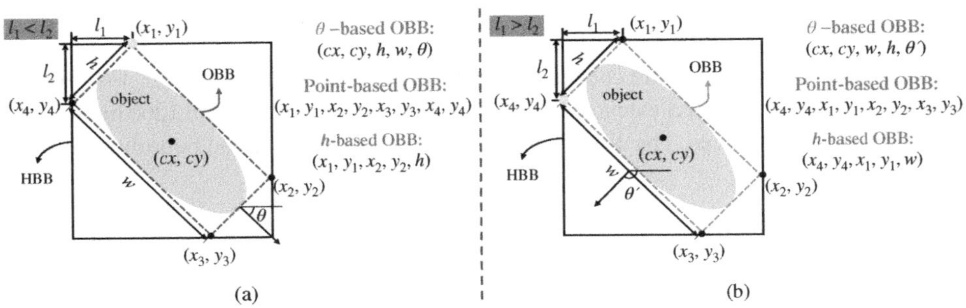

Figure 6.6 (a–b) Borderline states of regression-based OBB representations. The full line, dash line, and gray region represent horizontal bounding box, oriented bounding box and oriented object. The feature map of the left instance should be very similar to the right one. But by the definition in Xia et al. (2018) to choose the first vertex (x_1, y_1) in (a) and (x_4, y_4) in (b), the coordinates of θ-based OBB, point-based OBB, and h-based OBB representations differ greatly. The representation of Mask OBB can avoid the problem of ambiguity and obtain better detection results.

For h-based OBB and θ-based OBB, $\pi/4$ is still a discontinuity point. As shown in Figure 6.6 (a) and (b), with θ oscillating near $\pi/4$, the h-based OBB representation would switch between (x_1, y_1, x_2, y_2, h) and (x_4, y_4, x_1, y_1, w). The θ-based OBB representation would switch back and forth between (cx, cy, h, w, θ) and (cx, cy, w, h, θ') similarly.

Mask OBB for Oriented Object Detection For handling the ambiguity problem, Wang et al. (2019b) represents the oriented object as binary segmentation map which ensures the uniqueness naturally, and the problem of detecting oriented bounding box can be treated as the pixel-level classification for each proposal. Then the oriented bounding boxes are generated from the predicted masks by post-processing, and this kind of OBB representation is called mask oriented bounding box representation (Mask OBB). Under this representation, there is no discontinuity point and ambiguity problem.

Furthermore, aerial image datasets like DOTA (Xia et al. 2018) and HRSC2016 (Liu et al. 2016d) give the regression-based oriented bounding boxes as ground truth. Specifically, DOTA and HRSC2016 use point-based OBBs $\{(x_i, y_i)|i = 1, 2, 3, 4\}$ (Xia et al. 2018) and θ-based OBBs $\{(cx, cy, h, w, \theta)\}$ (Liu et al. 2016d) as the ground truth, respectively. However, for pixel-level classification, pixel-level annotations are essential. To handle this problem, pixel-level annotations are converted from original OBB ground truth. Specifically, pixels inside oriented bounding boxes are labeled as positive and pixels outside are labeled as negative. And then we obtain the pixel-level annotations which will be treated as the pixel-level classification ground truth. Figure 6.7 illustrates the point-based OBBs and converted Mask OBBs on DOTA images. The highlight points are original ground truth, and the highlight regions inside point-based OBBs are new ground truth for pixel-level classification, which is well known as instance segmentation problem. Being different from point-based OBB, h-based OBB, and θ-based OBB, Mask OBB is unique in the definition no matter how point-based OBB changes. Using Mask OBB, the problem of ground truth ambiguity can be solved in nature, and there are no discontinuity points allowed in this mode.

Figure 6.7 Samples for illustrating mask-oriented bounding box representation (Mask OBB). The corner points are original ground truth (point-based OBB), and the regions inside point-based OBBs are ground truth for pixel-level classification.

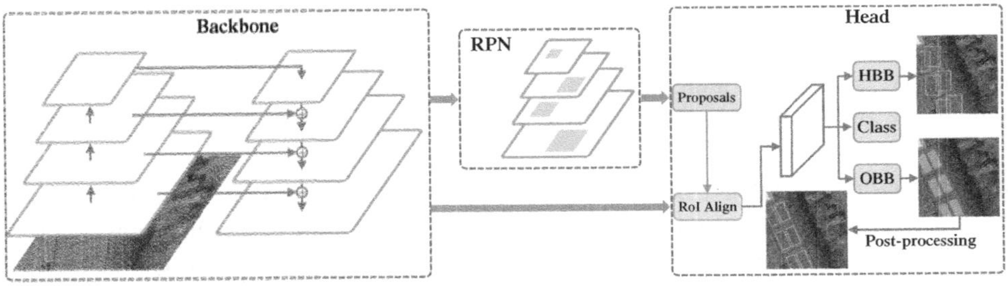

Figure 6.8 Overview of the pipeline for detecting oriented objects by Mask OBB. Horizontal bounding boxes and oriented bounding boxes are generated by HBB branch and OBB branch, respectively.

The overall architecture of Mask OBB is illustrated in Figure 6.8. Mask OBB is a two-stage method based on Mask R-CNN (He et al. 2017), which is well known as an instance segmentation framework. In the first stage, a number of region proposals are generated by a RPN (Ren et al. 2015). In the second stage, after the RoI Align (He et al. 2017) for each proposal, aligned features extracted from FPN (Lin et al. 2017) features are fed into the HBB branch and OBB branch to generate the horizontal bounding boxes and instance masks. Finally, the oriented bounding boxes are obtained by OBB branch based on predicted instance masks. Besides, Mask OBB applies FPN (Lin et al. 2017) with ResNet as the backbone to fuse low-level features and high-level features. Each level of the pyramid will be used for detecting objects at a different scale. We denote the output as $\{C_2, C_3, C_4, C_5\}$ for conv2, conv3, conv4, and conv5 of ResNet, and call the final feature map set of FPN as $\{P_2, P_3, P_4, P_5, P_6\}$. In this work, $\{P_2, P_3, P_4, P_5, P_6\}$ have strides of $\{4, 8, 16, 32, 64\}$ pixels with respect to the input image.

In the inference stage, we calculate the minimum area oriented bounding box of predicted segmentation map by *Topological Structural Analysis Algorithm* (Suzuki et al. 1985). The minimum area oriented bounding box has the same representation as θ-based OBB, which can be directly used by the HRSC2016 dataset for calculating mAP. For DOTA, the four vertexes of minimum area oriented bounding boxes can be used for evaluating performance.

Experiments In this section, we first study the different "first vertex" definition methods, which will affect the performance of point-based OBB and θ-based OBB in Table 6.2, and then, we study the effect of different OBB representations in Table 6.3. For a fair comparison, we re-implement the above three bounding box representations on the same basic network structure as Mask OBB.

For the first vertex definition, we compare two different methods. One is the same as Xia et al. (2018), which chooses the vertex closest to the "top left" vertex of the corresponding horizontal bounding box, and we call this method as "best point". The other one is defined by ourselves, which chooses the "extreme top" vertex of the oriented bounding box as the first vertex, then other vertexes are fixed in clockwise order, and we call this method the "extreme point". As shown in Table 6.2, the "best point" method significantly outperforms "extreme point" method on the OBB task of the DOTA dataset. We can learn that different "first vertex" definition methods will significantly affect mAPs of the OBB task. Thus if we want to obtain great performance on the OBB task by using point-based OBB and h-based

Table 6.2 Comparisons with different first vertex definition methods on the mAP of point-based OBB and *h*-based OBB representations. "Best point" method significantly outperforms the "extreme point" method on the OBB task of DOTA dataset.

dataset	first vertex	OBB represent.	backbone	OBB (%)	HBB (%)	gap (%)
DOTA	extreme point	point-based OBB	ResNet-50-FPN	64.40	68.72	4.32
		h-based OBB	ResNet-50-FPN	62.95	70.73	7.78
	best point	point-based OBB	ResNet-50-FPN	**69.35**	70.65	**1.30**
		h-based OBB	ResNet-50-FPN	**67.36**	70.46	**3.10**

Table 6.3 Comparison with different methods on the gap of mAP between HBB and OBB.

implementations	OBB representation	backbone	OBB (%)	HBB (%)	gap (%)
Ours	θ-based OBB	ResNet-50-FPN	69.06	70.22	1.16
	point-based OBB	ResNet-50-FPN	69.35	70.65	1.30
	h-based OBB	ResNet-50-FPN	67.36	70.46	3.10
	Mask OBB	ResNet-50-FPN	69.97	70.14	**0.17**
FR-O (Xia et al. 2018)	point-based OBB	ResNet-50-C4	54.13	60.46	6.33
ICN (Azimi et al. 2018)	point-based OBB	ResNet-50-FPN	68.16	72.45	4.29
SCRDet (Yang et al. 2019c)	θ-based OBB	ResNet-101-FPN	75.35	72.61	2.74
Li et al. (Li et al. 2019)	θ-based OBB	ResNet-101-FPN	75.38	73.28	2.10

OBB representations, we should design a special "first vertex" definition method which can represent OBB uniquely.

For different oriented bounding box representations, there is a higher gap between the HBB and OBB performance for both θ-based OBB, point-based OBB, and h-based OBB representation than Mask OBB. Theoretically, changing from the prediction of HBB to OBB should not affect the classification precision. But as shown in Table 6.3, the methods which use regression-based OBB representations have higher HBB task performance than OBB task performance. We argue that the reduction is due to the low-quality localization, which is caused by the discontinuity point. There should not be such a large gap between the performance of HBB and OBB task if the representation of OBB is defined well. The result of Mask OBB verified that. In addition, mAPs of HBB and OBB of Mask OBB are nearly all higher than the other three OBB representations in our implementations.

For other implementations, FR-O (Xia et al. 2018) uses point-based OBB and gets 60.46% HBB mAP and 54.13% OBB mAP, and the gap is 6.33%. ICN (Azimi et al. 2018) also uses point-based OBB and gets 72.45% HBB mAP and 68.16% OBB mAP, and the gap is 4.29%. SCRDet (Yang et al. 2019c) uses θ-based OBB and gets 72.61% OBB map and 75.35% HBB map, and the gap is 2.70%. Li et al. (2019) also uses θ-based OBB and gets 73.28% OBB map and 75.38% HBB map, and the gap is 2.10%. Note that the performances of ICN, SCRDet and Li et al. are obtained by using other modules and data augmentation technology. The gaps between HBB task and OBB task of these methods (6.33%, 4.29%, 2.70%, 2.10%) are all

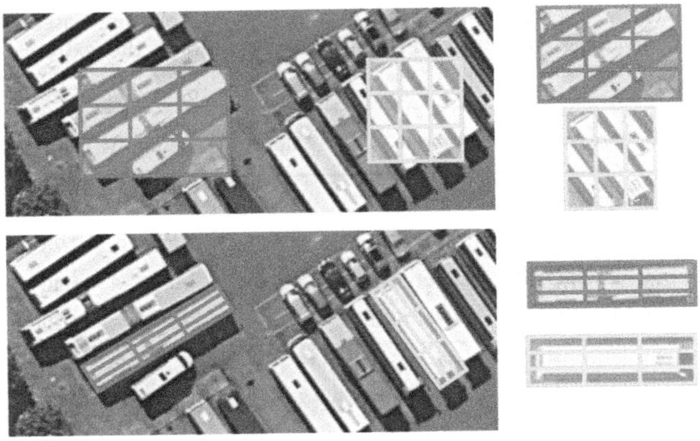

Figure 6.9 Horizontal RoI vs. Rotated RoI.

higher than Mask OBB (−0.16%). Therefore, we can draw the conclusion that Mask OBB is a better representation on the oriented object detection problem.

6.3.3.2 RoI Transformer

As shown in Figure 6.9, in the scenes where objects are rotated and densely packed, one horizontal RoI may contain several different object instances. So the features from horizontal RoI pooling are ambiguous and will influence the subsequent regression and classification. On the contrary, the oriented bounding box can provide more precise RoI and extract discriminative region features for object detection. To perform rotated RoI pooling, the rotated proposals are needed. As discussed before, SRBBS uses a hand-crafted way to generate rotated proposals, while RRPN has many redundant computations. To address this issue, we use a lightweight module, called RoI Transformer, to efficiently generate rotated proposals and extract rotated region features. With RoI Transformer, we can obtain more precise rotated proposals without increase the number of anchors. When matching between the rotated proposals and ground truth oriented bounding boxes, we directly use the IoU between OBBs as metric to avoid the problem of misalignment. The RoI Transformer contains two parts, namely, Rotate ROI Learner (RRoI Learner) and ROI Warping (RRoI Warping). Its architecture is shown in Fig. 6.10. For each horizontal region of interest (HRoI), it is passed to the RRoI Learner. The RRoI learner in the network uses position-sensitive ROI alignment, followed by a fully-connected layer. This layer will return the offset of the rotated true label relative to the horizontal RoI. The decoder is located at the end of the RRoI learner, which takes the horizontal RoI and offset as input and outputs the decoded RRoI. Then the feature map and the RRoI are feed into the RRoI Warping for geometrically robust feature extraction. The combination of the RRoI Learner and the RRoI Warping forms the ROI transformer. The RRoIs and robust geometric features from RoI Transformer are used for subsequent classification and bounding box regression.

RRoI Learner RRoI Learner aims to infer RRoIs from horizontal RoI features. Suppose we have obtained n horizontal RoI, represented as H_i. For each H_i, we use (x, y, w, h) to

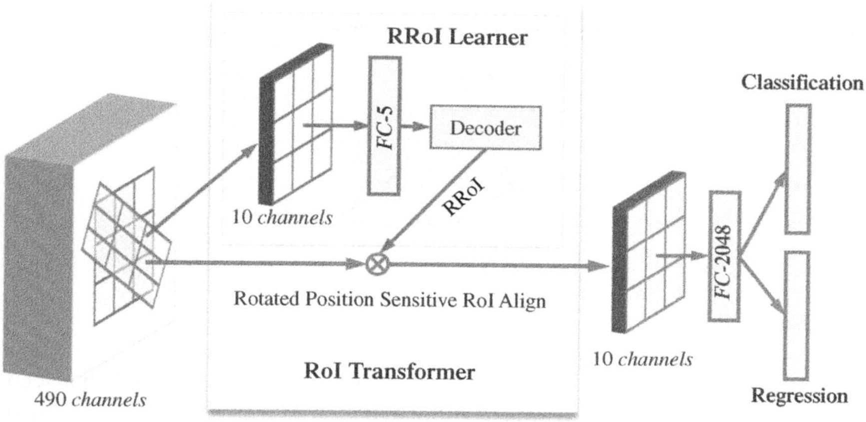

Figure 6.10 Network architecture of RoI Transformer.

represent the center, width, and height of horizontal RoI. The corresponding feature map is denoted as F_i. We can infer the geometry of RRoI from F_i followed by a fully-connected layer. The formulation of the learning targets are:

$$\begin{aligned}
t_x^* &= \frac{1}{w_r}\left((x^* - x_r)\cos\theta_r + (y^* - y_r)\sin\theta_r\right),\\
t_y^* &= \frac{1}{h_r}\left((y^* - y_r)\cos\theta_r - (x^* - x_r)\sin\theta_r\right),\\
t_w^* &= \log\frac{w^*}{w_r}, \quad t_h^* = \log\frac{h^*}{h_r},\\
t_\theta^* &= \frac{1}{2\pi}\left((\theta^* - \theta_r) \mod 2\pi\right),
\end{aligned} \tag{6.10}$$

where $(x_r, y_r, w_r, h_r, \theta_r)$ represent the center, width, height, and orientation of RRoI and $(x^*, y^*, w^*, h^*, \theta^*)$ represent the ground truth annotation. In fact, if $\theta^* = 3\pi/2$, the offset relative to the horizontal RoI is a particular case of Eq. (6.10). The general relative offset is shown in Figure 6.11. There are three coordinates: XOY is global coordinates binding to the image, and $x_1o_1y_1$ and $x_2o_2y_2$ are two local coordinates binding to the RRoIs. $(\Delta x, \Delta y)$ represent the offset between the oriented bounding box annotation and the RRoI. α_1 and α_2 represent the angles of two RRoIs. The yellow rotated rectangle represent the ground truth annotation. We can transform the left two rectangles to right two rectangles via translation and rotation, keeping the relative position unchanged. The $(\Delta x_1, \Delta y_1)$ and $(\Delta x_2, \Delta y_2)$ are the same if we observe them in $x_1o_1y_1$ and $x_2o_2y_2$ respectively. But they are not the same if we observe them in XOY. To derive Equation 6.10, we need to calculate the offsets in local coordinates such as $x_1o_1y_1$.

For each feature map F_i, the fully-connected layer is followed to output a vector $(t_x, t_y, t_w, t_h, t_\theta)$ by

$$\mathbf{t} = \mathcal{G}(\mathcal{F}; \Theta), \tag{6.11}$$

where $\mathcal{G}$, Θ denote the fully connected layer and its parameter. $\mathcal{F}$ denotes the feature map for each HRoI. During training, there is a matching process between input HRoIs and the

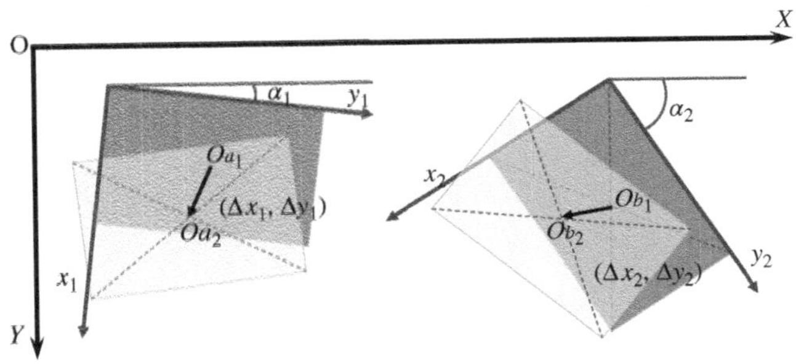

Figure 6.11 **Relative offsets.**

annotated oriented bounding boxes (OBBs). For efficiency, we calculate the IoU between input HRoIs and the corresponding HRoIs of annotated ground truths. For each matched HRoI, we assign the learning target by Equation 6.10. We use the Smooth L1 Loss (Girshick et al. 2014) for regression loss. For every predicted $\boldsymbol{t}$, we decode it to the RRoI.

RRoI Warping Once we get the RRoI, the rotation-invariant features can be extracted by RRoI Warping. Here, we implement the RRoI Warping by Rotated Position Sensitive (RPS) RoI Align in detail, because our baseline model is Light-Head R-CNN (Li et al. 2017). Given the input feature map $\mathcal{D}$ with shape of $(H, W, K \times K \times C)$ and a RRoI $(x_r, y_r, w_r, h_r, \theta_r)$, where (x_r, y_r) denotes the center of the RRoI and (w_r, h_r) denotes the width and height of the RRoI. The θ_r gives the orientation of the RRoI. Here, we implement the RRoI Warping by Rotated Position Sensitive (RPS) RoI Align in detail, because our baseline model is Light-Head R-CNN (Li et al. 2017). For the input feature $\mathcal{D}$ of shape $(H, W, K \times K \times C)$ and RRoI of shape $(x_r, y_r, w_r, h_r, \theta_r)$, where the (x_r, y_r) means the center and (w_r, h_r) are the width and height. θ_r is the angle of RRoI. We divide the feature map into $K \times K$ bins and output feature map of shape (K, K, C). For the bin at (i, j) location $(0 \leq i, j < K)$ and channel $c (0 \leq c < C)$, we have

$$\mathcal{Y}_c(i,j) = \sum_{(x,y)\in bin(i,j)} D_{i,j,c}(\mathcal{T}_\theta(x,y))/n, \tag{6.12}$$

where the $D_{i,j,c}$ represent one feature map of the $K \times K \times C$ output feature maps. The n represent the number of sampling ponits at one dimension. The $bin_{(i,j)}$ is the coordinates set $\{i\frac{w_r}{k} + (s_x + 0.5)\frac{w_r}{k\times n}; s_x = 0, 1, ...n-1\} \times \{j\frac{h_r}{k} + (s_y + 0.5)\frac{h_r}{k\times n}; s_y = 0, 1, ...n-1\}$. For each $(x, y) \in bin(i, j)$, it is transferred to (x', y') by $\mathcal{T}_\theta$, where

$$\begin{pmatrix} x' \\ y' \end{pmatrix} = \begin{pmatrix} cos\theta & -sin\theta \\ sin\theta & cos\theta \end{pmatrix} \begin{pmatrix} x - w_r/2 \\ y - h_r/2 \end{pmatrix} + \begin{pmatrix} x_r \\ y_r \end{pmatrix}. \tag{6.13}$$

RoI Transformer for Oriented Object Detection The RoI Transformer can be used to replace the regular RoI warping operation such as RoI Align and RoI Pooling. RoI Transformer outputs the rotation-invariant features and better initialization for subsequent regression. After RRoI warping, we add one 2048-dimension fully connected layer and two sibling fully connected layer for classification and regression respectively. The classification targets are calculated as normal. However, the regression targets are calculated different from

Table 6.4 Comparison between deformable RoI pooling and RoI Transformer. The DPSRP is the abbreviation of deformable position sensitive RoI pooling.

method	mAP	train speed	test speed	param
Light Head R-CNN OBB	58.3	**0.403 s**	**0.141s**	**273MB**
DPSRP	63.89	0.445s	0.206s	273.2MB
RoI Transformer	**67.74**	0.475s	0.17s	**273MB**

Table 6.5 Comparison with other methods.

method	backbone	W/FPN	test scales	mAP
FR-O (Xia et al. 2018)	resnet101		1	54.13
RRPN (Ma et al. 2018)	resnet101		1	61.01
R2CNN (Jiang et al. 2017)	resnet101		1	60.67
Yang et al. (Yang et al. 2018b)	resnet101	✓	1	62.29
ICN (Azimi et al. 2018)	dresnet101	✓	4	68.16
Baseline	resnet101		2	58.31
DPSRP	resnet101		2	63.89
RoITransformer	resnet101		2	67.74
Baseline	resnet101	✓	2	66.95
RoITransformer	resnet101	✓	2	**69.56**

the previous works. It must be rotation-invariant to maintain consistency. So we calculate the relative offsets as described in Figure 6.11. The core idea is to use the local coordinate system rather than the global image coordinate system for offsets calculation.

Experiments To conduct the experiments, we use the DOTA (Xia et al. 2018) dataset. To validate the performance is not from extra computation, we compare RoI Transformer with deformable position sensitive RoI pooling. Because both deformable RoI pooling and RoI Transformer are a kind of RoI Warping operation and can model the geometry transformation. To conduct experiments, we use the Light-Head R-CNN OBB (Li et al. 2017) as baseline. The deformable position sensitive RoI pooling and RoI Transformer are used to replace the position sensitive RoI align in Light-Head R-CNN OBB. The detailed results are shown in Table 6.4. It shows that the RoI Transformer is lightweight and efficient. We also compared the experiment results with other methods in Table 6.5. The model Light Head R-CNN + RoI Transformer outperforms other methods. The code of RoI Transformer is available[4]. Besides the original Mxnet implementation. There is another version implemented with Pytorch[5].

4 https://github.com/dingjiansw101/RoITransformer_DOTA
5 https://github.com/dingjiansw101/AerialDetection

6.4 Object Detection in SAR Images

6.4.1 Challenges of Detection in SAR Images

Synthetic Aperture Radar (SAR) is a kind of active remote sensing sensor which has been widely used in the monitoring of land surface, environment or the disasters for its all-time and all-weather acquisition, its sensitivity to different target properties, its penetration characteristics, and so on (Lee et al. 1994).

As a two-dimensional imaging system, SAR has opened up a new space for object detection tasks. In SAR images, we can not only measure the radar cross-section of the target (RCS), but also obtain the information of two-dimensional shape about the object. Object detection tasks based on SAR images are widely used in the military field and the civil field. In the military field, the current high-resolution SAR system whose resolution can reach the centimeter level can help to find and recognize all kinds of camouflaged targets. The famous program MSTAR (Moving and Stationary Target Recognition) has paid much attention to the research and development of object detection based on SAR images. In the civil field, object detection based on SAR images plays an important role in maritime surveillance, rescue, search, and so on. For example, compared with optical remote sensing image, SAR images have advantages in detecting ships, as the strong backward reflection of ships can form obvious bright spots in the background of low-intensity sea clutter.

As the imaging mechanism of SAR images is quite different from that of natural images, there are great differences between them. Therefore mostly using the existing object detection algorithms designed for natural image to process SAR images is impractical. However, applying these deep learning-based methods to SAR images directly often does not lead to satisfying performance since there are several challenges which limit the applications of deep learning methods in objects detection for SAR images.

6.4.2 Related Works

Object detection based on SAR images has been studied for a long time. Many methods have been proposed, among which automatic target recognition in SAR images (SAR-ATR) is the essential one (Khalid et al. 2018). It aims to detect and identify targets (such as aircraft, ships, vehicles, etc.) in many complex SAR scenarios quickly and effectively. The processing of SAR-ATR technology is divided into three stages: a detection processing stage, an identification processing stage, and a classification and recognition stage. The detection processing stage is mainly to detect the target and extract the region, which contains the interesting target, the identification stage is to remove false alarm areas caused by radar clutter, and the process of classification is to extract the features of the target and classify the different targets.

Traditional SAR target detection uses mathematical statistics theory to estimate the parameters of the image model. Two methods have been proposed: fixed threshold detection and Constant False Alarm Rate (CFAR) detection (D 1963). For the former, a threshold is calculated in advance and will not change once it is set, while the threshold of the CFAR can be adjusted adaptively according to the change of background clutter to ensure a constant false alarm rate. CFAR is at the pixel level, and the specific implementation

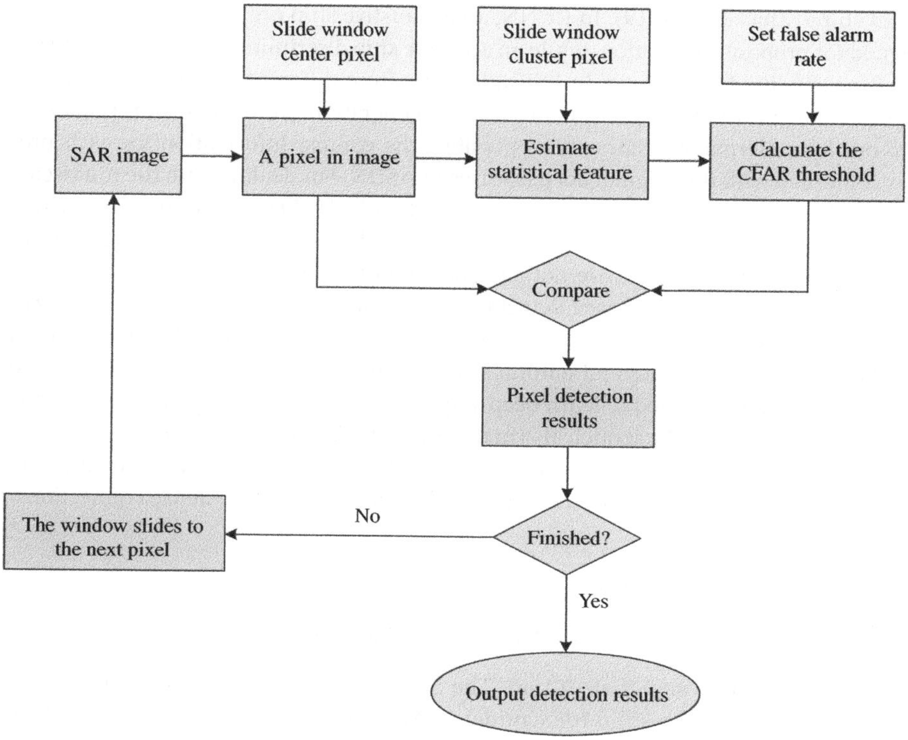

Figure 6.12 The flow chart of CFAR algorithm.

process is shown in Figure 6.12. Firstly, a false alarm probability value is set, and its statistical characteristics are calculated according to the background clutter pixels which near the pixel to be detected, then the threshold value of target detection is estimated adaptively according to this statistical feature; finally, the value of the pixel to be detected is compared with the estimated threshold value. If the value is larger than the threshold value, it is the target point; otherwise, it is the pixel background point. There are many kinds of CFAR algorithms. According to the different distribution of background clutter in SAR images, they can be divided into two categories: single-parameter CFAR (Finn and Johnson 1968) and multi-parameter CFAR (Leslie and Michael 1988). If the background clutter distribution is quantified by a single parameter(such as Rayleigh distribution, Exponential distribution, etc.) (Goldstein 1973), it is a single parameter CFAR, such as Cell Averaging CFAR and Greatest-Of CFAR. If the background clutter distribution contains two or more parameters (such as Gaussian distribution, Wilbur distribution, etc.), it is multi-parameter CFAR, such as double-parameter CFAR.

For more details, to solve the problem of detecting small objects, a ship detector composed of a RPN and an object detection network with contextual features has been proposed (Miao et al. 2017). The used strategy, fusing both the deep semantic and shallow high-resolution features helps to improve the detection performance for small-sized ships. A coupled convolutional neural network (CCNN) is also designed for small and densely clustered ships (Zhao et al. 2019). In CCNN, an exhaustive ship proposal network (ESPN) is designed for proposal generation, while an accurate ship discriminative network (ASDN) is used for excluding false alarms. In ESPN, features from different layers are reused and the representative intermediate layers are used to generate more reliable ship proposals. To rule out false alarms as accurately as possible, the context information for each proposal is combined with the original deep features in ASDN. For dealing with the multiscale problem, a densely connected multiscale neural network (DCMSNN) is proposed, which is a densely connected network. Clearly, the CNN-based methods have got huge success. However, for SAR images, there are not adequate annotated samples for model training. Therefore, most deep-learning-based detectors in SAR images have to fine-tune networks pre-trained on large-scale natural image datasets, such as ImageNet. But the huge domain gap exists between SAR images and natural images which will incur the learning bias. To solve this problem, the method to learn deep ship detector from scratch is proposed (Deng et al. 2019). Learning from the scratch also makes the redesign of the network structure possible. A condensed backbone network is designed, and several dense blocks are included to receive additional supervision from the objective function through the dense connections.

6.4.3 Datasets and Benchmarks

The followings are several typical SAR dataset which have been widely used.

- **SAR-Ship-Dataset** (Wang et al. 2019a). The SAR ship detection dataset contains 102 high-resolution SAR images and 108 sentinel-1 SAR images collected from Gaofen-3 SAR and sentinel-1 SAR. At present, the dataset has 43,819 ships in complex backgrounds and is suitable for many SAR image applications.
- **SSDD** (Li et al. 2017) It is the first publicly available dataset for SAR image ship detection. There are 1160 images and 2456 ships, with an average of 2.12 ships per image. These public SAR images were downloaded from the Internet and cropped into a size of 500 × 500 pixels. The dataset collects images from RadarSat-2, TerrasAR-X, and Sentinel-1 sensors with a resolution of 1–15 m and four polarizations (HH, HV, VV, and VH). It contains ships both in the sea and near shore. As a result, it is sufficient to train the ship detection model.
- **SpaceNet 6 Multi-Sensor All-Weather** (Shermeyer et al. 2020) Capella Space collected the data via an airborne sensor. Each image has four polarizations (HH, HV, VH, and VV) data and is preprocessed to show the backscattering intensity at a spatial resolution of 0.5 m. The entire dataset contains more than 48,000 high-quality architectural footprint notes, with extra quality control over the tags, removing incorrectly marked areas, and adding tags for unmarked or destroyed buildings. The dataset also contains a 3D component from the publicly available digital elevation model which is from airborne lidar. Therefore, for each annotation, we report the mean, median, and standard deviation of the height in meters for the 75th percentile. The height information will be valuable in detecting object height from the upper air.
- **AIR-SARShip-1.0** (Xian et al. 2019) The dataset contains 31 large images collected from Gaofen-3 satellite. The images consist of 1 m resolution and 3 m resolution, including bunching type and band type. The polarization mode is single-polarization, and the image

format is Tiff. Most of the images have a size of 3000 × 3000 pixels. The dataset also contains relevant information such as the surrounding sea, land, and port, which is closer to real-world applications.

6.5 Conclusion

In this chapter, we first gave the definition of object detection and introduced the evaluation metrics as well as applications. Object detection in remote sensing images is quite different from object detection in natural images and challenging. Because the objects can be an arbitrary orientation in remote sensing images, the oriented bounding box is more suitable for object detection in remote sensing images. Except for the arbitrary orientation, densely packed instances, scale variation, and large-size image inference are also challenges for object detection in remote sensing images. We categorized the previous works for object detection in remote sensing images according to the challenges they addressed. We also described two example algorithms and analyzed the experiment results to show some details to solve part of these challenges. With the large-scale datasets for object detection in optical remote sensing images available, there is a significant improvement. However, these challenges mentioned are still not well solved and remain open problems. Besides, large-scale datasets for object detection in SAR remote sensing images need to be established in the future.

7

Deep Domain Adaptation in Earth Observation

Benjamin Kellenberger, Onur Tasar, Bharath Bhushan Damodaran, Nicolas Courty, and Devis Tuia

7.1 Introduction

Environmental data, and in particular Earth Observation data, are not stationary in space and time. Signals acquired by sensors tend to vary (or *shift*) according to differences in acquisition conditions. For example, a crop field will return a very different signature if observed in the morning or at noon, after seeding, or at full growth of the crops or in different years of exploitation. This is due to different atmospheric effects, different relative positions of the sensor and the sun, or design differences in the sensors acquiring the measurements: for example, satellites have slightly different spectral ranges for bands, even if named identically (e.g. the first near infrared band of WorldView 2 covers the range of 770–895 nm, while the one of the French Pléiades covers 740–940 nm). All these factors lead to (spectral) signatures that are different when observing the same object. Since machine learning models base their decision on observed data only, these differences are oftentimes challenging. This problem is denoted as *dataset shift*, and particularly affects the generalization capability of a model. For example, a model trained to detect crops in Earth Observation data that has been acquired in the morning may learn to use the shadows cast by the crop plants. If such a model is applied on scenes acquired during mid-day, where such shadows are absent, it will likely fail.

A second challenge related to multiple data acquisition campaigns is that class concepts also drift in space and time and depend on where (and when) they are observed. Taking again crops as example, crops are dynamic, as they grow and change in time: a model trained on data acquired in an early growth stage will not be effective in recognizing crops at later stages, since leaf characteristics and biomass-to-soil ratios will be different, even when observed under the exact same experimental conditions. Shifts may be even worse in the case of classification across different geographical regions, for example when developing a model for building detection (as described in Chapter 5 of this book): models trained for suburban areas of Western cities will specialize in the detection of buildings with very different definitions compared to those trained for e.g. central business districts of mega-cities. Moreover, applying either of those models to a city in an Eastern country would probably be unsuccessful, because of differences in architecture, materials used and planning habits.

Deep Learning for the Earth Sciences: A Comprehensive Approach to Remote Sensing, Climate Science, and Geosciences, First Edition. Edited by Gustau Camps-Valls, Devis Tuia, Xiao Xiang Zhu, and Markus Reichstein.

This *concept drift* is also a serious problem hampering model generalization: when class definitions change between training and testing times, a model must be adapted accordingly to be successful.

A third challenge is to develop models that can generalize across different sensor types and configurations: this case arises when the same concept of interest is measured from different devices like sensors with different bands and spectral ranges or other data modalities (sensor networks, LiDAR, etc.). In this case, traditional models cannot work, as they expect a specific format of the input data (e.g. an image with a three-layered structure in case of RGB data) that would not be respected at inference time. This last type of domain difference has been referred to as *multi-modal domain shift* and is related to the domain shift case discussed above, but further requires models that are able to account for the different sensors configurations and specifications explicitly. Note finally that it is realistic to expect that most operational settings would suffer simultaneously from the three aforementioned challenges, making the domain adaptation at the same time particularly challenging, but omnipresent in modern applications.

In this chapter, we will discuss recent advances in methodologies addressing either of the three challenges discussed above and present different approaches to achieve generalization despite sensors or concepts differences, i.e. *domain adaptation* (Quiñonero-Candela et al. 2009). The field is actively studied in the communities of statistical learning (Redko et al. 2019), computer vision (Csurka 2017), and Earth Observation (Tuia et al. 2016b). Domain adaptation in the geo-spatial domain has strong implications on the ambition of building models that can be applied globally, at fine temporal scale, and with multiple sensors observing the planet from above. We will focus on methodologies addressing domain adaptation for deep learning models (Wang and Deng 2018), which, at least in the Earth observation community, is a young field with large potential for innovation.

7.2 Families of Methodologies

Similar to computer vision, also the Earth observation community nowadays increasingly employs deep learning-based methodologies for the majority of applications. Deep learning is a very powerful tool for a number of learning tasks, notably through its capacity to learn from a wealth of available data. For image data, particularly Convolutional Neural Networks (CNNs) allow for very precise predictions when trained with annotated data of high quality and quantity. Most of the problems arise whenever the amount of available annotated data is low, which in some cases may mean a complete absence of training data. A popular approach is to use an architecture that had been pre-trained on generic, large-scale datasets like ImageNet (Deng et al. 2009), and then fine-tune it on the available data, i.e. slightly modify the last layers of the network so that it can adjust to the observations at hand. Even though this approach is sometimes referred to as transfer learning, it can also be seen as re-tuning a model with a good initialization issued from another dataset with large variability, given high-quality labels are available in sufficient numbers: in this chapter we will focus on methodologies specifically designed for dealing with domain shifts and therefore fine-tuning will not be covered explicitly as a domain adaptation strategy.

In a domain adaptation context, the problem changes as the testing distribution (target domain) differs from the training distribution (source domain). Two variants of methodologies can be distinguished: **supervised domain adaptation** deals with the case where labeled data is also available in the target domain, although not in sufficient proportions to train an accurate model, and **unsupervised domain adaptation** considers cases where no labeled data is available for the target domain. Shallow (not deep learning-based) methods partially solve this problem by either *reweighing* source samples (Hidetoshi 2000), or by learning a common representation through a subspace projection that is invariant to domain shifts (Baktashmotlagh et al. 2013). In a deep learning setting, one can try to leverage the representational power of the network to mitigate the effect of domain shift. Deep learning-based domain adaptation methods generally belong to one of three families:

- **Adapting the inner representation**: methods attempt to minimize a statistical divergence criterion between the representations of the two domains at a given layer in the network. Popular choices for computationally tractable divergences include aligning second-order statistics (i.e. covariances) (Sun and Saenko 2016), contrastive domain discrepancy (Kang et al. 2019), maximum mean discrepancy (Mingsheng et al. 2015), or Wasserstein metrics (Damodaran et al. 2018). An alternative approach lies in adversarial training, which uses learning signals from a domain classifier to align inner source and target features (Ganin et al. 2016).
- **Adapting the input distribution**: other approaches align the input data distributions in the source and target domain before training the classifier. A first class of methods adapts the image statistics of the target domain, either using a common latent representation (as autoencoders) or using image-to-image translation principles (Zhu et al. 2017; Hoffman et al. 2018). A second class of methods focuses on generating adversarial examples that fit as close as possible into the target domain distribution, and then use these artificial data to train a domain-agnostic classifier. An example of this second strategy is CoGAN (Liu and Tuzel 2016).
- **Using (few, well-chosen) labels from the target domain**: sometimes the shift between domains is too severe or the class proportions vary too much, so that methods aligning domains in an unsupervised way cannot succeed. Supervised methods, i.e. methods using labels from the target domain, address those cases, but at the price of needing to annotate images from target. Strategies using selective sampling, or active learning (Settles 2012) can be used in this case to minimize the sampling effort.

It is worth mentioning that most of these methods are known to work in controlled settings, where the balance between classes is similar in the source and target domains. Variants of those methods, closer to real-world applications, are an active subject of research. This includes models for target shift, open set domain adaptation, partial domain adaptation, and multi-modal, sometimes called heterogeneous domain adaptation. Target shift occurs when source and target domains do not share the same class proportions. In open set domain adaptation, new classes are present in the target domain. Conversely, in partial domain adaptation, few classes are absent in the target domain, which is a specific case of the concept drift problem presented in the introduction. Finally, the multi-modal adaptation problem occurs when the source and target domains do not live in the same space. In the remainder of this chapter, we discuss examples of those three families of methods to give the reader an idea of concrete deep domain adaptation methods, without being exhaustive.

7.3 Selected Examples

7.3.1 Adapting the Inner Representation

One of the most frequently encountered families of domain adaptation modifies the machine learning model at hand to be able to generalize to data from previously unseen domains. For more traditional classifiers that operate on hand-crafted features, such as Random Forests or Support Vector Machines, the most straightforward way for domain adaptation is to fine-tune the classifier itself on the target domain, based on the final class predictions provided by the model. While this strategy can also be applied to deep learning models through a loss on the predicted class output, a more common strategy is to instead impose a loss on intermediate feature vectors as output by the hidden layers to generate similar feature vectors for the source and target samples. For example, Figure 7.1 shows intermediate and final outputs of a CNN with two convolutional layers ("conv1" and "conv2") and a fully-connected layer ("fc1"). The modularity of CNNs theoretically allows imposing domain adaptation losses on all feature vectors in addition to the final class predictions (right). In practice, however, a common strategy is to adapt the penultimate feature vectors only ("fc1"): these are expected to cover semantic concepts, which are more informative for the final task than lower-level image features learned in the earlier layers. Finally, note that most adaptation approaches keep a traditional classification loss for the source domain, in order to avoid catastrophic forgetting (Kirkpatrick et al. 2017), and also since the source labels are the only available semantic ground truth in an unsupervised adaptation scenario.

Several approaches in this direction have been proposed that work on deep learning models, among which the following examples will be briefly discussed and evaluated below:

- Maximum Mean Discrepancy (MMD)-based domain adaptation imposes a training loss on the model that tries to assimilate the source and target features by means of distances between their expected values (Mingsheng et al. 2015).
- Deep Correlation Alignment (DeepCORAL) follows a similar strategy, but instead attempts to minimize the covariances between predicted source and target feature vectors (Sun and Saenko 2016).

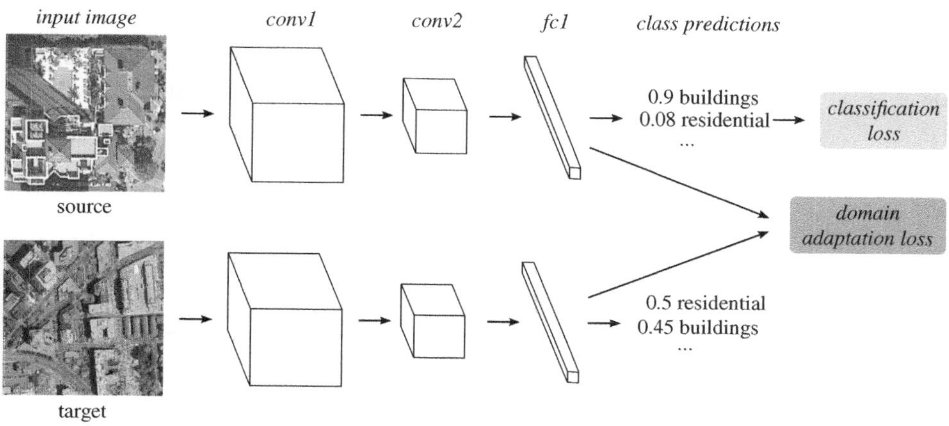

Figure 7.1 Domain adaptation loss imposed on a CNN's feature vectors produced by the penultimate layer ("fc1").

- Deep Joint Optimal Transport (DeepJDOT; Damodaran et al. (2018)) likewise attempts to minimize the discrepancy between the feature vector distributions of both domains, but also incorporates the label information associated to the features, thus aligning the joint distributions between feature and labels. It does so with Optimal Transport, which yields source-to-target couplings that provide the minimum Wasserstein distance between the two domains (Courty et al. 2017).

Experiments In the following, we evaluate the adaptation performance of all three methodologies on a remote sensing image classification problem (as described in Chapter 5 of this book). We employ two datasets named *UC Merced* (Yang and Newsam 2010) and *WHU-RS19* (Dai and Yang 2010). Both datasets contain satellite-derived RGB images, but obtained from different sources (USGS National Map Urban Area Imagery Collection for UC Merced, and Google Earth for WHU-RS19), with different image sizes (256×256 for UC Merced, 600×600 for WHU-RS19), different numbers of images per class (100 for UC Merced and 50 for WHU-RS19) and different class definitions. The two datasets were therefore limited to a set of ten overlapping classes, as described in Table 7.1. Both datasets were further divided by selecting 60% of the images of each class at random for training, 10% for validation, and 30% for testing. Example images are shown in Figure 7.2. As can be seen, the different resolutions, label classes, and geographic locations together pose a comparably strong domain shift. The experiments below will investigate adaptation performances of the three models above, and in both directions (i.e., UC Merced $\rightarrow$ WHU-RS19, and the inverse). Code to reproduce the experiments is provided on the dedicated GitHub page[1].

As a base classifier, we employ a ResNet-50 (He et al. 2016b) that has been pre-trained on the ImageNet classification dataset (Deng et al. 2009). We replace the last fully-connected layer with a new one with random initialization to match the number of classes in our

Table 7.1 Common label classes between the UC Merced and WHU-RS19 datasets.

Index	UC Merced	WHU-RS19
AG	agricultural	farmland
AP	airplane	airport
BE	beach	beach
DR	buildings, dense residential	commercial, industrial
FO	forest	forest
HA	harbor	port
MR	medium residential	residential
VI	overpass	viaduct
PA	parking lot	parking
RI	river	river

1 https://github.com/bkellenb/da-dl4eo

Figure 7.2 Examples from the UC Merced (top) and WHU-RS19 (bottom) datasets.

datasets. We initially train one model for each of the two datasets, with equal settings and hyperparameters for both: we draw minibatches of 32 images from the training set, resize the images to 128×128 pixels, and apply data augmentation in the form of random horizontal and vertical flips, as well as a slight color jitter. We use a softmax and multi-class cross-entropy loss for training, and employ the Adam optimizer (Kingma and Ba 2014) with an initial learning rate of 10^{-4} that gets divided by 10 after every ten epochs. We do not use weight decay for training. In order to ensure convergence, these base models are trained for 100 epochs on their respective dataset.

In a second step, we use the pre-trained models as a starting point for the three domain adaptation strategies presented above. We keep all settings constant for all strategies, but start with a learning rate of 10^{-5}. In addition to the cross-entropy loss on the predictions in the source domain, we add a domain adaptation loss from one of the three strategies on the 2048-dimensional feature vector output after the global average pooling layer in the ResNet-50 (i.e., the penultimate layer in the model). We train the respective model for another 100 epochs, with one epoch being defined as the maximum of the lengths of the two datasets, drawing 32 images per minibatch from each dataset at random.

Table 7.2 shows overall accuracies on the test sets for the unadapted baseline models (top), the three domain adaptation strategies (middle), as well as the target models (bottom). A first observation to make is that both source models yield perfect predictions on their respective datasets' test sets ("target only"), indicating that the separability of the datasets

Table 7.2 Overall accuracies for the discussed datasets and domain adaptation methods.

Method	**Adaptation: source → target**	
	UC Merced → WHU-RS19	**WHU-RS19 → UC Merced**
Source only	0.66	0.59
MMD	0.68	0.68
DeepCORAL	0.68	0.67
DeepJDOT	**0.75**	**0.73**
Target only	1.00	1.00
Source & target	1.00	1.00

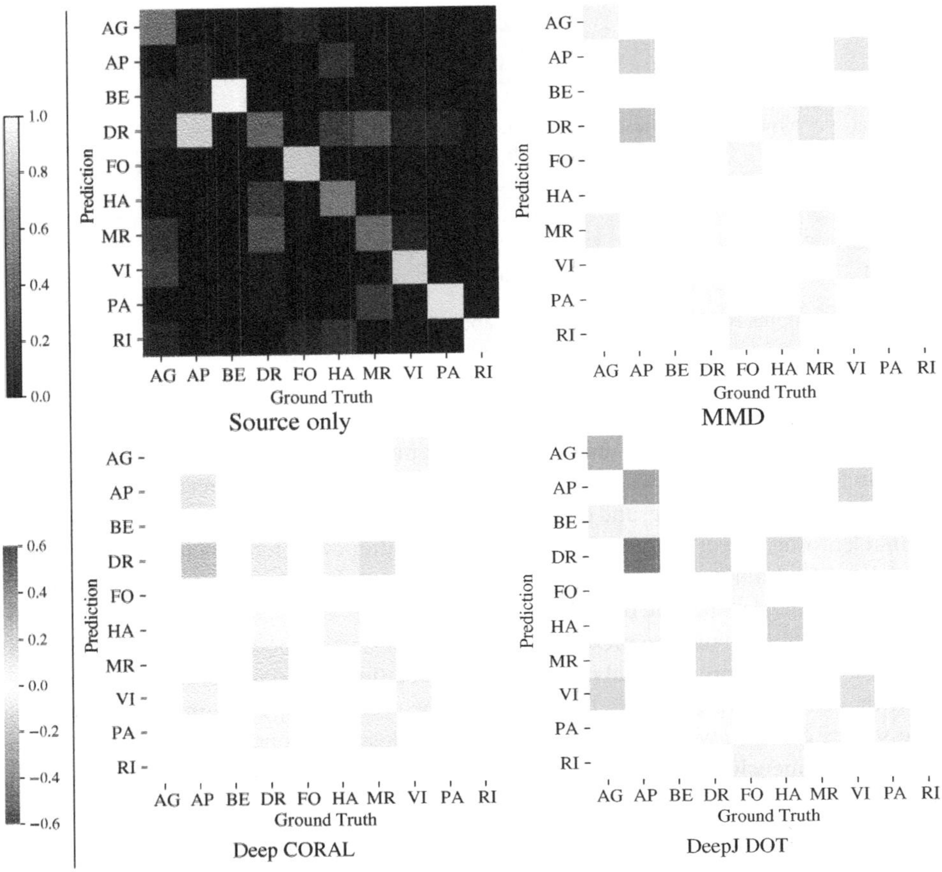

Figure 7.3 Confusion matrix of the source only model (top left) and differences to it for the domain adaptation strategies on the WHU-RS19 test set. Best viewed in color.

and learning capacity of the ResNet-50 are sufficient if enough labeled data from the specific domain are available. If applied without adaptation to the other domain's test sets ("source only"), accuracies drop by 34, resp. 41 absolute percent. A look at the per-class predictions (Figure 7.3) reveals that the primary confusion occurs between "airplane" and "commercial, industrial", with 88% of the WHU-RS19 "industrial" images misclassified as "airplane". This is likely to be attributable to WHU-RS19 showing entire airports rather than single airplanes, which looks similar to more industrial scenes. Other classes being confused are "buildings" and "industrial" (around 32% false positives), and "agricultural" and "viaduct" (around 20% false positives).

The three domain adaptation methods manage to slightly improve the overall accuracy in the UC Merced → WHU-RS19 case, and significantly raise the accuracy in the inverse adaptation experiment. As for the first adaptation direction, MMD lowers the confusion between "airplane" and "commercial, industrial" from 88% to 69%, but increases other confusions, such as between "industrial" and "residential". DeepCORAL does not significantly reduce the confusion between any two specific classes, but provides a more average result, decreasing confusion between some pairs, but increasing it in other cases. Finally, DeepJDOT significantly increases the true positive predictions or leaves them unaffected for all but two classes ("viaduct" and "parking"). It also significantly reduces the confusion between "dense residential" and "airplane". These improvements can in parts be attributed to the fact that DeepJDOT tries to minimize the feature vector distance between specific source-target samples, retrieved through Optimal Transport. This stands in contrast to MMD and DeepCORAL, which align samples that are close in feature space, but not necessarily with respect to the global distributions. This only works well if the source and target distributions lie in manifolds that are comparable with respect to global distribution characteristics. As soon as source and target samples of dissimilar classes lie closely together, those methods may force the model to consistently mis-predict target samples.

7.3.2 Adapting the Inputs Distribution

An alternative approach is to change the data distribution of either the source, or both source and target domains directly, before exposing them to a classification or regression model. The most straightforward approach to do so is to increase the number of samples and vary them by applying data augmentation methods, such as random contrast change (Buslaev et al. 2018), gamma correction (Tasar et al. 2019), or geometric deformations (Castelluccio et al. 2015; Marmanis et al. 2015). These strategies, while effective for the general task of fine-tuning models, fail if the shift between the source and target domains is large: in that case, dedicated strategies become necessary. In detail, the following two approaches have been proposed:

- *Generate synthetic data*: as described above, the intent behind synthetic data is to capture the distribution of the target domain. Those artificial data samples can then in turn be used to train or fine-tune a model on the fake source data. By proceeding this way, the distributions of the fake source and target data resemble each other, which is supposed to increase final model performance.
- *Standardize both domains*: a second strategy, often used in hyperspectral imaging (Gross et al. 2019) and multi-source image processing (Tuia et al. 2016a), maps the samples from both source and target domains into a common subspace. This is done in such a way that the samples belonging to the source domain are representative for the target domain. Then, good predictions can be obtained by training a model on the standardized source data and classifying the mapped target samples.

Methods Generative Adversarial Networks (GANs; Goodfellow et al. (2014a)) described in chapter 3 can be used for fake data generation (also see the dedicated chapter about GANs, chapter 3). The main challenge is to generate fake source images having similar distribution to the distribution of target domain, while keeping source and the fake source images semantically exactly the same. Image-to-Image translation approaches can be used to achieve this goal. We then can use the fake source images and the ground truth for source domain to fine-tune the already trained classifier. We compare the following translation methods:

- *CycleGAN* (Zhu et al. 2017) trains two GANs, with the first one generating target synthetic samples, and the second mapping the generated target samples back to the source domain. The aim is that the back-transformed, generated samples are realistic with respect to the source domain.
- *UNIT* (Liu et al. 2017b) maps both source and target data to a common latent space. Fake data are then sampled from the latent space.
- *MUNIT* (Huang et al. 2018d) combines the content code of a domain with the style code of another domain via Adaptive Instance Normalization (Huang and Belongie 2017).
- *DRIT* (Lee et al. 2018b): Similar to *MUNIT*, *DRIT* also combines content code of a domain with the style code of another domain. The difference is that combination is performed through concatenation, rather than Adaptive Instance Normalization.
- *ColorMapGAN* (Tasar et al. 2020) maps colors of the source domain to those of the target domain to correct the domain shift. Unlike the other GANs, the generator of ColorMapGAN performs only matrix multiplication and addition operations.

These deep learning-based methods will also be compared with two traditional image normalization methods:

- *Histogram matching* (Gonzalez and Woods 2006): The histograms for the spectral bands of the source data are matched with the histograms of the target data. This method is not based on GANs or any deep learning-based approaches.
- *Gray world* (Buchsbaum 1980): All the methods described above align the data distribution of the source domain to the distribution of the target domain. The *Gray world* algorithm, on the other hand, assumes that the color of the illuminant highly affects the colors of the objects and aims at removing this effect. We use the original *Gray world* algorithm to standardize both domains separately.

Experiments We consider here a semantic segmentation task (as described in Chapter 5 of this book). The source and the target images used in the experiments have been acquired over *Villach* and *Bad Ischl* in *Austria*, by the Pléiades satellite. The spatial resolution is 1 m in both cases. The images collected from *Bad Ischl* and *Villach* cover the total area of 27.71 km^2 and 43.59 km^2, respectively. The annotations for the *building*, *road*, and *tree* classes have been manually prepared. Given the changes in illumination and the differences in spatial structures, we can consider the shift between the domains as large.

We use *Bad Ischl* as the source and *Villach* as the target city. We first train a U-Net (Ronneberger et al. 2015a) on the source data for 2,500 iterations, where each iteration consists of batches of 32 randomly selected training patches. We then fine-tune the model on the synthesized training data, generated by the methods described above for 750 iterations. For the Gray world algorithm, we test the fine-tuned model on the standardized target data, whereas we test it on the real target data for the other approaches. Figures 7.4 and 7.5 illustrate source, target, and the fake data. Table 7.3 reports F1 scores. Finally, Figure 7.6 shows classification maps by the adapted models.

Discussion As can be seen in Figure 7.4, the fake source data generated by UNIT, MUNIT, and DRIT are semantically inconsistent with the real source data. Therefore, the fake source data and the ground truth for the real source data do not match, which results in a

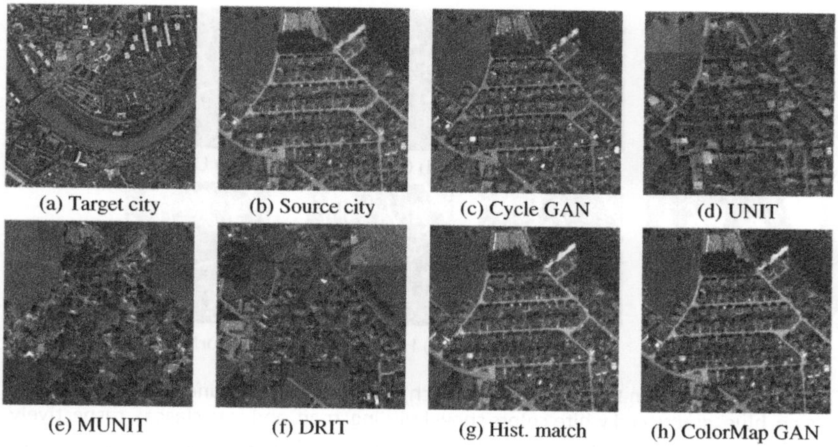

Figure 7.4 Source, target, and fake source images. Best viewed in color.

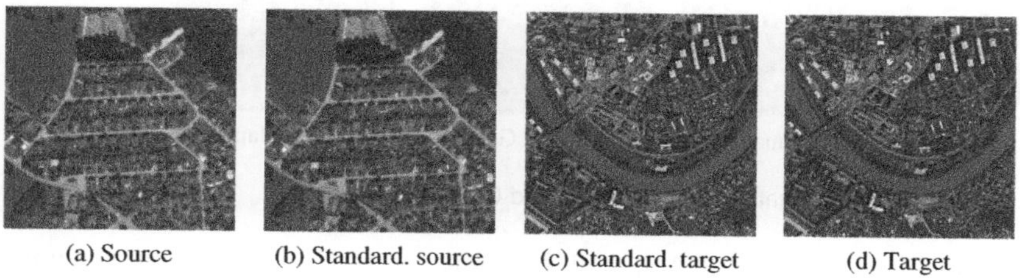

Figure 7.5 Real data and the standardized images by the *Gray-world* algorithm. Best viewed in color.

Table 7.3 F1 scores for the target city.

	Method	Building	Road	Tree	Overall
	U-net	32.74	32.92	73.62	46.43
U-net fine-tuned on data generated by	ColorMapGAN	**65.48**	**54.95**	74.64	**65.02**
	CycleGAN	60.55	44.83	**81.94**	62.44
	UNIT	46.31	26.66	77.57	50.18
	MUNIT	0.01	2.04	65.60	22.55
	DRIT	0.00	6.00	9.52	5.17
	Hist. matching	39.67	45.34	76.57	53.86
	Gray-world	39.95	41.65	72.37	51.32

poor performance. Hence, we conclude that these algorithms should not be considered for data augmentation. Figure 7.5 shows that the large gap between domains can be reduced by the Gray world algorithm. However, as confirmed by Table 7.3, the performance is not satisfactory.

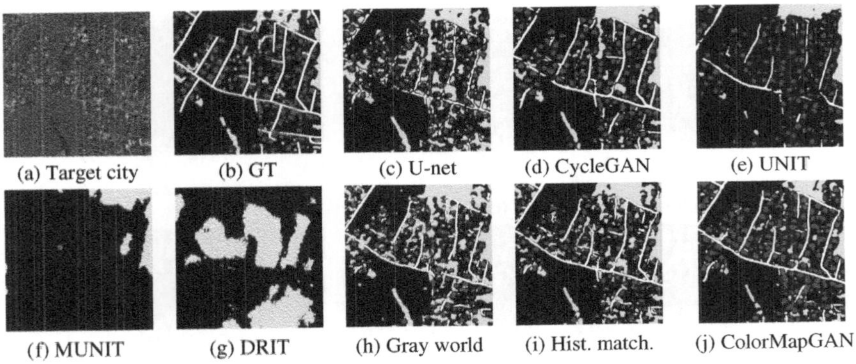

Figure 7.6 Classification maps on the target city (Villach) by the U-net fine tuned on the fake data. In the color version, blue, green, and white colors show building, road, and tree classes, respectively. When printed in black and white, these *correspond* to dark gray, white, and light gray, respectively. In both versions, black are pixels for which no class was predicted with more than 50% confidence.

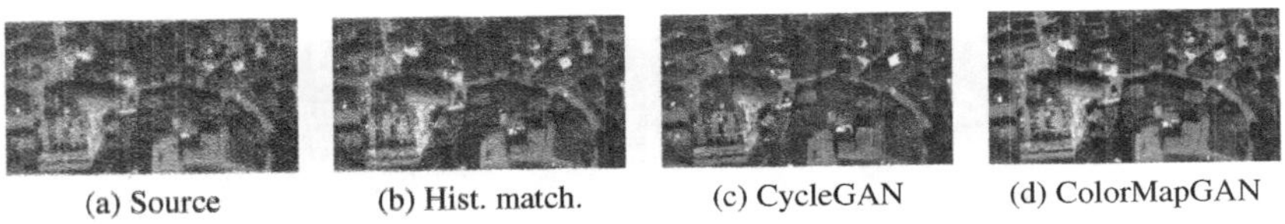

Figure 7.7 Limitations of hist. matching, CycleGAN, and ColorMapGAN.

Histogram matching, ColorMapGAN, and in some cases CycleGAN can generate semantically consistent fake data. However, each method has its own disadvantages. Figure 7.7 shows that histogram matching does not consider the contextual information: for example, it generates buildings with different rooftops colors, which do not exist in the target domain. In the same figure, we can see that CycleGAN adds a lot of artifacts. Since ColorMapGAN maps each color to another one, the fake image generated by this approach is slightly noisy. However, as it is a common practice to add noise (i.e., Gaussian noise) to training data to make the model to more robust, this may not be an issue. The detailed comparisons between these approaches can be found in Tasar et al. (2020).

7.3.3 Using (few, well-chosen) Labels from the Target Domain

Introduction Sometimes it is difficult to adapt inputs, or generate synthetic source domain samples, or else data distribution characteristics like heavy class imbalances disallow unsupervised domain adaptation strategies, as it is known to severely impair the adaptation process (e.g. Zhang et al. (2013)). In such cases, a third option can be employed that requires a small number of ground truth labels for the target domain, a procedure known as "semi-supervised domain adaptation". The rationale behind this is that labels for a small number of well-chosen target samples are sufficient to adapt a model appropriately. To this end, active learning (AL; Tuia et al. (2011); Settles (2012)) methods can be used.

The following case study employs semi-supervised domain adaptation for deep learning-based animal (object) detection in drone imagery (as considered in Chapter 6 of this book) and is taken from Kellenberger et al. (2019). Figure 7.8 shows images from the source and target datasets. Although both images were acquired with similar RGB

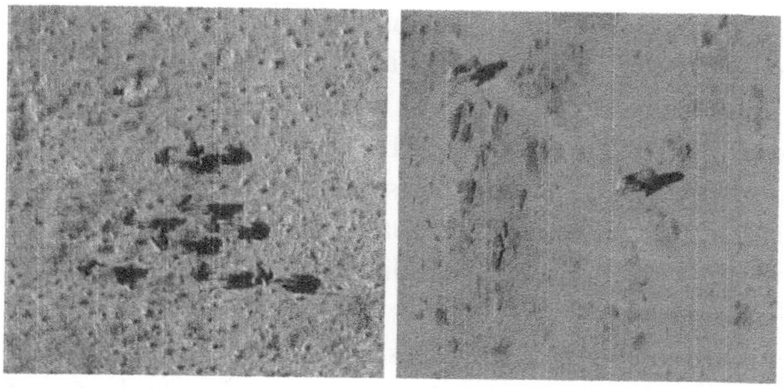

Figure 7.8 Example drone images from the source (left) and target (right) domains.

cameras, they exhibit domain shifts due to multiple causes, such as terrain properties, animal species and illumination conditions. The consequence of these shifts are that a CNN, trained on source, will generate a high number of false positives in target. Even worse, the number of objects of interest is minuscule in comparison to the vast amounts of background pixels, which makes this a challenging needle-in-the-haystack problem. This setting requires domain adaptation strategies that are robust to class imbalances.

Method One way to achieve robustness to imbalance is to focus on predictions that are most likely to be true positives, with respect to their location in the feature space. Figure 7.9 shows t-SNE embeddings (van der Maaten and Hinton 2008) of source (left) and target (right) domain predictions of a CNN detector. Since the model has been trained on source, it predicts a higher number of true positives (blue in the color version, black when printed in black and white) in that domain, compared to the significantly more false alarms (red in the color version, gray when printed in black white) in target. Of particular interest is the rightmost feature space area of the source domain, where the concentration of true

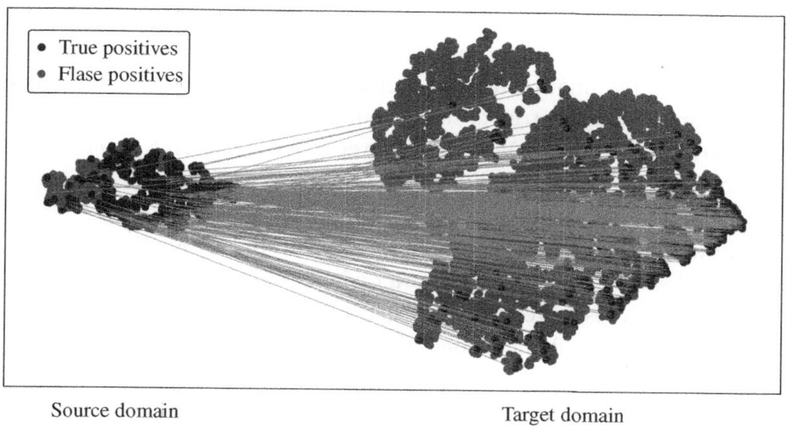

Figure 7.9 Feature space projections using t-SNE (van der Maaten and Hinton 2008) for predictions of the unadapted model in the source (left) and target (right) domains. True positives are shown in blue (black when printed in black and white), false positives in red (gray when printed in black and white). The gray lines show Optimal Transport correspondences between the target true positives and associated source samples.

positives is highest. The domain adaptation criterion of Kellenberger et al. (2019), named "Transfer Sampling" (TS), exploits the fact that this concentration in source exists, and tries to find the same hotspot of true positives in the (unlabeled) target domain. To this end, the work employs Optimal Transport (OT; Courty et al. (2017)), which is a distribution matching framework that finds correspondences between all samples of two distributions with respect to a minimal global cost. In essence, this means that source and target samples correspond well to each other if they are close according to a distance, such as an ℓ_2 norm, *and* lie in similar regions of the distribution.

These source-to-target associations are shown with gray lines in Figure 7.9. In the domain adaptation setting, OT essentially helps re-finding the hotspot of true positives in the target domain – note how most of the lines point from the source true positives hotspot to the target true positives. At runtime, the labels of the target domain are initially unknown, but the source ground truth is assumed to be present. Hence, by localizing the source true positives hotspot and establishing an OT source-to-target correspondence, it is theoretically possible to transfer the source labels to the target domain. For additional robustness to class imbalance, Kellenberger et al. (2019) instead transfers the "goodness" of the source samples, e.g. by their distance from the false positives hotspot (leftmost area in Figure 7.9), to the target domain and uses the scores as an AL criterion.

Experiments We evaluated the AL criterion on two sets of images collected in different years using an Unmanned Aerial Vehicle (UAV) over the Kuzikus wildlife reserve[2] in the African savanna. The datasets vary not only in acquisition year, but also in the area covered, background/soil composition, flying altitude, camera type, dataset size, and animal count (details in Kellenberger et al. (2019)). As a result, they are particularly challenging for animal detectors, which is also expressed in the high false positives count of the unadapted model on target.

We trained a CNN-based object detector of the 2014 dataset until convergence and then employed it in an AL loop with simulated user input: we sampled 50 image patches around those predictions with the highest transferred goodness score, updated the CNN for 12 epochs, and used the latest model state to re-predict the target samples. We repeated this cycle ten times and recorded the final model performance after adaptation to the target dataset, as well as the number of target true positives found. We compare TS with a series of conventional AL criteria, including Breaking Ties (Luo et al. 2005), the CNN's confidence value ("max confidence"), and random sampling.

Results Figure 7.10 shows precision-recall curves of the CNN in the source (left) and target (right) domains, and also model performances after adaptation. The unadapted model (right panel, black curve) clearly loses precision in target, but still reaches roughly the same recall of more than 80%. When adapted, the model significantly gains in precision, and does so the most when using TS.

In addition, Figure 7.11 shows the number of animals found in the target dataset during the ten AL iterations. Both TS and max confidence find about 80% of the total number of animals present (dash-dotted line), but TS manages to do so one entire AL iteration (50 label

2 https://kuzikus-namibia.de/xe_index.html

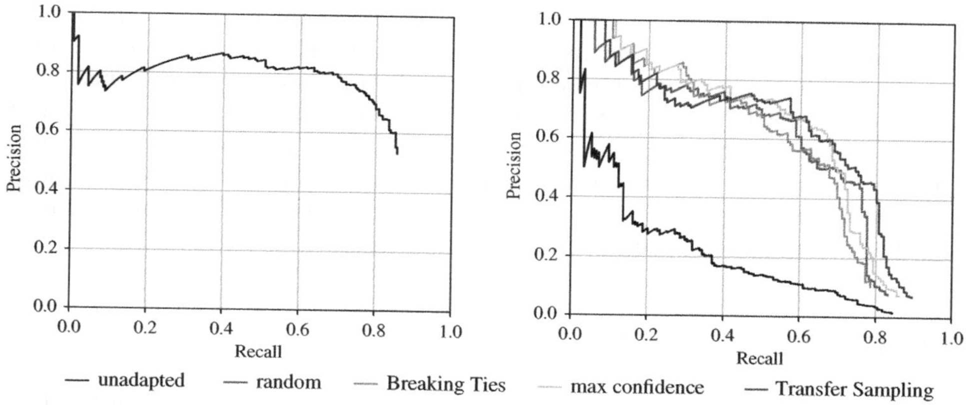

Figure 7.10 Precision-recall curves for the CNN on source (left) and target (right) datasets, without domain adaptation (black), and with (colored). Colored graphs show model performances on target after the final AL iteration. Best viewed in color.

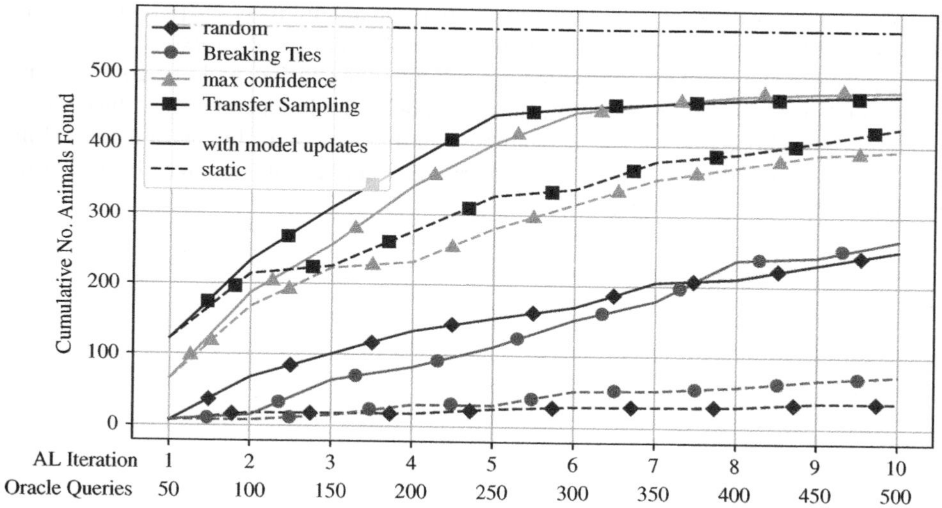

Figure 7.11 Cumulative number of animals found over the course of the ten AL iterations, with CNN updates at the end of every AL iteration (solid), and by simply sampling from the initial, unadapted CNN (dashed). The black dashed line denotes the total number of animals present in the dataset (upper bound).

queries) earlier. Breaking Ties and random sampling are not designed to focus on the most likely true positives, and hence retrieve less animals in the same span. Finally, the graph further shows that fine-tuning the CNN after every iteration (solid) yields significantly higher animal counts than just employing the criteria, without adapting the model (dashed).

Discussion One may find that unsupervised domain adaptation is not always suitable, depending on the data and application. In the example shown, the scarcity of the animals raises a significant class imbalance problem to the adaptation process, which requires a strategy that is virtually immune to such imbalances. Furthermore, the specific use case of animal detection not only requires satisfactory model performance in the target domain, but also an economic animal retrieval rate during adaptation. To this end, it may be required to obtain ground truth for the most relevant target domain samples, which can be achieved using AL. The presented TS criterion was designed explicitly for such purposes and, contrary to conventional AL criteria, focuses on the most probable target samples, which makes it robust to class imbalances and provides a high object retrieval rate already during the adaptation process.

7.4 Concluding Remarks

In this chapter, we presented recent advances in domain adaptation for geospatial data. Starting from a categorization of methods based on the stage they affect (inputs, inner representation or usage of labeled data), we presented a series of comparisons on remote sensing data and described pros and cons of the different approaches. We also provided reproducible code (on GitHub) to allow experimenting and better understanding of the properties of the model and further adoption of domain adaptation methods in geosciences.

The categories of approaches presented are clearly not exclusive, and one could potentially design hybrid methods using several aspects at once. But independently of the method of preference, we hope that we raised awareness on the need of dealing with dataset shift and the pitfalls that one could fall in if such distribution changes are not taken into account during model design and training.

8

Recurrent Neural Networks and the Temporal Component

Marco Körner and Marc Rußwurm

In the previous chapters, input data was assumed to be individual multi-dimensional measurements $\boldsymbol{x} \in \mathcal{R}^D, 0 < D \in \mathbb{N}$. If these signals come organized in a matrix-like structure, i.e. $\boldsymbol{x} \in \mathcal{R}^{M \times N}, 0 < M, N \in \mathbb{N}$, convolutional neural networks are, by design, able to process each individual measurement, i.e. each pixel, considering its spatial context and, thus, can be applied to images of various scales $M \times N$.

Earth observation data, in general, is mostly provided in the form of sequential data, i.e.

$$\mathbb{X} = \left\{\boldsymbol{x}_t = h(f(t)) \in \mathcal{R}^D\right\}_t \tag{8.1}$$

$$\text{or} \quad \mathbb{X} = \left\{\boldsymbol{x}_t = h(f(t)) \in \mathcal{R}^{M \times N}\right\}_t, \qquad 0 \leq t \leq T \in \mathbb{N} \tag{8.2}$$

representing a continuous-time dynamical system f sampled at discrete time-stamps t and projected into any observation space by an observation model h. As illustrated in Figure 8.1, regular DNNs can only process observations instance-wise, i.e.

$$y_t = f_{\text{DNN}}\left(\boldsymbol{x}_t\right), \tag{8.3}$$

or fixed-length concatenations of multiple observations, i.e.

$$y_t = \tilde{f}_{\text{DNN}}\left(\boldsymbol{x}_t, \boldsymbol{x}_{t-1}, \dots, \boldsymbol{x}_{t-p}\right). \tag{8.4}$$

On the contrary, it appears natural that the temporal context should be taken into account dynamically when processing this kind of data, i.e.

$$y_\tau = \hat{f}_{\text{DNN}}\left(\left\{\boldsymbol{x}_t\right\}_{t=0}^{\tau}\right). \tag{8.5}$$

For a long time, graphical models, e.g. hidden Markov models (HMMs) (Rabiner and Juang 1986), were considered the method of choice when dealing with time series of Earth observation data (Siachalou et al. 2015; Hoberg et al. 2015). These generative models update an unobservable *hidden state* of a dynamical system following the Markov assumption, i.e. solely by means of the state at the previous time step and the current observation, yielding the generic update rule

$$\boldsymbol{s}_t \leftarrow \pi\left(\boldsymbol{s}_{t-1}, \boldsymbol{x}_t\right). \tag{8.6}$$

Deep Learning for the Earth Sciences: A Comprehensive Approach to Remote Sensing, Climate Science, and Geosciences, First Edition. Edited by Gustau Camps-Valls, Devis Tuia, Xiao Xiang Zhu, and Markus Reichstein.

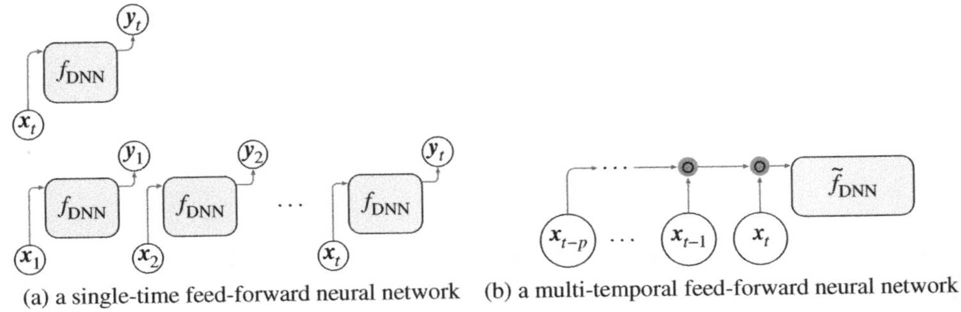

Figure 8.1 Applying deep feed-forward neural networks to multi-temporal data.

This abstract representation of the underlying processes producing the observations is further used to derive the final outputs, e.g. classification labels,

$$y_t = g\left(s_t\right). \tag{8.7}$$

Their straightforward formulation and tractable training regimes, like the Baum–Welch algorithm (Baum and Petrie 1966), made them handy to use and showed quite some success.

Despite that, HMMs show some major drawbacks, as their individual designs encode a lot of domain and expert knowledge. In particular, suitable state-space formulations and the correct temporal dependencies have to be modelled manually. Especially the extent of the temporal context used within the model needs to be hard-wired a-priori and remains unchanged throughout the entire process. This evidently contradicts the fundamental principle of representation learning that underlies the deep learning concept.

8.1 Recurrent Neural Networks

While the feed-forward deep learning models described so far required the involved computational graphs to be free of circles, the relaxation of this constraint gives rise to the entire family of feed-back or recurrent neural networks (RNNs). These recurrent connections allow the network to develop a set of internal dynamics over time that enable RNNs to learn to exploit the temporal context of individual data samples. This enables them to process data sequences of variable lengths.

In the purely continuous case, the dynamics of an RNN with n recurrent neurons at time step t can be described recursively by

$$\tau\dot{\boldsymbol{s}}_t = \tau\frac{\partial}{\partial t}\boldsymbol{s}_t \leftarrow f(\boldsymbol{s}_t, \boldsymbol{x}_{t+1}, \theta) \tag{8.8}$$

$$= -\boldsymbol{s}_t + \boldsymbol{W}_{\text{in}}\boldsymbol{x}_{t+1} + \boldsymbol{W}_{\text{rec}}\varsigma(\boldsymbol{s}_t) + \boldsymbol{b}_{\text{rec}} \tag{8.9}$$

with *weights* $\boldsymbol{W}_{\text{in}} \in \mathcal{R}^{n \times m}$ and $\boldsymbol{W}_{\text{rec}} \in \mathcal{R}^{n \times n}$, *biases* $\mathbf{b}_{\text{rec}} \in \mathcal{R}^n$, and where

$$\boldsymbol{h}_t = \varsigma\left(\boldsymbol{s}_t\right) \tag{8.10}$$

encodes the internal state after non-linear activation. A first-order Euler discretization of Equation 8.9 gives rise to the generic RNN update equation

$$\boldsymbol{s}_t = (1-\alpha)\boldsymbol{s}_{t-1} + \alpha\left(\boldsymbol{W}_{\text{in}}\boldsymbol{x}_t + \boldsymbol{W}_{\text{rec}}\varsigma(\boldsymbol{s}_{t-1}) + \boldsymbol{b}_{\text{rec}}\right). \tag{8.11}$$

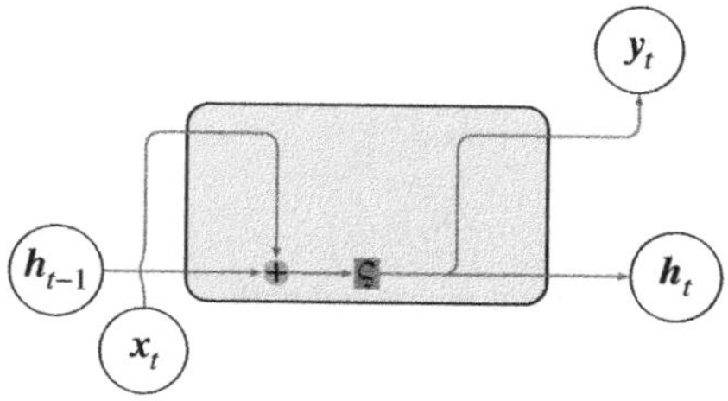

Figure 8.2 Schematic illustration of a single RNN cell that updates the cell state vector $\boldsymbol{h}_{t-1}$ when new data $\boldsymbol{x}_t$ becomes available.

Here, the temporal weighting parameter $\alpha = \frac{\Delta t}{\tau}$ tunes the influence of the system dynamics. One special border case, when $\alpha = 1$ and, thus, the new system state $\boldsymbol{s}_t$ is entirely inferred recurrently from only the previous state $\boldsymbol{s}_{t-1}$, gives rise to the standard formulation of RNNs, i.e.

$$\boldsymbol{s}_t = \boldsymbol{W}_{\text{in}}\boldsymbol{x}_t + \boldsymbol{W}_{\text{rec}}\varsigma(\boldsymbol{s}_{t-1}) + \boldsymbol{b}_{\text{rec}}, \tag{8.12}$$

or, its activated form, i.e.

$$\boldsymbol{h}_t = \varsigma\left(\boldsymbol{W}_{\text{in}}\boldsymbol{x}_t + \boldsymbol{W}_{\text{rec}}\boldsymbol{h}_{t-1} + \boldsymbol{b}_{\text{rec}}\right). \tag{8.13}$$

Figure 8.2 shows an illustration of a single RNN cell.

8.1.1 Training RNNs

Recurrent neural network architectures contain feedback loops. Hence, gradient-based parameter optimization based on back-propagation through the particular layers is neither applicable nor sufficient right away, as their node activations and outputs depend on data from different time steps. Unrolling such an RNN architecture through time, as exemplarily shown in Figure 8.3, visualizes this problem but also gives rise to a possible workaround. As shown, an RNN can be reformulated into a feed-forward NN with each layer representing another time step. Applying back-propagation on this dual architecture gives rise to the so-called *back-propagation through time (BPTT)* (Rumelhart et al. 1985; Werbos 1990b) algorithm. In order to exemplify that, let

$$\boldsymbol{W}_{\text{rec}} = \begin{pmatrix} \blacktriangleleft & 0 & 0 & \blacktriangleleft \\ \blacktriangleright & 0 & \blacktriangleleft & \blacktriangleleft \\ 0 & 0 & 0 & \blacktriangleleft \\ 0 & \blacktriangleright & 0 & \blacktriangleleft \end{pmatrix} \tag{8.14}$$

encode the connections of an RNN, as shown in Figure 8.3, where bold edges highlighted in grey and respective matrix elements $\blacktriangleleft$ correspond to recurrent feed-back loops.

8.1.1.1 Exploding and Vanishing Gradients

Figure 8.4 shows the computational graph of an unrolled recurrent network for the first three time steps. There, the temporal losses $\mathfrak{L}_t$ are combined into a total loss

$$\mathfrak{L} = \sum_{t=0}^{T} \mathfrak{L}_t. \tag{8.15}$$

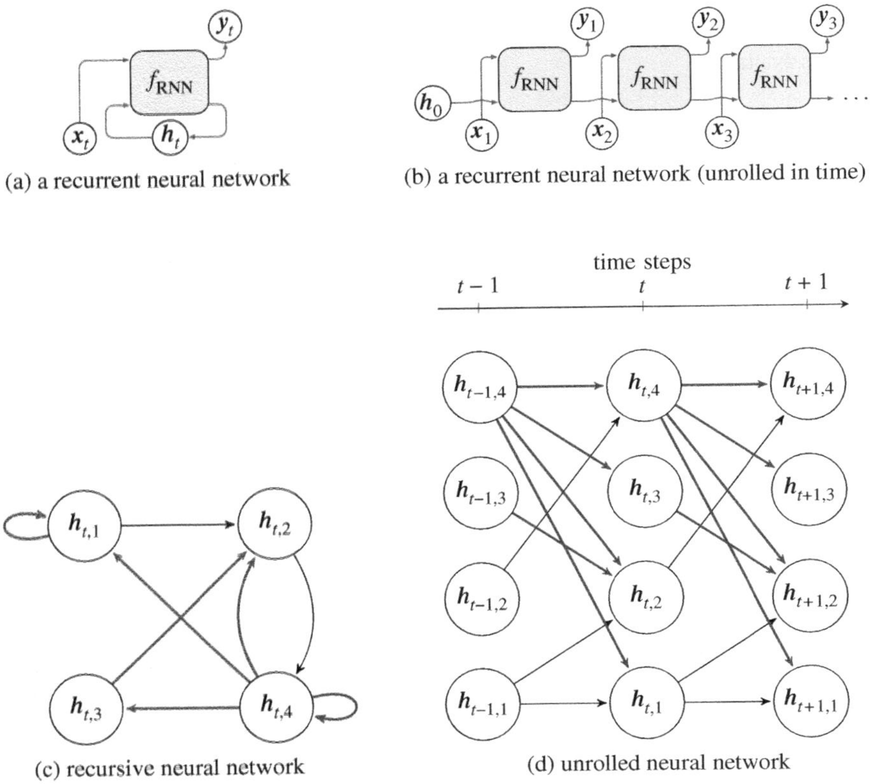

Figure 8.3 When a recursive, feed-back neural network is unrolled through time, it can be represented as a feed-forward neural network with layers corresponding with the individual time steps.

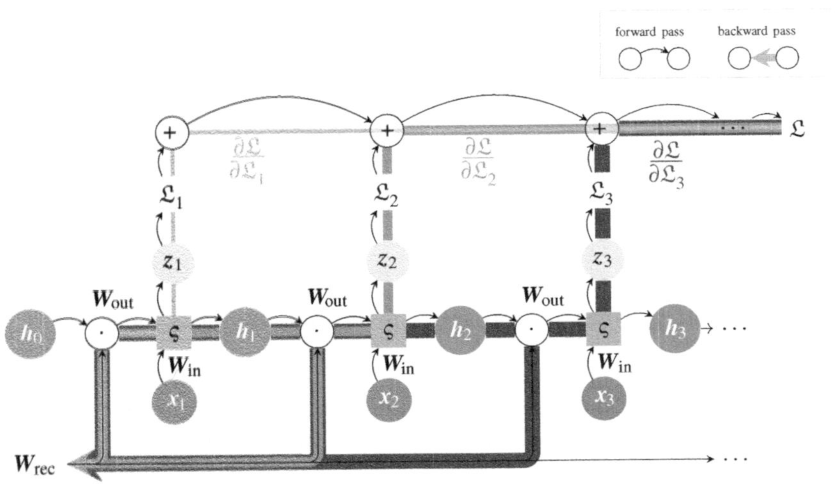

Figure 8.4 The computational graph of an unrolled RNN with forward (black arrows) and backward passes (grey arrows).

For updating, e.g. the shared recurrent weights $\boldsymbol{W}_{\text{rec}}$, the partial derivatives

$$\frac{\partial \mathfrak{L}}{\partial \boldsymbol{W}_{\text{rec}}} = \sum_{t=0}^{T} \frac{\partial \mathfrak{L}_t}{\partial \boldsymbol{W}_{\text{rec}}} \tag{8.16}$$

of the loss need to be computed. Similarly, for updating the internal states $\boldsymbol{h}$, gradients need to be propagated from $\boldsymbol{h}_{t+t}$ back to $\boldsymbol{h}_t$. For simplicity, let the nonlinear activation

$$\varsigma(x) = \varsigma_{\text{ReLU}}(x) = \max(0, x) \tag{8.17}$$

in Equation 8.13 be the rectified linear unit (ReLU), Equation 8.16 reformulates to

$$\frac{\partial \mathfrak{L}}{\partial \boldsymbol{W}_{\text{rec}}} = \sum_{t=0}^{T} \frac{\partial \mathfrak{L}_t}{\partial \boldsymbol{W}_{\text{rec}}} = \sum_{t=0}^{T} \frac{\partial \mathfrak{L}_t}{\partial \boldsymbol{h}_t} \cdot \prod_{\tau=0}^{t} \frac{\partial \boldsymbol{h}_{t-\tau}}{\partial \boldsymbol{h}_{t-\tau-1}} \cdot \frac{\partial \boldsymbol{h}_1}{\partial \boldsymbol{W}_{\text{rec}}} \tag{8.18}$$

which contains partial gradients

$$\frac{\partial \mathfrak{L}_t}{\partial \boldsymbol{h}_{t-k-1}} = \boldsymbol{W}_{\text{rec}}{}^{\mathcal{T}} \cdot \left(\boldsymbol{1}_{\geq 0} \left(\boldsymbol{h}_{t-k-1} \right) \odot \frac{\partial \mathfrak{L}_t}{\partial \boldsymbol{h}_{t-k}} \right) \tag{8.19}$$

iteratively defined by their upstream gradients $\frac{\partial \mathfrak{L}_t}{\partial \boldsymbol{h}_{t-k}}$. Here, it can be clearly seen that backpropagating gradients through $\tau \in \mathbb{N}$ time steps requires τ multiplications of the recurrent weight matrix $\boldsymbol{W}_{\text{rec}}{}^{\mathcal{T}}$, i.e. $\left(\boldsymbol{W}_{\text{rec}}{}^{\mathcal{T}}\right)^{\tau}$. Depending on their actual values – or, more precisely, their eigenvalues after eigenvalue decomposition [1] –, this yields an exponential growth or shrinkage of the recurrent weights, such that small perturbations in early iterations might manifest massive effects in later iterations. This effect is commonly referred to as *exploding* and *vanishing gradients* (Hochreiter 1991; Bengio et al. 1994), respectively, and poses a major problem when training very deep neural networks, and so RNNs, using gradient-based optimization, as the objective becomes effectively discontinuous. Thus, RNNs forfeit their ability to capture long-term temporal dependencies.

8.1.1.2 Circumventing Exploding and Vanishing Gradients

As the problem of exploding and vanishing gradients is fundamental for training deep neural network architectures, there are several countermeasures to avoid this phenomenon.

Real-time Recurrent Learning The probably most obvious countermeasure against exploding and vanishing gradients is to avoid backward passes through time entirely. The *real-time recurrent learning (RTRL)* algorithm (Williams and Zipser 1989), for instance, determines an optimal parameter update with only one complete forward pass and without memorizing elapsed hidden states. Thus, RTRL resembles a purely online learning procedure, as opposed to the batch-wise offline learning strategy in BPTT. Nevertheless, its extraordinarily high runtime and memory complexity make this procedure impractical to be used.

1

$$\left(\boldsymbol{W}_{\text{rec}}{}^{\mathcal{T}}\right)^{\tau} = \left(\boldsymbol{U\Lambda U}^{-1}\right)^{\tau} = \underbrace{\boldsymbol{U\Lambda} \overbrace{\boldsymbol{U}^{-1} \cdot \boldsymbol{U}}^{=1} \boldsymbol{\Lambda U}^{-1} \cdot \ldots}_{\tau \text{ times}} = \boldsymbol{U\Lambda}^{\tau}\boldsymbol{U}^{-1}$$

Truncated back-propagation through time In a similar way deep neural network architectures are suggested to be kept as shallow as possible, RNNs could be constrained to propagate back gradients not to the very beginning of the time series, but only for a fixed number of steps. Hence, Equation 8.18 becomes

$$\frac{\partial \mathfrak{L}}{\partial \boldsymbol{W}_{\text{rec}}} \approx \sum_{t=0}^{T} \frac{\partial \mathfrak{L}_t}{\partial \boldsymbol{h}_t} \cdot \prod_{\tau=0}^{k} \frac{\partial \boldsymbol{h}_{t-\tau}}{\partial \boldsymbol{h}_{t-\tau-1}} \cdot \frac{\partial \boldsymbol{h}_1}{\partial \boldsymbol{W}_{\text{rec}}}, \quad t > k \in \mathbb{N}. \tag{8.20}$$

Performing this temporal back-propagation only every κ forward passes, i.e.

$$\frac{\partial \mathfrak{L}}{\partial \boldsymbol{W}_{\text{rec}}} \approx \sum_{t \equiv_\kappa T} \frac{\partial \mathfrak{L}_t}{\partial \boldsymbol{h}_t} \cdot \prod_{\tau=0}^{k} \frac{\partial \boldsymbol{h}_{t-\tau}}{\partial \boldsymbol{h}_{t-\tau-1}} \cdot \frac{\partial \boldsymbol{h}_1}{\partial \boldsymbol{W}_{\text{rec}}}, \quad T > \kappa \in \mathbb{N}, \tag{8.21}$$

yields the *truncated back-propagation through time (tBPTT)* algorithm (Elman 1990; Williams and Peng 1990; Williams and Zipser 1995).

Temporal Skip Connections As the explosion and vanishing rates of a RNN are proportional to a function of the number of time steps τ to be back-propagated, temporal skip connections introduced to the recursive architecture might help to diminish this effect. If, for instance, a static time delay δ_t is added (Lin et al. 1996), the vanishing rates will now grow proportional to a function of $\frac{\tau}{\delta_t}$ instead, while gradient explosion is still possible.

Truncated Gradients When optimizing parameterized systems using gradient descent, large gradients $\mathbf{g} = \frac{\partial \mathfrak{L}}{\partial \boldsymbol{\omega}}$ with respect to certain parameters $\boldsymbol{\omega}$ yield large updates for these parameters. In awareness of the ability of gradients to explode, one commonly used strategy is to limit the gradient values, or the length of the gradient vectors $\mathbf{g}$, not to exceed a certain threshold $0 < c \in \mathcal{R}$, e.g. by rescaling

$$\mathbf{g} := \begin{cases} c \cdot \frac{\mathbf{g}}{\|\mathbf{g}\|} & \text{if } \|\mathbf{g}\| \geq c \\ \mathbf{g} & \text{else} \end{cases}. \tag{8.22}$$

Regularization Another commonly used way to restrict the parameter space during optimization is regularization, i.e.

$$\boldsymbol{\theta}^* = \text{argmin}_{\boldsymbol{\theta}} \mathfrak{L}(f, \boldsymbol{\theta}) + \lambda \cdot \Omega(\boldsymbol{\theta}), \tag{8.23}$$

where $\Omega(\boldsymbol{\theta})$ is a defined measure of complexity of the parameter set $\boldsymbol{\theta}$. In the context of training RNNs, Pascanu et al. (2013) proposed the regularizer

$$\Omega(\boldsymbol{\theta}) = \sum_{t=0}^{T-1} \left(\frac{\left\| \frac{\partial \mathfrak{L}}{\partial \boldsymbol{h}_{t+1}} \cdot \frac{\partial \boldsymbol{h}_{t+1}}{\partial \boldsymbol{h}_t} \right\|}{\left\| \frac{\partial \mathfrak{L}}{\partial \boldsymbol{h}_{t+1}} \right\|} - 1 \right)^2 \tag{8.24}$$

to enforce norm-preserving error updates.

Weight Initializations The problem when magnitudes of parameter updates grow without control is that a (local) optimum $\boldsymbol{\theta}^*$ could easily be omitted. As this is only problematic in non-convex optimization settings, a careful selection of initial parameterization $\boldsymbol{\theta}^{(0)}$ is often used to ensure stable convergence. For RNNs entirely based on ReLU activations, Le et al. (2015) propose to initialize the recurrent weights and biases with $\boldsymbol{W}_{\text{rec}} = \boldsymbol{I}$ and $\boldsymbol{b}_{\text{rec}} = \boldsymbol{0}$, respectively. In contrast, Talathi and Vartak (2015) propose to initialize $\boldsymbol{W}_{\text{rec}}$ in a way that its eigenvalues are normalized with respect to its largest one.

8.2 Gated Variants of RNNs

The aforementioned strategies to mitigate the problems of exploding and vanishing gradients all come with their particular advantages and downsides and are, thus, only rarely used in practice. Opposed to that, the idea of augmenting RNNs by internal gates that actively control the extent of temporal information to be memorized – or to be forgotten – to neatly address these problems brought the community quite a success.

8.2.1 Long Short-term Memory Networks

Instantiating the probably most popular gated variant of RNNs, the concept of *long short-term memory (LSTM)* cells – as first proposed by Hochreiter and Schmidhuber (1997) – introduces several of so-called *memory units* to steer the information flowing through the network containing cells of this type, as illustrated in Figure 8.5. These gates are, in particular, defined as

$$\text{the } \textit{forget gate} \quad \boldsymbol{f}_t = \sigma\left(\boldsymbol{W}_{\text{f}}\begin{bmatrix}\boldsymbol{h}_{t-1}\\ \boldsymbol{x}_t\end{bmatrix} + \boldsymbol{b}_{\text{f}}\right) \quad \in [0,1]^{N_{\text{f}}}, \tag{8.25}$$

$$\text{the } \textit{input gate} \quad \boldsymbol{i}_t = \sigma\left(\boldsymbol{W}_{\text{i}}\begin{bmatrix}\boldsymbol{h}_{t-1}\\ \boldsymbol{x}_t\end{bmatrix} + \boldsymbol{b}_{\text{i}}\right) \quad \in [0,1]^{N_{\text{i}}}, \tag{8.26}$$

$$\text{the } \textit{modulation gate} \quad \boldsymbol{v}_t = \tanh\left(\boldsymbol{W}_{\text{v}}\begin{bmatrix}\boldsymbol{h}_{t-1}\\ \boldsymbol{x}_t\end{bmatrix} + \boldsymbol{b}_{\text{v}}\right) \quad \in [-1,1]^{N_{\text{v}}}, \tag{8.27}$$

$$\text{and the } \textit{output gate} \quad \boldsymbol{o}_t = \sigma\left(\boldsymbol{W}_{\text{o}}\begin{bmatrix}\boldsymbol{h}_{t-1}\\ \boldsymbol{x}_t\end{bmatrix} + \boldsymbol{b}_{\text{o}}\right) \quad \in [0,1]^{N_{\text{o}}}, \tag{8.28}$$

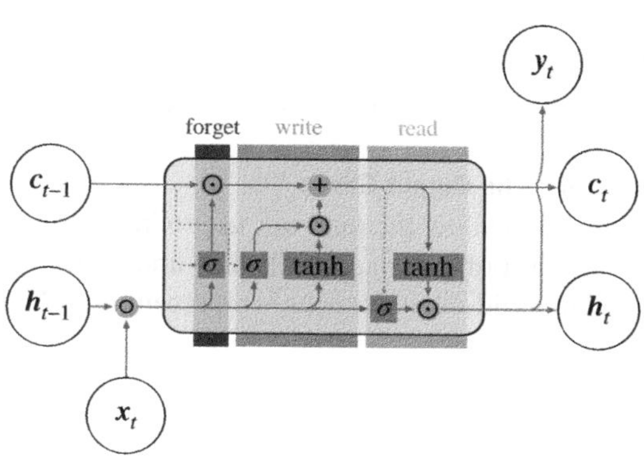

Figure 8.5 In a *long short-term memory (LSTM)* (Hochreiter and Schmidhuber, 1997) cell, several gates control the information flow during training and, thus, mitigate problems arising from vanishing for exploding gradients. When augmented with *peephole connections* (Gers and Schmidhuber, 2000) – shown as dotted paths –, training robustness can be improved remarkably.

which are used to update or create two state representations, i.e.

$$\text{the } \textit{cell state} \quad \boldsymbol{c}_t = \boldsymbol{f}_t \odot \boldsymbol{c}_{t-1} + \boldsymbol{i}_t \odot \boldsymbol{v}_t \qquad \in [-1, 1]^{N_c} \tag{8.29}$$

$$\text{and the } \textit{hidden state} \quad \boldsymbol{h}_t = \boldsymbol{o}_t \odot \tanh(\boldsymbol{c}_t) \qquad \in [-1, 1]^{N_h}. \tag{8.30}$$

Here, the

$$\textit{logistic sigmoid} \text{ function} \quad \sigma(x) = \frac{1}{1 + \exp(-x)} \tag{8.31}$$

$$\text{and } \textit{hyperbolic tangent} \text{ function} \quad \tanh(x) = \frac{\exp(x) - \exp(-x)}{\exp(x) + \exp(-x)} \tag{8.32}$$

are used for non-linear activation.

Intuitively, if the gates are closed, i.e. set to one, the involved gradients back-propagated through time can pass through these cells without being an influence for arbitrarily many time steps and thus are prevented from vanishing. In the following, the particular elements of LSTM cells will be briefly reviewed.

8.2.1.1 The Cell State c_t and the Hidden State h_t

The core concept of each recurrent network architecture is its ability to internalize a representation of the state of the system which is observed. This memory is iteratively updated as new evidence comes in, i.e. information can be written, read, and forgotten with every time step. In LSTMs, this memory is represented by the cell state $\boldsymbol{c}_t$ from which the final hidden state $\boldsymbol{h}_t$ is read out.

8.2.1.2 The Forget Gate f_t

The forget gate $\boldsymbol{f}_t$, as an update to the vanilla LSTM cell introduced later by Gers et al. (2000), is multiplied element-wise with the previous cell state in order to control the amount of historical information to be kept for future time steps before adding new evidence. The sigmoid activation σ restricts its values to a range between zero and one. While $\boldsymbol{f}_t \to \boldsymbol{1}$ results in full recall of previous state, $\boldsymbol{f}_t \to \boldsymbol{0}$ lets the LSTM cell forget any prior memorized state information.

8.2.1.3 The Modulation Gate v_t and the Input Gate i_t

LSTM cells assimilate all information extracted from the current signal into the modulation gate $\boldsymbol{v}_t$ which is activated by a hyperbolic tangent nonlinearity and thus contains values from the range -1 to 1. Then, the $[0, 1]$-valued input gate $\boldsymbol{i}_t$ controls how dominantly this new information should affect the updated cell memory.

8.2.1.4 The Output Gate o_t

In order to allow the network to produce predictions with each new piece of data coming in, the output gate $\boldsymbol{o}_t$ controls how and which information from the output gate should be used to build the hidden state representation $\boldsymbol{h}_t$ at that particular time step.

8.2.1.5 Training LSTM Networks

As already mentioned, LSTM cells have the ability to steer gradients through the entire networks without alteration and thus to avoid vanishing gradient problems for at least some cases. Still, whenever temporal information needs to be taken into account and, thus, the forget gates are not shut – i.e. $\boldsymbol{f}_t \rightarrow \boldsymbol{0}$ –, temporal gradients

$$\frac{\partial \mathfrak{L}}{\partial \boldsymbol{c}_{t-1}} = \frac{\partial \mathfrak{L}}{\partial \boldsymbol{c}_t} \odot \boldsymbol{f}_t \tag{8.33}$$

will still be dampened when propagated back through time. In order to avoid this phenomenon during early stages of LSTM training, it comes in practical to initialize especially the forget gate bias $\boldsymbol{b}_{\mathrm{f}}$ with sufficiently large values, such that the sigmoid-activated forget starts opened and can be shut only if this becomes necessary during optimization (Jozefowicz et al. 2015).

As modern deep learning frameworks offer very handy and highly optimized implementations for computing derivatives of such complex networks, training LSTM networks facilitated a lot during the last years and, thus, behaves better than training vanilla RNN architectures. As there is, especially in machine learning, no free lunch, this comes at the cost of four times the parameters to be optimized during training potentially resulting in a higher number of training iterations or larger datasets necessary to train such networks.

8.2.2 Other Gated Variants

Since their advent and in light of their proven computational power, manifold research has been carried out over the years to reduce the training complexity of LSTMs or to extend their capabilities.

In its original formulation, only the LSTM output gate has access to the internal cell state to derive the updated hidden state vector, while the other internal gates do not have access to this information. Gers and Schmidhuber (2000) proposed to relax this restriction by allowing also the input, forget, and output gates to take a glimpse of the current value of the cell state vector via so-called *peephole connections*, as illustrated by dotted paths in Figure 8.5. It has been shown that this architectural change results in more stable training and makes explicit teacher forcing dispensable (Gers et al. 2003).

These improved capabilities come at the cost of additional parameters to be optimized during training, which substantially increases the demand for training data. In contrast to that, researchers focused on developing more lightweight LSTM cell designs by removing or combining several LSTM gates. While there is no clear evidence on which of the LSTM gates are essential and which are dispensable – cf. Greff et al. (2017) *vs.* Jozefowicz et al. (2015) –, studies indicate that cells with gated information flows consistently show better performance than vanilla RNNs (Chung et al. 2014).

The *gated recurrent unit (GRU)* (Cho et al. 2014), for instance, couples the input and the forget gates into a combined *update gate*, in addition to a new *reset gate*. Additionally, a *minimal gated unit (MGU)* (Zhou et al. 2016) further reduces the number of gates to only one single *forget gate*. While these reformulations reduce the number of learnable parameters, they lose their ability to detect context-free languages and, hence, to count or to model the frequency of perceived events (Weiss et al. 2018).

8.3 Representative Capabilities of Recurrent Networks

A single recurrent cell (cf. Figure 8.6(a)) comes with a certain capacity and can, thus, store a limited amount of information. This capacity is architecture-specific and grows linearly with the number of its parameters (Collins et al. 2017). The particular gates of LSTM cells and their variants, which are introduced to direct the flow of information to improve the practical training properties, result in a reduced information storage capacity. For this reason, recurrent cells are commonly organized in various topologies to increase their representative capabilities.

8.3.1 Recurrent Neural Network Topologies

In the most straightforward case, recurrent cells can be stacked sequentially, as visualized in Figure 8.6. Depending on their particular organization, these stacked architectures can handle differently stated problems by generating various outputs. *Many-to-one* architectures, as shown in Figure 8.6(b), can be used to predict a single – scalar- or vector-valued – target $\boldsymbol{y}_t$ given a sequence of data samples $\mathbb{X}_{\rightarrow t} = \{\boldsymbol{x}_\tau\}_{0<\tau\leq t}$. The dual case, when a sequence of labels $\mathbb{X}_{t\rightarrow t+\delta t}$ has to be predicted from a single data sample $\boldsymbol{x}_t$, can be realized with *one-to-many* topologies (Figure 8.6(c)). *Many-to-many* architectures (Figure 8.6(e)), consequently, predict sequential targets from sequential input data.

As RNNs, in their canonical formulation, expect data to be organized in sequential order, these models can be used for *online* problem settings where targets should be

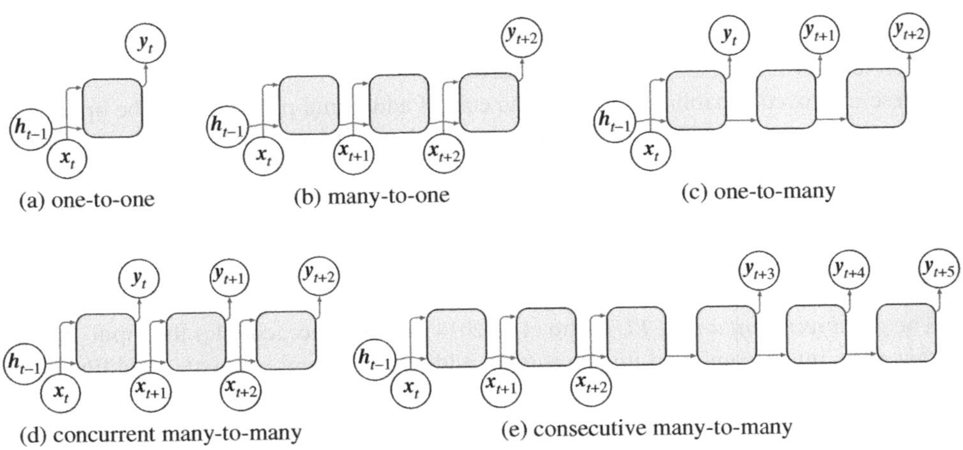

Figure 8.6 As the capacity of a single RNN cell is limited, several RNN cells can be organized in different sequential topologies, resulting in higher-capacitive networks.

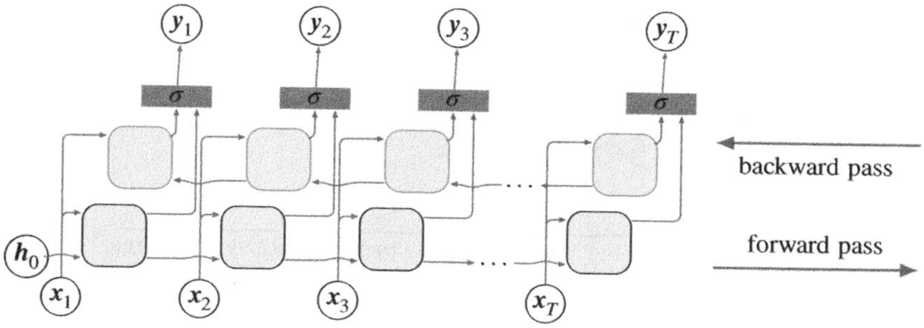

Figure 8.7 Bi-directional RNNs (Schuster and Paliwal, 1997) and LSTM networks (Graves and Schmidhuber, 2005) contain an additional backward path to process the input data in inverse sequential order. This allows them to access past and future data, which makes them suitable to be employed in *offline* problem settings.

predicted whenever new data $\boldsymbol{x}_{t+1}$ becomes available. For *offline* problems, when the entire data sequence $\mathbb{X}_{1\to T}$ is already available, this becomes an undesired yet unnecessary restriction. To mitigate this limitation, *bi-directional* models (Schuster and Paliwal 1997; Graves and Schmidhuber 2005), for instance, introduce an additional backward pass processing the input data sequence in inverted sequential order, i.e. $\mathbb{X}_{T\to 1}$., as illustrated in Figure 8.7.

Considering that Earth observation data, especially obtained by space-borne sensors, is mostly organized in matrix form, the vector-valued formulation of RNNs turns out to be another undesired restriction. *Multi-dimensional* network topologies (Fernández et al. 2007; Kalchbrenner et al. 2016) augment this purely sequential structure by further spatial dimensions and are, thus, able to process matrix-valued input signals. Irregular local neighborhood relations can be modeled with more general *graph RNN* (Goller and Küchler 1996) or *graph LSTM* (Liang et al. 2016) network architectures. In order to avoid full connectivity of recurrent LSTM cells, *ConvLSTMs* (Sainath et al. 2015) allow convolutional operations and, thus, local weight sharing and translation equivariance.

8.3.2 Experiments

As motivated earlier, global geophysical Earth system processes can be seen as dynamical systems that are governed by an as yet undiscovered multitude of sub-processes. These can be of chaotic or deterministic in nature and spread over various-length time spans. For that reason, computational models used to capture such processes based on time-discrete observations are required to show massive capacities in order to process the entire spatial and temporal variance of these Earth systems.

As recurrent neural networks are, formally, Turing-complete (Siegelmann and Sontag 1991, 1995) and can, thus, approximate any arbitrary program, they are theoretically suitable for such tasks. Their vanilla realizations, however, suffer from several severe practical limitations that restrict them in terms of their computational power. As described before, gated RNN variants aim to mitigate these shortcomings, mostly by counteracting the vanishing and exploding gradients phenomenon.

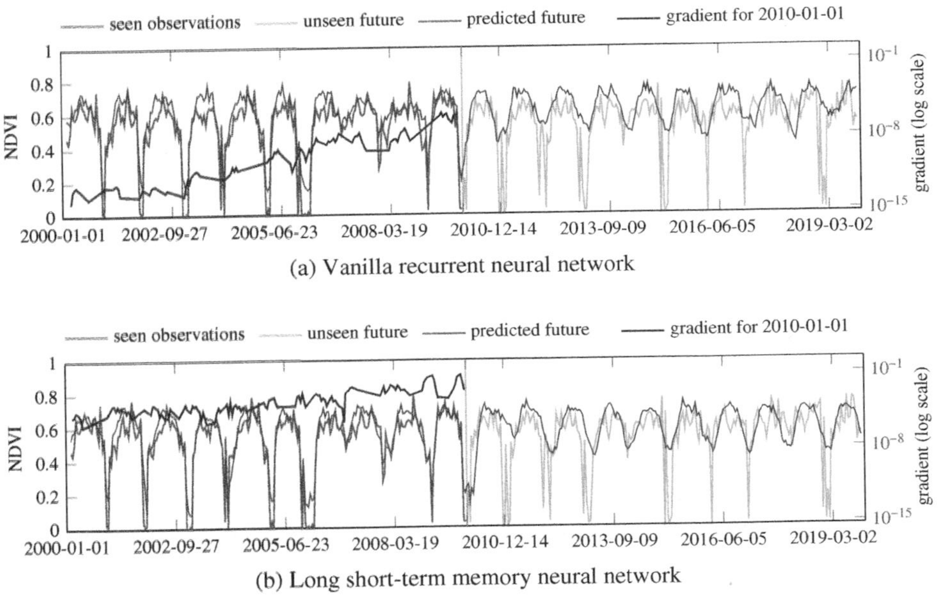

Figure 8.8 Two recurrent network models – i.e. (a) a vanilla recurrent neural network and (b) a long short-term memory neural network – have been trained to solve an auto-regressive NDVI data prediction task. The models have been shown NDVI time series acquired by MODIS from 2000 to 2010 over central Europe and predicted them further until 2020. The gradients are evaluated at the last known time point (2010) and indicate the influence of previous data (before 2010) on the observation in 2010. It can be seen that the vanishing gradients of the standard recurrent neural network restrict the ability to retrieve long-term temporal context while the LSTM network uses data from the previous five years.

In order to exemplify this effect, Figure 8.8 shows the results of a real-world experiment. For that purpose, a series of NDVI values has been derived from optical MODIS satellite observation from central Europe over the years from 2000 to 2010. Assuming that a high-capacity model is expected to be able to predict future data based on a sequence of past observations, this task was chosen as a proxy problem. The figure compares these prediction capabilities of a vanilla RNN and a LSTM network. As can be seen in the figure, while both models were able to estimate the unseen future data, i.e. the time span from 2010 to 2020, the LSTM model produced comparably smoother times series while still being able to reproduce periodical variation at different temporal scales. One reason for this behavior can be attributed to the evolution of their temporal gradients. The green curves (in log scale) show these gradients evaluated at the time step of the first prediction, i.e. 2010-01-01. It becomes evident that the gradient magnitudes decayed exponentially in the case of the RNN model, while they remained almost equally strong in the LSTM case. The ability to actively steer the gradient flows back through time enabled the LSTM network to consider a much longer temporal context and, thus, to predict the unseen data more stably. These stable predictions can build the basis for further classification tasks.

While LSTM networks have been shown to maintain the expressive power of RNNs, i.e. they can learn context-sensitive languages (Gers and Schmidhuber 2001), further design choices realize their further increased long-term stability at the cost of reduced computational capacity. It, thus, depends on the entirety of circumstances – i.e. the problem formulation, the used data, the available computational resources, etc. – which model variant performs best in a certain practical scenario.

8.4 Application in Earth Sciences

As motivated at the beginning of this chapter, the dynamical system Earth is governed by a multitude of processes at various spatial and temporal scales. Figure 8.9 shows the evolution of vegetation activity observed for two dedicated field parcels, located in the north of Munich, Germany, cultivated with meadow and corn crops, respectively. For this, the NDVI extracted from repeating Sentinel-2 observations has been used as a proxy for vegetation activity. These recordings nicely reflect the consequences of different influence processes, e.g. the onsets of plant growth, climatic and weather influences, as well as harvesting and cropping events. It can be safely assumed that the dynamics of phenological events and agricultural cultivation are crop-specific, while other disturbances like cloud cover are fairly uncorrelated with the particular crop type. Then, recurrent models can exploit these crop-specific patterns and make them accessible for further tasks, like crop classification or yield prediction.

Figure 8.10 summarizes the results of a crop classification experiment. We trained an LSTM network and, for comparison, a CNN baseline on Sentinel-2 data streams observing a region of interest located in the north of Munich, Germany, over the entire vegetative period of the year 2016. As can be seen in Figure 8.10(a), the recurrent model increased its classification performance with each newly presented observation. In comparison, the feed-forward baseline model showed a rather constant classification quality, fairly uncorrelated with the time of observation. It can further be shown that these recurrent models carry the ability to differentiate task-relevant temporal patterns from the irrelevant (Rußwurm and Körner 2018a).

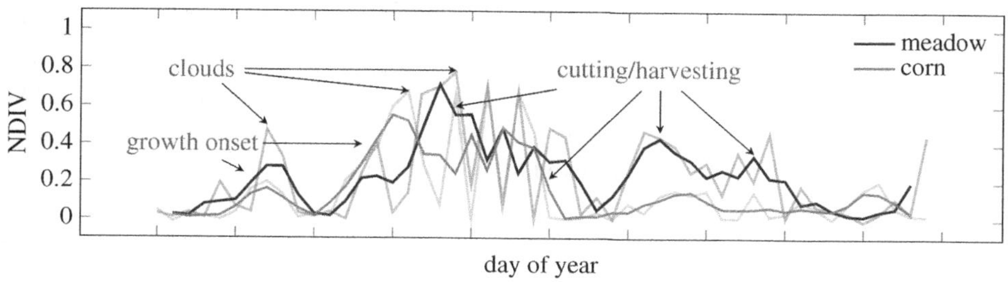

Figure 8.9 The vegetation activity of two field parcels – cultivated with meadow and corn crops, respectively – monitored over one entire season of 2016 by means of NDVI values derived from repeated Sentinel-2 observations. From these curves, it is possible to deduce the effects of various processes influencing crop growth, like, for instance, climatic and weather conditions (e.g. clouds), crop-specific phenological dynamics (e.g. growth onsets), and agricultural cultivation (cutting and harvesting events).

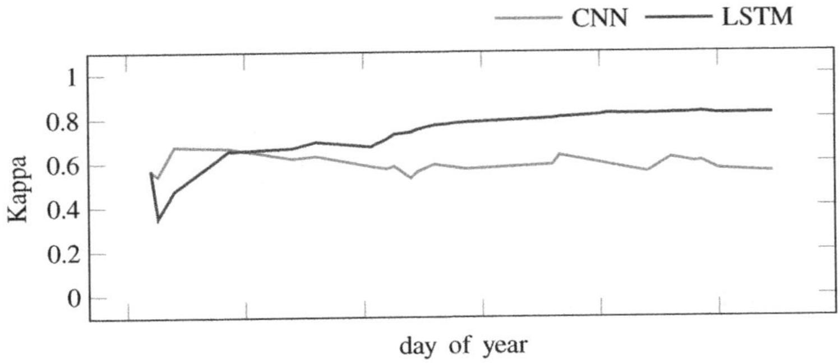

(a) Crop classification accuracy

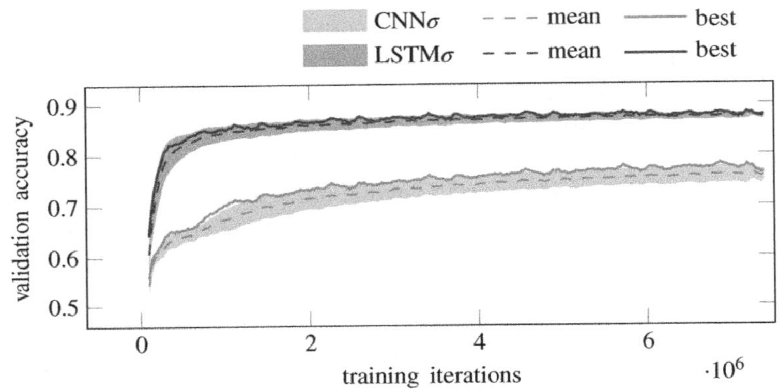

(b) Training process

Figure 8.10 Recurrent models can outperform feed-forward baselines in crop classification tasks, as they are able to extract crop-specific phenological patterns in Earth observation data. Furthermore, they show better training behavior, as they converge faster and more reliably to their optimal parameter settings.

Taking a closer look at the training process itself reveals another important observation. Figure 8.10(b) visualizes the validation accuracy of an LSTM and CNN model while training, as a function of the iterations performed for parameter optimization and accumulated over ten runs with varying initializations. It becomes evident that the recurrent LSTM model consistently converged faster to its final optimum, while the individual runs did, in general, show a smaller variance compared to the CNN baseline.

8.5 Conclusion

We have shown that recurrent neural network models, in their different formulations and variants, are able to capture the dynamics of Earth observation data that is assumed to be driven by complex latent dynamical systems. Most importantly, the active steering of gradient flows while training increases the representative power of these models which can, thus, produce more stable predictions over longer periods. These capabilities open the potential for tackling more complex machine learning tasks. Thus, such models can be employed in various Earth observation data processing systems, like, for instance, in land cover and crop type classification tasks (Rußwurm and Körner 2018, 2017a) or, at a global scale, for climate system analysis (see Chapter 18; (Kraft et al. 2020, 2019)).

Methodological research on this topic of recurrent data processing has gathered pace remarkably in recent years and these models have already been brought to numerous practical applications. They undoubtedly come with the potential to help to exploit the massive data stocks piled up since the rise of modern Earth observation satellite missions. Nevertheless, the research community is still facing unanswered questions. While feed-forward neural networks do already lack transparency, the information flows in trained recurrent neural networks are even harder to analyze and to interpret. Visualizing the information steering mechanisms of particular cells of gated variants already gives valuable insights, but the majority of these cells show complex, non-intuitive behavior. Further, active and prospective fields of research try to find approaches to integrate expert model knowledge into such data-driven models, to derive causal relationships between patterns present in Earth observation data, or to estimate the certainty and confidence at which predictions are made by such models.

9

Deep Learning for Image Matching and Co-registration

Maria Vakalopoulou, Stergios Christodoulidis, Mihir Sahasrabudhe, and Nikos Paragios

9.1 Introduction

Image matching and registration are some of the most important and popular problems for many communities, including Earth observation. Efficient and robust algorithms that can address such topics are essential for several other tasks including, but not limited to, optical flow, stereo vision, 3D reconstruction, image fusion, and change detection. Deep learning algorithms are becoming more and more popular, providing state-of-the-art performance for various problems including image matching and registration. These algorithms prove very efficient running time and robustness with variety of studies reporting their success in supervised and unsupervised settings.

Given a pair of images depicting the same area, image matching is the process of comparing the two images (source image S and target (or reference) image R) to obtain a measure of their similarity, while image registration is the process that aligns or maps these images by finding a suitable transformation between S and R. In particular, both image matching and registration are measuring or mapping identical pixels in the pair of images with the focus of the second to align S to R as accurately as possible. Although these problems seem to be easy conceptually, they are still an open research area for a variety of communities, considered as ill-posed problems that suffer from many uncertainties. This is mainly due to the nature of algorithms and images on which small changes in translation, illumination, or viewpoint can significantly affect these algorithms' performance even if the depicted areas are exactly identical. Therefore, numerous approaches have been proposed to address these problems and have been summarized in different surveys (Zitova and Flusser 2003; Sotiras et al. 2013; Leng et al. 2018; Burger and Burge 2016; Weng and He 2018). Nowadays, however, with the recent advances of deep learning, more and more techniques integrate these technologies for both matching and image registration, offering better performances especially in time requirements.

Some of the most common problems that image matching and registration algorithms need to address, especially for earth observation applications can be grouped in four main categories, namely (i) radiation distortions; (ii) geometric changes; (iii) areas including changes; and (iv) multimodal, large-scale sources of data. Starting with the first group,

Deep Learning for the Earth Sciences: A Comprehensive Approach to Remote Sensing, Climate Science, and Geosciences, First Edition. Edited by Gustau Camps-Valls, Devis Tuia, Xiao Xiang Zhu, and Markus Reichstein.

radiation distortions refer to the difference between the real emissivity of the ground objects and the one that is represented in the image level. This difference can be caused mainly by the imaging properties of the sensor itself or radiation transmission errors caused by the atmosphere during the acquisition of the objects' emissions. The later is also known as Bidirectional Reflectance Distribution Function (BRDF) with a lot of algorithms being proposed for its modeling (Montes and Ureña 2012). The second group refers to the geometric differences in the ground objects that have to do with differences in viewpoints of the sensors and the height of objects influencing mainly high-resolution images (with spatial resolution higher than 10 meters). The next group refers to the difficulty of these methods to work on places that contain changed areas. Both image matching and registration problems assume that the depicted regions are mainly identical, being unable by default to work properly on regions that contain changes, something that is quite common on earth observation and remote sensing datasets. Finally, it is also very challenging to create algorithms that are working robustly for multimodal datasets such as Synthetic-Aperture Radar (SAR) and multispectral or hyperspectral optical sensors or images produced by sensors with significantly different spatial resolutions. A variety of different sensor characteristics are available for earth observation and there is a significant need for matching and registration algorithms that can fuse their multitemporal information. All these cases make the problems of matching and co-registration very challenging and the use of algorithms proposed by other communities such as computer vision or robotics really challenging to be adapted to satellite data (Le Moigne et al. 2011; Nicolas and Inglada 2014).

Traditionally, image matching and image registration are two closely related problems, with the first providing usually reliable information for the second (Figure 9.1). Different sub-regions or pixels from the source and reference images are matched and used to define the best transformation model, G, to map S to R, thus resulting in a warped image D. Depending on the implemented strategy the matching algorithm can be applied globally, searching for the most similar regions in the entire image, or it can be applied locally

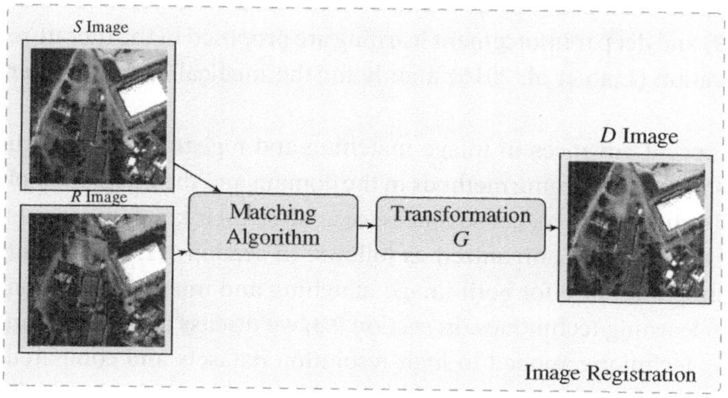

Figure 9.1 A schematic diagram of the image matching and image registration techniques for Earth observation. Image registration has two main components: the matching algorithm that is measuring how similar are different regions in the image; and the definition of the transformation G, which will be applied to the S image to generate the warped image D.

by searching for the best matching on predefined regions. The choice of the matching strategy also depends on the choice of transformation model used for the registration algorithm. Nowadays, various methods based on deep learning approaches are proposed from both the computer vision and earth observation fields. Most of the techniques that originate from the earth observation field are focusing on high-resolution datasets (Ma et al. 2019) due to the more challenging nature of this kind of images and their need for dense and more complex registration models.

Starting with image matching, traditionally two main components are usually applied – the feature extraction of keypoints or sub-regions of the images and the establishment of the proper correspondences. Feature extraction methods typically rely either on intensity-based methods using the raw image's intensities directly or on higher-level representations extracted from the pair of images. These representations are typically produced using either classical image descriptors or they are obtained using deep learning approaches. After feature extraction, optimal correspondences are found using a similarity function. Typical choices for this similarity function are mean squared error, normalized cross-correlation, and mutual information. The implementation of the similarity function in a deep learning framework is usually achieved using Siamese or triplet networks that share their weights (Kaya and Bilge 2019).

As far as the image registration task is concerned, depending on the transformation used, the methods can be categorized into: (i) rigid or linear; and (ii) deformable or elastic. Rigid methods define maps with transformations that include, e.g., rotations, scaling, and translations. They are global, and hence cannot model local geometric differences between images, which is usually the case in high-resolution datasets. However, they are very efficient and robust for the co-registration of satellite imagery. On the other hand, the deformable methods rely on a spatially varying model by associating the observed pair of images through non-linear dense transformations. After obtaining the optimal transformation G, the source image is resampled to construct the warped image which is in the same coordinate system as the reference image. Deep learning and convolutional neural networks have been used for image registration (Kuppala et al. 2020) while methods also based on generative models (Mahapatra and Ge 2019) and deep reinforcement learning are proposed in the literature for both 2D and 3D registration (Liao et al. 2016) mainly for the medical and computer vision communities.

This chapter focuses on recent advances in image matching and registration for earth observation tasks with emphasis on emerging methods in the domain and the integration of deep learning techniques. To study these recent advances we analyse their key components independently. The rest of the chapter is organized as follows. In section 9.2, we present a detailed overview of existing literature for both image matching and image registration focusing on the recent deep learning techniques. In section 9.3, we discuss and present an unsupervised deep learning technique applied to high-resolution datasets and compared with conventional image registration techniques. We describe the dataset used for this study in section 9.3.4, followed by experiments and results in 9.3.5. Finally, in 9.4 we summarize the chapter and enumerate future research directions for these algorithms.

9.2 Literature Review

9.2.1 Classical Approaches

Image matching has been dominated for long by hand-engineered and feature-based methods, with SIFT (Lowe 1999) being one of the most commonly used feature descriptor applied also in remote sensing with or without small variations (Vakalopoulou and Karantzalos 2014; Chen et al. 2018a). Additionally, descriptors such as SURF (Bay et al. 2008), DAISY (Tola et al. 2010), BRIEF (Calonder et al. 2012), the very recently proposed HOSS (Sedaghat and Mohammadi 2019), and other variations were equally popular. These descriptors are then used with intensity-based or more complex similarity functions such as mutual information (Viola and Wells 1995) or correlation-based methods (Pratt 1978) to establish correspondences. RANdom SAmple Consensus (RANSAC) (Fischler and Bolles 1981), a model-based technique, was also very commonly used to filter and establish proper matching of images or points. Although these features have alleviated the influences of radiometric and geometric deformations to some extent, their performance was significantly lower in the case of multi-sensor data, while their detection repeatability was still low. Tuytelaars and Mikolajczyk (2008) reports repeatability rates that were below 50% for datasets with three band image pairs.

Additionally, image registration techniques were based on the correspondences that were defined by the image matching algorithms to obtain the optimal transformation parameters, G that maps in the most optimal way S to R. Starting with the rigid methods, in remote sensing the transformations that are commonly used are translations, rotations, scaling, and shearing. These transformations can be captured by affine and homography mappings, which are used frequently in remote sensing applications (Zitova and Flusser 2003). In practice, these transformations can be described by a 3×3 matrix having 6 and 8 degrees of freedom respectively. To obtain the parameters of the transformation used, a number of correspondences have to be established (minimum 3 or 4 per dimension, respectively). The resulting system can be solved using least squares method to find the optimal values (Szeliski 2010). Numerous techniques (Wu et al. 2012; Vakalopoulou and Karantzalos 2014; Li and Leung 2007) fall into the category of methods using rigid transformations and have been tested using different spectral and spatial resolution satellite imagery.

While rigid methods are simple and efficient, they do not have the capacity to produce more complex transformations that can vary locally. To capture such locally-linear or non-linear transformations, deformable methods creating a dense transformation G are instead employed. Such methods construct a deformation grid G that defines transformations that locally express the correlation between the observations. Sotiras et al. (2013) presents a detailed survey of deformable registration methods and their categorization based on the geometric model they are using. These methods are commonly used in medical imaging and remote sensing datasets where the deformations between the images are not homogeneous (Karantzalos et al. 2014; Marcos et al. 2016). Some of the deformable strategies that have been proposed are based on correlation of objects (Marcos et al. 2016), contours (Hui Li et al. 1995), or intensity- and area-based similarity methods (Karantzalos et al. 2014; Vakalopoulou et al. 2016).

Table 9.1 Grouping of image matching techniques depending on the type of imagery they have been applied to.

Type of Imagery	Methods applied on Earth Observation
Optical to Optical	Altwaijry et al. (2016); Zhu et al. (2019a); En et al. (2018); Liu et al. (2018a); Chen et al. (2017b); Yang et al. (2018); He et al. (2018); Dong et al. (2019); Zhu et al. (2019); Jia et al. (2018); He et al. (2019a); Tharani et al. (2018); Wang et al. (2018)
SAR to SAR	Quan et al. (2016); Wang et al. (2018)
SAR to Optical	Merkle et al. (2017); Bürgmann et al. (2019); Merkle et al. (2018); Hughes et al. (2018); Quan et al. (2018); Ma et al. (2019); Merkle et al. (2017); Zhang et al. (2019)
Other	Ma et al. (2019); Zhang et al. (2019)

9.2.2 Deep Learning Techniques for Image Matching

Several deep learning methods for image matching in remote sensing images have been explored recently in the community. Most of the methods are based on Siamese architectures, extracting features from CNNs, and providing similarity scores for the input patches. Similar approaches were proposed by a variety of studies in the computer vision community for both the supervised and unsupervised settings (Revaud et al. 2016; Zagoruyko and Komodakis; Han et al. 2015). Even though these techniques are quite recent, deep learning-based methods have been shown to outperform traditional ones. A summary of deep learning-based methods applied to remote sensing is presented in Table 9.1. Methods have been grouped depending on the type of data they use, including supervised and unsupervised techniques, with the first being more dominant.

We begin with methods for matching optical to optical imagery. In Altwaijry et al. (2016), the authors propose an attention-based deep learning architecture trained with weak labels. In particular, the method provides local correspondence by framing the problem as a classification task, integrating an attention mechanism to produce a set of probable matches. To train their model, the authors used a dataset of urban high-resolution patches consisting of the labels "same" and "different". A similar weak annotation for different types of optical data is also used in En et al. (2018). A similar strategy is also presented in He et al. (2019a) for matching medium resolution multi-temporal imagery. In Zhu et al. (2019a) the authors proposed the use of densely-connected CNNs in a Siamese architecture to match RGB images with infrared images reporting very promising results for image pair matching. In Chen et al. (2017b) a deep hashing network is proposed to search for feature point matching. Concerning unsupervised methods for matching, adversarial networks (see Chapter 3) are mainly used for similar types of optical data. More specifically, in Tharani et al. (2018) an encoder-decoder architecture combined with a deep discriminator network to replace distance metrics is proposed. The authors report very promising results, while they also provide a comparative study of different commonly used convolutional architectures for the accurate registration of different land cover classes.

Siamese architectures are also popular for SAR to Optical image matching. In Merkle et al. (2017) a Siamese architecture is proposed to generate reliable matching points between TerraSAR-X and PRISM images. For the same type of images, a conditional generative adversarial network (see Chapter 3) is trained in Merkle et al. (2018) to generate SAR-like image patches from optical images to enhance the performance of known classical matching approaches. Moreover, a (pseudo-) Siamese network is proposed in Hughes et al. (2018). Medium- and high-resolution SAR and optical data are evaluated in Bürgmann et al. (2019), presenting an approach for matching ground control points (GCPs) from SAR to optical imagery. The training of conditional generative adversarial networks (see Chapter 3) is also proposed in Merkle et al. (2017) to generate artificial templates and the matching of optical to SAR data. Finally, a combination of deep and local features is used in Ma et al. (2019) to match and register multimodal remote sensing data.

Deep learning methods are also used to match images from completely different sources of data such as satellite images with maps. In Ma et al. (2019) the authors evaluated their methods using a pair of an optical image and the corresponding Tencent Map providing very promising results. A similar approach based on Siamese architectures is proposed in Zhang et al. (2019), evaluating its performance on multimodal data including optical to map matching.

9.2.3 Deep Learning Techniques for Image Registration

In computer vision, there are works that include ideas of modeling transformation (Hinton 1981) directly, learning transformation invariant representations (Kanazawa et al. 2014), and attention/detection mechanisms for feature selection (Gregor et al. 2015). The study presented in Jaderberg et al. (2015) was one of the first to introduce the idea of using a deep learning-based architecture to eliminate intra-object variance by transforming intermediate feature maps. The spatial transformer network proposed in this study is a trainable module that can be integrated and trained together with any deep learning architecture. The module estimates an optimal transformation of intermediate feature maps to remove variance due to intra-object shape differences, object placement, and object size, thus aiding in recognition by mapping objects to a canonical space. The estimated transformations can include rigid deformations such as scaling, cropping, rotations, as well as non-rigid deformations.

In remote sensing, the literature for methods that can obtain the parameters of the transformation directly from the developed models is not very vast. Most methods use deep learning-based models to match the images (as mentioned in the previous sub-section) and then they use these matching to obtain the parameters of the registration model independently. Recently, in Vakalopoulou et al. (2019) the authors proposed the use of the spatial transformer to regress rigid and non-rigid deformations directly from source and target images under a deep learning framework for the registration of high-resolution satellite datasets.

Moreover, a deep learning-based method that could output the displacement field directly from the networks was proposed in Zampieri et al. (2018) to register optical imagery to cadastrial maps of buildings and road polylines. The method was based on a fully convolutional architecture that learned scale-specific features predicting the deformations directly.

Additionally, the authors proposed an improvement to their previous work in Girard et al. (2019) by developing a multi-task scheme for simultaneous registration and segmentation, which improved the performance of the reported registration.

9.3 Image Registration with Deep Learning

In this section, we describe the approach presented in Vakalopoulou et al. (2019) in detail, which is one of the methods proposed recently for predicting the deformation maps in an end-to-end deep neural network. As discussed in the previous section, end to end deep learning architectures are more commonly used for matching than registration problems. It is for this reason that we have chosen to focus on the latter in this chapter. The method presented here is a modification of a recent work on accurate and efficient registration of 3D medical volumes (Christodoulidis et al. 2018). Both implementations are available online at https://github.com/stergioc/smooth-transformer.

The main advantages of this method are fourfold: (i) a completely unsupervised technique for regressing the dense deformation grid G from a pair of images, (ii) a modular formulation that couples rigid and deformable registration within a single optimization, (iii) a framework that is independent of the CNN architecture, (iv) fast inference allowing real-time applications even for very large-scale remote sensing datasets. The proposed framework can be divided into three different components – the transformation strategy, the CNN architecture, and the optimization procedure.

9.3.1 2D Linear and Deformable Transformer

The main component of the proposed CNN architecture is the 2D transformer layer, which enables the architecture to regress the spatial gradients. This layer warps the image S under a dense deformation G to create the warped image D that best matches R. This operation is defined by the equation

$$D = \mathcal{W}(S, G), \tag{9.1}$$

where $\mathcal{W}(\cdot, G)$ indicates a sampling operation $\mathcal{W}$ under the deformation G.

In our implementation, the deformation is fed to the transformer layer as sampling coordinates, which uses a backward bilinear interpolation sampler as $\mathcal{W}$, adapting a strategy similar to Shu et al. (2018). The backward sampling indicates that for every pixel of the warped image a coordinate in the original image S is computed indicating where the intensity value originates. Often backward sampling is preferred compared to forward due to the discrete nature of the images. The backward bilinear interpolation sampler is defined as

$$D(\vec{p}) \;= \mathcal{W}(S, G)(\vec{p}) = \sum_{\vec{q}} S(\vec{q}) \prod_{d} \max\left(0, 1 - \left|[G(\vec{p})]_d - \vec{q}_d\right|\right), \tag{9.2}$$

where $\vec{p}$ and $\vec{q}$ denote pixel locations, $d \in \{x, y\}$ denotes an axis, and $[G(\vec{p})]_d$ denotes the d-component of $G(\vec{p})$.

The formulation in this case consists of two different components – one which calculates a linear/affine transformation, and another that calculates a dense transformation. Depending on the application and the type of data, these two terms can be used and trained together or separately. In case, that these two operations are trained at the same time, they are applied one after the other, by integrating first the affine component and then the deformable for more fine transformations. Such scheme can be described by

$$\mathcal{W}(S, G) = \mathcal{W}(S, \mathcal{W}(G_N, G_A)) \tag{9.3}$$

where G_A represents the affine deformation grid, while G_N represents the deformable one. Here it should be mentioned that the network is trained end-to-end, optimizing both linear and deformable parts simultaneously. G_A is computed from six regressed affine transformation components, denoted by a 2×3 matrix A. For the deformable part G_N, an approach similar to Shu et al. (2018) is adopted. Instead of regressing the components of G_N directly, we regress instead a matrix Φ of spatial gradients along x- and y- axes. As is discussed in Shu et al. (2018), this approach helps generate smoother grids that render the deformable component, making it easier to train. The actual grid G_N can then be obtained by applying an integration operation on Φ along x- and y-axes, which is approximated by the cumulative sum in the discrete case. Adopting such a strategy enables us to draw some conclusions on the relative position of adjacent pixels in the warped image based on Φ. Concretely, two pixels $\vec{p}$ and $\vec{p}+1$ will have moved closer, maintained distance, or moved apart in the warped image, if $\Phi(p)$ is respectively less than 1, equal to 1, or greater than 1. Such an approach avoids self-crossings, while allows the control of maximum displacements among consecutive pixels.

9.3.2 Network Architectures

Such formulation is independent of the network architecture and according to the application and dataset used, different ones can be incorporated. To test this modular nature of the proposed approach, two different architectures were employed – one based on dilated filters and one based on maxpooling. The two architectures are presented in detail in Figure 9.2.

The network architecture is based on an encoder-decoder scheme. For the encoder part two different sets of experiments were constructed. The first is very similar to the one presented in Anthimopoulos et al. (2018) adopting for the encoder dilated convolutional kernels together with feature merging, while the decoder employs non-dilated convolutional layers. Specifically, a kernel size of 3×3 was set for the convolutional layers while LeakyReLU activation was employed for all convolutional layers. Each of the encoder-decoder parts contains four of these layer blocks with the feature maps starting from 16 and being doubled for each block, resulting in a 128 feature map. Before the decoder, all the feature maps were concatenated in order to create a more informative, multi-resolution feature space for the decoder.

The second architecture follows a U-Net like architecture (Ronneberger et al. 2015b) adopting consecutive layers of 2D convolutional layers with kernel size 3×3 followed by instance normalization, LeakyReLU and max-pooling that reduce the dimension of the input by half at each layer. The encoder part consists of four of these layer blocks with features maps from 16 to 128. The decoder part consists of a symmetric part where the max-pooling is replaced by upsampling to return the input to the initial dimensions. Moreover, skip connections flow information from the encoder to the decoder part.

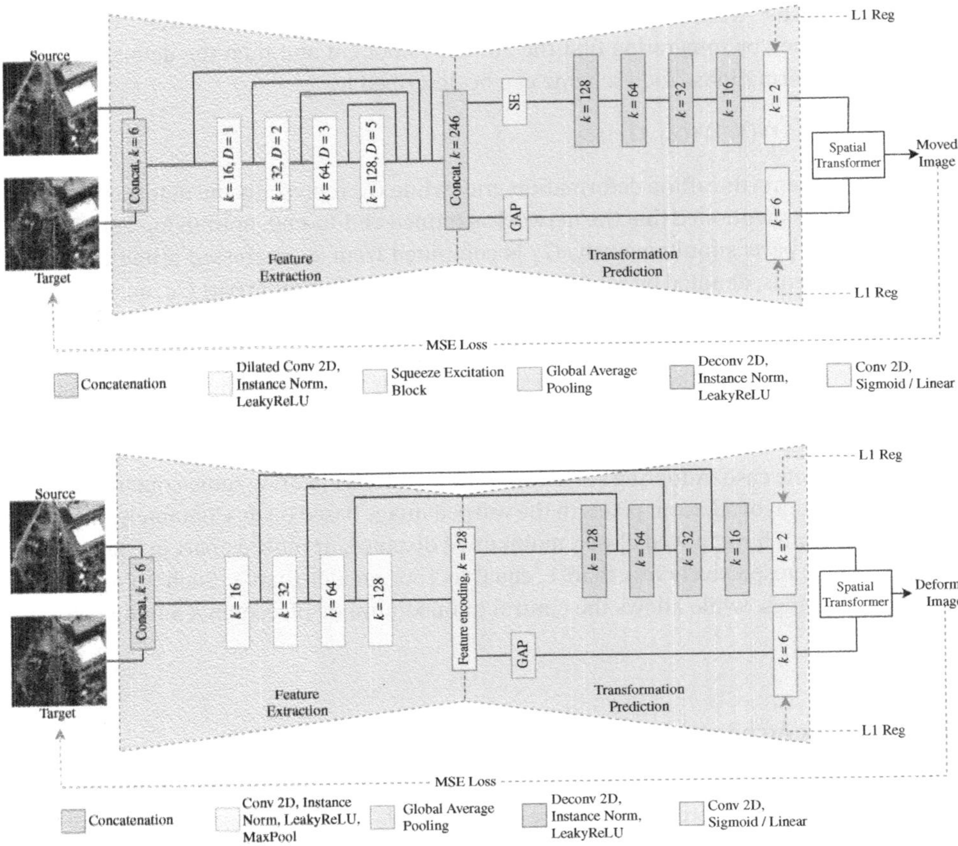

Figure 9.2 A schematic diagram of two different architectures presented in this chapter. The architecture consists of two different parts following an autoencoder scheme: the feature extraction part and the part for the prediction of the transformation.

The decoder part has two different branches, one that calculates the affine parameters and one the deformable ones. For the linear/affine parameters A, a linear layer was used together with a global average pooling to reduce the spatial dimensions, while for the spatial gradients Φ a sigmoid activation was employed. Finally, the output of the sigmoid activation was scaled by a factor of 2 to allow consecutive pixels to have larger displacements than the initial.

9.3.3 Optimization Strategy

The entire framework is trained in a completely unsupervised way, in that it does not require registered pairs of images to be trained. As similarity function between the R and D images, the mean squared error (MSE) is used and the overall loss is defined as

$$\text{Loss} = \|R - \mathcal{W}(S, G)\|^2 + \alpha\|A - A_I\|_1 + \beta\|\Phi - \Phi_I\|_1, \tag{9.4}$$

where A_I represents the identity affine transformation matrix, Φ_I the spatial gradients of the identity deformation, and α and β are regularization weights, controlling the influence of the regularization terms on the obtained displacements. The higher the values of α and β, the closer the deformation is to the identity. The regularization parameters are essential for the joint optimization, as they ensure that the predicted deformations will be smooth for both components. Moreover, the regularization parameters are very important in the regions of change, as they do not allow the deformations to become very large.

The most commonly employed reconstruction loss is the mean-squared error (MSE). However, MSE suffers from several drawbacks. Firstly, it cannot account for changes in contrast, brightness, tint, etc. Secondly, MSE tends to produce smooth images. Thirdly, MSE does not account for the perceptual information in the image (Wang et al. 2004). Recent papers have hence reported the use of other types of similarity functions, either instead of MSE or in combination with it, to construct more descriptive loss functions. One of this type of losses is the local cross correlation (LCC) presented in Balakrishnan et al. (2019).

9.3.4 Dataset and Implementation Details

To validate the method, a multimodal high-resolution dataset from Quickbird and WorldView-2 satellites was used. This dataset is a multitemporal dataset acquired in 2006 and 2007, covering a 14km^2 region in the East Prefecture of Attica in Greece. This particular dataset was challenging due to the very large size of the high-resolution satellite images, their complexity due to different acquisition shadows, angles, important height differences, numerous terrain objects and the sparse multitemporal acquisitions. To train the framework, patches of size 256×256 were created. In particular, 1350 patches were selected randomly for training, 150 for validation, and 150 for testing the proposed framework. Regions that are spatially independent were selected from the image to generate the training, validation and testing sets of pairs.

Concerning the implementation details of the framework, the initial learning rate was 10^{-3} and was divided by a factor of 10 if the performance on the validation set did not improve for 50 epochs, while the training procedure was stopped when there was no observed improvement for 100 epochs. The regularization weights α and β were both set to 10^{-6}. For the optimization, the Adam optimizer was selected, while the entire framework was implemented in tensorflow and keras. For all the experiments we used a GeForce GTX 1080Ti GPU. We noted that the training converges after around 140 epochs for the dilated architecture and 100 epochs for the maxpooling one. The overall training time was approximately 4 hours.

9.3.5 Experimental Results

Extensive experiments compare and benchmark the performance of the proposed method with other state-of-the-art algorithms that perform both rigid and deformable registration.

In particular, a framework similar to the one presented in Vakalopoulou and Karantzalos (2014) using SIFT, SURF and ASIFT descriptors, RANSAC and an affine transformation is developed to evaluate the performance of the non deep learning-based rigid transformation. Moreover, the method presented in Karantzalos et al. (2014) that applies a graph based method to obtain deformable transformations is also evaluated. Different similarity functions were used for comparison namely the Sum of Absolute Difference (SAD), the Sum of Absolute of Differences plus Gradient Inner Products (SADG), the Normalized Cross Correlation (NCC), and the Normalized Mutual Information (NMI).

Moreover, for the presented completely unsupervised CNN framework we benchmark each of its different components together with two different CNN architectures. In particular, experiments using only the linear A or deformable Φ components and also their ensemble were constructed. This enabled to better examine the framework and find the most optimal configuration for earth observation datasets. The performance with different network architectures are also benchmarked in this chapter.

Starting with the quantitative evaluation, \$5\$ different landmarks, mainly on the buildings' corners have been selected and their average errors in each of the axes are reported in Table 9.2. It should be noted that for all the methods the same landmarks have been selected and around 20 image pairs were used to extract the landmarks. These landmarks contained mainly roofs of buildings as they were the ones presenting the highest registration errors.

Both deep learning-based and classical techniques are evaluated in this study. All the methods recover the geometry of the pairs, achieving better performance than the

Table 9.2 Errors measured as average Euclidean distances between estimated landmark locations. *dx* and *dy* denote distances along *x*-, *y*-, respectively, while *ds* denotes the average error along all axes per pixel.

Method		***dx***	***dy***	***ds***	**Time (sec)**
Unregistered		**7.3**	**6.3**	**9.6**	**-**
Rigid (Vakalopoulou and Karantzalos 2014)	(SIFT)	3.0	2.8	4.1	~2
	(ASIFT)	4.0	3.5	5.3	~2.5
	(SURF)	4.7	3.0	5.6	~3
Deformable (Karantzalos et al. 2014)	(SAD)	1.5	2.7	2.9	~2
	(SADG)	1.5	2.6	2.8	~2
	(NCC)	1.3	2.3	2.6	~2
	(NMI)	1.4	2.4	2.7	~2
Dilated (Vakalopoulou et al. 2019)	A	2.5	2.8	3.7	~0.02
	Φ	1.2	2.0	2.3	~0.02
	A &Φ	**0.9**	**1.8**	**1.9**	**~0.02**
Maxpool (Vakalopoulou et al. 2019)	A	2.6	2.8	3.7	~0.02
	Φ	1.3	2.1	2.4	~0.02
	A &Φ	1.0	1.9	2.0	~0.02

unregistered case; however, the superiority of deformable methods compared to rigid ones is visible in this study, showing the need for more complex deformations for high-resolution imagery. The *ds* error for all the rigid methods is around 4 pixels with the ASIFT and SURF descriptor showing the highest errors. On the other hand, the deformable methods report *ds* errors around 2.5 pixels while the combination of rigid and deformable models reach an error of 2 pixels indicating that the combination of these two methods can boost even more the accuracy of the registration systems. Moreover, the use of different similarity metrics does not affect the performance a lot (Karantzalos et al. 2014); however, one can observe that area-based metrics like NCC and NMI perform slightly better. A similar conclusion was drawn for the influence of the deep neural network architecture in Vakalopoulou et al. (2019). The type of convolution did not really influence the performance of the algorithm, however, we should mention that the architecture with the maxpooling was converging slightly faster than the one with the dilated filters. Finally, it should be noted that the deep learning-based method is faster, with inference time for an image pair of size 256×256 less than half a second in comparison to the $2, 3$ seconds that the other methods need, giving a big advantage for large datasets such as the remote sensing ones, and allowing even real-time applications.

Comparing quantitatively the performance of the different methods, three different cases from the test set are presented in Figure 9.3 using checkerboard visualizations between the target R and warped image D before and after the registration. Regions of interest are indicated with red rectangles. For the different architectures in Vakalopoulou et al. (2019) and the different similarity metrics in Karantzalos et al. (2014) there were no differences in the visualizations and therefore the configuration with the lowest reported error is presented, namely the architecture with the dilated filters and the NCC metric, respectively. Even if the initial displacements were quite important all the methods recover the geometry and register the pair of images. However, the proposed method using only the A deformations fails to register accurately high buildings which have the largest deformations, due to the global nature of the transformation. On the other hand, all the deformable based methods report a good performance, registering very accurately the pair of images. One thing that should be mentioned is that the method in Vakalopoulou et al. (2019) reports easier convergence in the case that both the A and Φ parts are trained simultaneously, proving that the additional linear component is a valuable part of the proposed framework.

Moreover, in Figure 9.4 a comparison between the registration performance is provided for the explored methods after the application of the rigid (Vakalopoulou and Karantzalos 2014) using the SIFT descriptor, deformable (Karantzalos et al. 2014) using the NCC metric, and deep learning-based with deformable and affine (Vakalopoulou et al. 2019) methods. Again, the indicated problems are the same, with Vakalopoulou and Karantzalos (2014) failing to recover the local deformations, while the other two methods report very similar performance. In addition, Vakalopoulou et al. (2019) seem not to have a problem creating proper displacement fields for the different sensors, proving its power and potentials. However, experiments with sensors that have higher spectral and spatial differences should be performed.

Figure 9.3 Qualitative evaluation for three different pairs of images. From top to bottom: unregistered, (Karantzalos et al. 2014) with NCC, dilated (Vakalopoulou et al. 2019) only A, dilated (Vakalopoulou et al. 2019) only Φ, dilated (Vakalopoulou et al. 2019) A and Φ.

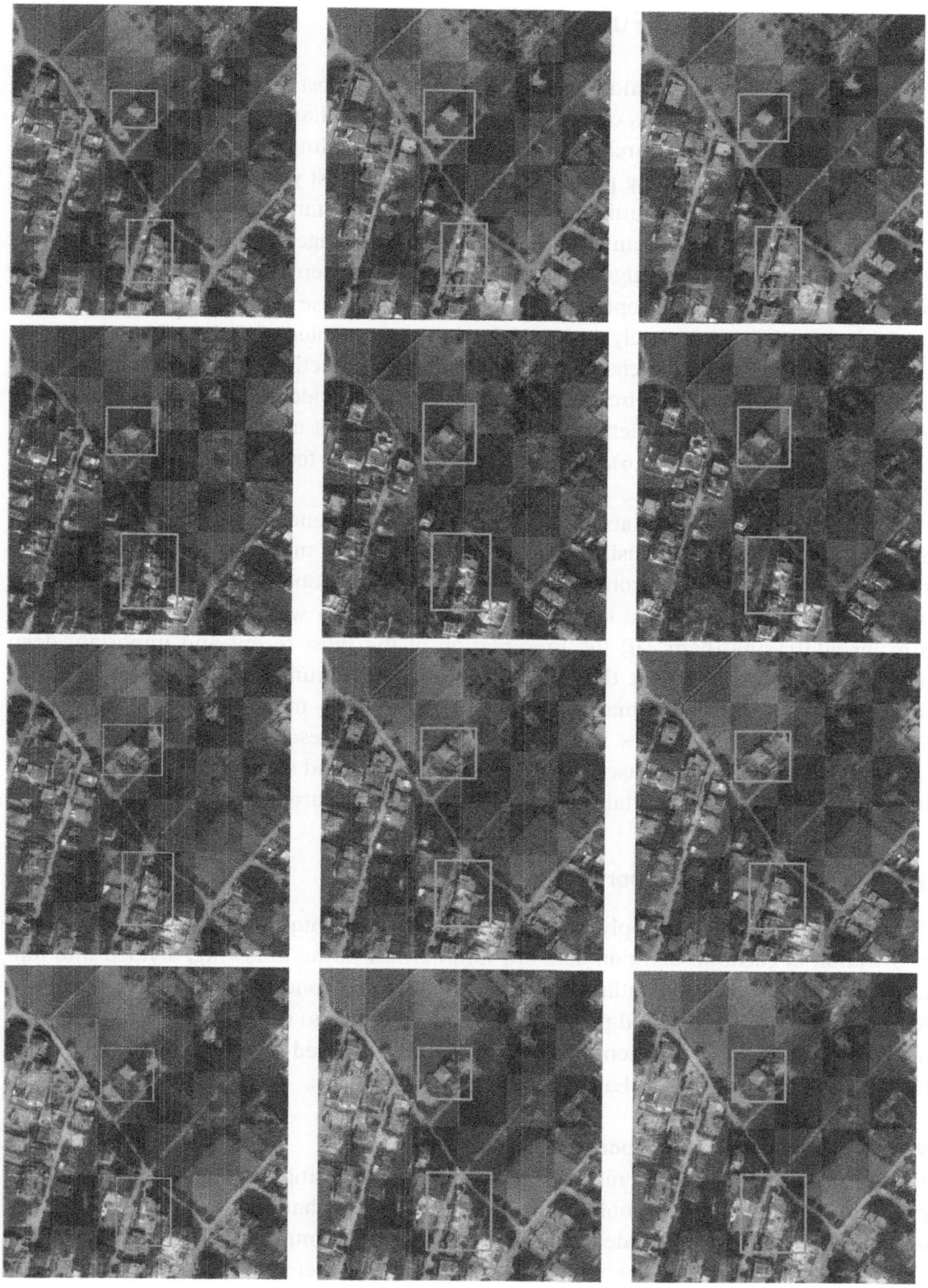

Figure 9.4 Qualitative evaluation for the different methods ((Vakalopoulou and Karantzalos 2014), (Karantzalos et al. 2014), (Vakalopoulou et al. 2019) respectively). From top to bottom: Quickbird 2006 - WorldView-2 2011, Quickbird 2007 - WorldView-2 2011, Quickbird 2009 - WorldView 2011, WorldView-2 2010 - WorldView-2 2011.

9.4 Conclusion and Future Research

Image matching and registration are two problems of utmost importance that have been extensively studied in various communities, including the Earth observation community. A significant amount of research has been devoted to providing the theory and the tools to address these two challenging problems properly. In the last years, both computer vision and earth observation communities have proposed ways to standardize the procedures and create generic methods that can properly address the challenges depending on the problem and the applications. Currently, with the development of deep learning techniques, multiple works in the literature propose approaches based on these techniques. However, these methods are not yet completely exploited for these two problems compared to other problems such as segmentation, classification, and change detection. This indicates the need to focus more towards this direction, especially if one considers that these two problems are very important for a variety of other problems such as image fusion, change detection (Vakalopoulou et al. 2016), 3D reconstruction, or even few shot learning (Sung et al. 2018).

In this chapter, we made an effort to provide a comprehensive survey of the recent advances in both fields focusing on deep learning-based methods. Our approach was structured around the key components of the problems, and in particular we focused on (i) the formulation of the two problems and the main ways proposed in the literature for addressing them, (ii) the presentation of the most recent and important deep learning-based methods that the earth observation community has proposed, and (iii) the comparison and benchmark of different registration methods summarizing their advantage and disadvantages using a challenging high-resolution dataset depicting a peri-urban region. Finally, based on these developments and state-of-the-art methods the present study highlighted certain issues and insights for future research and development.

9.4.1 Challenges and Opportunities

The current challenges of applying the deep learning techniques to image matching and registration of earth observation data are summarized in the following, divided into topics that have the most potential as future directions. To conclude, we believe that deep learning-based methods could provide very good solutions to the image registration problem and the future development in the field should be moved towards addressing the following challenges applied to Earth Observation applications.

9.4.1.1 Dataset with Annotations

Even if deep learning-based methods provide very promising directions for both image matching and registration problems, most of the methods that exist in the literature need annotations to train their models. Especially in the case of multimodal registration, annotations are most of the time essential. However, currently there are no available datasets to generate and evaluate these models, making the proper use and development of these algorithms very slow. It is very important to generate matching and registration datasets for earth observation while the community should also investigate unsupervised techniques for these two problems. Generative Adversarial Networks can provide very valuable tools towards developments in that direction.

9.4.1.2 Dimensionality of Data

One of the main problems of Earth observation data is their dimensionality both on the spatial and spectral domains, which is also an important difference compared to computer vision datasets. This is also one of the main problems for traditional algorithms to use all the available spectral information and efficiently provide matching or registration results. In particular, for rigid methods the computational time is not significant; however, for deformable methods the calculation of complex transformations is a bottleneck. Nowadays, with deep learning-based approaches the computational time even for deformable methods has considerably decreased, opening new opportunities for the community to design efficient solutions that exploit all the available information. Such methods will also be easily applicable for hyperspectral data, exploiting all their spectral information, something that currently is not easily achievable.

9.4.1.3 Multitemporal Datasets

Moreover, with the recent developments on satellites and the adaptation of open policies for many main space missions access to Earth observation data became easier. Moreover, due to these advances the community currently has access to multitemporal datasets with high temporal resolution that was unavailable earlier. Image matching and registration are the two methods that enable the use of these data for various applications and problems such as land monitoring, damage managing, environmental changes, and many others. There is a need for the community to propose methods and tools that efficiently solve these problems for two or even more images. The idea of group-wise registration has already been proposed in medical imaging (Kornaropoulos et al. 2016) and, with the recent deep learning advantages, it can provide a solution for registration of multitemporal datasets in the same coordinate system in an efficient way. The development of these algorithms will further boost the effectiveness and applicability of earth observation methods in large-scale monitoring of earth and environmental applications.

9.4.1.4 Robustness to Changed Areas

Earth observation datasets and especially the imagery based on optical sensors suffer from high radiometric, atmospheric changes and cloud coverage, making the application of matching and registration techniques really challenging. The use of deep learning-based methods could provide more robust algorithms on these changes as they rely on higher-level representations. Moreover, the development of algorithms that can handle regions with semantic changes would be an interesting direction for the future.

10

Multisource Remote Sensing Image Fusion

Wei He, Danfeng Hong, Giuseppe Scarpa, Tatsumi Uezato, and Naoto Yokoya

10.1 Introduction

Multisource remote sensing image fusion is used to obtain detailed and accurate information regarding the surface, which cannot be acquired from a single image, by fusing multiple image sources (Ghamisi et al. 2019). Typical examples of multisource image fusion used in remote sensing are the resolution enhancement tasks, which include (i) pansharpening to enhance the spatial resolution of multispectral imagery by fusing it with panchromatic imagery (Garzelli et al. 2007; Vivone et al. 2014; Loncan et al. 2015), and (ii) multiband image fusion to reconstruct high-spatial-resolution and high-spectral-resolution images (Lanaras et al. 2017; Yokoya et al. 2017; Ghamisi et al. 2019).

Since 1990, pansharpening has been actively researched to produce higher-level products for optical satellites that are composed of panchromatic and multispectral imagers. Furthermore, multiband image fusion has received great attention in the last decade with the emergence of space-borne hyperspectral sensors. Traditional approaches based on component substitution and multi-resolution analysis have been studied in detail and are commonly applied for practical applications. These approaches extract spatial details from a high-spatial-resolution image and inject them into an upsampled low-spatial-resolution image. These methods differs in how to extract and inject spatial details. As the next trend, researchers have formulated image fusion tasks as optimization problems and implemented priors of data structures in various models, such as low-rank, sparse, variational, and non-local modeling, to improve the quality of reconstruction. Even though these priors were useful for achieving significant improvement, the accompanying high computational cost has been a serious issue when applied to generate higher-level products of operational satellites that acquire large-scale data.

Deep learning (DL) has proven to be a powerful tool in many fields as well as various image processing tasks. Recently, DL-based methods have been proposed as a groundbreaking approach for pansharpening and multiband image fusion (Masi et al. 2016; Scarpa et al. 2018; Xie et al. 2019); these methods achieved state-of-the-art performance and computational efficiency at the inference phase. Such DL-based methods are capable of learning complex data transformations from input images to target images of training samples in an end-to-end manner. This book chapter provides an overview of state-of-the-art DL

Deep Learning for the Earth Sciences: A Comprehensive Approach to Remote Sensing, Climate Science, and Geosciences,
First Edition. Edited by Gustau Camps-Valls, Devis Tuia, Xiao Xiang Zhu, and Markus Reichstein.

methods for pansharpening and multiband image fusion as well as comparative experiments to demonstrate their advantages and characteristics. This is followed by a discussion of the future challenges and research directions.

10.2 Pansharpening

Pansharpening is a special type of super-resolution where, in addition to the multiband or multispectral (MS) image to be super-resolved, a high-resolution but single-band image, referred to as the panchromatic (PAN) band, which is spectrally overlapped with the MS, is also available in input. It can be considered as a single-sensor fusion task wherein spectral and spatial information are obtained by two distinct channels, i.e., MS and PAN, respectively. The goal of this technique is to obtain a spatial and spectral full-resolution data cube. A survey of the pansharpening methods proposed prior to the advent of the deep learning era can be found in Vivone et al. (2015). In the following sections, we first provide an overview of the most recent deep learning approaches for pansharpening and subsequently present and discuss a few related experimental results.

10.2.1 Survey of Pansharpening Methods Employing Deep Learning

To the best of our knowledge, the first pansharpening method based on convolutional neural networks (CNN), which are a special class of deep learning models suitable for image processing tasks, was proposed in Masi et al. (2016) and it was named PNN. The relative performance gain achieved by PNN, as compared to traditional methods, has encouraged the research community to adopt the same research idea, as testified by numerous works, such as Wei et al. (2017); Yang et al. (2017); Yuan et al. (2018); Shao and Cai (2018); Zhang et al. (2019) to mention a few.

The development of a DL-based method for pansharpening involves at least the following steps:

1. training dataset setup;
2. designing the network architecture;
3. definition of a proper measure for the error (*loss*) to guide the training;
4. training the network via any optimization method;
5. validating the network on a separate dataset.

As a peculiar trait, all CNN-based methods used for pansharpening share a common key problem, which is related to the collection of data for training (1), i.e., a sufficiently "rich" set of input-output examples. In practice, pansharpened images are in fact unavailable and must therefore be synthesized using an ad hoc generation procedure. Figure 10.1 (top) summarizes the main steps of such a process. The underlying idea is a resolution shift that enables the MS component to act as the output rather than the input. This is achieved by appropriately downgrading ($\downarrow_{4\times4}$) the resolution of the MS and PAN components of a training image by a factor equal to the resolution ratio between them, which was 4 in our experiment. Thus, we process input-output pairs in a reduced resolution domain. To complete the process, an image resize ($\uparrow_{4\times4}$) is used to match the size of the MS stack with that of

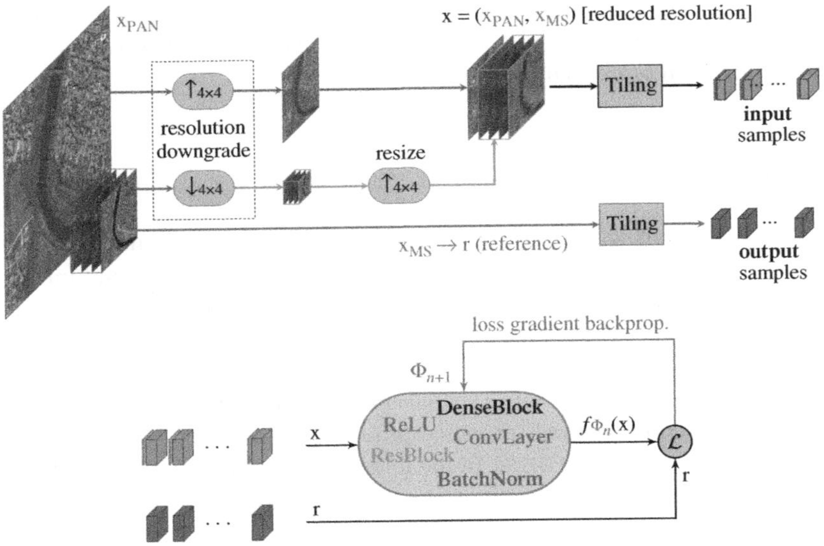

Figure 10.1 Training samples generation workflow (top) and iterative network parameters learning (bottom).

the PAN band, and a tiling phase is used to create mini-batches for the sake of computational efficiency during training. This generation process has quickly become a standard method for CNN-based pansharpening approaches following its introduction in Masi et al. (2016). The interested reader may refer to this for further details. However, it is worth to underline that such a resolution shift will come with a little price to pay as pansharpening is scale invariant only to a little extent. This is particularly evident for data-driven approaches since the same content of the images used for training will strictly depend on their ground sample distance (GSD).

Network architecture design (2) amounts to the definition of a directed acyclic graph (DAG), which defines the input-output information flow, thereby associating a specific task, such as convolution, point-wise nonlinearities (e.g., ReLU), batch normalization, concatenation, sum, and pooling to each DAG vertex. The specific sub-graph structures obtained by combining these elementary operations are also commonly employed (e.g., Residual, Dense, or Inception modules). The PNN model (Masi et al. 2016) is rather simple: a serial net composed of three sequential convolutional units interleaved by ReLU activations. In 2016, this simple architecture exhibited the significant potential of DL in contrast to traditional methods, and achieved state-of-the-art results. It has been further improved by converting it to a residual net named PNN+ (Scarpa et al. 2018). Residual learning, introduced by He et al. (2016a) as a very effective strategy for training very deep models, have quickly proved to be the optimal choice for resolution enhancement (Kim et al. 2016). Moreover, the desired super-resolved image can be viewed as a composition of its low and high frequency components, the former essentially being the input low-resolution image, and the latter being the missing (or *residual*) part that has to be restored. Owing to this partition, residual schemes naturally address problems associated with super-resolution or pansharpening, thereby avoiding an unnecessary reconstruction of the entire desired output and reducing the risk of altering the low-frequency contents of the image. The PanNet model proposed by Yang et al. (2017) is a further improvement in this direction; it eliminates the low-frequency contents from the input stack as well. Additionally, a majority of the recent DL pansharpening models include residual modules (Wei et al. 2017; Yang et al. 2017; Scarpa et al. 2018; Liu et al. 2018c; Shao and Cai 2018; Zhang et al. 2019), although they can be significantly different in terms of complexity. While Scarpa et al. (2018) keep a relatively shallow architecture with only three convolutional layers, other models employ tens of them (Wei et al. 2017; Yang et al. 2017; Liu et al. 2018c). Shao and Cai (2018) have presented a two-branch architecture wherein the MS and PAN components follow different convolutional paths before they are combined, and the influence of the number of convolutional layers per branch is analyzed. Their main conclusion was that the PAN branch should be relatively deeper than the MS one; in particular, the optimal values for the proposed model were eight and two layers, respectively. Finally, it is also worth mentioning a complementary approach such as the U-Net-like model proposed by Yao et al. (2018).

Once the model architecture is fixed, say $f_\Phi(\mathbf{x})$, its parameters Φ have to be learned (4), with the help of a suitably chosen guiding loss functions $\mathcal{L}(f_\Phi(\mathbf{x}), \mathbf{r})$ (3) that quantifies the network error at each iteration over a training mini-batch in order to adjust the parameters by moving in the opposite direction of their gradient:

$$\nabla_\Phi \mathcal{L}(f_{\Phi_n}(\mathbf{x}), \mathbf{r}) \longrightarrow \Phi_{n+1}.$$

The top-level workflow of this process is summarized in the bottom part of Figure 10.1. Multiple optimizers, which are the variants of the stochastic gradient descent algorithm, can be used to this purpose. For the sake of brevity, we skip this point as is not highly relevant to our specific problem, and refer the interested reader to the pansharpening papers discussed above. However, the so-called generative adversarial learning paradigm (Liu et al. 2018c) is an important exception where it is used, which is worth mentioning.

A more interesting aspect is the selection of the loss function. The L2-norm can be considered a standard option as its minimization corresponds to the minimization of the mean squared error. It has been actually adopted in many cases (Masi et al. 2016; Wei et al. 2017; Yao et al. 2018; Shao and Cai 2018; Yuan et al. 2018) for its simplicity and convergence properties. However, the pansharpening quality assessment is still an open issue, likewise the intimately related choice of a proper loss function for an optimal training. In fact, in addition to the classical mean squared (MSE) and the mean absolute (MAE) errors, both spectral oriented measurements with slightly different properties, many other quantitative quality measurements have been proposed during the last decades (see Vivone et al. (2015)). Some of these, like the *erreur relative globale adimensionnelle de synthèse* (ERGAS), a revisited MSE measure with bandwise weighting, or the Spectral Angle Mapper (SAM), are more related to the spectral fidelity, whereas measures such as the spatial cross-correlation, average cross-correlation between gradient images, focus on spatial consistency. Moreover, different loss functions may present different convergence properties during training. For these reasons Scarpa et al. (2018) decided to use the L1-norm achieving better results (less smooth than using L2-norm). The same selection was adopted by Liu et al. (2018c), while (Zhang et al. 2019) have recently proposed to use ERGAS that relates to the root mean squared error (RMSE) as follows:

$$\text{ERGAS} \triangleq 100\frac{1}{R}\sqrt{\frac{1}{L}\sum_{l=1}^{L}\left(\frac{\text{RMSE}(l)}{\mu(l)}\right)^2},$$

where $l \in \{1, \dots, L\}$ indexes the generic band, $\mu(l)$ is the l-th band average, and R is the resolution ratio.

Finally, a validation process (5) completes the network design cycle. Samples reserved for validation, hence "unseen" by the training process, are used both for detecting overfitting and for comparison of different hyper-parameters configurations. Once the training-validation process is stopped, the hyper-/parameters are frozen and the net is ready to be used.

10.2.2 Experimental Results

10.2.2.1 Experimental Design

We conduct the pansharpening experiments by adopting the remote sensing data from WorldView-2 (WV-2) satellites, and evaluate quantitatively and qualitatively the performance in terms of six full-reference measures: average universal image quality index (Q) (Wang and Bovik 2002), 8-band extension of Q (Q8) (Alparone et al. 2004), peak signal-to-noise ratio (PSNR), spectral angle mapper (SAM) (Hong et al. 2019), *erreur relative globale adimensionnelle de synthèse* (ERGAS) (Wald 2000), and spatial correlation coefficient (SCC) (Zhou et al. 1998), and three no-reference indices: quality with no-reference index (QNR) (Alparone et al. 2008), spectral component of QNR ($\mathbf{D}_\lambda$), and spatial component of QNR($\mathbf{D}_S$).

Moreover, three classic non-DL methods (e.g., component substitution based BDSD (Garzelli et al. 2007), multiresolutional analysis based MTF-GLP (Vivone et al. 2014), and variational method based SIRF (Chen et al. 2015)) and four representative DL methods (e.g., PNN (Masi et al. 2016), DRPNN (Wei et al. 2017), PanNet (Yang et al. 2017), and PNN+ (Scarpa et al. 2018)) are selected for the performance comparison. Note that the codes for these methods are available from the websites[1,2,3,4,5,6,7]. The trained models provided by the original developers were fine-tuned for our experiments to maximize the performance. We conduct the experiments with four WV-2 scenes using one fourth region of each scene for training and the rest for testing. The experimental setups are identical to the original setups used in the experiments of the compared methods.

10.2.2.2 Visual and Quantitative Comparison in Pansharpening

Evaluation at reduced-resolution WV-2 images: Owing to the fact that the MS images having the same resolution as the PAN images are not available in the real case, we assess

1 BDSD: http://openremotesensing.net/wp-content/uploads/2015/01/pansharpeningtoolver_1_3.rar
2 MTF-GLP: http://openremotesensing.net/wp-content/uploads/2015/01/pansharpeningtoolver_1_3.rar
3 SIRF: http://cchen156.web.engr.illinois.edu/code/CODE_SIRF.zip
4 PNN: http://www.grip.unina.it/download/prog/PNN/PNN_v0.1.zip
5 DRPNN: https://github.com/Decri/DRPNN-Deep-Residual-Pan-sharpening-Neural-Network
6 PanNet: https://xueyangfu.github.io/paper/2017/iccv/ICCV17_training_code.zip
7 PNN+: https://github.com/sergiovitale/pansharpening-cnn

Table 10.1 Performance comparison of three non-DL and four DL methods at reduced (reference) and full (no-reference) resolutions on the WV-2 datasets. The best results are shown in bold.

Method	Reference						No-reference		
	Q8	Q	PSNR	SAM	ERGAS	SCC	QNR	D_λ	D_S
BDSD	0.888	0.893	28.719	6.319	4.899	0.878	0.894	0.045	0.064
MTF-GLP	0.907	0.910	30.114	5.745	4.163	0.902	0.918	0.040	0.049
SIRF	0.894	0.901	31.041	5.925	3.984	0.897	0.911	0.053	0.039
PNN	0.907	0.921	30.777	6.559	4.155	**0.913**	0.927	0.032	0.043
DRPNN	0.922	0.933	31.016	5.822	3.792	0.913	0.921	0.031	0.050
PanNet	0.916	0.916	30.392	**5.448**	3.976	0.887	0.939	**0.021**	0.042
PNN+	**0.923**	**0.933**	**31.598**	5.796	**3.743**	0.905	**0.949**	0.025	**0.027**
Ideal value	↑ 1	↑ 1	↑	↓ 0	↓ 0	↑ 1	↑ 1	↓ 0	↓ 0

the model performance at a reduced-resolution dataset. Therefore, Wald's protocol (Wald et al. 1997) is applied for simulating this experimental process.

Table 10.1 details the quantitative comparison of the three non-DL and the four DL-based pansharpening methods on the WV-2 datasets. Overall, the pansharpening performance of the non-DL approaches is inferior to that of the DL approaches. To some extent, this demonstrates the powerful ability of DL-based techniques in learning the complex processes of image upsampling. We provide additional detailed discussions and analyses for these DL-based methods as follows. The result of image fusion when using baseline PNN is competitive, particularly SCC, wherein PNN achieves the best performance. Pansharpening MS remote sensing images is a challenging task, wherein additional domain-related knowledge or priors should be considered. Hence, PanNet works on image residuals (high-pass components) to preserve spectral and spatial information, thereby yielding a desirable performance for several important indices (e.g., Q8 and SAM). Owing to its deeper network architecture and residual learning strategy, DRPNN achieves a significant improvement in the quality of the pansharpened image. It is worth noting that PNN+ extends PNN by processing with image residuals. This enables PNN+ to pansharpen the new MS images with a higher generalization ability, as listed in Table 10.1, where PNN+ outperforms the others for a majority of the assessment indices. In addition, we also visualize the pansharpened products and the corresponding residual images with the ground truth (GT) on an example of the WV-2 image, as shown in Figure 10.2.

Evaluation at full-resolution of WV-2 images: In the pansharpening task, we expect to observe the performance of these developed methods at the original resolution. Therefore, we directly use these resulting models trained at the reduced resolution images to resolve the WV-2 images at the full-resolution scale.

As the high-resolution MS images are not available, a reference-free index: QNR and its variants are used to quantitatively assess the performance, as listed in Table 10.1. As expected, although these traditional pansharpening methods based on multiresolution analysis or variational inference maintain a relatively good performance, they are evidently

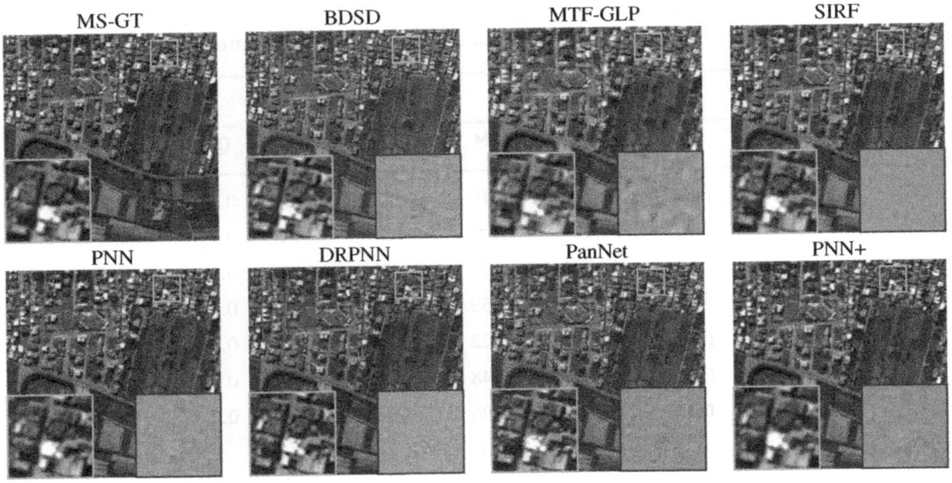

Figure 10.2 Pansharpening results with different compared methods at a reduced resolution WV-2 image. An enlarged region is framed on the bottom-left of each image and the corresponding residual image between the fused image and MS-GT is framed on the bottom-right of each image.

incomparable to DL-based methods. Moreover, there is a significant performance gain for the PanNet and PNN+ methods at the original scale, with the no-reference measures, as compared to the other methods. This can be attributed to the use of processing with image residuals to jointly preserve spatial-spectral information. Furthermore, Figure 10.3 presents the pansharpening results on a given WV-2 image, similar to Figure 10.2, where a similar trend of the quantitative assessment exists. As compared to the others, the fused images obtained by PanNet and PNN+ exhibit better visual effects that are sharper at the

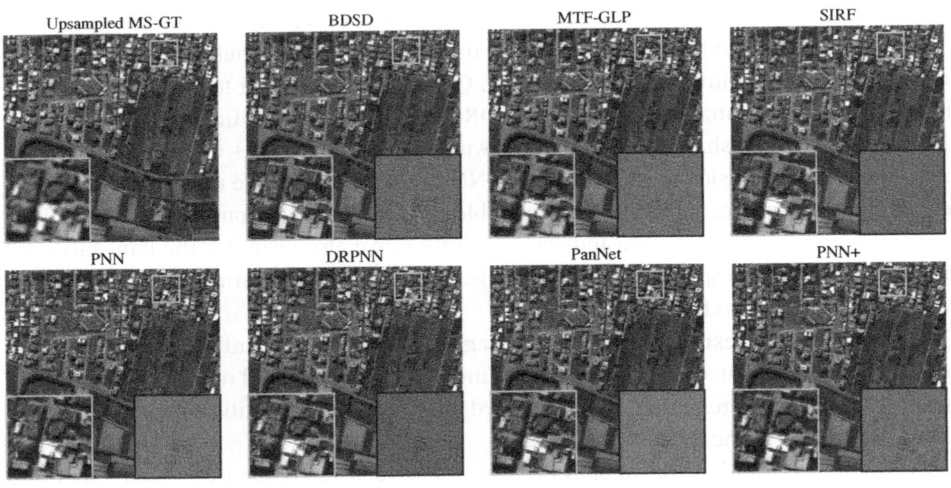

Figure 10.3 Pansharpening results with different compared methods at a full-resolution WV-2 image. An enlarged region is framed on the bottom-left of each image and the corresponding residual image between the fused image and MS-GT is framed on the bottom-right of each image.

Table 10.2 Processing time comparison of three non-DL and four DL methods in the test phase.

Method	BDSD	MTF-GLP	SIRF	PNN	DRPNN	PanNet	PNN+
Running Time (s)	0.1278	0.1729	9.3736	0.1095	0.1184	0.1036	0.1101

edges and also yield lower residual errors between the MS-GT up-sampled via bi-cubic interpolation.

Computational time: All test experiments in this chapter were implemented on a Windows 10 operating system and conducted on Intel Core i7-8700K 3.70GHZ desktop with 64GB memory. With the setting, the running time for those compared non-DL and DL methods is given in Table 10.2. Overall, the running time of DL-based techniques is basically superior to that of non-DL ones, particularly optimization-based approaches (e.g., SIRF). Remarkably, those DL-based methods hold a similar and faster running speed in practice, owing to the linear processing time of their feed-forward mechanism in the inference process.

10.3 Multiband Image Fusion

Hyperspectral (HS) and multispectral (MS) data fusion is a typical example of multiband image fusion and has been extensively studied in the field of remote sensing. Conventional methods formulate the fusion problem as an optimization problem by using hand-crafted priors. The performance of the methods significantly rely on these hand-crafted priors. However, the hand-crafted priors require prior knowledge of the latent high-resolution HS (HR-HS) image. A few HS and MS data-fusion methods based on deep learning have been developed; these methods require fewer assumptions with respect to hand-crafted-based methods on the latent HR-HS image. The DL-based methods can be categorized into two main approaches: (i) supervised DL-based approaches (Xie et al. 2019; Dian et al. 2018; Palsson et al. 2017; Han et al. 2018) and (ii) unsupervised DL-based approaches (Qu et al. 2018; Fu et al. 2019). The supervised approaches assume that the reference HR-HS images are available and minimize the training loss between the reference HR-HS and the estimated HR-HS images in an end-to-end manner. The unsupervised approaches do not require the reference HR-HS image. In this chapter, the term "unsupervised approaches" refers to the DL-based methods that do not require training data. These approaches consider the reconstruction loss between the HS and MS images, derived from the estimated HR-HS and the observed HS and MS images. In the following sub-section, the most recent supervised and unsupervised DL-based methods are briefly discussed.

10.3.1 Supervised Deep Learning-based Approaches

The supervised approaches mainly comprise the following two steps.

1. training a CNN model using paired training data;
2. computing an HR-HS image by using the trained CNN model.

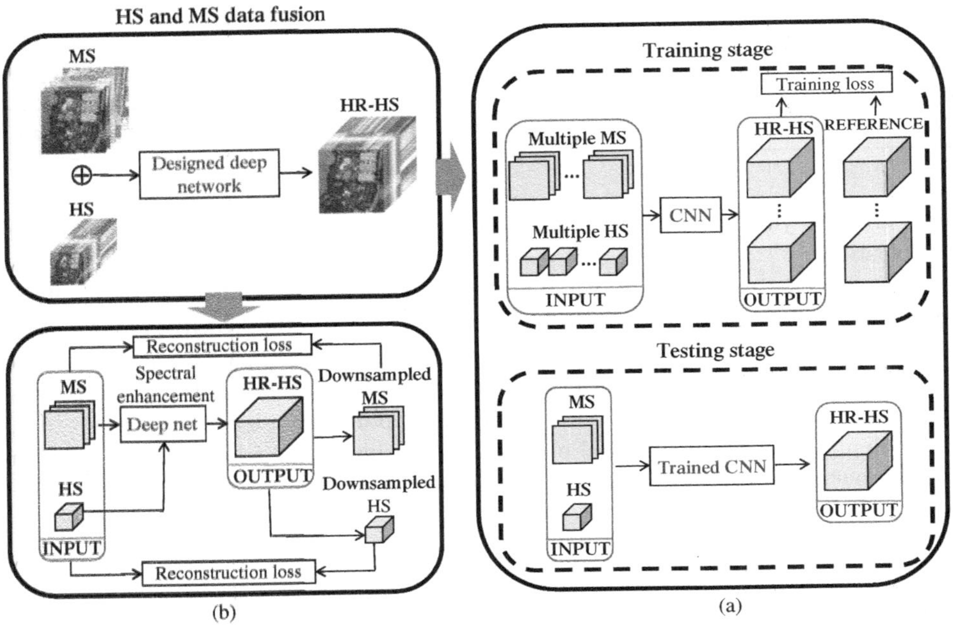

Figure 10.4 Example of HS and MS data fusion: (a) supervised approaches and (b) unsupervised approaches.

The supervised approaches commonly assume that the paired training data are available. The training data used in the supervised approaches include low spatial resolution HS (LR-HS) and MS images as the inputs and the HR-HS images as the outputs. In the methods (Han et al. 2018; Palsson et al. 2017), the LR-HS and MS images are simply concatenated as a single input after applying spatial resampling. In the other methods (Xie et al. 2019; Dian et al. 2018), the LR-HS and MS images are separately incorporated as inputs in the optimization process.

The architecture of the network differs significantly, depending on the method. The spectral spatial fusion architectures of CNN (SSF-CNN) (Han et al. 2018) are aimed at learning the nonlinear relationship between the concatenated LR-HS and MS images and the HR-HS images, using CNN. Additionally, the training loss between the estimated and reference HR-HS images is considered. Once the model has been trained, the HR-HS images are computed using the trained CNN for a new given input (i.e., the concatenated LR-HS and MS images).

A 3D convolutional neural network (3D-CNN) (Palsson et al. 2017) also adopts a similar approach. One noticeable difference is that 3D-CNN uses different training data. The input data are spatially decimated HS and MS images, while the observed HS image is used as the target HR-HS images. The similar trick is also explored in the pansharpening. The relationship is learned by using 3D-CNN in an end-to-end manner. Once the model has been trained, the observed HS and MS images are provided as inputs to compute HR-HS images. This is based on the assumption that the relationship learned by 3D-CNN at a low resolution can also be applicable at higher resolutions.

The aforementioned two models possess a simple network architecture and perform well in the experiments. However, the CNN models used are not specifically designed for the MS/HS fusion problem. MS/HS Fusion Net (MHF-net) (Xie et al. 2019) was proposed to incorporate the following observation models:

$$\mathbf{Y} = \mathbf{XR} + \mathbf{N}_y, \tag{10.1}$$

$$\mathbf{Z} = \mathbf{CX} + \mathbf{N}_z, \tag{10.2}$$

where $\mathbf{Y}$ is the observed MS image, $\mathbf{X}$ is the HR-HS image, $\mathbf{R}$ is the spectral response of the multispectral sensor, $\mathbf{Z}$ is the observed HS image, $\mathbf{C}$ is the linear operator that is composed of a cyclic convolution operator and a downsampling operator, and $\mathbf{N}_y$ and $\mathbf{N}_z$ represent noise present in the MS and HS images, respectively. MHF-net formulates a new optimization problem from the observation models and shows that the optimization problem can be solved by a specifically designed deep network. MHF-net automatically estimates the parameters related to the downsampling operator within a supervised deep learning framework. MHF-net can exploit the general prior structure of the latent HS images (e.g., low-rankness) and also enables each step of the network architecture to be interpretable.

A deep HSI sharpening method DHSIS which can incorporate the observation model has been developed (Dian et al. 2018). DHSIS comprises three steps. The first step estimates the initial HR-HS image from a conventional optimization problem derived by the observation model. In the second step, it learns image priors representing the mapping between the initialized HR-HS image and the reference HR-HS image using deep residual learning. Finally, the learned image priors are incorporated into an image fusion optimization framework to reconstruct the final HR-HS image.

10.3.2 Unsupervised Deep Learning-based Approaches

Although the aforementioned methods have achieved state-of-the-art performances, the models assume that training data are available, which may not be true in remote sensing applications. In addition, the learned nonlinear mapping function is only suitable for one combination of specific sensors. In this sub-section, the unsupervised DL-based approaches that have been developed to address these problems are discussed. Unsupervised approaches commonly consider the reconstruction loss between the downsampled HR-HS images and the observed MS or HS images because the ground truth HR-HS image is not available. To solve the highly ill-posed HS/MS fusion problem, a few network architectures have been previously developed.

An unsupervised CNN-based method has been proposed in Fu et al. (2019). The method assumes that the observed RGB and HS images act as the inputs, and the HR-HS image as the output. However, the HR-HS is unknown in advance. Therefore, this method proposed an additional spectral network to learn the spectral response function from the HR-HS values to RGB values. The loss function is built as the similarity between the simulated RGB image and the input original RGB image. The parameters of CNN are optimized in the end-to-end network.

An unsupervised sparse Dirichlet net (uSDN) (Qu et al. 2018) has been proposed to address the HS and MS image fusion problem. uSDN is based on an unsupervised encoder-decoder architecture. The architecture comprises two encoder-decoder networks. One network encodes and decodes an HS image, whereas the other network encodes and decodes an MS image. The decoder is shared between the two networks, and the network activations derived from the encoders are promoted to have a similar pattern. In this architecture, the HS and MS images are encoded as proportional coefficients, and the decoder represents spectral signatures corresponding to the coefficients. The parameters of the encoder and the decoder are alternatively optimized until it converges to an HR-HS image.

10.3.3 Experimental Results

10.3.3.1 Comparison Methods and Evaluation Measures

We select the following low-rank matrix/tensor related methods: coupled non-negative matrix factorization (CNMF) (Yokoya et al. 2012)[8], FUSE (Wei et al. 2016)[9], HySure (Simões et al. 2015)[10], coupled CP factorization (STEREO) (Kanatsoulis et al. 2018)[11], and coupled Tucker factorization (CSTF) (Li et al. 2018a)[12], to compare with the DL-based methods, unsupervised uSDN (Qu et al. 2018)[13], and supervised MHF-net (Xie et al. 2019)[14]. We use the peak signal-to-noise ratio (PSNR), root mean square error (RMSE), relative dimensional global error in synthesis (ERGAS) (Wald 2000), spectral angle mapper (SAM), and structure similarity (SSIM) (Wang et al. 2004) as evaluation criteria for the SR results of different methods.

10.3.3.2 Dataset and Experimental Setting

We select two datasets for the experiments. The first is the Chikusei dataset[15], which was obtained at Chikusei, Ibaraki, Japan, on 29 July 2014. The size of the Chikusei dataset is $2517 \times 2335 \times 128$. We select a subimage of size $500 \times 2210 \times 128$ for the training of the MHF-net method, and a subimage of size $448 \times 544 \times 128$ for its testing. The second is the CAVE multispectral dataset[16], from which we selected 20 images for the training of the MHF-net method, and the rest 12 CAVE images for the testing. The size of the simulated HSIs and MSIs of the test Chikusei dataset and the CAVE images are presented in Table 10.3. The spectral downsampling in the Chikusei dataset is the same as that in Xie et al. (2019), and the spatial downsampling scale is set as 4. For the CAVE dataset, the spatial downsampling is the same as that in Xie et al. (2019), and the spectral downsampling matrix is the spectral response matrix of Nikon D700 (Qu et al. 2018).

---

8 http://naotoyokoya.com/Download.html
9 https://github.com/qw245/BlindFuse
10 https://github.com/alfaiate
11 https://sites.google.com/site/harikanats/
12 https://sites.google.com/view/renweidian/
13 https://github.com/aicip/uSDN
14 https://github.com/XieQi2015/MHF-net
15 http://naotoyokoya.com/Download.html
16 http://www1.cs.columbia.edu/CAVE/databases/

Table 10.3 The size of the image used for HSR experiments.

Image name	HR-HSI	HSI	MSI
Chikusei	448×544×128	112×128×128	448×544×3
CAVE	512×512×31	16×16×31	512×512×3

Table 10.4 Quantitative comparison of different algorithms on two different images. The best results are marked in bold.

Method	Chikusei					CAVE				
	PSNR	RMSE	ERGAS	SAM	SSIM	PSNR	RMSE	ERGAS	SAM	SSIM
CNMF	35.21	5.09	6.54	5.591	0.906	34.40	5.84	0.77	8.05	0.942
FUSE	34.98	5.79	6.83	6.418	0.859	34.88	5.94	0.71	12.24	0.902
HySure	35.85	5.17	6.23	**5.698**	0.904	34.45	6.50	0.79	19.43	0.897
STEREO	34.66	6.52	7.40	10.272	0.832	34.56	5.10	0.80	17.09	0.907
CSTF	36.39	5.37	6.11	7.616	0.855	38.46	4.75	0.54	11.86	0.939
uSDN	34.04	6.30	7.48	7.055	0.888	33.17	6.38	0.83	14.21	0.905
MHF-net	**39.52**	**4.09**	**4.52**	5.896	**0.932**	**39.22**	**3.34**	**0.39**	**6.38**	**0.978**
Ideal vaule	↑	↓0	↓0	↓0	↑1	↑	↓0	↓0	↓0	↑1

10.3.3.3 Quantitative Comparison and Visual Results

Table 10.4 presents the quantitative comparison between DL-based methods and low-rank matrix/tensor related methods for the Chikusei and CAVE toy images. From the table, it can be observed that MHF-net achieves good reconstruction results for both cases and outperforms uSDN. It is worth noting that MHF-net requires a sufficient training dataset to train the model; thus, the quality and the similarity between the training and test datasets have a significant influence on the final results of the test. For the CAVE test image, MHF-net achieves the best results in difference indices. In summary, the DL-based methods are promising for the HSI and MSI fusion task. Figure 10.5 presents the SR results of different methods, and the difference images between the different SR image and the original image. From the figure, we conclude that MHF-net achieves the best visual results.

Computational time: Non-DL methods are programmed in Matlab R2018b on a laptop with CPU Corei7-8750H 32G memory, and the DL methods are programmed in Python on a laptop with a single GTX1080 GPU. The running time of those comparison methods are presented in Table 10.5. For MHF-net, the test time is reported in the table, meanwhile, the training time is about 20 hours. From the table, it can be concluded that the supervised MHF-net is really fast in the test stage, indicating the possibility of real-time fusion via DL methods in the future.

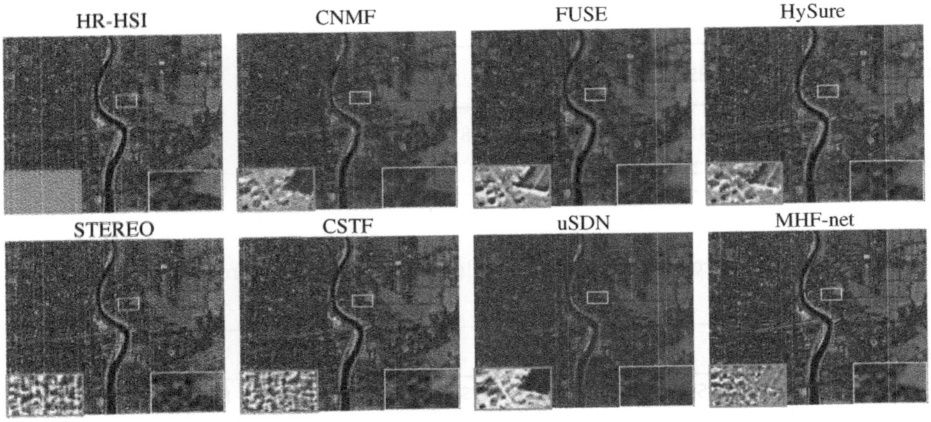

Figure 10.5 HSR results of different methods with Chikusei image. We choose bands 70, 100, 36 for illustration. An enlarged region is framed on the bottom-right of each image and the corresponding residual image between the fused image and MS-GT is framed on the bottom-left of each image.

Table 10.5 Processing time compassion of five non-DL and two DL methods in the test phase.

Method	CNMF	FUSE	HySure	STEREO	CSTF	uSDN	MHF-net
Running Time (s)	31	16	303	21	135	273	1

10.4 Conclusion and Outlook

This chapter presents a review of state-of-the-art DL-based image fusion techniques for two spatial-spectral resolution enhancement tasks, namely, pansharpening and multiband image fusion. The DL-based methods exhibit the capability to learn complex data transformations from input image sources to the targets. Unlike the conventional approaches that are based on hand-crafted filters and priors (or regularizations), DL-based methods have the potential to learn priors from the training samples in an end-to-end manner. However, a careful design of the network architecture and a definition of the loss function are required for the DL-based methods. The popular concepts of conventional approaches, such as the injection of spatial details and spatial-spectral preservation based on observation models, can facilitate more efficient learning. As demonstrated in the comparative experiments, the DL-based algorithms achieved higher reconstruction quality as compared to the conventional approaches, with a relatively fast speed at the inference. Hence, DL-based image fusion is suitable for processing large-scale optical satellite images.

A limitation of DL-based approaches is that a majority of these methods require numerous input-output training samples for each combination of the sensors, which cannot be easily obtained. Owing to this limitation, generalization so as to accept a pair of images acquired by any combination of sensors as the input remains a key challenge from a practical point of view. Unsupervised DL approaches are a potential solution to address this challenge because they can be trained by only using the input data; however, their computational cost is high and there is still room for improvement in their reconstruction performance. Transfer learning is a possible direction to improve computational efficiency and reconstruction performance of unsupervised DL approaches. A majority of the architectures developed for multisource remote sensing image fusion have been manually designed by humans. An automated neural architecture search can be another direction for future research.

11

Deep Learning for Image Search and Retrieval in Large Remote Sensing Archives

Gencer Sumbul, Jian Kang, and Begüm Demir

11.1 Introduction

With the unprecedented advances in the satellite technology, recent years have witnessed a significant increase in the volume of remote sensing (RS) image archives. Thus, the development of efficient and accurate content-based image retrieval (CBIR) systems in massive archives of RS images is a growing research interest in RS. CBIR aims to search for RS images of the similar information content within a large archive with respect to a query image. To this end, CBIR systems are defined based on two main steps: (i) image description step (which characterizes the spatial and spectral information content of RS images); and (ii) image retrieval step (which evaluates the similarity among the considered descriptors and then retrieve images similar to a query image in the order of similarity). A general block scheme of a CBIR system is shown in Figure 11.1.

Traditional CBIR systems extract and exploit hand-crafted features to describe the content of RS images. As an example, bag-of-visual-words representations of the local invariant features extracted by the scale invariant feature transform (SIFT) are introduced in Yang and Newsam (2013). In Aptoula (2014), a bag-of-morphological-words representation of the local morphological texture features (descriptors) is proposed in the context of CBIR. Local Binary Patterns (LBPs), which represent the relationship of each pattern (i.e., pixel) in a given image with its neighbors located on a circle around that pixel, are found very efficient in RS. In Tekeste and Demir (2018), a comparative study that analyzes and compares different LBPs in RS CBIR problems is presented. To define the spectral information content of high-dimensional RS images the bag-of-spectral-values descriptors are presented in Dai et al. (2018). Graph-based image representations, where the nodes describe the image region properties and the edges represent the spatial relationships among the regions, are presented in Li and Bretschneider (2007); Chaudhuri et al. (2016, 2018). Hashing methods that embed high-dimensional image features into a low-dimensional Hamming (binary) space by a set of hash functions are found very effective in RS (Demir and Bruzzone 2016; Li and Ren 2017; Reato et al. 2019). By this method, the images are represented by binary hash codes that can significantly reduce the amount of memory required for storing the RS images with respect to the other descriptors. Hashing methods differ from each other on how the hash functions are generated. As an example, in Demir and Bruzzone (2016);

Deep Learning for the Earth Sciences: A Comprehensive Approach to Remote Sensing, Climate Science, and Geosciences, First Edition. Edited by Gustau Camps-Valls, Devis Tuia, Xiao Xiang Zhu, and Markus Reichstein.

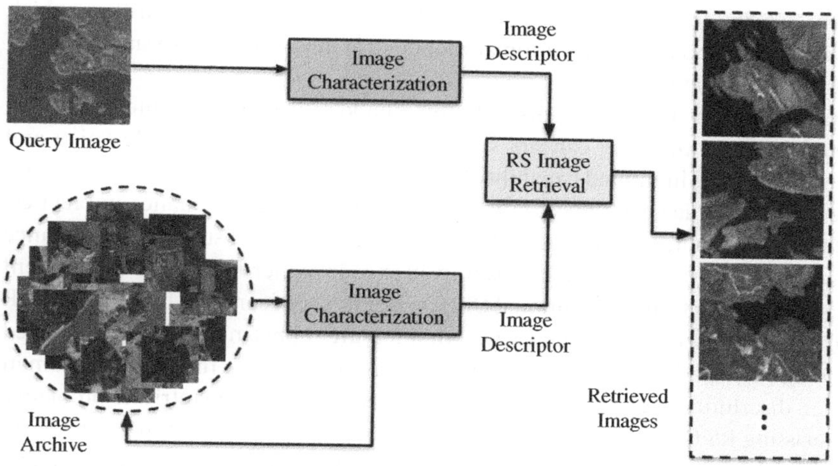

Figure 11.1 General block scheme of a RS CBIR system.

Reato et al. (2019) kernel-based hashing methods that define hash functions in the kernel space are presented, whereas a partial randomness hashing method that defines the hash functions based on a weight matrix defined using labeled images is introduced in Li and Ren (2017). More details on hashing for RS CBIR problems are given in section 11.3.

Once image descriptors are obtained, one can use the k-nearest neighbor (k-NN) algorithm, which computes the similarity between the query image and all archive images to find the k most similar images to the query. If the images are represented by graphs, graph matching techniques can be used. As an example, in Chaudhuri et al. (2016) an inexact graph matching approach, which is based on the sub-graph isomorphism and spectral embedding algorithms, is presented. If the images are represented by binary hash codes, image retrieval can be achieved by calculating the Hamming distances with simple bit-wise XOR operations that allow time-efficient search capability (Demir and Bruzzone 2016). However, these unsupervised systems do not always result in satisfactory query responses due to the semantic gap, which is occurred among the low-level features and the high-level semantic content of RS images (Demir and Bruzzone 2015). To overcome this problem and improve the performance of CBIR systems, semi-supervised and fully supervised systems, which require user feedback in terms of RS image annotations, are introduced (Demir and Bruzzone 2015). Most of these systems depend on the availability of training images, each of which is annotated with a single broad category label that is associated to the most significant content of the image. However, RS images typically contain multiple classes and thus can simultaneously be associated with different class labels. Thus, CBIR methods that properly exploit training images annotated by multi-labels are recently found very promising in RS. As an example, in Dai et al. (2018) a CBIR system that exploits a measure of label likelihood based on a sparse reconstruction-based classifier is presented in the framework of multi-label RS CBIR problems. Semi-supervised CBIR systems based on graph matching algorithms are proposed in Wang et al. (2016); Chaudhuri et al. (2018). In detail, in Wang et al. (2016) a three-layer framework in the context graph-based learning is proposed for query expansion and fusion of global and local features by using the label information of query images. In Chaudhuri et al. (2018) a correlated label propagation algorithm, which operates on a neighborhood graph for automatic labeling of images by using a small number of training images, is proposed.

The above-mentioned CBIR systems rely on shallow learning architectures and hand-crafted features. Thus, they can not simultaneously optimize feature learning and image retrieval, resulting in limited capability to represent the high-level semantic content of RS images. This issue leads to inaccurate search and retrieval performance in practice. Recent advances in deep neural networks (DNNs) have triggered substantial performance gain for image retrieval due to their high capability to encode higher-level semantics present in RS images. Differently from conventional CBIR systems, deep learning (DL)-based CBIR systems learn image descriptors in such a way that feature representations are optimized during the image retrieval process. In order words, DNNs eliminate the need for human effort to design discriminative and descriptive image descriptors for the retrieval problems. Most of the existing RS CBIR systems based on DNNs attempt to improve image retrieval performance by: (i) learning discriminative image descriptors; and (ii) achieving scalable image search and retrieval. The aim of this chapter is to present different DNNs proposed in the literature for the retrieval of RS images. The rest of this chapter is organized as follows. Section 11.2 reviews the DNNs proposed in the literature for the description of the complex information content of RS images in the framework of CBIR. Section 11.3 presents the recent progress on the scalable CBIR systems defined based on DNNs in RS. Finally, section 11.4 draws the conclusion of this chapter.

11.2 Deep Learning for RS CBIR

The DL-based CBIR systems in RS differ from each other in terms of: (i) the strategies considered for the mini-batch sampling; (ii) the approaches used for the initialization of the parameters of the considered DNN model; (iii) the type of the considered DNN; and (iv) the strategies used for image representation learning. Figure 11.2 illustrates the main approaches utilized in DL-based CBIR systems in RS. In detail, a set of training images is initially selected from the considered archive to train a DNN. Then, the selected training images are divided into mini-batches and fed into the considered DNN. After initializing the model parameters of the network, the training phase is conducted with an iterative estimation of the model parameters based on a loss function. The loss function is selected on the basis of the characteristics of the considered learning strategy.

During the last years, several DL-based CBIR systems that consider different strategies for the above-mentioned factors are presented. As an example, in Zhou et al. (2015) an unsupervised feature learning framework that learns image descriptors from a set of unlabeled RS images based on an autoencoder (AE) is introduced. After random selection of mini-batches and initialization of the model parameters, SIFT-based image descriptors are encoded into sparse descriptors by learning the reconstruction of the descriptors. The learning strategy relies on minimization of a reconstruction loss function between the SIFT descriptors and the reconstructed image descriptors in the framework of the AE. A CBIR system that applies a multiple feature representation learning and a collaborative affinity metric fusion is presented in Li et al. (2016b). This system randomly selects RS images

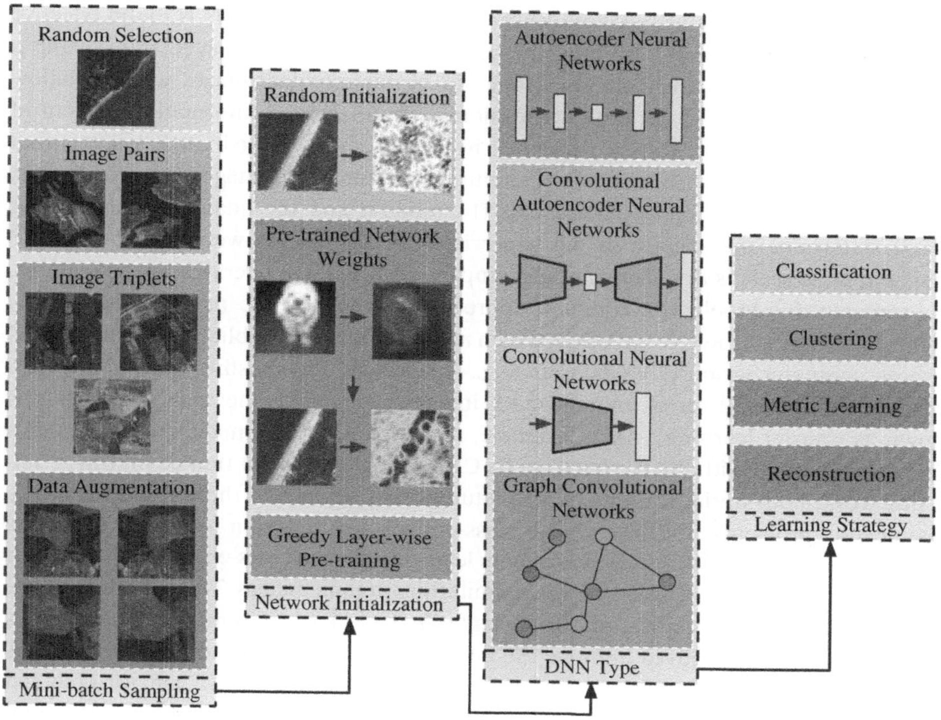

Figure 11.2 Different strategies considered within the DL-based RS CBIR systems.

for mini-batches and initializes the model parameters of a Convolutional Neural Network (CNN). Then, it employs the CNN for k-means clustering (instead of classification). To this end, a reconstruction loss function is utilized to minimize the error induced between the CNN results and the cluster assignments. Collaborative affinity metric fusion is employed to incorporate the traditional image descriptors (e.g., SIFT, LBP) with those extracted from different layers of the CNN. A CBIR system with deep bag-of-words is proposed in Tang et al. (2018b). This system employs a convolutional autoencoder (CAE) for extracting image descriptors in an unsupervised manner. The method first encodes the local areas of randomly selected RS images into a descriptor space and then decodes from descriptors to image space. Since encoding and decoding steps are based on convolutional layers, a reconstruction loss function is directly applied to reduce the error between the input and constructed local areas for the unsupervised reconstruction based learning. Since this system operates on local areas of the images, bag-of-words approach with k-means clustering is applied to define the global image descriptor from local areas. Although this system has the same learning strategy as Zhou et al. (2015), its advantages are two-fold compared to Zhou et al. (2015). First, model parameters are initialized with greedy layer-wise pre-training that allows more effective learning procedure with respect to the random initialization approach. Second, the CAE model has better capability to characterize the semantic content of images since it considers the neighborhood relationship through the convolution operations. The reader is referred to Chapter 2 for the detailed discussion on unsupervised feature learning in RS.

Reconstruction based unsupervised learning of RS image descriptors is found effective particularly when annotated training images are not existing. However, minimizing a reconstruction loss function on a small amount of unannotated images with a shallow neural network limits the accurate description of the high-level information content of RS images. This problem can be addressed by supervised DL-based CBIR systems that require a training set that consists of a high number of annotated images to learn effective models with several different parameters. The amount and the quality of the training images determine the success of the supervised DL models. However, annotating RS images at large scale is time-consuming, complex, and costly in operational applications. To overcome this problem, a common approach is to exploit DL models with proven architectures (such as ResNet or VGG), which are pre-trained on publicly available general purpose computer vision (CV) datasets (e.g., ImageNet). The existing models are then fine-tuned on a small set of annotated RS images to calibrate the final layers (this is known as transfer learning). As an example, in Hu et al. (2016) model parameters of a CNN are initialized with the parameters of a CNN model that is pre-trained on ImageNet. In this work, both initial training and fine-tuning are applied in the framework of the classification problems. To this end, the cross-entropy loss function is utilized to reduce the class prediction errors. Image descriptors learned with the cross-entropy loss function encode the class discrimination instead of similarities among images. Thus, it can limit the performance of CBIR systems. A data augmentation technique for mini-batch sampling is utilized in Hu et al. (2016) to improve the effectiveness of the image descriptors. To this end, different scales of RS images in a mini-batch are fed into the CNN. Then, the obtained descriptors are aggregated by using different pooling strategies to characterize the final image descriptors. A low-dimensional convolutional neural network (LDCNN) proposed in Zhou et al. (2017a) also utilizes parameters of a pre-trained DL model to initialize the network parameters. However, it randomly selects RS images for the definition of mini-batches and adopts a classification based learning strategy with the cross-entropy loss function. This system combines convolutional layers with cross channel parametric pooling and global average pooling to characterize low-dimensional descriptors. Since fully connected (FC) layers are replaced with pooling layers, LDCNN significantly decreases the total number of model parameters required to be estimated during the training phase. This leads to significantly reduced computational complexity and also reduced risk of over-fitting (which can occur in the case of training CNNs with a small amount of training images). A CBIR system based on a CNN with weighted distance is introduced in Ye et al. (2018). Similar to Hu et al. (2016) and Zhou et al. (2017a), this system also applies fine-tuning on a state-of-the-art CNN model pre-trained on ImageNet. In addition, it enhances the conventional distance metrics used for image retrieval by weighting the distance between a query image and the archive images based on their class probabilities obtained by a CNN. An enhanced interactive RS CBIR system, which extracts the preliminary RS image descriptors based on the LDCNN by utilizing the same mini-batch sampling, network initialization and learning strategy (based on the cross-entropy loss function), is introduced in Boualleg and Farah (2018). Labeled training images are utilized to obtain the preliminary image descriptors. Then, a relevance feedback scheme is applied to further improve the effectiveness of the image descriptors by considering the user feedbacks on the automatically retrieved images. The use of aggregated deep local features for RS image retrieval is proposed in Imbriaco et al. (2019). To this end, the VLAD representation of local convolutional descriptors from multiplicative and additive attention mechanisms are considered to characterize the descriptors of the most relevant regions of the RS images. This is achieved based on three steps. In the first step, similar to Ye et al. (2018) and Zhou et al. (2017a), the system operates on randomly selected RS images and applies fine-tuning to a state-of-the art CNN model while relying on a classification based learning strategy with the cross-entropy loss function. In the second step, additive and multiplicative attention mechanisms are integrated into the convolutional layers of the CNN and thus are retrained to learn their parameters. Then, local descriptors are characterized based on the attention scores of the resized RS images at different scales (which is achieved based on data augmentation). In the last step, the system transforms VLAD representations with Memory Vector (MV) construction (which produces the expanded query descriptor) to make the CBIR system sensitive to the selected query images. In this system, the query expansion strategy is applied after obtaining all the local descriptors. This query-sensitive CBIR approach further improves the discrimination capability of image descriptors, since it adapts the overall learning procedure of DNNs based on the selected queries. Thus, it has a huge potential for RS CBIR problems.

Most of the above-mentioned DL-based supervised CBIR systems learn an image feature space directly optimized for a classification task by considering entropy-based loss functions. Thus, the image descriptors are designed to discriminate the pre-defined classes by taking into account the class based similarities rather than the image based similarities during the training stage of the DL models. The absence of positive and negative images with respect to the selected query image during the training phase can lead to a poor CBIR performance. To overcome this limitation, metric learning is recently introduced in RS to take into account image similarities within DNNs. Accordingly, a Siamese graph convolutional network is introduced in Chaudhuri et al. (2019) to model the weighted region adjacency graph (RAG) based image descriptors by a metric learning strategy. To this end, mini-batches are first constructed to include either similar or dissimilar RS images (Siamese pairs). If a pair of images belongs to the same class, they are assumed as similar images, and vice versa. Then, RAGs are fed into two graph convolutional networks with shared parameters to model image similarities with the contrastive loss function. Due to the considered metric learning strategy (which is guided by the contrastive loss function) the distance between the descriptors of similar images is decreased, while that between dissimilar images is increased. The contrastive loss function only considers the similarity estimated among image pairs, i.e., similarities among multiple images are not evaluated, which can limit the success of similarity learning for CBIR problems.

To address this limitation, a triplet deep metric learning network (TDMLN) is proposed in Cao et al. (2020). TDMLN employs three CNNs with shared model parameters for similarity learning through image triplets in the content of metric learning. Model parameters of the TDMLN are initialized with a state-of-the-art CNN model pre-trained on ImageNet. For the mini-batch sampling, TDMLN considers an anchor image together with a similar (i.e., positive) image and a dissimilar (i.e., negative) image to the anchor image at a time. Image triplets are constructed based on the annotated training images (Chaudhuri et al. 2019). While anchor and positive images belong to the same class, the negative image is associated to a different class. Then, similarity learning of the triplets is achieved based on the triplet loss function. By the use of triplet loss function, the distance estimated between

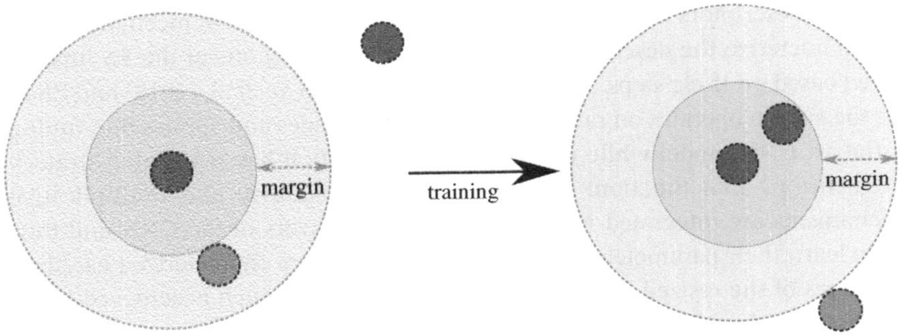

Figure 11.3 The intuition behind the triplet loss function: after training, a positive sample is moved closer to the anchor sample than the negative samples of the other classes.

the anchor and positive images in the descriptor (i.e., feature) space is minimized, whereas that computed between the anchor and negative images is separated by a certain margin. Figure 11.3 illustrates intuition behind the triplet loss function. Metric learning guided by the triplet loss function learns similarity based on the image triplets and thus provides highly discriminative image descriptors in the framework of CBIR. However, how to define and select image triplets is still an open question. Current methods rely on the image-level annotations based on the land-cover land-use class labels, which do not directly represent the similarity of RS images. Thus, metric learning-based CBIR systems need further improvements to characterize retrieval specific image descriptors. One possible way to overcome this limitation can be an identification of image triplets through visual interpretation instead of defining triplets based on the class labels. Tabular overview of the recent DL-based CBIR systems in RS is presented in Table 11.1.

11.3 Scalable RS CBIR Based on Deep Hashing

Due to the significant growth of RS image archives, an image search and retrieval through linear scan (which exhaustively compares the query image with each image in the archive) is computationally expensive and thus impractical. This problem is also known as large-scale CBIR problem. In large-scale CBIR, the storage of the data is also challenging as RS image contents are often represented in high-dimensional features. Accordingly, in addition to the scalability problem, the storage of the image features (descriptors) also becomes a critical bottleneck. To address these problems, approximate nearest neighbor (ANN) search has attracted extensive research attention in RS. In particular, hashing-based ANN search schemes have become a cutting-edge research topic for large-scale RS image retrieval due to their high efficiency in both storage cost and search/retrieval speed. Hashing methods encode high-dimensional image descriptors into a low-dimensional Hamming space where the image descriptors are represented by binary hash codes. By this way, the (approximate) nearest neighbors among the images can be efficiently identified based on the Hamming distance with simple bit-wise operations. In addition, the binary codes can significantly reduce the amount of memory required for storing the content of images. Traditional hashing-based RS CBIR systems initially extract hand-crafted image descriptors and then utilize hash functions that map the original high-dimensional

Table 11.1 Main characteristics of the DL-based CBIR systems in RS.

Reference	Mini-batch Sampling	Network Initialization	DNN Type	Learning Strategy	Loss Function
Zhou et al. (2015)	Random selection	Random initialization	Auto-encoder	Reconstruction (unsupervised)	Reconstruction
Hu et al. (2016)	Data augmentation	Pre-trained network weights	Convolutional neural network	Classification (supervised)	Cross-entropy
Li et al. (2016b)	Random selection	Random initialization	Convolutional neural network	Clustering (unsupervised)	Reconstruction
Zhou et al. (2017a)	Random selection	Pre-trained network weights	Convolutional neural network	Classification (supervised)	Cross-entropy
Ye et al. (2018)	Random selection	Pre-trained network weights	Convolutional neural network	Classification (supervised)	Cross-entropy
Tang et al. (2018b)	Random selection	Greedy layer-wise pre-training	Convolutional auto-encoder	Reconstruction (unsupervised)	Reconstruction
Boualleg and Farah (2018)	Random Selection	Pre-trained network weights	Convolutional neural network	Classification (supervised)	Cross-entropy
Imbriaco et al. (2019)	Random selection Data augmentation	Pre-trained network weights	Convolutional neural network	Classification (supervised)	Cross-entropy
Chaudhuri et al. (2019)	Image pairs	Random initialization	Graph convolutional network	Metric learning (supervised)	Contrastive
Cao et al. (2020)	Image triplets	Pre-trained network weights	Convolutional neural network	Metric learning (supervised)	Triplet

representations into low-dimensional binary codes, such that the similarity to the original space can be well preserved (Demir and Bruzzone 2016; Li and Ren 2017; Reato et al. 2019; Fernandez-Beltran et al. 2020). Thus, descriptor extraction and binary code generation are applied independently from each other, resulting in sub-optimal hash codes. Success of DNNs in image feature learning has inspired research on developing DL-based hashing methods (i.e., deep hashing methods).

Recently, several deep hashing-based CBIR systems that simultaneously learn image representations and hash functions based on the suitable loss functions are introduced in RS (see Table 11.2). As an example, in Li et al. (2018b) a supervised deep hashing neural network (DHNN) that learns deep features and binary hash codes by using the contrastive and quantization loss functions in an end-to-end manner is introduced. The contrastive loss function can also be considered as the binary cross-entropy loss function, which is optimized to classify whether an input image pair is similar or not. One advantage of the contrastive loss function is its capability of similarity learning, where similar images can be grouped together, while moving away dissimilar images from each other in the feature space. Due to the ill-posed gradient problem, the standard back-propagation of DL models to directly optimize hash codes is not feasible. The use of the quantization loss mitigates the performance degradation of the generated hash codes through the binarization

Table 11.2 Main characteristics of the state-of-the-art deep hashing-based CBIR systems in RS.

Reference	Loss Functions	Learning Type	Hash Layer
Li et al. (2018b)	Contrastive, Quantization	supervised	linear
Li et al. (2018a)	Contrastive, Quantization	supervised	linear
Roy et al. (2020)	Triplet, Bit balance, Quantization	supervised	sigmoid
Song et al. (2019)	Contrastive, Quantization, Cross-entropy	supervised	linear
Tang et al. (2019)	Cross-entropy, Contrastive, Reconstruction, Quantization, Bit balance	semi-supervised	linear
Liu et al. (2019)	Adversarial, Quantization, Contrastive, Cross-entropy	supervised	sigmoid

on the CNN outputs. In Li et al. (2018a) the quantization and contrastive loss functions are combined in the framework of the source-invariant deep hashing CNNs for learning a cross-modality hashing system. Without introducing a margin threshold between the similar and dissimilar images, a limited image retrieval performance can be achieved based on the contrastive loss function. To address this issue, a metric-learning based supervised deep hashing network (MiLaN) is recently introduced in Roy et al. (2020). MiLaN is trained by using three different loss functions: (i) the triplet loss function for learning a metric space (where semantically similar images are close to each other and dissimilar images are separated); (ii) the bit balance loss function (which aims at forcing the hash codes to have a balanced number of binary values); and (iii) the quantization loss function. The bit balance loss function makes each bit of hash codes to have a 50% chance of being activated, and different bits to be independent from each other. As noted in Roy et al. (2020), the learned hash codes based on the considered loss functions can efficiently characterize the complex semantics in RS images. A supervised deep hashing CNN (DHCNN) is proposed in Song et al. (2019) in order to retrieve the semantically similar images in an end-to-end manner. In detail, DHCNN utilizes the joint loss function composed of: (i) the contrastive loss function; (ii) the cross-entropy loss function (which aims at increasing the class discrimination capability of hash codes); and (iii) the quantization loss. In order to predict the classes based on the hash codes, a FC layer is connected to the hash layer in DHCNN. As mentioned above, one disadvantage of the cross-entropy loss function is its deficiency to define a metric space, where similar images are clustered together. To address this issue, the contrastive loss function is jointly optimized with the cross-entropy loss function in DHCNN. A semi-supervised deep hashing method based on the adversarial autoencoder network (SSHAAE) is proposed in Tang et al. (2019) for RS CBIR problems. In order to generate the discriminative and similarity preserved hash codes with low quantization errors, SSHAAE exploits the joint loss function composed of: (i) the cross-entropy loss function; (ii) a reconstruction loss function; (iii) the contrastive loss function; (iv) the bit balance loss function; and (v) the quantization loss function. By minimizing the reconstruction loss function, the label vectors and hash codes can be obtained as the latent outputs of the AEs. A supervised deep hashing method based on a generative adversarial network (GAN) is proposed in Liu et al. (2019). For the generator of the GAN, this method introduces a joint loss function that

Table 11.3 Comparison of the DL loss functions considered within the deep hashing-based RS CBIR systems. Different marks are provided: "×" (no) or "✓" (yes).

Loss Function	Similarity Learning Capability	Mini-batch Sampling Requirement	Bit Balance Capability	Binarization Capability	Annotated Image Requirement
Contrastive	✓	Image pairs	×	×	✓
Triplet	✓	Image triplets	×	×	✓
Adversarial	×	Random selection	✓	×	×
Reconstruction	×	Random selection	×	×	×
Cross-entropy	×	Random selection	×	×	✓
Bit balance	×	Random selection	✓	×	×
Quantization	×	Random selection	×	✓	×

composed of: (i) the cross-entropy loss function; (ii) the contrastive loss function; and (iii) the quantization loss function. For the discriminator of the GAN, the sigmoid function is used for the classification of the generated hash codes as true codes. This allows the learned hash codes following the uniform binary distribution to be restricted. Thus, the bit balance capability of hash codes can be achieved. It is worth noting that the above-mentioned supervised deep hashing methods preserve the discrimination capability and the semantic similarity of the hash codes in the Hamming space by using annotated training images.

In Table 11.3, we analyze and compare all the above-mentioned loss functions based on their: (i) capability on similarity learning, (ii) requirement on the mini-batch sampling; (iii) capability of assessing the bit balance issues; (iv) capability of binarization of the image descriptors; and (v) requirement on the annotated images. For instance, the contrastive and triplet loss functions have the capabilities to learn the relationship among the images in the feature space, where the semantic similarity of hash codes can be preserved. Regarding to the requirement of mini-batch sampling, pairs of images should be sampled for the contrastive loss function, image triplets should be constructed for the triplet loss function. The bit balance and adversarial loss functions are exploited for learning the hash codes with the uniform binary distribution. It is worth noting that an adversarial loss function can be also exploited for other purposes, such as for image augmentation problems to avoid overfitting (Cao et al. 2018). The quantization loss function enforces the produced low-dimensional features by the DNN models to approximate the binary hash codes. With regard to the requirement on image annotations, the contrastive and triplet loss functions require the semantic labels to construct the relationships among the images.

11.4 Discussion and Conclusion

In this chapter, we presented a literature survey on the most recent CBIR systems for efficient and accurate search and retrieval of RS images from massive archives. We focused our attention on the DL-based CBIR systems in RS. We initially analyzed the recent DL-based CBIR systems based on: (i) the strategies considered for the mini-batch sampling; (ii) the approaches used for the initialization of the parameters of the considered DNN models; (iii) the type of the considered DNNs; and (iv) the strategies used for image representation learning. Then, the most recent methodological developments in RS related to scalable image search and retrieval were discussed. In particular, we reviewed the deep hashing-based CBIR systems and analyzed the loss functions considered within these systems based on their: (i) capability of similarity learning, (ii) requirement on the mini-batch sampling; (iii) capability of assessing the bit balance issues; (iv) capability of binarization; and (v) requirement on the annotated images. Analysis of the loss functions under these factors provides a guideline to select the most appropriate loss function for large-scale RS CBIR problems.

It is worth emphasizing that developing accurate and scalable CBIR systems is becoming more and more important due to the increased number of images in the RS data archives. In this context, the CBIR systems discussed in this chapter are very promising. Despite the promising developments discussed in this chapter (e.g., metric learning, local feature aggregation, and graph learning), it is still necessary to develop more advanced CBIR systems. For example, most of the systems are based on the direct use of the CNNs for the retrieval tasks, whereas the adapted CNNs are mainly designed for learning a classification problem and thus model the discrimination of pre-defined classes. Thus, the image descriptors obtained through these networks cannot learn an image feature space that is directly optimized for the retrieval problems. Siamese and triplet networks are defined in the context of metric learning in RS to address this problem. However, the image similarity information to train these networks is still provided based on the pre-defined classes, preventing to achieve retrieval specific image descriptors. Thus, CBIR systems that can efficiently learn image features optimized for retrieval problems are needed. Furthermore, the existing supervised DL-based CBIR systems require a balanced and complete training set with annotated image pairs or triplets, which is difficult to collect in RS. Learning an accurate CBIR model from imbalanced and incomplete training data is very crucial and thus there is a need for developing systems addressing this problem for operational CBIR applications. Furthermore, the availability of an increased number of multi-source RS images (multispectral, hyperspectral and SAR) associated to the same geographical area motivates the need for effective CBIR systems, which can extract and exploit multi-source image descriptors to achieve rich characterization of RS images (and thus to improve image retrieval performance). However, multi-source RS CBIR has not been explored yet (i.e., all the deep hashing-based CBIR systems are defined for images acquired by single sensors). Thus, it is necessary to study CBIR systems that can mitigate the aforementioned problems.

Acknowledgement

This work is supported by the European Research Council (ERC) through the ERC-2017-STG BigEarth Project under Grant 759764.

Part II

Making a Difference in the Geosciences With Deep Learning

12

Deep Learning for Detecting Extreme Weather Patterns

Mayur Mudigonda, Prabhat Ram, Karthik Kashinath, Evan Racah, Ankur Mahesh, Yunjie Liu, Christopher Beckham, Jim Biard, Thorsten Kurth, Sookyung Kim, Samira Kahou, Tegan Maharaj, Burlen Loring, Christopher Pal, Travis O'Brien, Kenneth E. Kunkel, Michael F. Wehner, and William D. Collins

12.1 Scientific Motivation

Anthropogenically-forced changes in the number and character of extreme storms have the potential to significantly impact human and natural systems. Current high-performance computing technologies enable multidecadal simulation of global climate models at resolutions of 25 km or finer (Wehner et al. 2017). Such high-resolution simulations are demonstrably superior in simulating extreme storms such as tropical cyclones than the coarser simulations available in the Coupled Model Intercomparison Project (CMIP5) and provide the capability to more credibly project future changes in extreme storm statistics and properties (Oouchi et al. 2006; Strachan and Camp 2013; Walsh et al. 2013; Wehner et al. 2014; Murakami et al. 2012; Murakami 2014; Scoccimarro 2016). The High Resolution Model Intercomparison Project (HighResMIP), a subproject of the forthcoming CMIP6 (Eyring et al. 2016a), is an opportunity to advance understanding of extreme storms and precipitation (Haarsma et al. 2016).

These high-resolution climate models are inherently better able to emulate observations of strong gradients of temperature and moisture than their coarser counterparts. Hence, simulated storms of many types including tropical cyclones exhibit greater realism in high-resolution, multidecadal simulations. A challenge in analyzing high-resolution, multidecadal simulations is posed by the identification and tracking of storms in the voluminous, TB-PB sized model output. In contrast to meteorological feature tracking in the real world, it is impractical to manually identify storms in such simulations due to the enormous size of the datasets and therefore automated procedures are used. Traditionally, these procedures are based on a multi-variate set of physical conditions based on known properties of the class of storms in question.

For instance, tropical cyclones (TC) have been identified and tracked by using criteria of co-located high values of low level vorticity, low surface pressure values, elevated temperatures aloft, and high 10m wind speeds maintained for a specified duration of time (Knutson et al. 2007; Ullrich and Zarzycki 2017). The Tropical Cyclone Climate Model Intercomparison Project (TCMIP) summarized its findings as follows: "The results herein indicate

Deep Learning for the Earth Sciences: A Comprehensive Approach to Remote Sensing, Climate Science, and Geosciences,
First Edition. Edited by Gustau Camps-Valls, Devis Tuia, Xiao Xiang Zhu, and Markus Reichstein.

moderate agreement between the different tracking methods, with some models and experiments showing better agreement across schemes than others. When comparing responses between experiments, it is found that much of the disagreement between schemes is due to differences in duration, wind speed, and formation-latitude thresholds. After homogenization in these thresholds, agreement between different tracking methods is improved. However, much disagreement remains, accountable for by more fundamental differences between the tracking schemes. The results indicate that sensitivity testing and selection of objective thresholds are the key factors in obtaining meaningful, reproducible results when tracking tropical cyclones in climate model data at these resolutions, but that more fundamental differences between tracking methods can also have a significant impact on the responses in activity detected." (Horn et al. 2014a)

Extratropical cyclones (ETC) are often identified by conditions of locally maximal vorticity and minimal pressure but are considered more difficult to identify than tropical cyclones due to their larger and more asymmetric physical characteristics, faster propagation speeds, and greater numbers. The Intercomparison of Mid-Latitude Storm Diagnostics (IMILAST) project examined 15 different ETC identification schemes applied to a common reanalysis and found profound sensitivities in annual global counts, ranging from 400 to 2600 storms per year (Neu et al. 2013). Atmospheric rivers (AR) are characterized by "a long, narrow, and transient corridor of strong horizontal water vapor transport that is typically associated with a low-level jet stream ahead of the cold front of an extratropical cyclone" (American Meteorological Society 2017). As a non-cyclonic event, AR identification schemes are even more heterogeneous and diverse, and are based upon a wide variety of criteria involving total precipitable water, integrated water transport, and other several variables. The AR community has recently self organized the Atmospheric River Intercomparison Project (ARTMIP) along similar lines to the IMILAST project (Shields et al. 2018). Some of the recent conclusions of the ARTMIP project are: (i) AR frequency, duration, and seasonality exhibit a wide range of results; and (ii) "understanding the uncertainties and how the choice of detection algorithm impacts quantities such as precipitation is imperative for stakeholders such as water managers, city and transportation planners, agriculture, or any industry that depends on global and regional water cycle information for the near term and into the future. Understanding and quantifying AR algorithm uncertainty is also important for developing metrics and diagnostics for evaluating model fidelity in simulating ARs and their impacts. ARTMIP launched a multitiered intercomparison effort designed to fill this community need. The first tier of the project is aimed at understanding the impact of AR algorithm on quantitative baseline statistics and characteristics of ARs, and the second tier of the project includes sensitivity studies designed around specific science questions, such as reanalysis uncertainty and climate change."

Other types of weather events are less amenable to such heuristics-based automated identification. Blocking events are obstructions "on a large scale, of the normal west-to-east progress of migratory cyclones and anticyclones" (American Meteorological Society 2017) and are associated with both extreme precipitation and extreme temperature events. Three objective schemes were compared by Barnes et al. (2012), who found that differing block structure affects the robustness of identification. Objective identification of fronts, "interface or transition zone between two air masses of different density" (American Meteorological Society 2017), is even less developed. Location of fronts can often be detected visually from maps of pressure and temperature, but a clear identification of the boundary usually requires the synthesis of multiple variables. Hewson (1998) summarizes efforts to objectively detect fronts.

The climate science community is yet to fully exploit the capabilities of modern machine learning and deep learning methods for pattern detection. Self-organizing maps (SOMs) have been used to a limited extent for visually summarizing patterns in atmospheric fields (Sheridan and Lee 2011). For example, Hewitson and Crane (2002) and Skific et al. (2009) used this technique to detect patterns in surface pressure fields in 4the eastern U.S. and the Arctic, respectively. Loikith et al. (2017) found large-scale meteorological patterns associated with temperature and precipitation extremes for the northeastern U.S. by utilizing SOMs. Gibson et al. (2017) applied the method to examination of extreme events in Australia. While SOMs are a powerful tool to visualize the variations in atmospheric patterns, these patterns often represent different locations of the same meteorological phenomenon. Thus, they are not relevant to the purposes of our study.

These heuristic schemes can be abstracted into two distinct parts. The first part (detection) scans each time step of high frequency model output for candidate events according to the specified criteria while the second part (tracking) implements a continuity condition across space and time (Prabhat et al. 2012, 2015a). Details, of course, vary significantly. The detection step is a massive data reduction but may still contain many candidates that will not satisfy the tracking criteria. This is especially the case for tropical cyclones but much less so for atmospheric rivers.

Supervised machine learning techniques tailored to identify extreme weather events offer an alternative to these objective schemes as well as provide an automated method to implement subjective schemes. The latter is critical to understanding how climate change affects weather systems, such as frontal systems, for which objective identification schemes have not been developed. In both cases, the construction of suitable labeled training datasets is necessary and the details of how they are constructed are important. We discuss towards the end some progress on this front as well.

The heuristics-based identification schemes described above present quantitative definitions of the weather events in question. Hence, it should come as no surprise that different definitions yield different results. Supervised machine learning techniques trained on datasets produced by these schemes should, at the very least, mimic the original identification scheme.

Modern day deep learning presents opportunities for computational modeling that are to-date unparalleled. Deep learning has shown remarkable successes on a wide range of problems in vision, robotics, natural language processing and more. A large category of problems in climate science can be posed as supervised learning problems such as weather front classification, identifying atmospheric rivers and tropical cyclones, etc. There are many parallel applications to these problems in computer vision.

In summary, identifying, detecting, and localizing extreme weather events is a crucial first step in understanding how they may vary under different climate change scenarios. Pattern recognition tasks such as classification, object detection, and segmentation have remained challenging problems in the weather and climate sciences. While there exist many empirical heuristics for detecting extreme events, the disparities between the output of these different methods even for a single event are large and often difficult to reconcile. Given the success of deep learning in tackling similar problems in computer vision, we advocate a DL-based approach. Leveraging the efforts in other scientific communities and combining them with physics-based constraints provides a unique opportunity to approach these problems in a way that has not been before. We will look at some problems and solutions in following sections.

12.2 Tropical Cyclone and Atmospheric River Classification

Tropical cyclones (TCs) are strongly rotating weather systems that are characterized by a core of low pressure and warm temperature with high circulating winds. However, the defining dynamic and thermodynamic characteristics of TCs vary by region, mechanism and impacts; therefore no universal agreement on criteria that best characterize TCs exists (Nolan and McGauley 2012). Very different, and sometimes contradictory, results are reported by employing these varied detection algorithms on the same dataset (Horn et al. 2014a).

Atmospheric Rivers (ARs) are increasingly recognized as the cause of heavy precipitation over mid-latitude landmasses (Neiman et al. 2008; Lavers et al. 2012). These narrow channels of enhanced moisture transport occur in the lower troposphere in the low-level jet region of extra-tropical cyclones. Landfall making ARs often result in heavy precipitation, especially in regions with mountainous topography because of the forced vertical uplift of the moisture-laden air. Despite the negative impacts of flooding, the moisture that ARs transport and deliver is also essential for water resources and supply (Lavers et al. 2012). Most popular approaches for AR detection use thresholds for the integrated water vapor (IWV) or for the vertically-integrated horizontal water transport (IVT). Sometimes wind speeds and their directions are also used. However, the detection of ARs remains challenging, especially because the thresholds used are not well established and vary regionally. Further, the coupling between a narrow AR region of moisture transport with extra-tropical cyclones and jet streams complicates their detection and analyses.

Advances in supervised deep learning have demonstrated promising success on pattern recognition in natural images (Krizhevsky et al. 2012; Simonyan and Zisserman 2015; Szegedy et al. 2015) and speech (Graves et al. 2013; Sutskever et al. 2014). These successes have demonstrated the potential transformative power of deep learning to replace hand-crafted feature engineering with direct feature learning from training data. Most of the state-of-art deep learning architectures for visual pattern recognition are build on the hierarchical feature learning convolutional neural network. Modern convolutional neural network tend to be deep and large with many hidden layers and millions of hidden units, making them very flexible in learning a broad class of patterns simultaneously from training data. In this section, we formulate the problem of detecting extreme weather events such as TCs and ARs as a classic supervised visual pattern recognition problem that consists of two components: classification and localization. We develop deep convolutional architecture and primarily address the task of classification in this section.

12.2.1 Methods

We collected ground truth labeling of TCs and ARs obtained via application of heuristic multivariate threshold-based criteria (Prabhat et al. 2015a; Knutson et al. 2007) and manual

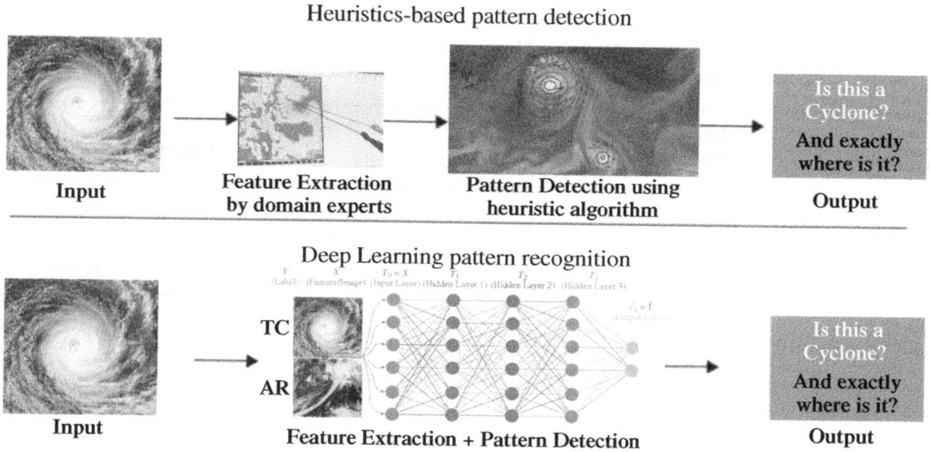

Figure 12.1 Contrasting traditional heuristics-based event detection versus deep learning-based detection.

Table 12.1 Data sources used for TC and AR binary classification.

Climate Dataset	Time Frame	Temporal Resolution	Spatial Resolution (lat × lon degree)
CAM5.1 historical run	1979–2005	3 hourly	0.23 × 0.31
ERA-Interim reanalysis	1979–2011	3 hourly	0.25 × 0.25

classification by expert meteorologists (Neiman et al. 2008; Lavers et al. 2012). Training data for these two types of events consist of image patches, defined as a prescribed geometrical box that encloses an event, and a corresponding spatial grid of relevant variables extracted from global climate model simulations or reanalyses. The size of the box is based on domain knowledge – for example, a 500 × 500 km box is likely to contain most tropical cyclones. To facilitate model training, an image patch is centered over the event. Because the spatial extent of events vary and the spatial resolution of simulation and reanalysis data is non-uniform, final training images differ in their size across the two types of events. This is one of the key limitations that prevents developing one single convolutional neural network to classify both types of storms simultaneously. The images are classified as those that contain events and those that do not contain events.

A summary of the attributes of training data is listed in Table 12.1, and attributes of original reanalysis and model simulation data are documented in Table 12.2. The datasets are split 80% and 20% for training and testing respectively.

12.2.2 Network Architecture

Figure 12.2 and Table 12.3 present the architecture used for addressing the supervised, binary classification problem of predicting TCs and ARs. We use two sets of convolutional

Table 12.2 Dimension of image, diagnostic variables (channels) and labeled dataset size for classification task (PSL: sea surface pressure, U: zonal wind, V: meridional wind, T: temperature, TMQ: vertical integrated water vapor, Pr: precipitation).

Events	Image Dimension	Variables	Total Examples
Tropical Cyclone	32 × 32	PSL,V-BOT,U-BOT, T-200,T-500,TMQ, V-850,U-850	10,000 +ve 10,000−ve
Atmospheric River	148 × 224	TMQ, Land Sea Mask	6,500 +ve 6,800−ve

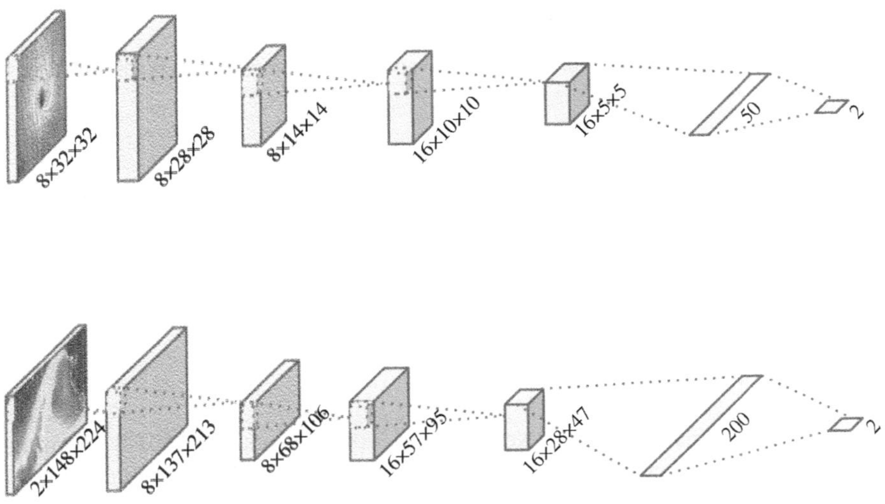

Figure 12.2 Top: architecture for tropical cyclone classification. Right: architecture or atmospheric rivers. Precise details can be found in Table 12.3.

Table 12.3 Classification CNN architecture and layer parameters. The convolutional layer parameters are denoted as <kernel size> − <number of feature maps> (e.g. 5 × 5 − 8). The pooling layer parameters are denoted as <pooling window> (e.g. 2 × 2). The fully connected layer parameter are denoted as <number of units> (e.g. 2). Non-linear activation function of hidden unit is shown in parentheses.

	Conv1 (ReLu)	Pool1	Conv2 (ReLu)	Pool2	Fully (ReLu)	Fully (Sigmoid)
Tropical Cyclone	5 × 5 − 8	2 × 2	5 × 5 − 16	2 × 2	50	2
Atmospheric River	12 × 12 − 8	3 × 3	12 × 12 − 16	2 × 2	200	2

Table 12.4 Accuracy of deep learning for TC and AR binary classification task.

Event Type	Train (%)	Test (%)
Tropical Cyclone	99.3	99.1
Atmospheric River	90.5	90.0

and pooling layers, followed by a fully connected layer and a sigmoid unit, that predicts the binary label.

Training deep architecture is known to be difficult (Larochelle et al. 2009; Glorot and Bengio 2010) and requires carefully tuning of parameters. In this study, we employ a Bayes' framework of hyper-parameter optimization technique to facilitate parameter selecting. Referring to AlexNet (Krizhevsky et al. 2012), we build a classification system with two convolutional layers followed by two fully connected layers. Details of the architecture and layer parameters can be found in Table 12.3 and Figure 12.2.

12.2.3 Results

The distinct characteristics of tropical cyclones, such as a low pressure center with strong winds circulating the center and a warm temperature core in the upper troposphere, make their patterns relatively easy to learn to represent with a CNN. Our deep CNNs achieved nearly perfect (99%) classification accuracy with failures associated with weakly developed storms that did not have the distinct features described above. Table 12.5 shows the confusion matrix for TCs. The confusion matrix reports accuracies obtained by a procedure in predicting labels compared to ground truth. Further details on the statistics and accuracy, including examples of failure modes, are presented in Liu et al. (2016b, c).

In contrast, deep CNNs achieve 90% classification accuracy for ARs. The challenges faced by deep CNNs in AR detection are their relative weakness and/or disjointed (broken) features of IWV and the presence of other weather systems in the vicinity, such as extra-tropical cyclones and jet streams. Table 12.6 shows the confusion matrix for ARs. Further details on the statistics and accuracy, including examples of failure modes, are in Liu et al. (2016b, c).

The results in this section suggest that deep convolutional neural networks are a powerful and novel approach for extreme event detection that does not rely on cherry picking features and thresholds. They also motivate the application of deep learning for a broader class of pattern detection problems in climate science.

Table 12.5 Confusion matrix for tropical cyclone classification.

	True TC	True Non_TC
Predict TC	0.989	0.003
Predict Non_TC	0.011	0.997

Table 12.6 Confusion matrix for atmospheric river classification.

	True AR	True Non_AR
Predict AR	0.93	0.107
Predict Non_AR	0.07	0.893

12.3 Detection of Fronts

Fronts are centrally important because of the variety of significant weather events associated directly with them, including severe thunderstorms and a wide spectrum of precipitation types and amounts. In a recent study, fronts were found to be the direct cause of more than half of observed extreme precipitation events in the contiguous U.S. (Kunkel et al. 2012).

Among the types of weather events discussed herein, weather fronts have the most complex spatial pattern. While tropical and extra-tropical cyclones (ETC) can be characterized by a point representing the pressure minimum, fronts are extended in space with one dimension much larger than the other. Thus, a front requires representation at a minimum as a line, but that line can have a complex shape. In middle latitudes most weather fronts are associated with extra-tropical cyclones. Fronts are identified visually based on the approximate spatial coincidence of a number of quasi-linear localized features – a trough in air pressure in combination with gradients in air temperature and/or humidity and a shift in wind direction (Stull 2015). Fronts are categorized as cold, warm, stationary, or occluded, with each type exhibiting somewhat different characteristics.

12.3.1 Analytical Approach

A supervised 2D CNN was implemented to investigate whether it could imitate the visual fronts recognition task. The goal for our front classification CNN is to estimate the likelihood that a given *pixel* in the image lies within a front.

The CNN architecture (Figure 12.3) is trained by optimizing the values of the pixels in the convolution filters to minimize the difference between the truth dataset and the output of the network applied to the input dataset as measured using a cost, or loss, function. Our categorical cross-entropy loss function has the form

$$H(p,t) = -\sum_{i=1}^{I}\sum_{c=1}^{C} w_c * \log(p_{ic}) * t_{ic}$$

where p is a set of output pixels, t is a set of truth pixels, w is a per-category weight, I is the number of pixels, and C is the number of categories. There are five possible categories including four for different types of fronts and a fifth for the absence of a front (or "no-front"). Each pixel is assigned one and only one category. The lower the likelihood value for the corresponding output category, the larger the contribution to the loss. The per-category weights are used to adjust the relative significance of the contributions from the different categories.

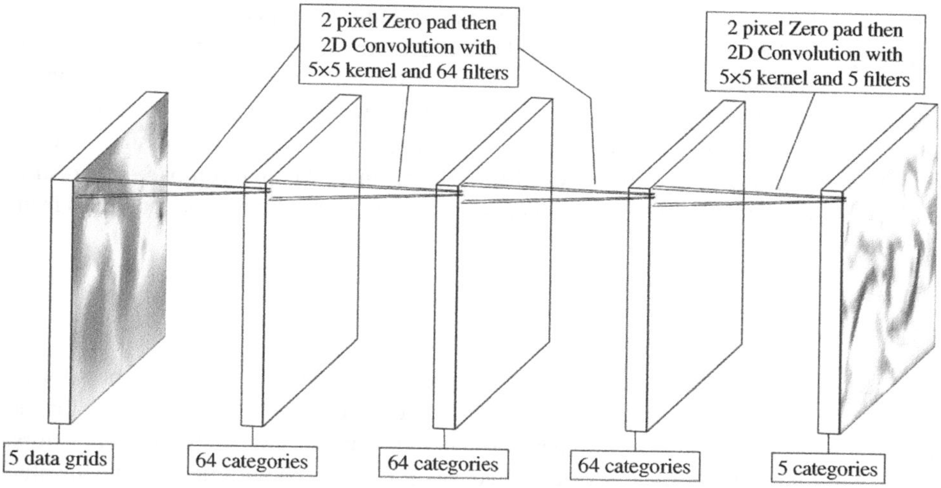

Figure 12.3 4-layer front detection CNN architecture with 64 5 × 5 filters per layer in the first three convolutional layers and 5 5 × 5 filters in the last layer. All convolutional layers are padded to have the same output shape. Source: Biard, J.C. and Kunkel, K.E., 2019: Automated detection of weather fronts using a deep learning neural network. Adv. Stat. Clim., Meteorol. Oceanogr, 5, 147–160. Licensed under CC BY-4.0

12.3.2 Dataset

The input dataset consisted of gridded fields of five surface variables taken from the MERRA2 https://gmao.gsfc.nasa.gov/reanalysis/merra2. The variables were 3-hourly instantaneous values of 2 m air temperature, 2 m specific humidity, sea level pressure, and the 10 m wind velocity.

Our truth dataset was extracted from the Coded Surface Bulletin (CSB) (http://www.nws.noaa.gov/noaaport/html/noaaport.shtml). Each text bulletin contains latitudes and longitudes specifying the locations of pressure centers, fronts, and troughs identified visually. Each front and trough is represented by a polyline We obtained all the bulletins possible for 2003–2016 and produced a 5-channel (categories of cold, warm, stationary, occluded, and none) image for each time step by drawing the front lines into latitude/longitude grids with one degree cell size. The image is filtered so that only one channel is set in each pixel, with a front-type preference order of warm over occluded over cold over stationary over none. Each front is drawn with a transverse extent of three degrees to account for the fact that a front is not a zero-width line and to add tolerances forslight lateral differences in position between the CSB and MERRA-2 derived fronts. In addition, the quantitative evaluation was restricted to regions where the frequency of fronts is at least 40 per year. The network was implemented in Keras and Theano, and trained with the data for 2003–2016 using an 80%-20% training-test split.

Each front is drawn with a transverse extent of three degrees to account for the fact that a front is not a zero-width line and to add tolerances for slight lateral differences in position between the CSB and MERRA-2 derived fronts. In addition, the quantitative evaluation was restricted to regions where the frequency of fronts is at least 40 per year. The network was implemented in Keras and Theano, and trained with the data for 2009 using an 80%–20% training-test split.

We then tested and calculated the confusion matrix and per-category IOU (ratio of correctly categorized pixels to total pixels in that category in either truth or CNN data, computed over each category) for the entire set of images. We also extracted polylines describing the fronts in each timestep by tracing out the lines following the maxima for each type of front. These were used to calculate the annual average number of front crossings at each 96 km × 9 kmm grid cell in a Lambert Conformal Conic map. We then compared the results with the annual average number of front crossings found using the original polylines from the CSB.

12.3.3 Results

Figure 12.4 shows an example of the output from the CNN using MERRA2 inputs for 20 March, 2010 at 12:00 UTC along with the fronts produced by NWS meteorologists for the same date and time. All four front types have been drawn together using different colors, with the color intensity in the CNN-generated image correlating with the likelihood value produced for that front type. Areas of mixed color correspond with regions where more

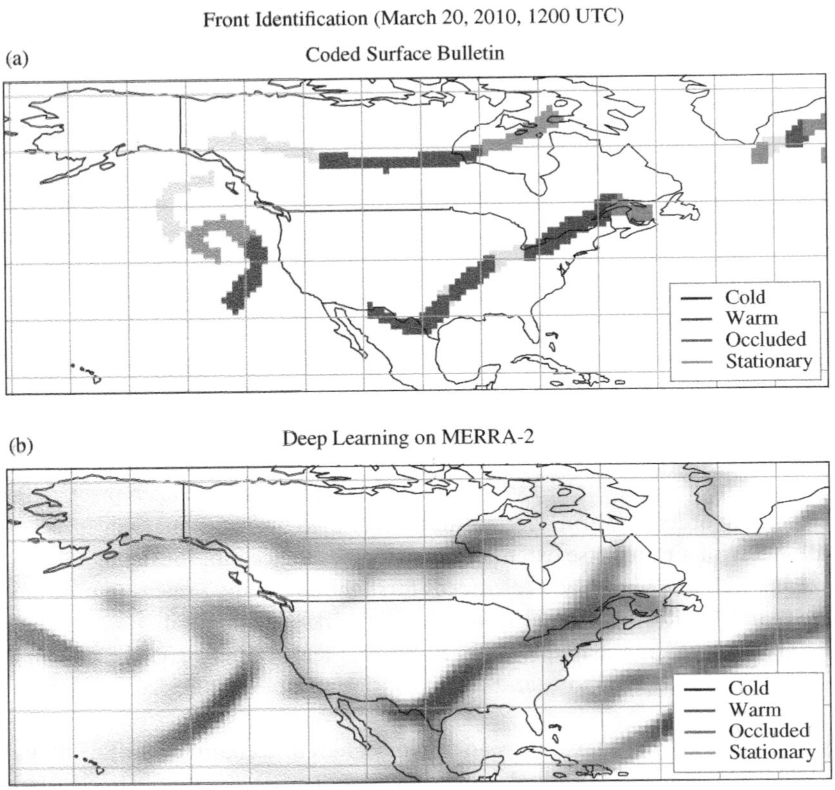

Figure 12.4 Coded Surface Bulletin fronts and CNN-generated front likelihoods.

Table 12.7 Per-category counts and IOU front detection metrics for 2013–2016.

Front Type	Truth Counts	CNN Counts	IOU
Warm	2,812,209	1,477,776	0.15
Occluded	2,593,094	2,961,205	0.24
Cold	6,964,750	6,358,892	0.28
Stationary	7,910,374	6,005,437	0.19
None	152,275,957	155,753,074	0.89

Table 12.8 Front category confusion matrices for 2013–2016.

		Percentages of Truth totals		
Truth/CNN	Warm	Occluded	Cold	Stationary
Warm	42.0%	16.8%	13.3%	27.9%
Occluded	4.9%	78.7%	9.9%	6.4%
Cold	1.7%	4.6%	79.3%	14.5%
Stationary	4.6%	4.2%	19.6%	71.6%
		Percentages of CNN totals		
Truth/CNN	Warm	Occluded	Cold	Stationary
Warm	67.6%	14.2%	4.7%	11.7%
Occluded	8.0%	67.2%	3.6%	2.7%
Cold	7.2%	10.3%	75.6%	16.3%
Stationary	17.2%	8.2%	16.0%	69.3%

than one front type produced a significant response. This image pair shows that the CNN is capturing a large majority of the CBS features. There are some, mostly low likelihood, fronts found by the CNN that may not be physical, and there are some regions of disagreement.

Table 12.7 shows the per-category IOU metric for the front likelihoods prediction. The IOU values indicate low agreement between the truth and predicted data, but inspection of the images (as seen in Figure 12.4) suggests that this is largely due to spatial differences between CSB and CNN in the location of fronts. The counts and IOU values show that the performance of the CNN is best for cold and occluded fronts, and worst for warm fronts.

Table 12.8 shows the confusion matrices for the cases when both the truth and CNN datasets indicated the presence of some type of front. Warm fronts are shown to have the greatest confusion, with true warm front pixels being categorized by the CNN as a different type of front almost 60% of the time. The other three types of fronts have much better performance, with cold and occluded fronts nearing 80% accuracy relative to the truth counts for those front types.

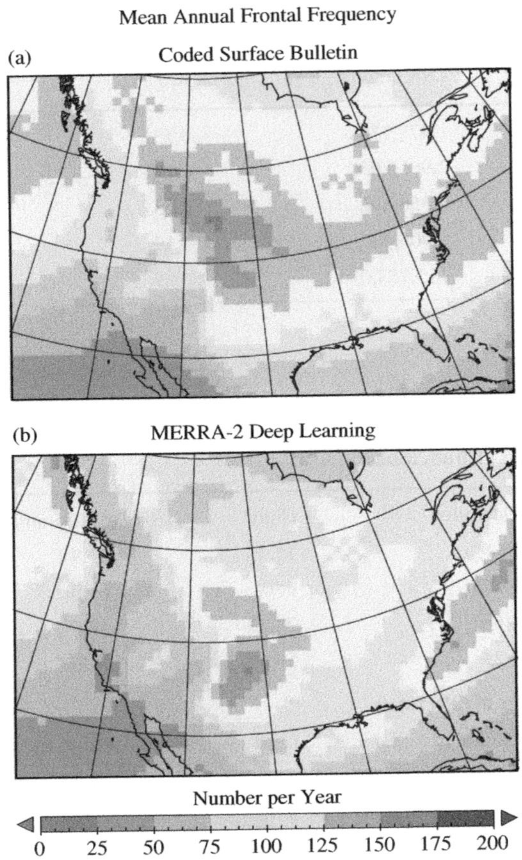

Figure 12.5 Mean annual frontal frequencies for Coded Surface Bulletin and CNN-generated fronts.

Figure 12.5 displays the results of taking the per-pixel mean of the number occurrences of fronts of all types over the entire 2003–2016 period covered by both of our datasets. The spatial pattern for the CSB fronts (Figure 12.5a) shows the highest values off the east and west Coasts and in the central U.S. The absolute numbers for the MERRA2 fronts (Figure 12.5b) are lower than the CSB numbers almost everywhere, but the spatial pattern is very similar (Figure 12.5b) with maxima in the same locations. The overall difference between the frequencies indicates that the CNN is detecting approximately 80% of the fronts found by NWS meteorologists.

12.3.4 Limitations

The current front detection CNN consistently under-detects fronts by 20%, particularly warm fronts, and tends to conflate cold and stationary fronts. Improvement is likely possible by experimenting with a larger number of layers, with different numbers of filters, and with adjustments to the relative weights given to the different front types in the loss function used for training. The CNN currently treats each 3-hourly image independently and does not take advantage of the high temporal correlations in front locations. Building a CNN or a hybrid LSTM architecture that makes use of multiple time steps has potential to produce significant improvements. The current system also only uses surface variables as inputs. Adding one or more variables at some elevation above the surface also has the potential to produce improvements.

12.4 Semi-supervised Classification and Localization of Extreme Events

As is the case in many scientific domains, there are vastly larger amounts of *unlabeled* climate data relative to the amount of *labeled* data. Ideally we would like to be able to learn from both sources at once by using labeled data to inform learning of the unlabeled data and vice-versa. This approach is called semi-supervised learning. In order to explore methods for semi-supervised learning of climate data, we use a 1979–2005 run of the Community Atmospheric Model v5 (CAM5) (Wehner et al. 2014) at 25 km resolution with unlabeled output every 3 h and labeled output every 6 h. The test set consists of 365 days from 1984. For 30 atmospheric levels, the output is a stack of 30 "images" consisting of vectors of data at the 768×1152 grid cells, with each vector including 16 scalar variables and labels for Tropical Depression (TD), Tropical Cyclones (TC), Extra-Tropical Cyclones (ETC), and Atmospheric Rivers (AR). In this study we consider only surface quantities over time. In total this results in a dataset of $27 \times 365 \times 8 = 78{,}840$ "images", of which 39,420 are labeled. Train, test, and validation splits, are summarized in Table 12.9.

More precisely, the "labels" are bounding boxes drawn around each extreme weather event. There may be multiple, overlapping events in each image. The dataset thus poses the problem of multiclass *detection* (identifying which events are in the image) and *localization* (identifying the coordinates of the bounding box).

12.4.1 Applications of Semi-supervised Learning in Climate Modeling

This work differs from our previous efforts in four distinct ways. First, we utilize the entire, high-resolution climate simulation images as the input, rather than cropped images only containing the event to be classified. Second, we perform event localization and classification, which means that the network must learn to scan through the image and correctly detect and classify any number of events. Third, we employ semi-supervised learning in order to show the plausibility of harnessing unlabeled data in addition to the labeled data for climate detection. Fourth, we utilize a 3D convolutional network operating on temporally

Table 12.9 Class frequency breakdown for Tropical Cyclones (**TC**), Extra-Tropical Cyclones (**ETC**), Tropical Depressions (**TD**), and Atmospheric Rivers (**AR**). Raw counts in parentheses. Table reproduced with permission from (Racah et al. 2016).

Data	TC %	ETC%	TD%	AR%
Train	42.32 (3190)	46.57 (3510)	5.74 (433)	5.36 (404)
Test	39.04 (2882)	46.47 (3430)	9.44 (697)	5.04 (372)

consecutive simulation frames in order to extract spatiotemporal features that can potentially assist with climate event detection.

12.4.1.1 Supervised Architecture

For the 2D supervised architecture, we use a convolutional neural network that takes as input CAM5 images of 768×1152 cells each containing 16 state variables. The 3D architecture uses 3D convolutions so each example is 8 images consecutive in time. Besides these differences, the architectures are very similar. Both networks are designed as a modified variant of "YOLO" ('You Only Look Once') (Redmon et al. 2016) detailed in Racah et al. (2016). YOLO is a popular object detection method that uses a single convolutional neural network to find objects in an image by predicting the location and class of the object as well as the shape and size of the box around it. This method is in contrast to more complex multi-stage pipelines (such as R-CNN (Ren et al. 2015)), where multiple neural networks are used, e.g., one to propose regions of interest of the image and another to classify the object within the region. Our network differs from YOLO in a few ways. We trade full context for translational invariance and for spatial refinement of features. That is, while YOLO uses features from the entire image to predict each box by use of fully connected layers, we only use features from the region where the box is predicted to be. While YOLO has the advantage of using features from the entire image as context for every object, it suffers from the fact that some spatial information is lost in combining all the features in the image. Moreover, our network predicts bounding box offsets to a 64×64 box, which leads to faster convergence for tropical cyclones and extratropical cyclones compared to YOLO, which predicts box offsets from a box the size of the image (Racah et al. 2016).

12.4.1.2 Semi-supervised Architecture

Our semi-supervised architecture (illustrated in Figure 12.6) consists of two parts: an encoder and a decoder. The encoder is identical to the supervised architecture. The decoder, however, takes as input the features extracted from the encoder and uses Transposed Convolutional (Deconvolutional) layers to upsample the features back to the original image size (Dumoulin and Visin 2016). This network of an encoder and decoder has two main objectives: first, to minimize the error between the predicted and true bounding boxes for the labeled data, and second, to reconstruct the original input for the unlabeled data using the features used for bounding box detection. In situations where there is limited labeled data, unlabeled data can inform the encoder not only to extract features useful for reconstructing the input, but also to use these features for localizing bounding boxes. Our decoder uses tied weights, i.e. the parameters of the decoder are the same as the ones in the encoder after transposition.

12.4.2 Results

12.4.2.1 Frame-wise Reconstruction

Before bounding box prediction, we first trained a 2D convolutional autoencoder on the data by treating each time-step as an individual training example, in order to visually assess reconstructions and to ensure the unsupervised part of the architecture was extracting useful features. Figure 12.8 shows the original and reconstructed feature maps for the 16 climate variables of one image in the dataset. We are able to achieve excellent reconstructions

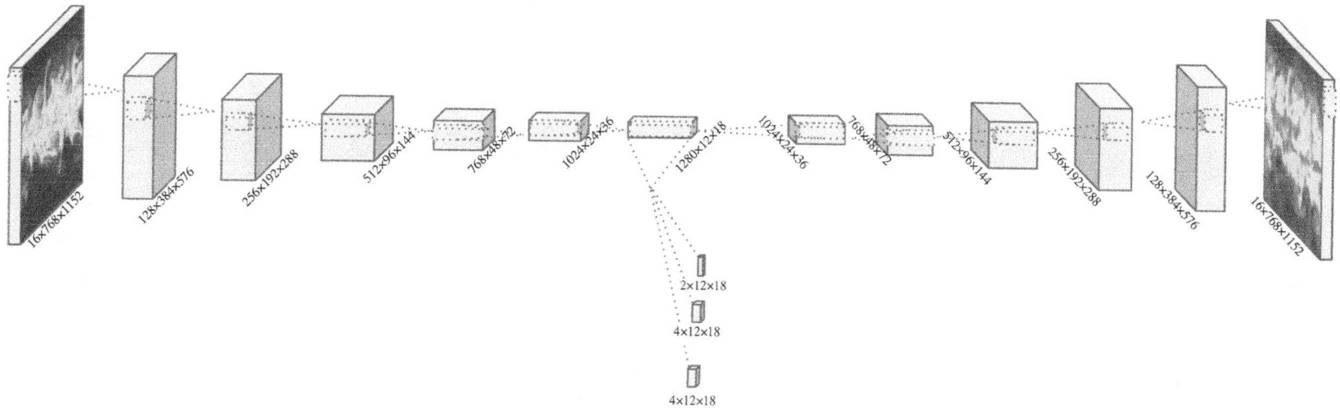

Figure 12.6 Diagram of the 3D semi-supervised convolutional network architecture. Output shapes of the various layers are denoted below the feature maps. The small network at the bottom of the diagram is responsible for predicting the bounding boxes, while the decoder part of the network (right), which is symmetric to the encoder, is responsible for reconstructing the output. Because both the bounding box predictor and decoder feed off of the encoder part of the network (left), they both contribute useful error signals to shape the underlying representation of the network.

Table 12.10 Semi-Supervised Accuracy Results: Mean AP for the models. Table reproduced with permission from (Racah et al. 2016).

Model	Mode	No. params (million)	λ	mAP (%) IOU=0.1	mAP (%) IOU=0.5
2D	Supervised	66.53	0	51.42	**16.98**
2D	Semi-Supervised	66.53	10	48.85	9.24
2D	Semi-Supervised	66.53	1	51.11	6.21
2D	Supervised	16.68	0	49.21	15.49
2D	Semi-Supervised	16.68	1	44.01	7.71
3D	Supervised	50.02	0	51.00	11.60
3D	Semi-Supervised	50.02	1	**52.92**	7.31

using an extremely compressed bottleneck representation (slightly less than 1% of the original input size).

12.4.2.2 Results and Discussion

In Table 12.10 we present 2D and 3D supervised and semi-supervised results for various settings of λ, which is the ratio of reconstruction loss to bounding box loss. We also show these results on a per-class basis in Table 12.11. Average Precision (AP) was calculated in the manner of ImageNet (Russakovsky et al. 2015), where we integrated the precision-recall curve for each class. Results are shown for two performance metrics by counting as true positives each occurrence of a bounding box prediction with intersection-over-union (IOU) with the ground truth box of at least 0.1 or 0.5. Furthermore, in Figure 12.7 we provide bounding box predictions shown on 2 consecutive 6-hourly simulation frames comparing the 3D supervised vs. 3D semi-supervised model predictions.

In Table 12.11 we see that some gains were achieved with the use of semi-supervised learning for both 2D and 3D methods. We note, however, that for our most sophisticated 3D semi-supervised architecture, AP tapers off dramatically for the IOU=0.5 case. Although incomplete sampling of the hyperparameter space might be a cause, another potential reason for this reduction in AP can be explained with Figure 12.7. As shown in the figure, the 3D models are able to locate the general position of the event, but are unable to adjust the size of the boxes since our network predicts offsets from "reference" boxes. Because we only used reference boxes of size 64×64, most predicted boxes are of that size as shown in Figure 12.7. For events much smaller than 64×64 (Tropical Depressions) or events much bigger than 64×64 (ARs), this can lead to detection errors. For an IOU threshold of 0.1, a 64×64 box around a TD or AR can still be classified as a true positive, but for a more IOU stringent threshold such as 0.5, these phenomena may not be detected. This leads to a sharp decrease in performance for ARs and TDs and also for some sizes of TCs as depicted in Table 12.11. One solution to this issue would be to have "reference" boxes of varying shapes and sizes and even to use features of different resolutions like in Liu et al. (2015).

Table 12.11 AP for each class. Frequency of each class in the test set shown in parentheses. First number is at IOU = 0.1 and second number after the semicolon is at IOU = 0.5 as the criteria for a true positive. In each column, highlighted in bold is the best result for that particular class and IOU setting. Table reproduced with permission from (Racah et al. 2016).

Model	Mode	Parameters (millions)	λ	ETC (46.47%) AP (%)	TC (39.04 %) AP %	TD (9.44 %) AP (%)	AR (5.04 %) AP (%)
2D	Sup	66.53	0	21.92; 14.42	52.26; 9.23	95.91; 10.76	35.61; 33.51
2D	Semi	66.53	1	18.05; 5.00	52.37; 5.26	97.69; 14.60	36.33; 0.00
2D	Semi	66.53	10	15.57; 5.87	44.22; 2.53	98.99; **28.56**	**36.61**; 0.00
2D	Sup	16.68	0	13.90; 5.25	49.74; **15.33**	97.58; 7.56	35.63; **33.84**
2D	Semi	16.68	1	15.80; 9.62	39.49; 4.84	**99.50**; 3.26	21.26; 13.12
3D	Sup	50.02	0	22.65; **15.53**	50.01; 9.12	97.31; 3.81	34.05; 17.94
3D	Semi	50.02	1	**24.74**; 14.46	**56.40**; 9.00	96.57; 5.80	33.95; 0.00

12.5 Detecting Atmospheric Rivers and Tropical Cyclones Through Segmentation Methods

Analyzing extreme events in large datasets poses a significant challenge in climate science research. Conventional tools to analyze extreme events are built upon human expertise, and they require subjective thresholds of relevant physical variables to define specific events. Tropical Cyclones (TCs), Extra Tropical Cyclones (ETCs), and Atmospheric Rivers (ARs) are important and impactful extreme weather events. Current methods to detect storms rely on sequential processing of the same data to detect each class of storm (TCs, ETCs, ARs, etc.). It would be significantly more efficient to detect all types of extreme weather events based on features/patterns that exist in multivariate climate datasets. Deep learning methods could achieve this goal when they are applied to physical variables such as integrated water vapor, surface pressure, and wind speed. Furthermore, traditional detection methods resort to subjective, arguably arbitrary thresholds, which may change with global warming. Accurate, efficient, and automatic tracking of extreme events can play a critical role in weather prediction if the network can learn precursors to these events. Deep neural networks may serve as an automated detector and tracker of extreme weather that relies on spatiotemporal patterns, not thresholds, in climate model simulations. With this tool, scientists can better study the environmental drivers that control the frequency, intensity, and location of extreme weather events and how they may change in a warming world.

12.5.1 Modeling Approach

In this work, we work at the finest possible resolutions. That is, pixel-level output masks are generated. To do this we employ the power of segmentation architectures. Recent progress in computer vision and machine learning on segmenting large natural images for various categories has been remarkable. Here, we explore the possibility of such work being applied to climate data.

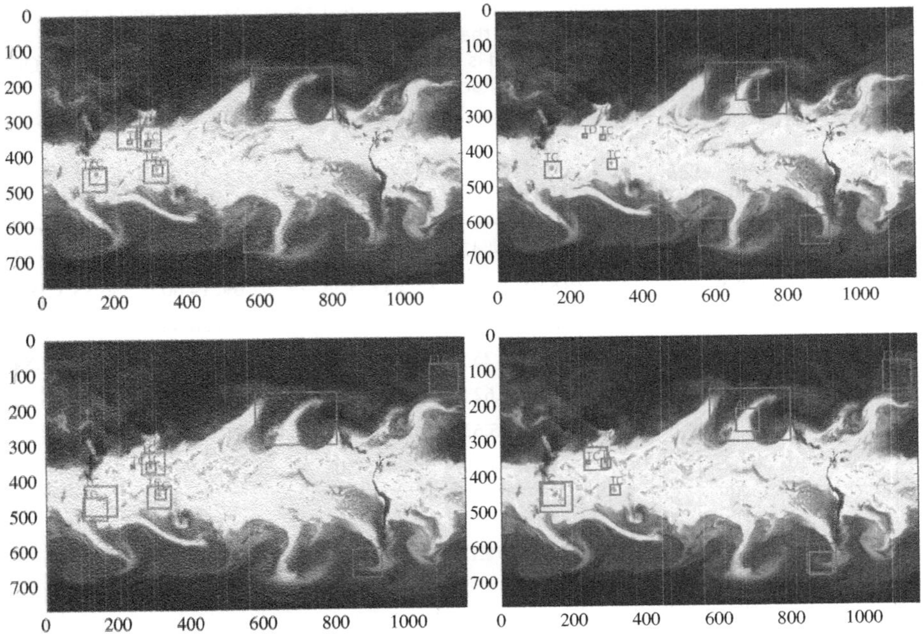

Figure 12.7 Bounding box predictions shown on 2 consecutive (6 hours in between) simulation frames (integrated water vapor column). Green = ground truth, red = high confidence predictions (confidence above 0.8 IoU). Left: 3D supervised model, right: 3D semi-supervised model. Figure reproduced with permission from (Racah et al. 2016).

12.5.1.1 Segmentation Architecture

High-level frameworks like TensorFlow make it convenient to experiment with different networks. We evaluated two very different networks for our segmentation needs. The first is a modification of the *Tiramisu* network (Jégou et al. 2017). Tiramisu is an extension to the Residual Network (ResNet) architecture (He et al. 2016b) which introduced skip connections between layers to force the network to learn residual corrections to layer inputs. Where ResNet uses addition, Tiramisu uses concatenation to combine the inputs of one or more layers of the network with their respective outputs. The Tiramisu network is comprised of a down path that creates an information bottle-neck and an up path that reconstructs the input. To perform pixel-level segmentation, Tiramisu includes skip connections spanning the down and up paths to allow re-introduction of information lost in the down path. Our Tiramisu network uses five dense blocks in each direction, with 2, 2, 2, 4, and 5 layers respectively (top to bottom). We then train the model using adaptive moment estimation (ADAM) (Kingma and Ba 2014).

The second network we evaluated is based on the recent DeepLabv3+ network (Chen et al. 2018a) and is shown in Figure 12.9. DeepLabv3+ is an encoder-decoder architecture that uses well-proven networks (in our case ResNet-50) as a core. The encoder performs a function similar to Tiramisu's down path but avoids loss of information by replacing some of the downscaling with atrous convolution. Atrous convolutions sample the input sparsely according to a specified dilation factor to detect larger features. This simplifies the decoder

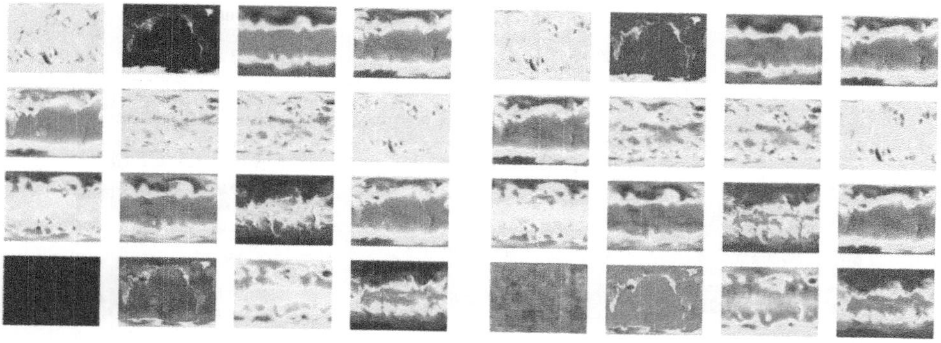

Figure 12.8 Feature maps for the 16 channels for one of the frames in the dataset (left) versus their reconstructions from the 2D convolutional autoencoder (right). The dimensions of the bottleneck of the encoder are roughly 0.8% of the size of the input dimensionality, which demonstrates the ability of deep autoencoders to find a robust and compressed representation of their inputs. Figure reproduced with permission from (Racah et al. 2016).

(corresponding to Tiramisu's up path) considerably. Our modifications to these existing networks are described in section 12.2.2.

12.5.1.2 Climate Dataset and Labels

We utilize 0.25-degree Community Atmosphere Model (CAM5) output for this study. Climate variables are stored on an 1152 × 768 spatial grid, with a temporal resolution of 3 hours. Over 100 years of simulation output are available in HDF5 files. Ideally, climate scientists would hand-label pixel masks corresponding to events. In practice, scientists currently use a combination of heuristics to produce masks on large datasets. The first step is to process climate model output with the Toolkit for Extreme Climate Analysis (Prabhat et al. 2012, 2015a) to identify TCs. A floodfill algorithm is used to create spatial masks of ARs (Shields et al. 2018), which provides the labels for our training process.

There are about 63K high-resolution samples in total, which are split into 80% training, 10% test and 10% validation sets. We use all available 16 variables (water vapor, winds, precipitation, temperature, pressure, etc.). The pixel mask labels correspond to three classes: Tropical Cyclone (TC), Atmospheric River (AR), and background (BG) class.

12.5.2 Architecture Innovations: Weighted Loss and Modified Network

The image segmentation task for climate analysis is challenging because of the high class imbalance: about 98.2% of the pixels are BG and about 1.7% of the overall pixels are ARs. Pixels labeled as TCs make up less than 0.1% of the total. With an unweighted loss function, each pixel contributes equally to the loss function, and a network can (and did, in practice) achieve high accuracy (98.2% in our case) by simply predicting the dominant background class for all pixels. To improve upon this situation, we use a weighted loss calculation in which the loss for each pixel is weighted based on its labeled class. The per-pixel weight map is calculated as part of the input processing pipeline and provided to the GPU along with the input image. Our initial experiments used the inverse of the class frequencies for weights, attempting to equalize the collective loss contribution from each class. We found

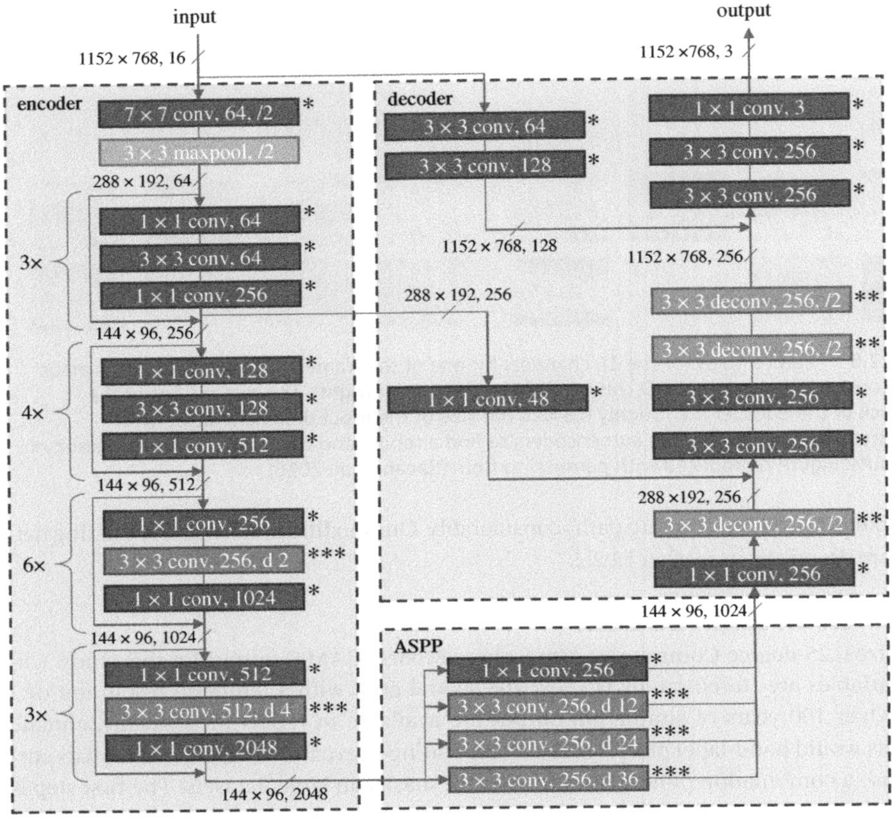

Figure 12.9 Schematic of the modified DeepLabv3+ network used in this work. The encoder (which uses a ResNet-50 core) and atrous spatial pyramid pooling (ASPP) blocks are changed for the larger input resolution. The DeepLabv3+ decoder has been replaced with one that operates at full resolution to produce precise segmentation boundaries. Standard convolutions*; deconvolutional layers**; Atrous convolution layers*** (specifying *dilation* parameter used).

that this approach led to numerical stability issues, especially with FP16 training, due to the large difference in per-pixel loss magnitudes. We examined more moderate weightings of the classes and found that using the inverse square root of the frequencies addressed stability concerns while still encouraging the network to learn to recognize the minority classes (see Figure 12.10).

The developers of the original Tiramisu network advocate the use of many layers with a relatively small growth rate per layer (e.g. 12 or 16) (Jégou et al. 2017) and our initial network design used a growth rate of 16. This network learned well, but performance analysis of the resulting TensorFlow operations on Pascal and Volta GPUs found considerable room for improvement and we determined that a growth rate of 32 would be significantly more efficient. To keep the overall network size roughly the same, we reduced the number of layers in each dense block by a factor of two and changed the convolutions from 3×3 to 5×5 to maintain the same receptive field. Not only was the new network much faster

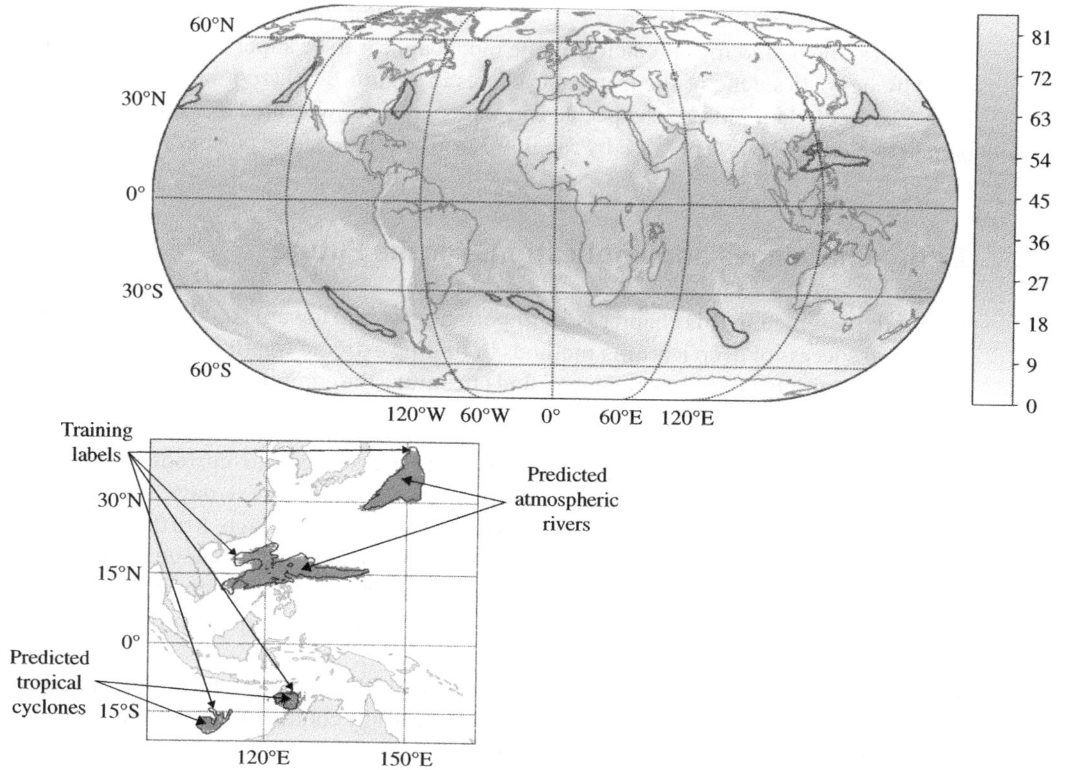

Figure 12.10 Top: Segmentation masks overlaid on a globe, indicating IWV (integrated water vapor, kg/m^2), one of the 16 input channels used by the network. Bottom: Detailed inset showing predictions vs. labels used in training. Segmentation results from modified DeepLabv3+ network.

to compute, we found that it trained faster and yielded a better model than our original network.

For DeepLabv3+, the atrous convolutions result in a more computationally expensive network than Tiramisu. The standard DeepLabv3+ design makes the compromise of performing segmentation at one-quarter resolution (i.e. 288×192 rather than 1152×768) to keep the computation tractable for less-powerful systems, at the cost of fidelity in the resulting masks. The irregular and fine-scale nature of our segmentation labels requires operating at the native resolution of the dataset. With the performance of *Summit* available for this work, we were able to replace the standard DeepLabv3+ decoder with one that operates at full resolution, thereby benefiting the science use case.

12.5.3 Results

Segmentation accuracy is often measured using the intersection over union (IoU) metric. The Tiramisu network obtained an IoU of 59% on our validation dataset, while our modified DeepLabv3+ network was able to achieve 73% IoU. Visually, this translates into qualitatively pleasing masks as seen in Figure 12.10. Not only does the network find the same atmospheric features, it makes a very good approximation of their exact boundaries. In some cases, the boundaries predicted by the model appear to be superior to the labels provided by heuristics. One of the tropical cyclones in Figure 12.10 (zoom inset) does suffer from overprediction. This is an expected consequence of our weighted loss function, which penalizes a false negative on a TC by roughly 37× more than a false positive.

12.6 Challenges and Implications for the Future

The work presented in this chapter is an important first step towards establishing the relevance and success of deep learning methods in finding extreme weather patterns. We now enumerate a number of open challenges, and encourage the community to work with us in addressing these problems in the future.

- *Training time*: Deep learning is computationally expensive. Our current front detection and supervised classification implementations take several days to train; the semi-supervised architectures currently take 1–2 weeks to converge. Kurth et al. (2017, 2018) and Mudigonda et al. (2018) present early results targeting this issue with impressive results with training times less than a day in many cases. It is very important that the climate science community have access to optimized, multi-node implementations of deep learning libraries;
- *Hyper-parameter optimization*: Determining the right DL network architecture for a given problem is currently an art. Practitioners have to typically conduct some amount of exploration in the space of number of layers, type of layers, learning rates, training schedules, etc. If each hyper-parameter combination requires a few days to converge, this exploration quickly becomes infeasible. We would like to request the mainstream AI research community to develop efficient software and techniques for solving the problem of finding optimal DL architectures.
- *Extension to temporal patterns*: Our current results are largely based on processing instantaneous snapshots of 2D fields. Climate patterns often span 3D fields and persist for extended periods of time. Ideally, we would train hybrid convolutional + LSTM architectures (Xingjian et al. 2015);
  The challenges of modeling time at high-dimensional scales is non-trivial. That said, there are many physical constraints that we are aware of (such as conservation of energy, etc.) that can help constrain the manifold for learning.
  While we present some work (based on Kurth et al. (2018) and Mudigonda et al. (2018)) on higher-dimensional grids, more work is required to understand larger-scale phenomena.
- *Interpretability*: Deep Networks are complicated functions with several layers of linear and non-linear filtering. While some effort has been spent in understanding ImageNet architectures (Zeiler and Fergus 2013), there is currently a gap in mapping the extracted feature hierarchy to semantic concepts from climate science. Recent approaches targeted at interpreting climate data (Toms et al. 2019b) demonstrate promising results; however, much remains to be accomplished towards the goal of developing interpretable, explainable DL networks.

- *Lack of training data*: Commercial ImageNet-style architectures operate on *millions* of labeled images. We hypothesize that deep learning works reasonably well in our application context because of the relatively small number of classes and lack of visual complexity encountered in natural scenes (e.g. occlusion, perspective foreshortening, illumination, material properties). Nevertheless, in order to improve the accuracy of deep learning-based classifiers, the climate science community will need to conduct coordinated labeling campaigns to create curated datasets which are broadly accessible to researchers.

Some progress towards these problems is being addressed by work in our group with ClimateNet (https://www.nersc.gov/research-and-development/data-analytics/big-data-center/climatenet/; Prabhat et al. (2018); Mudigonda et al. (2018)) and the climatecontours tool (http://labelmegold.services.nersc.gov/climatecontoursgold/tool.html). While these tools and limited labeling campaigns run thus far aim to address the challenge of curating and providing a standard dataset for the AI and climate communities, we essentially need the mainstream climate science community to provide labeled datasets for dozens of extreme weather patterns. We believe that such a dataset would probably result in the near-complete solution of supervised pattern recognition problems in climate science, enabling the community as a whole to move towards much harder problems involving causal mechanistic discovery.

12.7 Conclusions

This article presents an assessment of deep learning for extracting extreme weather patterns from climate datasets. We have demonstrated the application of supervised convolutional architectures for detecting tropical cyclones and atmospheric rivers in cropped, centered image patches. Subsequently, we demonstrated the application of similar architectures to predicting the type of weather front at the granularity of a grid-cell. Finally, we developed a unified architecture for simultaneously localizing, as well as classifying tropical cyclones, extra-tropical cyclones and atmospheric rivers. The benefit of the semi-supervised approach lies in the possibility of detecting other coherent fluid-flow structures that may not yet have a semantic label attached to them.

This work also highlights a number of avenues for future work motivated by the pragmatic challenges associated with improving the performance and scaling of deep learning methods and hyper-parameter optimization. Extending the methods to 3D space-time grids is an obvious next step. However, this will require creation of large training datasets, requiring the climate science community to conduct labeling campaigns. Finally, improving the interpretability of these methods will be key to ensure adoption by the broader climate science community.

Note, this chapter reviews work published in the following paper:
https://arxiv.org/abs/1612.02095

13

Spatio-temporal Autoencoders in Weather and Climate Research

Xavier-Andoni Tibau, Christian Reimers, Christian Requena-Mesa, and Jakob Runge

13.1 Introduction

Understanding and predicting weather and climate is one of the main concerns of humankind today. This is more true now than ever before due to the urgent need to understand how climate change will affect the Earth's atmosphere with its severe societal impacts (Stocker et al. 2013).

The main tools in weather and climate research are physics-based models and observational data analysis. In contrast to other complex systems, such as the human brain, a large body of knowledge on the underlying physical processes governing weather and climate exists. Since the 1960s numerical weather and climate models that simulate these processes have greatly progressed (Simmons and Hollingsworth 2002), with each new generation of models bringing better resolution and more accurate forecasts. Physics-based models are also used to understand particular processes, for example, by targeted model experiments to evaluate how the atmosphere is coupled to the ocean. A major challenge of such models today lies in the computational complexity of simulating all relevant physical processes, from atmospheric turbulence to the biosphere, where dynamics are chaotic and occur on multiple scales. Deep learning could make simulations much faster by augmenting physical climate models with faster deep learning-based parametrization schemes for such processes, termed hybrid modeling, with the challenge to preserve physical consistency (Reichstein et al. 2019).

On the other hand, in the last decades satellite systems and ground-based measurement stations lead to vast amounts of observational data of the various subsystems of Earth. Such datasets, together with increasing computational power, pave the way for novel data-driven learning algorithms to better understand the underlying processes governing the Earth's climate. Two prominent data analysis approaches, causal inference (Runge et al. 2019) and deep learning (Reichstein et al. 2019), both bear great promise in this endeavor. However, typical Earth science datasets pose major challenges for such methods, from dataset sizes and nonlinearity to the spatio-temporal nature of the underlying system.

Deep Learning for the Earth Sciences: A Comprehensive Approach to Remote Sensing, Climate Science, and Geosciences, First Edition. Edited by Gustau Camps-Valls, Devis Tuia, Xiao Xiang Zhu, and Markus Reichstein.

In this chapter we present a particular deep learning method, autoencoders (AE), as a tool well-suited for Earth data challenges (Vincent et al. 2010; Lusch et al. 2018). We cover different application scenarios that AEs have in weather and climate science and relate to applications presented in Chapter 15. However, here we will focus on why and how AEs help in specific tasks. The structure is as follows: In the first part (section 13.2), we dive into the theoretical background of AEs, what they are, and for which purposes they can be used. In the second part (section 13.3), we discuss some specific applications in weather and climate research. In the last part (section 13.4), we discuss some possible future applications.

13.2 Autoencoders

The goal of an AE is to encode information in an efficient way. The basic idea is to build two neural networks, one encoder and one decoder. The encoder contains multiple layers, each with less neurons than the layer before. The decoder is inverse to the encoder in the sense that every layer contains more neurons than the layer before. The two networks do not have to be symmetric. They connect at the layer with the smallest number of neurons, the bottleneck. The general architecture of an autoencoder can be seen in Figure 13.1.

To fit the parameters of the AE, the standard back-propagation (Rumelhart and Williams 1986) is used to minimize the loss function $\mathfrak{L}(X, X')$, for example, the mean squared error, between the input X and output X'. In this process the parameters of the AE converge in the direction of steepest descent. Formally, the encoder ϕ and the decoder φ are given by

$$\begin{aligned} \phi : \mathcal{X} &\rightarrow \mathcal{H} \qquad & \varphi : \mathcal{H} &\rightarrow \mathcal{X} \\ X &\mapsto H & H &\mapsto X'. \end{aligned} \tag{13.1}$$

We adapt the parameters of these functions to minimize the distance between $\varphi(\phi(x))$ and x'. The idea is that during this training the shrinking layers of the AE will encode the information in an efficient way in the bottleneck.

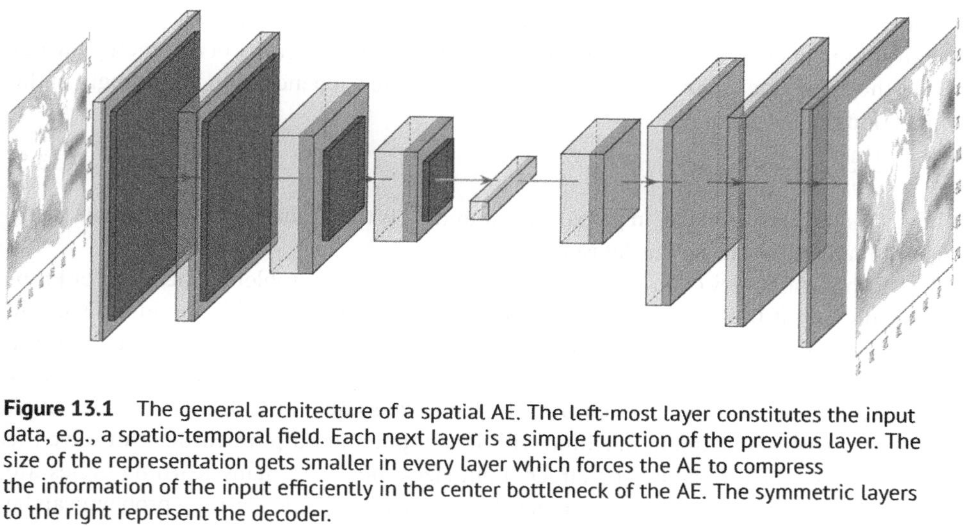

Figure 13.1 The general architecture of a spatial AE. The left-most layer constitutes the input data, e.g., a spatio-temporal field. Each next layer is a simple function of the previous layer. The size of the representation gets smaller in every layer which forces the AE to compress the information of the input efficiently in the center bottleneck of the AE. The symmetric layers to the right represent the decoder.

13.2.1 A Brief History of Autoencoders

In this section, we provide a brief introduction to the roots of AEs. Readers who are mainly interested in the practical applications of AEs can skip this section and continue reading in section 13.2.2.

Autoencoders emerged from the auto-associative problem to approximate the identity mapping between inputs and outputs. The first idea based on neural networks were Boltzmann Machines by Ackley et al. (1985). The proposed machine has many units that are in one of two states and connected through weighted connections. A positive weight indicates that the two units are often in a similar state, while a negative weight indicates that the neurons are often in opposite states. More specifically, the global state of a Boltzmann Machine is associated with the energy E:

$$E = -\sum_{i<j} w_{ij} s_i s_j + \sum_i \theta_i s_i. \tag{13.2}$$

Here w_{ij} is the weight between the units i and j, by s_i we denote the state of unit i, and θ_i is a threshold. Some of the units are associated to observables while some of the states are associated to latent variables. The machine is trained by fixing the units associated with observables and then computing the other states such that the energy is minimized. The advantage of this approach is that the optimal state of a unit can be determined locally by the difference in energy between a unit being on or off,

$$\Delta E_k = \sum_i w_{ki} s_i - \theta_k. \tag{13.3}$$

The actual state of all units is decided randomly to avoid getting stuck in local minima. The level of randomness is given by a temperature constant T. The probability P_α to be in global state α compared to the probability P_β to be in global state β is given by the Boltzmann distribution

$$\frac{P_\alpha}{P_\beta} = \mathrm{e}^{-(E_\alpha - E_\beta)/T} . \tag{13.4}$$

Boltzmann machine are trained by adapting the weights such that the expected energy in the minimum of energy for any observation is minimal. The linearity of the energy further allows the derivative of the log probability to be computed depending on the strength of the connections as

$$\frac{\partial \ln}{\partial w} = \frac{1}{T}(s_i^\alpha s_j^\alpha - p_{ij}) , \tag{13.5}$$

where s_i^α is the state of unit i in global state α and p_{ij} is the probability of both units being on at the same time across the dataset.

The authors of Ackley et al. (1985) call the abovementioned problem "The Encoder Problem" and credit it to Sanjaya Addanki. They use an architecture with one input and one output layer of size v and $\log v$ hidden units. Due to the symmetric connections of the Boltzmann machine, the hidden units are not ordered in layers, but every hidden unit is connected to every other hidden unit. In their experiments the authors (Ackley et al. 1985) show that a Boltzmann machine with four input and output neurons can solve the auto-associative task reliably. For larger input-sizes, however, many learning cycles are needed to find at least a close to optimal solution. While reliable learning problem can be solved by allowing for more hidden units, the problem that training Boltzmann machines is slow remains.

Today, the term "neural network" is mainly used for feed forward networks that are trained using backpropagation. This kind of network was first presented by Rumelhart and Williams (1986); Rumelhart et al. (1985). They used the same problem setting as presented above and found that their AE consistently managed to learn the task while being simpler and faster in training than Boltzmann machines.

In the following, several authors presented examples that indicated that AEs could solve difficult and interesting problems. For example, Cottrell et al. demonstrated (Cottrell and Willen 1988) that AEs with only three layers reach results on image compression that are comparable with the state-of-the-art of their time. They achieved this using an architecture similar to a modern convolutional neural network with just one filter. They divided the image in patches of 8×8 pixels and used a network consisting of 64 input neurons, 64 output neurons and 16 hidden neurons to encode and decode every single patch of the image individually. Their approach did not just learn to encode images with information loss close to the state-of-the-art of that time but also generalized to different, unseen images. One of the main drawbacks was that the AEs as described above converge towards a principal component analysis (PCA). This phenomenon is discussed, for example, in Cottrell and Willen (1988); Bourlard and Kamp (1988); Baldi and Hornik (1989). We discuss the similarities between AEs and standard machine learning methods in subsection 13.2.4.

In the following, the development of AEs was closely related to the general development of deep neural networks, which enabled to fit more complex and expressive encoders and decoders. We now mention some steps of this development. Firstly, the development of convolutional neural networks, an idea proposed by Fukushima (1980) and further considered by Rumelhart and Williams (1986); Rumelhart et al. (1985), demonstrated impressive effectiveness on real-world problems (see, e.g., LeCun et al. (1989)). Secondly, the work of Nair and Hinton (2010) and Krizhevsky et al. (2012) demonstrated that these methods can be scaled up to large image datasets and outperform state-of-the-art methods at that time. Thirdly, Rumelhart and Williams (1986); Rumelhart et al. (1985), Hochreiter and Schmidhuber (1997), and Cho et al. (2014) developed recurrent neural networks and made them feasible for large datasets. Recurrent neural networks have the advantage that they can handle inputs of different size, for example time series of different length. Additionally, such networks can couple the weights and create models with fewer parameters to optimize. The scope of this chapter does not allow for a discussion of the progress that enabled deep learning to reach the results and applications that it has achieved today.

13.2.2 Archetypes of Autoencoders

In the standard implementation of an AE, the fundamental objective is to reconstruct the original data from a smaller dimensional representation. However, several implementations pursue other tasks than mere reconstruction. Each of the different variants shares the same intent of identifying meaningful features to better approximate the data manifold. Below, some are described (for a summary of their loss functions see Table 13.1).

Table 13.1 This table summarizes all discussed variations of the standard AE. We describe shortly how they approximate the data manifold better than a standard AE and which form of regularization is used. We use $\mathfrak{L}$ to denote the standard loss function of the AE and $F(\cdot)$ stands for the function fitted by the AE (i.e., $F(x) = \varphi(\phi(x))$.

Name	Additional loss term	Goal
GAE	$\mathfrak{L}(X^{\sim}, F(X))$	Reproduce not only the input but also similar inputs
RAE	$\mathfrak{L}(R(X), R(F(X)))$	Reproduce not only the input but also the relation between inputs
SAE	$\sum_{j=1}^{m} \rho \log \frac{\rho}{\hat{\rho}_j} + (1-\rho) \log \frac{1-\rho}{1-\hat{\rho}_j}$	Reproduce not only the inputs but also enforce sparse activations
DAE	loss replaced by$\mathfrak{L}(X, F(\tilde{X}))$	Reproduce the input from a noisy version of it
CAE	$\|\|J_F(X)\|\|_F^2$	Force the derivative of F to be small near the inputs

Generalized Autoencoder (GAE) The GAE was introduced by Wang et al. (2014). A GAE is based on the common assumption in dimensionality reduction that the data is forming a lower-dimensional manifold in the input space. Its main goal is to reconstruct this manifold and the relations between examples that it represents. This is achieved by training the AE to not only reconstruct the input but a set of similar input samples. The loss for all these reconstructions is weighted by the similarity of these samples to the original input as measured by k-nearest neighbor distance or by all images belonging to the same class. The authors argue that the GAE captures the structure of the dataset better than an AE because the encoder is forced to map similar input examples closer on the manifold.

Relational Autoencoder (RAE) Presented in 2017 by Meng et al. (Meng et al. 2017), this AE is designed to not only reconstruct the input, but additionally preserve the relation between different samples in the dataset. While the loss function for the AE is

$$\min_{\theta} \mathfrak{L}(X, F(X)), \tag{13.6}$$

where θ denotes all network parameters, the loss for the RAE is

$$\min_{\theta} (1-\lambda)\mathfrak{L}(X, F(X)) + \lambda\mathfrak{L}(R(X), R(F(X))). \tag{13.7}$$

Here R is a measure of some relation between the data points (e.g., the variance) and F is the function realized by the AE. The authors claim that considering the relationships between different inputs decreases the reconstruction error and generate more robust features.

Sparse Autoencoder (SAE) The underlying idea of the SAE is to find a sound and meaningful representation of the inputs that allow transferring knowledge between similar datasets.

The SAE was introduced by Deng et al. (2013). The standard loss function of the AE, Equation 13.6, is enhanced by a term

$$\lambda \sum_{j=1}^{m} \rho \log \frac{\rho}{\hat{\rho}_j} + (1-\rho) \log \frac{1-\rho}{1-\hat{\rho}_j} \tag{13.8}$$

that enforces sparsity (Lee et al. 2008). Here, λ is a constant to weigh the importance of sparsity compared to reconstruction performance. The parameter ρ is a fixed sparsity level, for example, 0.01. The dimension of the embedding is denoted by m and $\hat{\rho}_j$ is the average activation of hidden unit j averaged over all inputs. Notice that in AEs, the compression effect emerges from the bottleneck being much smaller than the input, while in SAE, the compression is given by the redundant nature of sparse representations. Indeed, the bottleneck of a SAE does not have to be smaller than the input size.

Denoising Autoencoder (DAE) The core assumption leading to a DAE is that the data is located on a low dimensional manifold in the input space. Additionally, it is assumed that one can only observe noisy versions of these inputs. To allow an AE to learn from noisy inputs and identify useful features for classification, Vincent et al. (2010) introduced the DAE.

The DAE is related to the idea of data augmentation. For training, instead of the original input X, a corrupted version $\tilde{X}$ is used. The target to reconstruct is the clean input X, which has not been observed by the network. The goal of this training method is that the latent space captures the areas of high probability, i.e., the data manifold, and maps inputs of low probability onto these areas by deleting the noise.

While this method was used for feature extraction in the paper mentioned above, the actual classification is trained using clean inputs only. Hence, denoising is different from data augmentation. Interestingly, the authors compare a three-layer stacked DAE with a network build of three stacked Boltzmann Machines and could not outperform the latter on some datasets, showing that the main drawback of Boltzmann Machines is the complicated training and not the resulting performance.

Contractive Autoencoder (CAE) To determine the data manifold in the input space, this AE uses a penalty on the Frobenius norm of the Jacobian of the encoder. The idea was first presented by Rifai et al. (2011). In comparison to the loss function of the AE (Equation 13.6), the loss function of the CAE is given by

$$\min_{\theta} \mathfrak{L}(X, F(X)) + \lambda ||J_F(X)||_F^2.$$

The penalty term on the Jacobian matrix J_F leads to the representation being robust to small perturbations. It also enforces that the eigenvectors corresponding to large eigenvalues of J_F point into directions of high variance of the data points, which indicates a more meaningful representation.

13.2.3 Variational Autoencoders (VAE)

The idea behind a VAE is different from the idea of an AE. Instead of reproducing single inputs, the goal is to map the unknown distribution of the inputs onto a d-dimensional multivariate Gaussian distribution and then map it back to the original one. The

Gaussian distribution is centered at zero with unit variance (identity) matrix I_d, where d is the dimensionality of the compressed feature space. The VAE was first described by Kingma and Welling (2013).

We denote probability distributions by $P(X)$ and the corresponding density functions by lowercase $p(X)$. The objective is to ensure two things. Firstly, the distribution of the output data is intended to approximate the distribution of the input data and secondly, to force the latent distribution to be multivariate standard normal. The first goal is addressed by maximizing the likelihood of the inputs under the output distribution. The main problem is that such a distribution is unknown. The VAE approach solves this problem by using the evidence lower bound, also called ELBO,

$$\log p(x) \geq \mathbb{E}_{Hq}(\log p(x, H)) + \mathbb{H}(H) = \log p(x) - D_{KL}(q(H)||p(H|x)) .$$

Here $\mathbb{H}$ denotes the entropy of H and $\mathbb{E}_{Hq(\cdot)}$ denotes the expectation over any density function q. The quality of this bound depends on the density function q. The better $q(H)$ approximates $p(H|x)$, the tighter the bound. Therefore, by finding a q such that the Kullback–Leibler divergence $D_{KL}(q(H)||p(H|x)$ is minimized, one can push both distributions to be close. The calculation of the ELBO is mostly an application of Jensen's inequality (see Yang (2017)).

The second goal is to force the latent distribution to be multivariate standard normal, i.e., $q = \mathcal{N}(0, I_d)$. The VAE is trained as follows: The encoder maps from the original distribution to the latent distribution (and not a representation as in a standard AE). The reconstruction of the input is performed by sampling from that distribution and feeding the decoder with that sample. Now the problem is that it is not possible to calculate the derivative of the sampling process. This problem is solved by the so-called reparametrization trick. Essentially, the encoder learns a distribution and the network tries to determine the parameters of that distribution. Hence, the latent distribution is restricted to be a factorized Gaussian, and the encoder only derives the mean μ and the variance σ^2 of every latent component. Then the latent sample h is obtained by sampling $\tilde{h}$ from a standard normal distribution and compute

$$h = \sigma\tilde{h} + \mu.$$

This representation allows derivatives for μ and σ to be calculated and backpropagation to be used train the network. An example of an architecture for a VAE can be seen in Figure 13.2.

13.2.4 Comparison Between Autoencoders and Classical Methods

The ability of neural networks to automatically encode meaningful features is well known (Rumelhart and Williams 1986). For example, in Baldi and Hornik (1989) the authors show that a linear AE with an n-dimensional encoding space is converging towards a projection onto the first n principal components.

For nonlinear AEs, the situation is different. As mentioned by, for example, Tibau et al. (2018), nonlinear AEs can be thought of as kernel-PCA. PCA selects the subspace that contains most variability of the data. Therefore, a kernel-PCA can be understood as first applying an invertible feature function to the data and afterward selecting the subspace of

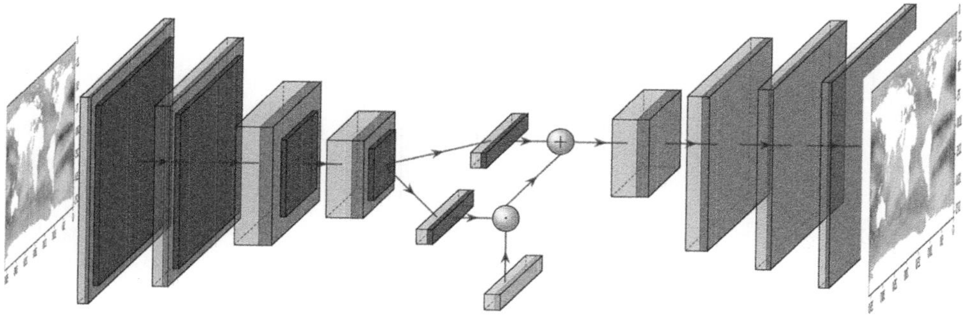

Figure 13.2 The architecture of a variational autoencoder. The main difference to an autoencoder is the random sampling process in the center.

the highest variance. From this point of view, the nonlinear encoder and decoder of the AE approximate this feature function, and hence optimizing an AE on data can be thought of as finding the optimal kernel for a kernel-PCA.

Note that the search space of this optimization does not span all possible kernels. The dimension of the latent space has to be decided before starting the optimization. This implies that, in general, it is only possible to optimize over kernels with the given latent dimension. For example, it is not possible to approximate the RBF-Kernel which has an infinite latent dimensionality. If the encoder is a close approximation to the optimal kernel, then the solution found to the dimensionality reduction problem is also optimal. Ham et al. (2004) show how many standard dimensionality reduction methods can be understood as a kernel-PCA.

13.3 Applications

In this section, we suggest several different applications that AEs can have in weather and climate sciences, divided into the common uses of AEs. The applications are divided into two main categories "Use of the latent space" (section 13.3.1) and "Use of the decoder" (section 13.3.2). In the first, it is shown how the lower-dimensional latent variables can be used to improve predictions and for knowledge extraction. In the second, we describe the capabilities of a neural network based decoder for denoising, sampling generation, and anomaly detection. Figure 13.3 synthesizes this classification.

13.3.1 Use of the Latent Space

High-dimensional data is difficult to grasp. For this reason, methods have been developed for projecting data onto a lower-dimensional and meaningful representation. If explainability of the original data is preserved, this lower-dimensional space allows us to use other classical methods to unravel the underlying dynamics and relations that exist in the data.

The first attempt to represent data in fewer dimensions came from H. Hotelling, who in 1933 presented PCA as a method for reducing results of psychological tests into "components" (Hotelling 1933). He claimed that "A few mental characters [...] are sufficient to

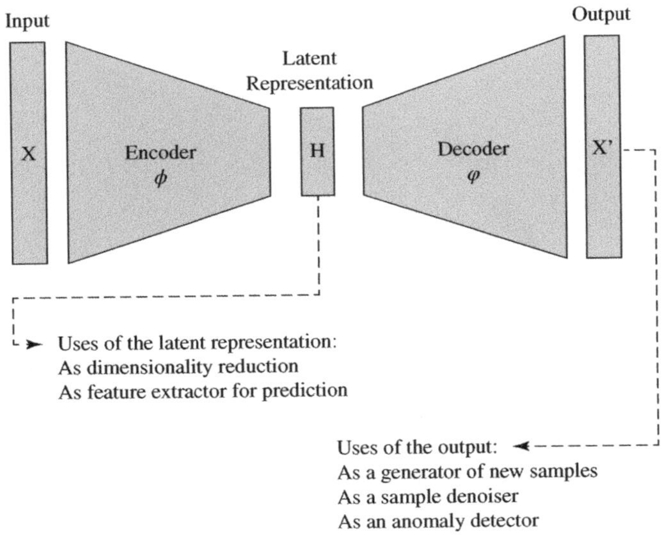

Figure 13.3 Summary of the use of an AE for weather and climate. These can be divided in (a) direct utilization of the latent space and (b) the use of the decoder function. (a) can be subdivided into (a.1) to use extracted features for better understanding the encoded data and (a.2) to use the latent representation for prediction. (b) can be subdivided into the use of the decoder function for (b.1) generating new samples, (b.2) denoising, and (b.3) as an anomaly detector.

account for virtually all the variance among individuals…." In climate and weather science, the first approach came from Eduard N. Lorenz, who, in 1959, used PCA (in his works called Orthogonal Empirical Functions, EOFs) as a method for reducing the data dimension before prediction (Lorenz 1956). Later on, to improve interpretability of these lower-dimensional variables, some modifications to PCA were introduced, again, first in the psychological (Kaiser 1958) and later in the climatology domain (Richman 1986).

Next to the methods used to visualize linear dependencies among the data, there have been several approaches to capture nonlinear dependencies, motivated by the fact that real-world data, and especially weather, often has nonlinear dependencies. One of the leading methods is Kernel-PCA (Schölkopf et al. 1998), which uses the kernel-trick to account for nonlinearities in the data. However, two main problems arise with Kernel-PCA: First, the new space is often no longer interpretable, and, second, it is not easy to know before-hand which kernel is suited (Rasmussen 1999).

In this context, AEs can address both problems, finding a suitable kernel for dimensionality reduction and understanding the latent space. It is well known that AEs are able to represent original data in a lower-dimensional space, often called *hidden representation*, *latent space*, *embedding*, *code*, *bottleneck*, or *probabilistic latent space* (in case of the VAE). One key feature of this mapping is that the samples that are close to each other in the latent representation are also close in the original space. However, this property is not ensured by a plain AE and this issue can be addressed with the introduction of additional loss terms.

The latent representation aims to be a good lower-dimensional description of latent variables, and therefore one can use it for different purposes; the first, and most straightforward one, is to predict future stats of the system or classify the input data from these extracted features, as is done with other dimensionality reduction methods like PCA. In section 13.3.1.2 we show several examples. Another way to make use of these extracted latent variables is to help scientists to understand the governing dynamics of the system better.

13.3.1.1 Reduction of Dimensionality for the Understanding of the System Dynamics and its Interactions

One of the main pitfalls in using the latent space of an AE to understand the data dynamics is the loss of interpretability in that space. One example to overcome this problem can be found in Tibau et al. (2018). There, Tibau et al. present an architecture where a VAE was used as an unsupervised kernel approximation of kernel-PCA[1]. The considered AE encodes temporal relations and PCA reduces the spatial dimension of those embeddings. The encoder function $\phi : X \mapsto H$ (where $X \in \mathbb{R}^{nxmxt}$) represents the data in the original space, $\mathcal{H} = \mathbb{R}^{nxmxf}$ represents the feature space, n, m, t, and f are indices for latitude, longitude, time, and the latent variable, respectively. While the entire dataset is used as a training set to learn ϕ, the mapping is performed independently for each grid-point ($m \times n$) and does not take into account the dynamics of the entire dataset.

Then PCA is applied in $\mathcal{H}$ to create orthogonal representations (features) of the latent dynamics according to the variance of H. How relevant those features are in each grid-point can then be visualized. Points with similar PCA weights share a common dynamic. The decoder (φ) is trained to map from $\mathcal{H}$ to $\mathcal{X}$ as an approximation of the inverse of ϕ, which can be used to transform a specific combination of principal components into a signal. This property allows the same decoder to understand to be used the dynamics or create noise-free versions of the original data. The idea behind the latter is similar to a Fourier filter, but instead of removing frequencies, feature components are removed.

As a proof of concept, the authors performed a set of experiments where they aimed to find a hidden driving pattern of different chaotic and stochastic dynamics. They used two synthetic datasets, the first one based on a Lorenz '96 dynamical model (Lorenz 1996). For every grid point, a set of 10 variables was simulated, and only the first variable was recorded while the others are treated as unobserved variables. The forcing pattern at every grid-point is the ground truth that they aimed to recover from the observed variable. Notice that in the model equations, the forcing parameter $\mathcal{F}$ defines the chaoticity of the dynamics of the system. The second dataset was a cellular automaton where the value of each grid point depends on the previous time step and its neighbors' values through a set of rules. To introduce stochastic behavior, the authors allowed those rules to change according to some degree of randomness defined by the hidden pattern $\mathcal{F}$ (See Figure 13.4). In summary, they presented two experiments with a hidden spatial pattern that governs the temporal dynamics with the objective to recover that hidden pattern. As a baseline method, they used PCA, Kernel-PCA (with different kernels: linear, polynomic, RBF, sigmoid and cosine), and the first 12 moments of the distribution at each grid point. For validation, the correlation value between each grid point of the weights in the principal components and the real pattern $\mathcal{F}$ was computed. The results show that baseline methods fail to approximate the hidden pattern while the VAE based kernel-PCA well captured them (see Table 13.2).

---

1 Code available in https://github.com/ClimateInformatics/SupernoVAE

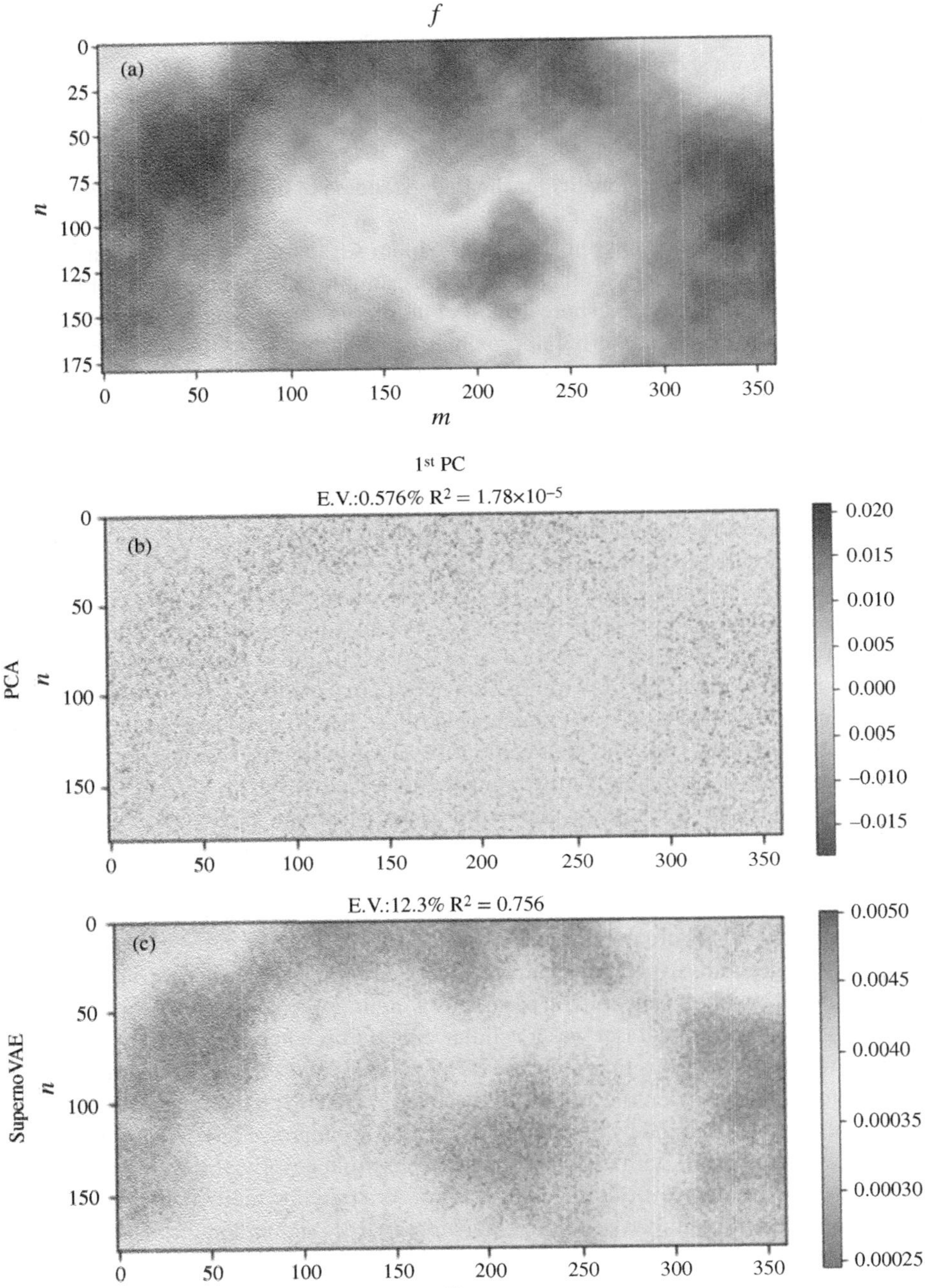

Figure 13.4 An example of the results in Tibau et al. (2018). Plot of (a) $\mathcal{F}$, (b) the 1st PC for the Lorenz '96 dataset and (c) the 1st PC for Lorenz '96 after applying SupernoVAE. E.V.: Explained Variation by the first principal component.

Table 13.2 Summary of results in Tibau et al. (2018). The column *Reconstruction* shows the coefficient of determination of the VAE reconstructions and the input time series, the columns 1st, 2nd and 3rd PC show the coefficient of determination between the kth principal component and the forcing pattern $\mathcal{F}$. θ stands for the relation of the bottleneck dimension size and the input dimensional size. The PC marked * is represented in Figure 13.4 (c).

	Lorenz '96				Cellular automata			
θ	Reconstruction	1st PC	2nd PC	3rd PC	Reconstruction	1st PC	2nd PC	3rd PC
0.01	0.129	0.000	0.001	0.001	0.513	0.009	0.002	0.005
0.10	0.643	0.476	0.000	0.022	0.968	0.627	0.003	0.005
1.00	0.865	0.756*	0.851	0.001	0.981	0.287	0.005	0.767
		time-permuted				*time-permuted*		
0.10	0.446	0.001	0.000	0.000	0.939	0.598	0.027	0.007
1.00	0.988	0.006	0.000	0.000	0.997	0.397	0.328	0.074
Other methods								
EOF	—	$5.94 \cdot 10^{-6}$	$5.22 \cdot 10^{-8}$	$1.36 \cdot 10^{-7}$	—	$7.13 \cdot 10^{-5}$	$4.81 \cdot 10^{-6}$	$7.42 \cdot 10^{-5}$
Linear Kernel	—	$8.15 \cdot 10^{-4}$	$3.24 \cdot 10^{-4}$	$3.84 \cdot 10^{-5}$	—	$7.80 \cdot 10^{-5}$	$1.63 \cdot 10^{-6}$	$5.16 \cdot 10^{-5}$
Poly Kernel	—	$2.12 \cdot 10^{-4}$	$2.54 \cdot 10^{-4}$	$1.87 \cdot 10^{-6}$	—	$4.72 \cdot 10^{-4}$	$1.36 \cdot 10^{-4}$	$1.31 \cdot 10^{-5}$
RBF Kernel	—	$5.12 \cdot 10^{-4}$	$1.17 \cdot 10^{-5}$	$9.53 \cdot 10^{-5}$	—	$4.06 \cdot 10^{-4}$	$9.65 \cdot 10^{-5}$	$1.63 \cdot 10^{-4}$
Sigmoid Kernel	—	$9.42 \cdot 10^{-4}$	$2.48 \cdot 10^{-7}$	$3.83 \cdot 10^{-4}$	—	$6.65 \cdot 10^{-4}$	$1.13 \cdot 10^{-3}$	$1.13 \cdot 10^{-5}$
Cosine Kernel	—	$1.06 \cdot 10^{-3}$	$6.35 \cdot 10^{-5}$	$2.76 \cdot 10^{-5}$	—	$1.17 \cdot 10^{-4}$	$4.50 \cdot 10^{-5}$	$2.60 \cdot 10^{-5}$

As mentioned above, one of the main problems with this approach is that linearity in the latent space is not ensured. Another interesting work where this issue is faced is Lusch et al. (2018). Lusch et al. rely on the fact that the eigenfunctions of the Koopman operator $\mathcal{K}$ provide intrinsic coordinates that globally linearize the dynamics. Since the identification and representation of such functions is complex for highly nonlinear dynamics, the authors propose an AE for embedding it into a lower-dimensional manifold. The authors aim to satisfy:

$$\phi(X_{k+1}) = \mathcal{K}\phi(X_k) \tag{13.9}$$

where k denotes the different states of the system over time. They expect ϕ to project the data in a space where the Koopman operators are linear, that is, $\phi : \mathbb{R}^n \mapsto \mathbb{R}^p$ and then, $H_{k+1} = \mathbf{K}H_k$, where n and p are the dimensions in the original and in the latent space, respectively. To train ϕ, a regular AE with the MSE loss function $\mathcal{L}_1 = ||X_k - \varphi\left(\phi(X_k)\right)||_2^2$ is used. The final network is built with this encoder-decoder function, where the encoder encodes X_k and the decoder decodes H_{k+1} through a dense layer with weights $\mathbf{K}$ and a

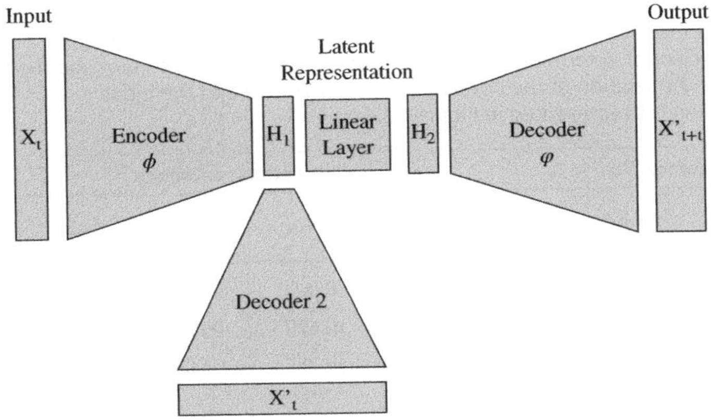

Figure 13.5 Schematic view of the architecture used in Lusch et al. (2018).

linear activation between H_k and H_{k+1}. See Figure 13.5 for a depiction of the architecture. To train the entire network they add two more losses: one for the linear dynamics in the latent space $\mathcal{L}_2 = ||\phi(X_{k+1}) - \mathbf{K}\phi(X_k)||_2^2$, and the other for future state predictions $\mathcal{L}_3 = ||X_{k+1} - \varphi\left(\mathbf{K}\phi(X_k)\right)||_2^2$.

As a proof of concept, the authors conducted three experiments. In the first, they used a well-studied model with a discrete spectrum of eigenvalues. In the second, the system was based on a nonlinear pendulum with a continuous spectrum of eigenvalues and increasing energy. Finally, in the third experiment, they used a high-dimensional model of nonlinear fluid flow. In all three cases, the authors achieved linear representations of the dynamics in $\mathcal{H}$ space.

The representation of the latent space using a dimensionality reduction method is also explored in Racah et al. (2017). The goal was to detect and track extreme events, using an AE as a method to extract features. The architecture proposed by the authors consisted of an AE where a bottleneck connects to three CNNs to predict the type of event, its location, and the confidence interval. As is often done with dimensionality reduction methods, the authors use t-SNE to visualize the different representations of the original data in the latent space. The resulting visualization showed the data grouped by the original categories, even if they were unknown to the AE. The authors stress the importance of using these visualization techniques to better understand weather data and how features interact with each other. For a more detailed explanation of how to use deep learning for extreme event detection and tracking, see Chapter 12.

Another example of the use of the latent space to understand data is found in Krinitskiy et al. (2019). There, the authors employed a sparse variational autoencoder (SVAE) to cluster polar vortex states. The authors pose that the idea behind the use of an SVAE is that the variational inference ensures that similar examples in the original space are close in the latent space. The sparsity constraint enforces features in the latent space to be Bernouilli-distributed, allowing a clustering that produces more valuable results than standard normal distributed latent spaces (see subsection 13.2.2). They also applied the commonly used techniques for convolutional networks, such as transfer learning, deploying an encoder derived from a powerful network previously trained on ImageNet. In the latent space, they applied Lance–Williams hierarchical agglomerative clustering, a method frequently used for clustering atmospheric and stratospheric states. They found their classification of polar vortex states to be physically valid and consistent with recent studies.

13.3.1.2 Dimensionality Reduction for Feature Extraction and Prediction

We have shown works where the latent space is used as a representation of the data and its dynamics in a lower-dimensional space. Good representations can also be used to make predictions.

McAllister and Sheppard presented two works where an AE is used as a feature extractor for wind vector determination (McAllister and Sheppard (2017) and McAllister and Sheppard (2018)). In these papers, the authors used a SAE to extract characteristics from each of the points on a lattice. They then used each of them, with a fully connected layer, to determine the wind speed at the surrounding locations.

The same approach is taken by hernandez2016rainfall to predict the daily accumulation of rain. In their work, they used an AE to extract features and then a dense layer to estimate the amount of rainfall accumulated the next day. To assess the results, they compared them with other works and with different alternatives as a baseline.

In Klampanos et al. (2018), the authors model how a radioactive plume evolves over time. They created multiple simulations with different setups to classify the pattern of pollution dispersion in 15 specific categories based on the observations (10 or 30).

For the input, the authors perform simulations of a real dispersion over Europe and sample only 10 or 30 points, standing for real measurement points. They also feed the network with two classes of cluster measurements of atmospheric characteristics. One based on some spatial-base transformation (referred to as km^2) and the other based on some temporal aggregation (referred to as density). They also used a scenario where only the most probable origins are given.

For the experiments, several methods were tested; k-means, a shallow DAE (two hidden layers), a Deep DAE (seven hidden layers), deep convolutional AE, Deep multichannel convolutional AE, and PCA with the first 16 principal components. The results (Table 13.3) show that depending on how the variables are given, the complexity of the best architecture is different.

13.3.2 Use of the Decoder

The decoder is capable of mapping from the latent distribution back into the space of the original data $\varphi : \mathcal{X} \mapsto \mathcal{H}$. While previously covered methods make use mainly of the latent space, the decoder also has many applications in Earth system science.

In Li and Misra (2017) the idea was to map from standard measurements of soil composition to nuclear magnetic resonance (NMR) distributions (also known as T2), which are expensive to obtain. NMR is a tool widely used in geological characterization to investigate the structure of geomaterials, possibly filled with hydrocarbons. In the first step, the authors trained a VAE to map from the resulting T2 distributions of NMR to a latent representation and then back to T2, as is usual with VAEs. After training, the weights of the decoder were frozen and the encoder is then exchanged by a fully connected layer that maps from the

Table 13.3 Summary of results in Klampanos et al. (2018). Accuracy of the different methods used to classify clusters of dispersion of a nuclear plume. Data input was as a function of km^2 or density, and for 1 or 3 likely origins.

	10 reading points				30 reading points			
	1 estimated origin		*3 estimated origins*		*1 estimated origin*		*3 estimated origins*	
Methods	**km^2**	**density**	**km^2**	**density**	**km^2**	**density**	**km^2**	**density**
Raw k-means	0.287	0.290	0.584	0.581	0.377	0.385	0.699	0.696
Shallow DAE	**0.310**	0.296	**0.609**	0.601	**0.409**	0.384	**0.739**	0.723
Deep DAE	0.301	**0.303**	0.599	**0.616**	0.406	0.393	0.727	0.727
Deep Conv AE	0.305	0.288	0.603	0.583	0.399	0.389	0.721	0.710
Deep MC Conv AE	0.266	0.301	0.559	0.602	0.359	**0.416**	0.675	**0.745**
PCA (16)	0.297	0.291	0.592	0.595	0.402	0.376	0.722	0.713
PCA^T	0.294	0.251	0.584	0.568	0.394	0.328	0.702	0.683

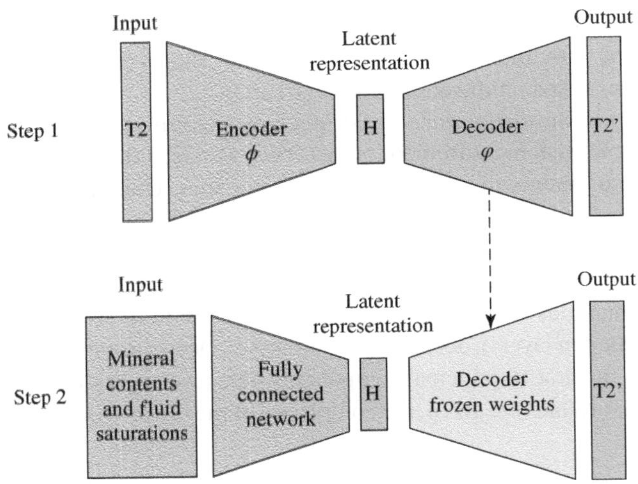

Figure 13.6 Schematic view of the architecture used by Li and Misra (2017). In a first step, a VAE is trained on the results of NMR (T2), then the encoder is exchanged by a fully connected layer that maps from standard measures of soil composition to the latent representation and the decoder maps to T2'. In this way the network learns how to obtain the results of a NMR from another source.

standard measurements of fluids and mineral content to the main NMR features as previously represented by the VAE in the latent space. See the architecture used in Figure 13.6. After the second step training is complete and the model can generate the desired NMR T2 distributions from the fluids and mineral contents measurements.

Scher (2018) implemented an autoencoder-like network to emulate a simple General Circulation Model (GCM). While the network he implemented does not attempt a self-reconstruction of the inputs, it shares a similar bottlenecked architecture. He used as input a complete set of atmospheric fields of the GCM and as target the set of fields at a later time. Given the spatial nature of both inputs and outputs, he made use of a convolutional encoder and a deconvolutional decoder. While one can not expect such a model to be able to generate cohesive time series that respect the physics of the system in the long run, in his study, the network was able to generate similar time series.

13.3.2.1 As a Random Sample Generator

Trained VAEs can be used as a generative model. The latent distribution allows sampling from a prior distribution and decoding that sample back to the original space by using the decoder (for more details see section 13.2.3). Among other uses, the ability to generate new samples can potentially be used to emulate new runs of physical models.

An ensemble of runs of Coupled General Circulation Models are our best approximation to future states of Earth's climate, see Figure 13.7. Producing a sufficient amount of runs to build ensemble forecasts is, however, very computationally costly, especially when large ensembles are needed as in Murphy et al. (2004). Deep learning approaches, such as spatio-temporal VAEs, allow learning the behavior of simulations from previous runs. Ideally, this results in speeding up climate models. While there will be a penalty on prediction accuracy, the speed-up might be worth in some cases. In addition, the same unsupervised autoencoder-like models could be used to approximate the subgrid processes.

As the architecture of the decoder can be freely adapted, implementing recurrent decoders allow for the generation of novel temporal sequences. A spatio-temporal decoder, like the one seen in the work of Lee et al. (2018a), can generate sequences of two-dimensional arrays such as videos. Such architectures allow for the stochastic modeling of the Earth system from raw observations. Currently, no use of this approach for the data-driven modeling of Earth dynamics has been published. In many cases, such a VAE network is aided with an auxiliary discriminative network (forming what became known as a VAEGAN). Adversarial models are explained in more detail in Chapter 3.

13.3.2.2 Anomaly Detection

The encoder-decoder function F is trained on a specific distribution. When used on a sample that does not belong to that distribution, the AE will fail in reconstructing it. This is the basic idea of the use of AEs as anomaly detectors. This anomaly detection approach considers the reconstruction error $\varepsilon = \mathfrak{L}(X, X')$ as a score for anomalies. The farther from the original distribution the input is, the larger ε. Samples exceeding a specific threshold can be considered anomalies.

Anomalies can be of many different types. The main advantage of using unsupervised methods to detect anomalies, such as an autoenconder, relies on the ability to detect unseen anomalies. The supervised counterpart methods are trained with labeled data for known anomalies and have the advantage of being able to classify the anomaly type in a single step. However, they cannot detect unseen anomalies. Recent work by Kawachi et al. (2018) extends the use of VEAs for supervised anomaly detection.

Anomaly detection using the reconstruction loss is a very popular application of deep learning in other fields of science. However, we are not aware of applications in weather

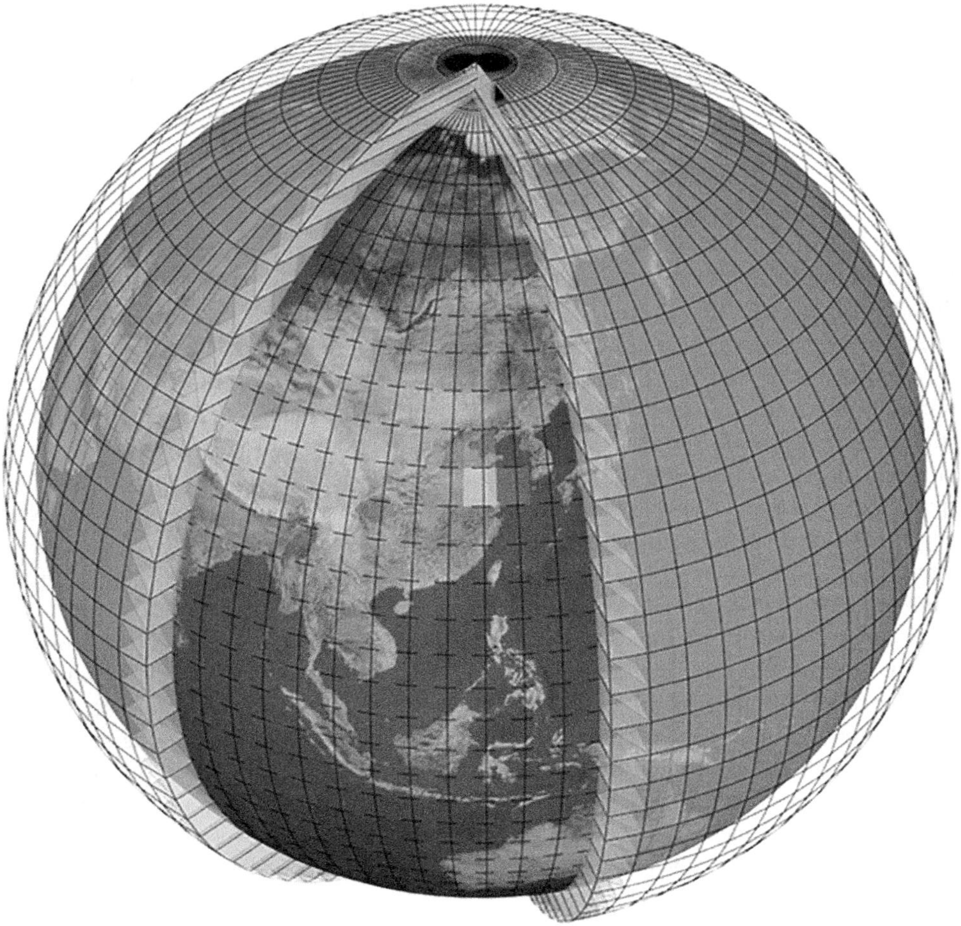

Figure 13.7 Climate data is typically represented on a grid at different levels in both satellite data and the output of numerical models. The spatial structure makes these datasets an ideal case for the use of convolutional architectures. Unsupervised models such as spatio-temporal VAEs can unravel climatic processes and different regimes by analyzing the latent representation of those datasets.

and climate. For a more detailed overview of the use of deep learning for anomaly detection, see Chapter 6.

13.3.2.3 Use of a Denoising Autoencoder (DAE) Decoder

The DAE (Vincent et al. 2008) was conceived as a robust feature extractor. However, its unique training scheme, with pairs of noisy inputs and clean targets, makes them able to clean noise-corrupted images. Variations of DAEs have been used to improve atmospheric river forecasts by Chapman et al. (2019). The forecasts are treated as noisy images, where the noise represents the prediction error, and the network corrects the forecast toward a "clean" image. Using this scheme, the authors achieved state-of-the-art performance on the forecasting of atmospheric rivers.

13.4 Conclusions and Outlook

Machine learning has made significant progress in weather and climate sciences in recent years. The use of AEs has increased as deep learning was widely introduced in the climate community. The use of AEs seems to be particularly appropriate, given the characteristics of the data. In particular, AEs can better represent complex spatio-temporal nature of the underlying datasets that is currently often summarized with indices or convoluted features. Additionally, there are usually no labels making unsupervised AEs useful.

The motivation for AEs comes from their successes in other fields and the diverse utilities that a neural network with an AE structure offers: (1) the unsupervised extraction of useful features to predict, (2) their potential to generate new samples from the original distribution, (3) the removal of noise, and (4) the identification of anomalies. Some of these applications can be used as intermediate steps in larger networks or other machine learning algorithms.

AEs are still not widespread in climate research. We think this is due to, among others, two factors. Firstly, there is still a certain reluctance in the use of models that are not guided by physical principles. Secondly, there is no clear framework in the literature among all the possible configurations of AEs (stacked, variational, LSTM-based, CNN-based, etc.) that researchers can use to determine what architecture is best suited for their particular problem. At the same time, the baseline methods are not clearly defined. Concerning the first obstacle, in recent years, new methods based on deep learning have been developed that allow the incorporation of physical knowledge (Reichstein et al. 2019). Incorporating knowledge into deep learning promotes the acceptance of data-driven models in the community.

We believe that the climate and weather communities will benefit from the different types of AEs. However, the adoption of such techniques by the community is not without challenges. On the one hand, it is necessary to compare climate models and methods. Most of the works presented in this chapter consist of the application of a single deep learning model on spatio-temporal climate or weather data, lacking a comparison with baseline models. Although it is undoubtedly informative that the results make sense, it would be more useful to compare them with baseline methods and other competitive deep learning architectures. In other disciplines, such as image classification (Deng et al. 2009) or causal inference (Runge et al. 2019), there are established standard benchmark databases. There is a need to continue with the effort of Racah et al. (2017) and establish more benchmark databases for climate and weather challenges: prediction, event classification, clustering, etc. This would help researchers to know the state-of-the-art and what architectures are best suited for their particular problem. The availability of high-dimensional multivariate spatio-temporal datasets, often unlabeled, grows every day. AEs constitute a very promising dimensionality reduction method and we expect that they can lead to significant improvements in our understanding of the Earth system and our ability to forecast weather and climate.

14

Deep Learning to Improve Weather Predictions

Peter D. Dueben, Peter Bauer, and Samantha Adams

14.1 Numerical Weather Prediction

Why is it so hard to predict the weather? Generating an accurate and reliable weather forecast is a difficult exercise that requires a complex workflow. In a first step, the initial conditions for forecasts are created, for which **observations** need to be collected that are required to be as accurate as possible, with a global spatial coverage that is as good as possible, across national borders, land and ocean. Unfortunately, it is not feasible to collect observations that describe all processes of the Earth System or to update important boundary conditions – such as land use – in real time. Still, the amount of observations that are collected is impressive. Forecast centers are collecting hundreds of millions of observations every day that include observations from satellites, ships, planes, weather balloons, ground stations, and many more. These observations are evaluated and filtered and a subset of the observations – tens of millions per day – are used for Data Assimilation (DA) and to feed information into the forecast model.

Data assimilation (DA) is a process that combines observations and information from model simulations to generate initial conditions for weather predictions. This process is difficult and error-prone since the quantities that are observed are not the same as the physical fields that are represented within forecast models. To give an example, a satellite radiometer measures radiation at the top of the atmosphere in a specified wavelength which is not the same as the temperature in a broad layer in the atmosphere, as represented within the model, along the path of the satellite orbit. Still, for DA, this radiation signal needs to be compared to model temperatures at the collocated grid point. Furthermore, DA needs to estimate both the errors in the forecast model and of observations that are fed into the model framework.

Once the initial conditions have been fixed, **model simulations** are performed that propagate the model's initial conditions into the future to make a prediction. The underlying dynamical systems of the atmosphere and ocean exhibit non-linear, chaotic behavior which implies that errors in the initial conditions will grow exponentially with time even if the forecast model would be perfect. Therefore, forecast models show significant errors if they are run for long forecast lead times. On the other hand, errors that are caused by

Deep Learning for the Earth Sciences: A Comprehensive Approach to Remote Sensing, Climate Science, and Geosciences, First Edition. Edited by Gustau Camps-Valls, Devis Tuia, Xiao Xiang Zhu, and Markus Reichstein.

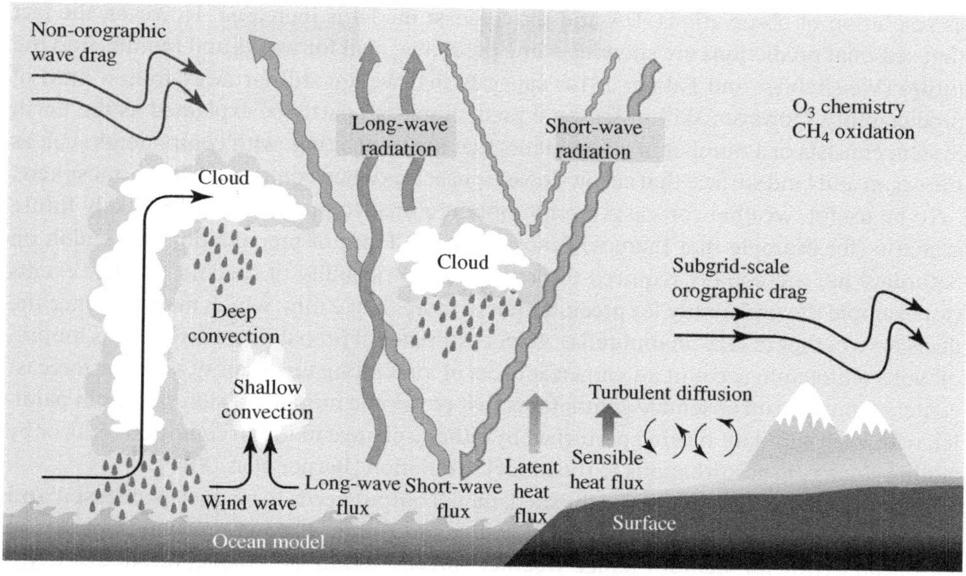

Figure 14.1 Processes that influence weather and climate. The figure is reproduced from Bauer et al. (2015).

insufficiencies of the model tend to grow linearly over time (Magnusson and Källén 2013) and will eventually make a weather prediction useless even if the model could have been started from perfect initial conditions.

Furthermore, the Earth is large and current computing resources restrict the resolution that is available within models to a typical grid-spacing of around 10 km for global weather predictions and around 2 km for limited area models. As illustrated in Figure 14.1, the Earth is also complex, with many processes interacting with each other, which makes it difficult to represent all important physical processes within numerical models. The individual components of the Earth system will also show non-linear behavior and respond on different timescales from seconds to decades. Due to the complexity of the underlying system, forecast model software is also very complex, often with more than one million lines of code. Even so, many of the important processes cannot be represented explicitly within model simulations since they can either not be resolved due to the lack of spatial resolution – the horizontal extent of clouds is, for example, typically smaller than 10 km – or since the equations of the model components are unknown – for example for soil physics. Sub-grid-scale processes need to be described by so-called parametrization schemes, which need to be based on model fields at spatial scales that can be resolved within simulations.

Edward Lorenz has argued that the forecast horizon of weather predictions is limited since the error doubling time (which is the time that it takes for an errors to double in size on average) reduces if smaller and smaller scales of the weather are considered (Lorenz 1969; Palmer et al. 2014). To increase resolution of a weather model by a factor of two will increase the computational cost of the model significantly (a factor of two for each dimension in space and time). However, if the error doubling time is reducing towards smaller scales, this will result in an ever-decreasing increase of the forecast horizon as resolution is increased. This may eventually cause a stagnation of improvements of weather predictions as resolution of observations, DA and the forecast model is increased. However, the fact that seasonal predictions are showing some predictive skill for weeks and months into the future (Weisheimer and Palmer 2014) suggests that we are still far away from a limit of predictability. However, skill in seasonal predictions can partly be explained as the Earth System consists of a number of components that are interacting, with components such as the ocean and land surface that act on slower timescales when compared to the atmosphere.

To be useful, weather forecasts are not only required to provide the most likely future scenario (for example that there will most likely be 1 mm of precipitation in London on Saturday) but are equally required to produce the probability of specific weather events (for example the probability for precipitation of more than 2 mm, which may influence the decision whether to take an umbrella). To get estimates of probability distributions for predictions taking into account all known sources of forecasting uncertainty, weather forecast centers typically run ensemble simulations that perform a number of simulations in parallel, with each simulation being perturbed by either a change in initial conditions and/or by adding stochastic perturbations to the model simulation (Berner et al. 2017).

Once the model simulations are finished, model output needs to be **post-processed and disseminated**. This typically requires selection and compression of model output. At the European Centre for Medium-Range Weather Forecasts (ECMWF) model variables are typically stored with 16 bits per variable while the forecast model is using 64 bits per variable. The forecast results are then distributed to end-users and will eventually reach the general public. The data output is huge, the ensemble forecast of ECMWF is producing more than 70 terabytes of output data during a single day.

This complex workflow needs to be completed end-to-end within about three hours to be useful to the public audience since a weather prediction can only be useful if it is timely. To run through the workflow processing as many observations as possible and with a model at optimal resolution requires supercomputing facilities and therefore weather prediction centers like ECMWF or the UK Met Office have supercomputers that rank within the fastest 50 supercomputers of the world[1].

To build DA and modeling frameworks that scale efficiently to peta-scale supercomputers requires a significant investment in software engineering infrastructure. Examples are the Scalability Programme at ECMWF[2] and the LFRic project at the UK Met Office (Adams et al. 2019). International collaborations between the Weather and Climate community across Europe and worldwide are also required (Lawrence et al. 2018). To optimize models, the community is investigating a number of directions of research including the development of new dynamical cores, the use of domain-specific languages to improve portability, improvements of workflow management, and mixed precision (Biercamp et al. 2019).

Next to the computing challenge, there is also a challenge to manage Earth system data. This includes observations but also the output of weather models and distribution of forecast data. ECMWF's data archive is growing by 233 terabytes per day and has a total storage of 210 petabytes of primary data[3]. This amount of data will likely increase further and it will be more and more challenging to process the data in a timely fashion and to make it

1 https://www.top500.org/
2 https://www.ecmwf.int/en/about/what-we-do/scalability
3 https://www.ecmwf.int/en/computing/our-facilities/data-handling-system

accessible to users, in particular since data storage capacity has not been growing at the same pace as supercomputing processing power.

Despite these challenges, there have been significant improvements in weather forecasts during the past decades for many reasons, including: an increase in the performance of supercomputers, the increase in the number of observations that can be assimilated, the increase in resolution of forecast models, improvements in the efficiency of models, and the increase of complexity of modeling frameworks (Bauer et al. 2015).

14.2 How Will Machine Learning Enhance Weather Predictions?

Many techniques that are already part of the standard toolbox of Numerical Weather Prediction (NWP) scientists can be considered as "Machine Learning" (ML). For example, linear regression, the calculation of teleconnections and correlation, and principal component analysis. The process of DA could also be counted as machine learning. It is therefore not fundamentally new to use Machine learning (ML) to improve predictions. However, there are many other ML tools such as deep neural networks and complex decision trees that are applicable to non-linear systems but have not been used very frequently for weather predictions in the past. New ML tools allow for the representation of non-linear systems that are much more complex when compared to the complexity that could be addressed a decade ago. They also allow scientists to model and interpret processes that would be too complex for the evaluation with standard (linear) techniques. Two examples are the adjustment of model parameters using Approximate Bayesian Computation or unsupervised learning to understand complex, non-linear relationships using self-organizing maps (SOMs; see also chapter 12).

Once considered a niche academic discipline, ML has become increasingly popular over the last 10 years in many domains and achieved human-level performance in many challenging application areas such as Machine Vision and Natural Language Processing. The toolbox of methods, and in particular of neural network architectures, is growing at a breath-taking pace, which allows for customized ML solutions for specific application areas. The largest neural networks that are trained today comprise millions of trainable parameters, allowing for the representation of very complex systems. The full potential of ML has been realized due to various factors. Firstly, developments in computing hardware mean that a single scientist can now easily train a fairly complex deep learning network from gigabytes of data on a single GPU of a laptop. Freely available, open source frameworks that are easy to use (e.g. TensorFlow, Caffe, Keras, MXNet) as well as many free online training resources have helped to democratize ML knowledge. These frameworks allow the development of complex ML applications that can run efficiently on modern supercomputers based on a couple of hundred lines of Python code.

When using these techniques, there is, in principle, no need for an understanding of the underlying physics. ML tools can enable representation of processes where there is no theory available or where the underlying theory is too complex or chaotic to be solved within a numerical model given the computational resources that are available. The community of weather forecasting is only starting to explore the new tool box.

As outlined in the previous section, the Earth system is complex and consists of a number of components that show non-linear dynamics. Furthermore, a lot of data is available for training (both observations and model data). Such an environment therefore offers a large number of potential application areas for ML tools. In fact, applications for ML potentially cover the entire workflow of numerical weather predictions as described in the following section.

14.3 Machine Learning Across the Workflow of Weather Prediction

This section aims to summarize the most evident application areas of ML within the workflow of numerical weather predictions (see Figure 14.2). Some links to the existing literature will be provided but this list is not meant to be complete as the literature of ML applications in numerical weather predictions is growing very fast.

Observations: ML has been used for deriving environmental information from satellite data for several decades (Krasnopolsky and Schiller (2003) and references therein). In this application, ML performs multivariate, non-linear regression of selected geophysical parameters such as precipitation or wind speed over oceans from multi-wavelength radiance observations.

Another obvious use of ML for observations is feature detection and real-time error correction when measurements are performed. This includes, for example the use of ML to reduce data volume for data transmission from satellites. The next step is the processing of observation data once they have reached weather forecast centers. Observations carry systematic biases that can often not be understood from theory but can potentially be learned using ML tools. ML could also be applied to perform real-time anomaly detection. Here, different sources of observations are compared and unreasonable observations are filtered. ML techniques could maybe also be used to correct historical observations (Leahy et al. 2018) and are also discussed for gap-filling strategies (Krasnopolsky et al. 2016).

ML has already been used frequently for “nowcasting” applications. Here, observations are used to predict the future with only a couple of hours lead time. This forecast horizon only allows for very short simulation times and typically does not allow to perform proper DA or to run a complex forecast model. Nowcasting techniques are very commonly used for precipitation forecasting since very short forecast lead times are needed to predict potentially damaging extreme weather events. Precipitation nowcasting direct from radar or satellite data involves optical flow techniques. This is a two-stage method: firstly the computation of a flow field based on a sequence of input 'frames' and secondly advecting forward in time to generate predictions. For this technique, Lagrangian persistence is assumed - i.e. that the predicted field mass only moves and does not change shape, shrink, grow or split. This works well for local analysis, but it is unrealistic when considering larger areas. To overcome these limitations, researchers have begun to explore Machine Learning approaches to learn the underlying equations of motion from observations (e.g. De Bézenac et al. (2017)). Most of the examples in the literature predict image frames from either radar or satellite data, many of them using the newer deep learning neural network architectures such as U-NETs, CNNs, LSTMs, GRUs or hybrids of these which are all suitable for

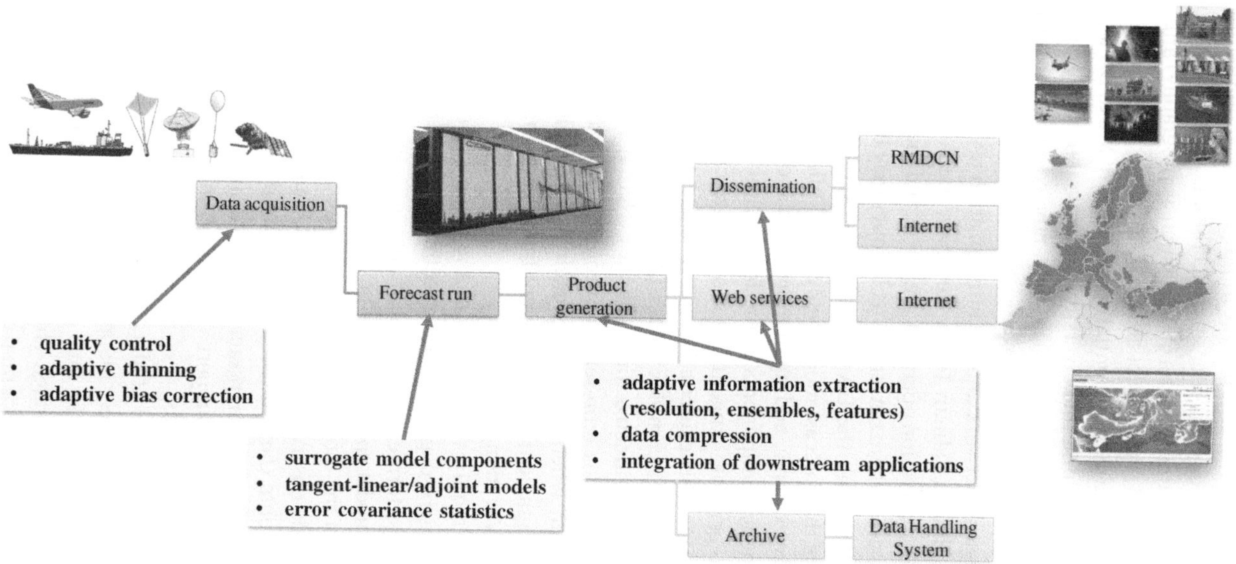

Figure 14.2 Workflow of weather prediction from observations to forecast dissemination and the impact of machine learning on the different workflow components (dark grey boxes). The figure is reproduced from Dueben et al. (2019).

spatio-temporal forecasting problems (Shi et al. 2015b; Heye et al. 2017; Zhang et al. 2017; Foresti et al. 2019; Lebedev et al. 2019b). There are, however, challenges in applying these techniques as they have been designed for quite different applications. Although initial results are promising, ML solutions are not yet improving on traditional methods. More research into probabilistic methods (in order to model the chaotic nature of convective evolution), incorporating other sources of information (such as orography) and incorporating physical mechanisms is needed (Prudden et al. 2020).

Data assimilation: There are many potential applications of ML to DA that could help to make better use of increasing volume and diversity of observations data thus leading to more accurate weather predictions. Actually, DA as such can be considered as being a machine learning process. There are many similarities for example between the minimization process during training of neural network and the minimization used during 4DVar DA or for applications of Kalman Filters (Kalman 1960). However, standard methods often assume that model dynamics and observation operators are linear and that probability distributions are Gaussian. This is one area where newer ML techniques may be able to improve matters.

ML could be used to speed up the DA process in the same ways as suggested for the forecast (see discussion of the Forecast model below), thus allowing for more time to spend on model simulation (for example, adding in more science or increasing resolution). If it is possible to emulate model components of the forward model using ML, it should also be comparably simple to generate tangent linear or adjoint code for ML emulators which are required for 4DVar DA. This exercise is usually difficult and consumes a significant amount of a scientist's time for complex parts of the non-linear forward model, and in particular for the physical parametrization schemes. Neural network emulators could therefore allow for more complex representations of the forward model as tangent linear and adjoint code. However, whether this approach is applicable for complex networks still needs to be shown as gradients of the tangent linear code may become too steep and irregular for use in the 4DVar framework.

Since conventional DA requires estimation of the error covariance matrix, ML could be used to learn this matrix dependent on specific weather situations. Furthermore, bias correction could be performed during the mapping from the spatial distribution in the observation space to model space. ML could also be used to learn model bias when comparing the tendency of the model with analysis increments (this is the forcing that is pushing the model towards observations during DA). Some recent work (Poterjoy et al. 2017, 2019) shows that Monte Carlo methods can be used to create 'local' particle filters without assumptions on the prior or posterior error distributions. Gilbert et al. (2010) have looked at kernel methods and Support Vector Machines as an alternative to standard Kalman filter and variational methods.

One promising application by Hall et al. (2018) uses Generative Adversarial Networks (GANs; see chapter 3) to learn a direct mapping between GOES-15 satellite upper tropospheric water vapor observations to the total column precipitable water (pwat) variable in the Global Forecast System (GFS) model. This shows that ML techniques can help when the observed (sensed) quantities do not map exactly to model variables, in particular if the relationships are non-linear. GANs have also been used by Finn et al. for DA in a Lorenz'96 model (the Lorenz'96 model is a dynamical system formulated by Edward Lorenz (Lorenz 1996) that often serves as toy model for atmospheric dynamics), where they concluded that although the technique is much faster than standard methods, the error is about the same and success is very dependent on the stability of GAN training. Another possible direction is to combine ML with standard DA techniques. Moosavi et al. (2019) have, for example, used ML for adaptive tuning of localization in standard ensemble Kalman filter methods. It has recently been shown for the Lorenz'96 model that ML can be used within a DA framework to either learn the equations of motion or to develop model emulators (Bocquet et al. 2019; Brajard et al. 2019).

Forecast model: In order to make use of accelerator technologies such as GPUs, and potentially also programmable hardware such as FPGAs, many weather and climate centers have been working to port parts of their code bases to these types of hardware (Fuhrer et al. 2018). To port an entire forecast model onto accelerators is a difficult exercise since it requires specialist knowledge and very often code refactoring since traditional weather and climate codes have not been designed with these hardware architectures in mind. Physical parametrization schemes within the model are particularly difficult since they often comprise a very large fraction of the model code, are typically written by domain scientists and are very heterogeneous regarding the code and underlying analytic structures. In contrast, ML techniques such as Neural Networks are ideally suited to running on GPUs and the use of dense linear algebra makes them efficient on almost all hardware. So if it is possible to emulate some of the model components via neural networks they will be able to run efficiently on GPUs. In order to do this, the original model would be run for a long time and input/output pairs of a specific part of the model would be stored. In a second step, the data pairs would be used to train a neural network which would eventually allow replacement of the original model component within forecast simulations. This approach has now been tested by a number of groups and results show great potential (Chevallier et al. 1998; Krasnopolsky et al. 2010; Pal et al. 2019). As well as emulating parametrizations, ML could also be used to develop parametrization schemes which are better in comparison to the schemes that are used with forecast models. If, for example, neural network emulators are trained from super-parametrized simulations, which use a two-dimensional large eddy simulation model to mimic sub-grid-scale cloud features, the neural network emulator may become more realistic in comparison to existing parametrization schemes, in particular for cloud physics and convection (Brenowitz and Bretherton 2019; Gentine et al. 2018; Rasp et al. 2018). For an emulator of the radiation scheme, the neural networks could be trained from schemes that allow to represent 3-dimensional cloud effects or a more detailed evaluation of gas optics which are currently too expensive to be used as the default option. Observations could also be taken as reference (Ukkonen and Mäkelä 2019). Another example where a neural network has been used to emulate and improve on the representation in a model is ozone (He et al. 2019b).

Finally, it will also be interesting whether weather and climate models will be able to make use of the newer accelerators developed specifically for deep learning applications that will be available on the next generation of supercomputers, e.g. in form of tensor processing units (TPUs) or accelerators of low-precision matrix-matrix multiplications. Concerning the latter, a recent study has shown that TensorCores accelerators of NVIDIA Volta GPUs could, in principle, be used to calculate the most expensive kernel of spectral atmosphere models – the Legendre Transformation (Hatfield et al. 2019). The peak performance of half precision matrix-matrix multiplications is sixteen times higher when compared to the speed of double precision matrix-matrix multiplications on the same hardware.

Finally, ML could also open opportunities for the detection of complex features within simulations, such as tropical cyclones and atmospheric rivers (Mudigonda et al. (2017); Kurth et al. (2018); chapter 12). This would potentially allow to trigger more-detailed model output in areas of interest or to run diagnostic or attribution tools online within simulations with no need to tackle the big data challenge when studying huge model output files. However, this area of research typically requires labeled datasets which are often not available or based on existing tools.

Finally, instead of improving the complex non-linear model, neural network tools could also be used to replace forecast models altogether. The equations of motion could simply be learned from data such as observations or re-analysis datasets. For global models, this approach was suggested and tested for the first time in Dueben and Bauer (2018) when performing global forecasts of geopotential height at 500 hPa and the approach was picked up and improved by others (Scher and Messori 2019; Weyn et al. 2019). This approach would be very transformative but many scientists in the community are very skeptical whether neural networks will be sufficient to represent the interaction of complex processes better than existing models. However, for short-term predictions or nowcasting this may be a valid alternative. Physics-guided neural networks may provide an acceptable hybrid solution where constraints are applied to neural networks guided by knowledge about the real physical processes (De Bézenac et al. 2017; Karpatne et al. 2017c; Beucler et al. 2019).

The application of ML is also interesting for very long lead times and seasonal predictions. Nooteboom et al. (2018) have, for example, claimed that a neural network correction combined with a simple dynamical system prediction model of El-Nino indices can be competitive with the same predictions of large weather centers that are based on high-dimensional forecast models.

Post-processing and dissemination: ML has many potential applications that could result in greater accuracy and speed in the post-processing chain. Automated data mining for anomalies or interesting patterns could help forecasters deal with ever-increasing volumes of data and spend more time using their expertise. This could allow for faster dissemination of severe weather event warnings (for example McGovern et al. (2017)). The same way in which ML could be used to compress observational data via feature detection, ML could also be used to compress model output. If, for example, tropical cyclones can be detected within simulations, the position and a couple of variables to define their strength could be diagnosed. This would reduce the amount of data that users who are interested in tropical cyclones would need to evaluate by several orders of magnitude.

It may also be possible to perform regional downscaling and to project coarse resolution model output onto a fine resolution grid using a neural network to increase the quality of local features in the forecast (Rocha Rodrigues et al. 2018). This could, for example, help to improve predictions for valleys in the mountains or for fog at airports when neural networks are trained from model forecasts and high-resolution fields for boundary conditions such as topography and local observations.

Finally, neural networks could also be used for bias correction of forecast errors or the correction of ensemble spread (Krasnopolsky and Lin 2012; Rasp and Lerch 2018; Scher and Messori 2018; Grönquist et al. 2019; Watson 2019). These corrections could be applied both within a running model and as post-processing and allow to take specific weather regimes into account. ML approaches to uncertainty quantification in order to produce better probabilistic forecasts are another possibility since many of the modern deep learning models incorporate probabilistic mechanisms (e.g. Deep belief networks, Variational Autoencoders, GANs) and the idea of quantifying predictive uncertainty in NNs is already being considered in general ML research (Lakshminarayanan et al. 2017). Wang et al. (2019a) have already explored a deep neural network ensemble approach to weather forecasting that incorporates uncertainty quantification.

14.4 Challenges for the Application of ML in Weather Forecasts

There is no fundamental reason why tools that act as a black box should not be used within weather simulations. However, there are a number of open scientific questions that will need to be addressed in the future. The following will provide an (incomplete) list and a preliminary discussion of the most fundamental questions:

- **How can we use our knowledge about the Earth system to build customized ML tools?** We have been able to gather a lot of knowledge about the dynamics of the Earth system which was used to design prediction models for weather and climate. If ML can be used to replace some of the existing tools, the new tools will need to compete with physically informed conventional models. For many approaches in ML, and in particular for deep neural networks, it is still unknown how to use our knowledge about the underlying physical system to improve the ML tools. Designing physically informed neural networks is an active area of research (Reichstein et al. (2019); see also Chapter 23).
- **How can we fix ML models that fail to capture essential dynamical properties?** If ML tools are trained to enhance our models, it is difficult to interpret the dynamics of these black-box tools. However, if one of the fundamental links of atmospheric dynamics, such as the generation of gravity waves from clusters of deep convection, is not represented correctly by a ML solution, it is not clear how to fix such a problem. Given the lack of a physical understanding of what the ML model is doing, it will be difficult for domain scientists to remove or address biases and systematic errors. This is different in conventional models that are based on a physical interpretation. In particular quantitative diagnostic tools are missing.
- **How can we enforce conservation properties?** There are a couple of approaches to introduce a notion of conservation properties into ML products, for example by using cost functions that take these properties into account or by training for cost-functions that are based on predictions of longer lead times (Brenowitz and Bretherton 2018). The work of Beucler et al. (2019) has also suggested building conservation into the ML model architectures. However, these are small steps that cannot be considered as a solution of the problem.
- **How can we introduce ML tools in weather and climate models which suffer from compensating errors?** Modern weather and climate models have been hand-tuned for a couple of decades to provide optimal prediction scores that are visualized using complex score cards (Figure 14.3). The simultaneous tuning of components

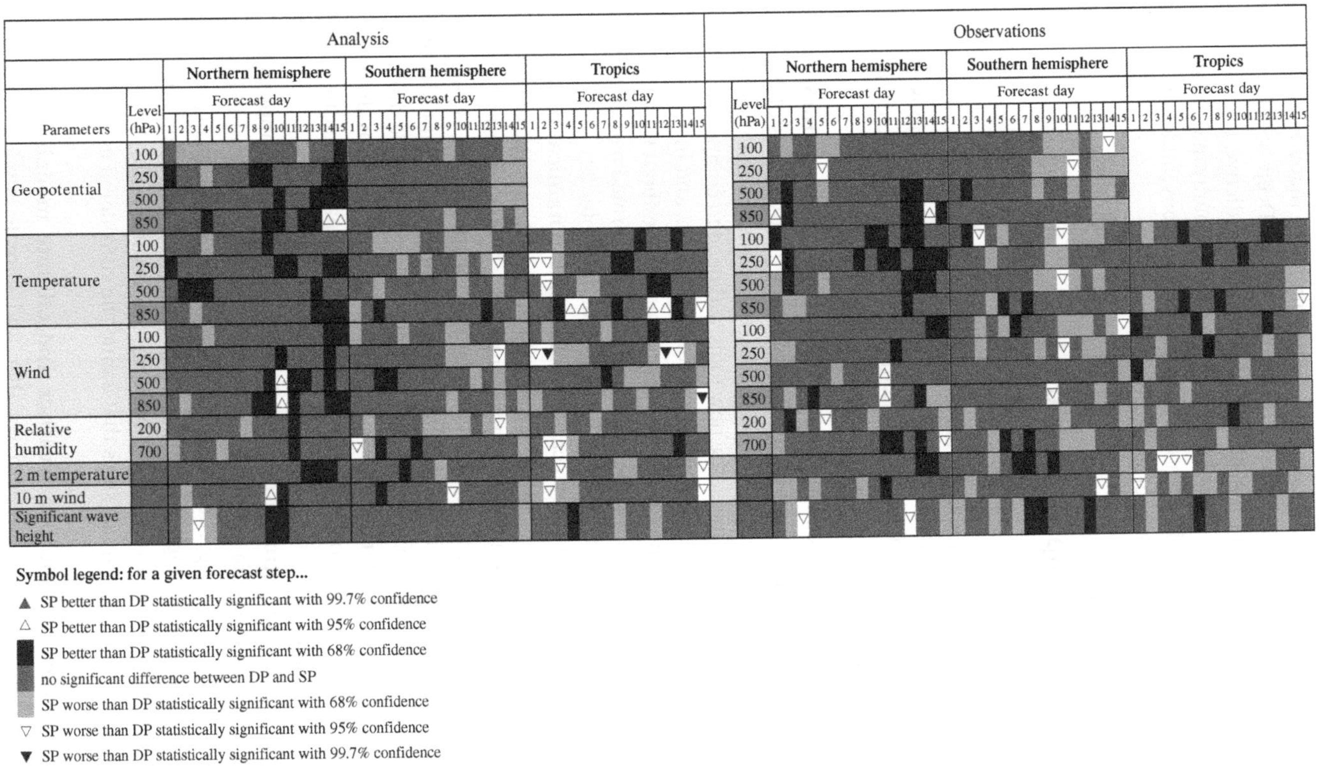

Figure 14.3 Score-card for ensemble simulations at ECMWF (reproduced from Dueben et al. (2019)) that shows the differences between single precision (SP) and double precision (DP) simulations. In a tightly linked system such as the Earth system, a change in one model component (such as parametrization of deep convection) can have unforeseen effects on many components (such as surface temperature at the pole). If the score-card shows a strong negative impact on any of the scores, it is unlikely that the change will be adopted for operational predictions.

to optimize global tendencies often results in compensating errors between the different model components. If individual model components are replaced by ML tools that are, for example, trained from observations or high-resolution simulations, this may generate a degradation of forecast scores for some quantities due to un-compensated errors. This is not a new problem caused by ML and is also a challenge for other approaches that replace individual model components. Furthermore, it would be beneficial in the long term to remove compensating errors from the model. However, it will still make it difficult to introduce ML tools within complex prediction models.

- **How can we ensure reliability in a changing climate?** As the climate is changing, new weather situations will happen and ML tools that are trained on the current climate may fail if, for example, the Arctic is suddenly ice-free in summer. If model components are based on physical understanding, they will be more likely to provide a reliable response to an unforeseen weather situation.
- **How can we ensure reliability when changing the design of parametrization schemes?** Weather and climate models grow and develop. It will therefore be important that the tool-chain is adjustable. If, for example, a neural network emulator of the radiation scheme is developed for the current operational setting, it will be essential that the neural network will also be able to perform a proper emulation if vertical resolution is increased or if the underlying conventional scheme is improved. At the moment, we are still missing the confidence that this will be possible.
- **How can we optimize hyper parameters?** To find the optimal setting of ML tools, and in particular deep neural networks, requires the optimal setting of hyper-parameters to be found – such as the number of layers, the number of neurons per layer, the activation function, optimizer, the use of recurrent or convolutional networks, the number of epochs, etc. Today, this optimization will require a very large number of trial-and-error tests and the optimization may well have step changes when changing one of the hyper-parameters. To have a lot of training cost and a comparably small cost for forecast applications is, in principle, great for weather and climate predictions due to the critical time window of forecasts. However, training costs may become prohibitive as data and network sizes are increasing.
- **How can we prepare supercomputers of weather centers to balance different needs between ML and conventional models?** Today, most weather and climate models are still unable to run on hardware accelerators such as GPUs despite their introduction in HPC in 2006. On the other hand, ML tools, and in particular neural networks, are most efficient on GPUs. This will make the allocation of hardware difficult if a model configuration is relying on both ML and conventional methods.
- **How can we scale neural networks to more complexity?** At the moment, most of the ML solutions that are investigated are still not supercomputing applications but rather networks that are trained on single GPUs with a couple of gigabytes of training data. For challenging ML applications, such as the use for global weather and climate predictions, it will require a complexity of millions of input and output parameters which is still not possible.
- **How can we prepare for different use of data, data mining and data fusion in the future with more/larger data requests?** Deep learning can make use of a sheer unlimited amount of data and still improve the quality of result. As more and more users

are working with complex deep learning tools for both model output (for example of CMIP simulations) or observations, weather and climate prediction centers will need to prepare for changes of data requests and larger data needs.

- **How can we embed user products within the operational forecast model and how can we interface ML code with legacy code?** To run ML tools live within forecast models (as described above) may allow for a new family of use-cases for numerical weather and climate predictions. However, it is still unclear how to couple conventional models with new ML applications and how to make sure that load-balancing and the critical time window for predictions remain unaffected. It is also surprisingly difficult to use common libraries for ML, such as TensorFlow, within the framework of conventional forecast models that are typically based on Fortran.
- **How can we design good training data and labeled datasets?** While there is an enormous amount of Earth system data available, this data is not suitable for many ML applications mainly since the time-frequency of data snapshots is not sufficient to resolve the evolution of important dynamic features. If ML models are trained from observations, these observations will often have biases between products and satellite measurements are only available since a couple of decades. This may often be insufficient for training of complex ML solutions. There is also a lack of large labeled datasets for many applications. If a feature detection algorithm is trained from data that a conventional algorithm for feature detection has labeled, the usefulness of ML is reduced significantly. However, there are first initiatives to generate large labeled datasets (Rasp et al. 2019).
- **How should we train the next generation of weather and climate scientists?** Traditionally, numerical weather predictions have required skills in the domain sciences of the different model components (for example Meteorology, Physics, Chemistry, or Applied Mathematics) but also Software Engineering and HPC. The skill of ML Engineering may need to be added into this mix in the future and it will become even more important to foster collaborations across the borders of the individual domains.

14.5 The Way Forward

As outlined above, the application of new ML tools provides promising opportunities throughout the workflow of weather predictions. However, there are also a number of challenges that will need to be addressed to enable the efficient application of these tools in the future.

The scientific developments should follow both a top-down and bottom-up approach (see Figure 14.4). A top-down approach since it will help to tackle challenges that are relevant for weather predictions as soon as possible. This will require realistic and not idealized applications to be investigated, meaning for example to emulate parametrization schemes at the resolution that is actually used in operational predictions rather than for idealized cases. This is important since the community will require "beacon" applications that show that ML tools are ready for operational use within the next 2–5 years. If this is not achieved, much of the momentum that ML is gathering at the moment will get lost and ML will eventually be considered as a wave that went past. To develop these beacons requires ML applications to be picked wisely and to always evaluate progress regarding usefulness to learn real-world applications. It is also important that scalability of ML tools and the

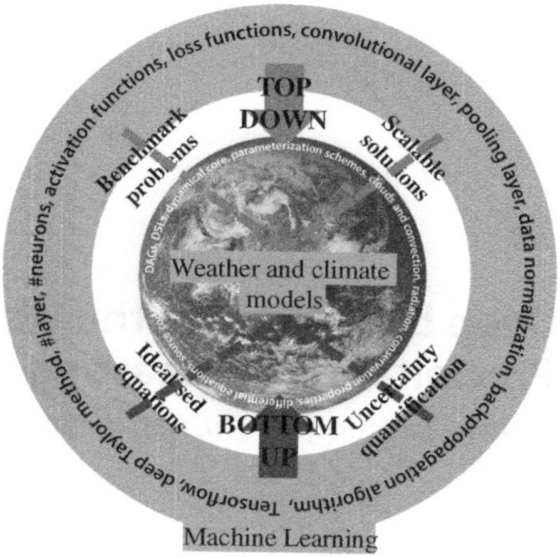

Figure 14.4 A visualization of the way forward. It will require a concerted effort between a top-down approach to develop a scalable, production-ready ML solution for weather predictions and a bottom-up approach that is investigating basic scientific questions within idealized systems. It will furthermore require the study of idealized equations to learn basic rules for ML applications in physical systems, the study of uncertainty quantification for ML tools, the development of benchmark problems to tie ML to weather and climate models as well as the development of scalable solutions that are ready for implementation on modern supercomputers.

high-performance computing framework of today's models is always kept in mind during the development phase.

On the other hand, a bottom-up approach is required at the same time which will perform more idealized studies to answer more general questions such as how to achieve physical interpretability or ways to project physical laws and physical knowledge into the architecture of neural networks. This will require to study how to represent basic differential equations using ML tools and how to develop ML methods that obey conservation laws. It may also be useful to study whether the structure of model source code, and the underlying connectivity of model fields, can be used to generate blue prints of neural network emulators at least for idealized applications. Finally, the application of Bayesian neural networks may allow new possibilities for the representation of model uncertainty and in particular stochastic parametrization schemes that should be explored.

Operational forecast centers will need to provide a healthy environment to support the community of ML. This includes to prepare for a change in data use and data retrievals but also work on infrastructure. There is a need for standardized solutions to couple conventional models to Python libraries that are used for ML. It would also give the community a substantial push if benchmark datasets and problems would be developed for ML in weather and climate predictions (see also Rasp et al. (2020)). Many developments of ML tools were triggered by benchmark problems such as the MNIST dataset[4] for image recognition that allow a quantitative comparison between different ML solutions. This will be more difficult for weather and climate applications when compared to other domains due to the complexity of the underlying system and one benchmark can not possibly cover all application cases in the weather predictions workflow. However, it would help to form a vibrant community and to foster quick developments towards customized ML tools for weather and climate applications.

4 https://en.wikipedia.org/wiki/MNIST_database

15

Deep Learning and the Weather Forecasting Problem: Precipitation Nowcasting

Zhihan Gao, Xingjian Shi, Hao Wang, Dit-Yan Yeung, Wang-chun Woo, and Wai-Kin Wong

15.1 Introduction

Precipitation nowcasting refers to the forecasting of rainfall and other types of precipitation up to 6 hours ahead (as defined by the World Meteorological Organization)[1]. Since rainfall can be localized and highly changeable, users of precipitation nowcast typically demand to know the exact time, location and intensity of rainfall. It is therefore necessary to make very high resolution, both spatially and temporally, precipitation nowcast products in a timely manner, typically in the order of minutes. The most important use of precipitation nowcast is to support the operations of rainstorm warning systems managed by meteorological services around the world. Rainstorm warning systems provide early alerts to the public, disaster risk reduction agencies, government departments in particular those related to public security and works, as well as managers of infrastructures and facilities. Upon the issuance of rainstorm warnings, these parties take actions according to their own standard operating procedures with a view to saving lives and protecting properties. It has tremendous impact on various areas from aviation service and public safety to people's daily life.

For example, commercial airlines rely on precipitation nowcasting to predict extreme weather events and ensures flight safety. On land, heavy rainfall can severely affect the road conditions and increases the risk of traffic accidents, which can be avoided with the help of precipitation nowcasting. For local businesses, the number of customers and their feedback about a restaurant are largely related to the weather (Bujisic et al. 2019), especially the rain rate. Thus, accurate and timely prediction of rainfall helps restaurants predict and adjust their sales strategies. Therefore, the past years have seen an ever-growing need for real-time, large-scale, timely and fine-grained precipitation nowcasting (Xingjian et al. 2015; Shi et al. 2017; Lebedev et al. 2019b; Agrawal et al. 2019). Due to the inherent complexities of the atmosphere and relevant dynamical processes, the problem imposes new challenge to the meteorological community (Sun et al. 2014).

Traditionally, precipitation nowcasting is approached by either *Optical Flow* (OF)-based methods (Li et al. 2000; Reyniers 2008) or numerical methods (Weisman et al. 2008; Sun et al. 2014; Benjamin et al. 2016). OF-based methods first estimate the flow field, which

1 See WMO-No.1198, accessible from https://library.wmo.int/doc_num.php?explnum_id=3795

Deep Learning for the Earth Sciences: A Comprehensive Approach to Remote Sensing, Climate Science, and Geosciences, First Edition. Edited by Gustau Camps-Valls, Devis Tuia, Xiao Xiang Zhu, and Markus Reichstein.

represents the convective motion of the precipitation, with the observed weather data (e.g., the *Constant Altitude Plan Position Indicator* (CAPPI) radar echo maps (Douglas 1990)) and then use the flow field for extrapolation (Woo and Wong 2017). The numerical methods build mathematical models of the atmosphere on top of the physical principles such as the dynamic and thermodynamic laws. Future rainfall intensities are predicted by numerically solving partial differential equations within the mathematical models. However, both approaches have deficiencies that limit their success. The OF-based methods attempt to identify the convective motion of the cloud, but they fail to represent cloud initiation or decay and lack the ability of expressing strong nonlinear dynamics. In addition, the flow field estimation step and the radar echo extrapolation step are separated, making it challenging to determine the best model parameters. Numerical methods can provide reliable forecast but require meticulous simulation of the physical equations. The inference time of numerical models usually take several hours and they are therefore not suitable for generating fine-grained predictions required by precipitation nowcasting.

Recently, a new approach, *deep learning for precipitation nowcasting*, has emerged in the area and shown promising results. Shi et al. (Xingjian et al. 2015) first formulated precipitation nowcasting as a spatiotemporal sequence forecasting problem and proposed a DL-based model, dubbed *Convolutional Long Short-Term Memory* (ConvLSTM), to directly predict the future rainfall intensities based on the past radar echo maps. The model is learned end-to-end with a large amount of historical weather data and performs substantially better than the OF-based algorithm in the operational *Short-range Warning of Intense Rainstorms in Localized Systems* (SWIRLS) developed by the *Hong Kong Observatory* (HKO) (Li et al. 2000; Woo and Wong 2017). After this seminal work, researchers start to explore DL-based methods for precipitation nowcasting and have built models with state-of-the-art performance (Hernández et al. 2016; Shi et al. 2017; Qiu et al. 2017; Tran and Song 2019; Lebedev et al. 2019b; Chen et al. 2019; Agrawal et al. 2019). In essence, precipitation nowcasting is well-suited for DL for three reasons. Firstly, the problem satisfies the big data requirement of DL. Numerous amount of weather data are generated on a daily basis and can be used to train the nowcasting model. For example, in *National Oceanic and Atmospheric Administration* (NOAA), tens of terabytes of data are generated in a single day (Szura 2018). Secondly, DL is suitable for modeling complex dynamical systems (Goodfellow et al. 2016); a single-hidden-layer *Multi-Layer Perceptron* (MLP), which is the most basic form of DL models, is a universal functional approximator (Csáji et al. 2001). Thirdly, the inference speed of DL models is faster than numerical methods (Agrawal et al. 2019). Moreover, in the inference stage, we can dynamically update the DL model with the newly observed data (Shi et al. 2017), making the model more adaptive to the emerging weather patterns.

In this chapter, we introduce the current progress of DL-based methods for precipitation nowcasting. In section 15.2, we describe how to mathematically formulate precipitation nowcasting as a spatiotemporal sequence forecasting problem. In section 15.3, we review the high-level strategies for constructing and learning DL models for precipitation nowcasting; because precipitation nowcasting requires predicting rainfall intensities for multiple timestamps ahead, we introduce various strategies to learn such a multi-step forecasting model. In section 15.4.1 and section 15.4.2, we introduce the DL models in two categories: *Feed-forward Neural Network* (FNN)-based models and *Recurrent Neural Network* (RNN)-based models. In section 15.5, we describe the first systematic benchmark of the DL models for precipitation nowcasting, the HKO-7 benchmark. We conclude this chapter and discuss the potential future works along this area in section 15.6.

15.2 Formulation

Precipitation nowcasting can be formulated as a spatiotemporal sequence forecasting problem. Suppose that the meteorological system is defined over an $M \times N$ grid, in which there are M rows and N columns. Within each cell (i, j) of the grid, there are D measurements that vary over time. By taking snapshots of the system at timestamps $t_1, t_2, ..., t_T$, we get a spatiotemporal sequence that can be denoted as a sequence of tensors $\mathbf{X}_{t_1:t_T} = [\mathbf{X}_{t_1}, \mathbf{X}_{t_2}, \dots, \mathbf{X}_{t_T}]$. Here, $\mathbf{X}_{t_i} \in \mathcal{R}^{D\times M\times N}$ is the observed meteorological data at timestamp t_i. In most DL models for precipitation nowcasting, the meteorological observations $\mathbf{X}_{t_i}$s are $2D$ CAPPI radar echo maps (Xingjian et al. 2015; Shi et al. 2017), satellite images (Lebedev et al. 2019b), or data from another *Quantitative Precipitation Estimation* (QPE) product (Agrawal et al. 2019; Zhang et al. 2016a). Thus, in most cases, the grid is regular and each pixel covers a region in the geographical map, e.g., a 1 km×1 km local area. Some works (Hernández et al. 2016; Qiu et al. 2017) mainly explore the impact of multi-modal meteorological data without considering the spatial correlations. In this case, $M = N = 1$ and the spatiotemporal sequence forecasting problem degenerates to sequence forecasting problem, in which $\mathbf{X}_{t_i}$ degenerates to a vector $\boldsymbol{x}_{t_i} \in \mathcal{R}^D$. In addition, in most scenarios, the time difference between two consecutive snapshots, i.e. $t_{i+1} - t_i$, is always the same. Thus, we are able to simplify the definition and denote $\mathbf{X}_{t_i}$ as $\mathbf{X}_i$.

The spatiotemporal sequence forecasting problem is to predict the most likely length-L sequence in the future given the previous J observations including the current one (Xingjian et al. 2015). The mathematical definition is given in Equation 15.1, in which $\tilde{\mathbf{X}}_{t+1:t+L}$ are the predictions, $\mathbf{X}_{t-J+1:t}$ are the observations, and $p(\mathbf{X}_{t+1:t+L} \mid \mathbf{X}_{t-J+1:t})$ is the model. In the terminology of precipitation nowcasting, the goal is to use the previously observed sequence to predict short term rainfalls of a local region (e.g., Hong Kong, Shanghai, New York, or Tokyo) in the future. In most nowcasting systems, radar echo map is the mainstay of the meteorological observations due to its high spatial and temporal resolutions. The radar maps are usually taken from the weather radar every 6–10 minutes and predictions are given for the following 0–6 hours. If we record one radar frame every six minutes, the task is to predict for 0–60 frames ahead:

$$\tilde{\mathbf{X}}_{t+1:t+L} = \underset{\mathbf{X}_{t+1:t+L}}{\operatorname{argmax}}\, p(\mathbf{X}_{t+1:t+L} \mid \mathbf{X}_{t-J+1:t}). \tag{15.1}$$

The spatiotemporal sequence forecasting problem is different from the conventional multi-variate time series forecasting problem because the prediction target of our problem is a sequence that contains both spatial and temporal structures. Because the number of possible sequences grows exponentially with respect to both the spatial and temporal dimensionality, we have to, in practice, exploit the structure of the spatiotemporal space to reduce the dimensionality and hence make the problem tractable.

15.3 Learning Strategies

Precipitation nowcasting is intrinsically a multi-step forecasting problem. Learning a model for multi-step forecasting is challenging because the elements in the predicted sequence $\hat{\mathbf{X}}_{t+1:t+L}$ are not independent and identically distributed (i.i.d.). Nevertheless, predicting the rainfall for multiple timestamps ahead is a crucial requirement of precipitation nowcasting and DL-based methods adopt different ways to solve the issue. In this section, we introduce the learning strategies for multi-step forecasting. We first explain and compare two basic strategies called *Iterative Multi-step Estimation* (IME) and *Direct Multi-step Estimation* (DME) (Chevillon 2007) and then introduce one extension called *Scheduled Sampling* (SS) that bridges the gap between IME and DME.

Iterative Multi-Step Estimation The IME strategy trains a single-step forecasting model and iteratively feeds the generated samples to the forecaster to get multi-step-ahead predictions. The IME model can either be deterministic or probabilistic. Here, we denote the model as $p(\mathbf{X}_{t+1} \mid \mathbf{X}_{1:t}; \boldsymbol{\theta})$ to cover both cases, in which the deterministic model has a delta distribution. In the terminology of precipitation nowcasting, $\mathbf{X}_{1:t}$ is the sequence of past weather data, $\mathbf{X}_{t+1}$ is the rainfall intensity that the model will predict at timestamp $t+1$, and $\boldsymbol{\theta}$ is the parameter of the model. To train the model, we factorize the distribution $p(\mathbf{X}_{t+1:t+L} \mid \mathbf{X}_{t-J+1:t})$ as $\prod_{i=1}^{L} p(\mathbf{X}_{t+i} \mid \mathbf{X}_{t-J+1:t+i-1}; \boldsymbol{\theta})$. The optimal parameter $\boldsymbol{\theta}^\star$ can be estimated by maximizing the likelihood:

$$\boldsymbol{\theta}^\star = \arg\max_{\phi} \mathbb{E}_{\hat{p}_{\text{data}}} \left[\sum_{i=1}^{L} \log p(\mathbf{X}_{t+i} \mid \mathbf{X}_{t-J+1:t+i-1}; \boldsymbol{\theta}) \right]. \tag{15.2}$$

There are two advantages of the IME approach: (i) The objective function in equation 15.2 is easy to train because it only requires optimizing for the one-step-ahead forecasting error and (ii) we can predict for an arbitrary horizons in the future by recursively applying the basic forecaster. However, there is an intrinsic discrepancy between training and testing in IME. In the training phase, we use the ground-truths from $t+1$ to $t+i-1$ to predict the regional rainfall at timestamp $t+i$, which is also known as teacher-forcing (Goodfellow et al. 2016). While in the testing phase, we feed the model predictions instead of the ground-truths back to the forecaster. This makes the model prone to accumulative errors in the forecasting process (Bengio et al. 2015). Usually, the optimal forecaster for timestamp $t+i$, which is obtained by maximizing $\mathbb{E}_{\hat{p}_{\text{data}}}\left[\log p(\mathbf{X}_{t+i} \mid \mathbf{X}_{t-J+1:t}; \boldsymbol{\theta})\right]$, is not the same as recursively applying the optimal one-step-ahead forecaster when the model is nonlinear. This is because the forecasting error at earlier timestamps will propagate to later timestamps (Lin and Granger 1994).

Direct Multi-Step Estimation The main motivation behind DME is to avoid the error drifting problem in IME by directly minimizing the long-term prediction error. Instead of training a single model, DME trains a different model $p(\mathbf{X}_{t+i} \mid \mathbf{X}_{t-J+1:t}; \boldsymbol{\theta}_i)$ for each forecasting horizon i, in which $\boldsymbol{\theta}_i$ is the parameter. There can thus be L models in the DME approach. The set of optimal parameters $\{\boldsymbol{\theta}_1^\star, \ldots, \boldsymbol{\theta}_L^\star\}$ can be estimated from the following optimization problem:

$$\theta_1^\star, \ldots, \theta_L^\star = \underset{\theta_1,\ldots,\theta_L}{\arg\max}\, \mathbb{E}_{\hat{p}_{\text{data}}} \left[\sum_{i=1}^{L} \log p(\mathbf{X}_{t+i} \mid \mathbf{X}_{t-J+1:t}; \theta_i) \right] \tag{15.3}$$

To disentangle the model size from the number of forecasting steps L, we can also construct $p(\mathbf{X}_{t+i} \mid \mathbf{X}_{t-J+1:t}; \theta_i)$ by recursively applying the single-step forecaster $p(\mathbf{X}_{t+1} \mid \mathbf{X}_{1:t}; \theta)$. In this case, the model parameters $\{\theta_1, \ldots, \theta_L\}$ are shared. For example, when the single-step forecasting model is deterministic and predicts $\tilde{\mathbf{X}}_{t+1}$ as $m(\mathbf{X}_{1:t}; \theta)$, we can obtain the second step prediction by feeding in the predicted rainfall intensity, i.e., $\tilde{\mathbf{X}}_{t+2} = m\left(\mathbf{X}_1, \mathbf{X}_2, \ldots, \mathbf{X}_t, m(\mathbf{X}_{1:t}; \theta); \theta\right)$. By repeating the process for L times, we obtain the predictions $\tilde{\mathbf{X}}_{t+1:t+L}$. The optimal parameter $\theta^\star$ can be estimated by minimizing the distance between the prediction and the ground-truth, i.e., $\theta^\star = \arg\min_\theta \mathbb{E}_{\hat{p}_{\text{data}}} d(\tilde{\mathbf{X}}_{t+1:t+L}, \mathbf{X}_{t+1;t+L})$, in which $d(\cdot, \cdot)$ is a distance function. We need to emphasize here that the aforementioned objective function directly optimizes the multi-step-ahead forecasting error and is different from Equation 15.2, which only minimizes the one-step-ahead forecasting error.

Scheduled Sampling According to Chevillon (2007), DME leads to more accurate predictions when (i) the model is misspecified, (ii) the sequences are non-stationary, or (iii) the training set is too small. However, DME is more computationally expensive than IME. For DME, if the θ_hs are not shared, we need to store and train L models. If the θ_hs are shared, we need to recursively apply the basic forecasting model for $O(L)$ steps (Chevillon 2007; Bengio et al. 2015; Lamb et al. 2016). Both cases require larger memory storage or longer running time than solving the IME objective. Overall, IME is easier to train but less accurate for multi-step forecasting, while DME is more difficult to train but more accurate.

Schedules sampling (SS) Bengio et al. (2015) tries to bridge the gap between IME and DME. The idea of SS is to first train the model with IME and then gradually replace the ground-truths in the objective function with samples generated by the model itself. When all ground-truth samples are replaced with model-generated samples, the training objective falls back into the DME objective. The generation process of SS is described in Equation 15.4:

$$\begin{aligned}
&\forall 1 \leq i \leq L, \\
&\tilde{\mathbf{X}}_{t+i} \sim p(\mathbf{X}_{t+i} \mid \hat{\mathbf{X}}_{t-J+1:t}, \overline{\mathbf{X}}_{t+1:t+i-1}; \theta), \\
&\overline{\mathbf{X}}_{t+i} = (1 - \tau_{t+i})\hat{\mathbf{X}}_{t+i} + \tau_{t+i}\tilde{\mathbf{X}}_{t+i}, \\
&\tau_{t+i} \sim \text{Binomial}(1, \epsilon_k).
\end{aligned} \tag{15.4}$$

Here, $\tilde{\mathbf{X}}_{t+i}$ and $\hat{\mathbf{X}}_{t+i}$ are correspondingly the generated sample and the ground-truth at timestamp $t + i$. $p(\mathbf{X}_{t+i} \mid \hat{\mathbf{X}}_{t-J+1:t}, \overline{\mathbf{X}}_{t+1:t+i-1}; \theta)$ is the basic single-step forecasting model. Meanwhile, τ_{t+h} is generated from a binomial distribution and controls whether to use the ground-truth or the generated sample. ϵ_k is the probability of choosing the ground-truth at the kth iteration. In the training phase, SS minimizes the distance between $\tilde{\mathbf{X}}_{t+1:t+L}$ and $\hat{\mathbf{X}}_{t+1:t+L}$. In the testing phase, τ_{t+i}s are fixed to 0, meaning that the model-generated samples are always used.

SS lies in the mid-ground between IME and DME. If ϵ_k equals to 1, the ground-truths are always chosen, and the objective function will be the same as in the IME strategy. If ϵ_k is 0, the generated samples are always chosen, and the optimization objective will be the same as in the DME strategy. In practice (Bengio et al. 2015; Wang et al. 2019b), ϵ_k is gradually decayed during the training phase to make the optimization objective shift smoothly from IME to DME, which is a type of *curriculum learning* (Bengio et al. 2009).

When applied for precipitation nowcasting, existing DL models adopt either of these three learning strategies. We will introduce the detailed architectures of these two types of models in section 15.4.1 and section 15.4.2 and give an overview of the learning strategy that each model uses in section 15.6.

15.4 Models

15.4.1 FNN-based Models

FNN refers to deep learning models that construct the mapping $\mathbf{Y} = f(\mathbf{X}; \theta)$ by stacking various basic blocks such as the *Fully-Connected* (FC) layer, the convolution layer, the deconvolution layer, and the activation layer. Common types of FNNs include *Multilayer Perceptron* (MLP) (Rosenblatt 1962; Goodfellow et al. 2016), which stacks multiple FC layers and nonlinear activations and *Convolutional Neural Network* (CNN) (Krizhevsky et al. 2012), which stacks multiple convolution layers, pooling layers, deconvolution layers, FC layers, activation layers, normalization layers (Ioffe and Szegedy 2015; Wu and He 2018) and other transformations. The parameters of FNN are usually estimated by minimizing the loss function plus some regularization terms, i.e., $\theta^\star = \arg\min_\theta \mathbb{E}_{\hat{p}_{\text{data}}} \left[l(\mathbf{Y}, f(\mathbf{X}; \theta)) \right] + \Omega(\theta)$ where $l(\cdot,\cdot)$ is the loss function and $\Omega(\cdot)$ is the regularization function such as the L_2 or L_1 loss (Goodfellow et al. 2016). Usually, the optimization problem is solved via stochastic-gradient-based methods (Goodfellow et al. 2016), in which the gradient is computed by backpropagation (Nocedal and Wright 2006).

The convolution layer takes advantage of the translational invariance property of image data. The convolution layer computes the output by scanning over the input and applying the same set of linear filters. Although the input can have an arbitrary dimensionality (Tran et al. 2015), we mainly focus on 2D convolution, since for precipitation nowcasting, the convolution layer is mainly used for extracting the spatial correlation in meteorological images. For input $\mathbf{X} \in \mathcal{R}^{C_i \times H_i \times W_i}$, the output of the convolution layer $\mathbf{H} \in \mathcal{R}^{C_o \times H_o \times W_o}$, which is also known as feature map. Ayzel et al. (2019) proposed DozhdyaNet for precipitation nowcasting. DozhdyaNet consists of only convolution layers and is a type of all convolutional network (Springenberg et al. 2014). The input of the model is a sequence of radar images $\mathbf{X}_{t-J+1:t}$. To facilitate the 2D convolution layer to deal with the sequence of 3D tensors, the author concatenates all frames along the temporal dimension and treats them as different channels:

$$\mathbf{X}_{in} = \text{concat}(\mathbf{X}_{t-J+1}, \mathbf{X}_{t-J+2}, \ldots \mathbf{X}_t), \tag{15.5}$$

where $\mathbf{X}_{in} \in \mathcal{R}^{C'_i \times H_i \times W_i}$ is the network input and the number of channels C'_i is the product of the number of observations within each local grid and the input sequence length, i.e.,

$C_i' = C_i \times J$. Notice that in this manner, the channel C_i' is determined by the input length, hence unlike the RNN-based models, which will be explained in detail in section 15.4.2, the input length must be fixed for the FNN-base models. The radar image at the next timestamp $\mathbf{X}_{t+1}$ is predicted by feeding the preprocessed input sequence $\mathbf{X}_{in}$ into the FNN: $\mathbf{X}_{t+1} = f(\mathbf{X}_{in}; \theta)$. Also, to predict for multiple steps ahead, the author adopted the IME strategy by feeding the predicted radar image back to the network in the inference phase. Also, the author compared different transformation techniques for preprocessing the 2D radar images and used two radar images taken with the 5min interval as the input to predict the next radar image.

Agrawal et al. (2019) also concatenates the input images in the temporal dimension and used the U-Net (Ronneberger et al. 2015a) architecture for prediction. U-Net combines down-sampling, up-sampling and skip connections to learn better hidden representations. Figure 15.1 illustrates how these building blocks are organized. The iterative down-sampling part extracts more global and more abstract representations and the up-sampling part gradually refines the representation and adds the finer details to the generated output. The skip connection helps preserve high-resolution details and facilitates gradient backpropagation. The author used QPE data from the *Multi-Radar Multi-Sensor* (MRMS) system (Zhang et al. 2016a) for training and testing the model. The input is a sequence of radar images taken with 2min interval for one hour and the output is the sequence of radar images for the next several hours. Experiments show that the U-Net based DL model outperforms OF-based model and the HRRR model by NOAA (Benjamin et al. 2016).

Klein et al. Klein et al. (2015) designed the dynamic convolution layer to replace the conventional convolution layer. Instead of using a data-independent filter, the dynamic

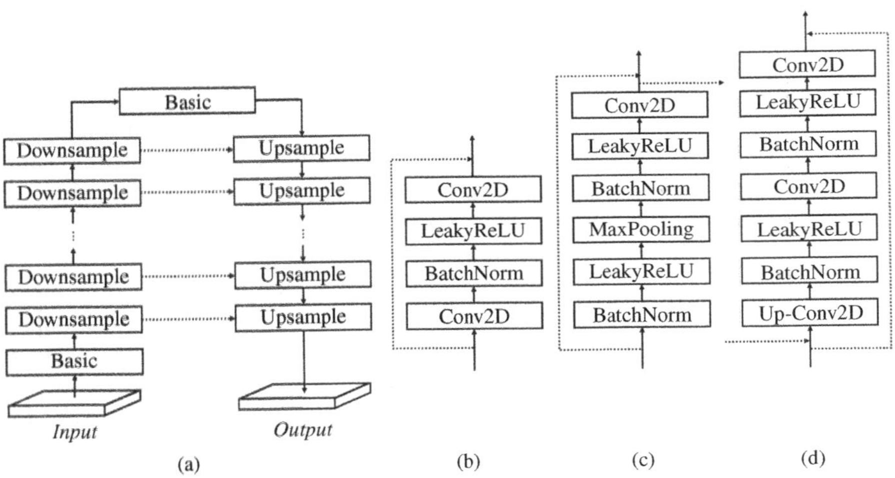

Figure 15.1 (a) The overall structure of the U-NET in Agrawal et al. (2019). Solid lines indicate input connections between layers. Dashed lines indicate skip connections. (b) The operations within the basic layer. (c) The operations within our down-sample layers. (d) The operations within the up-sample layers.

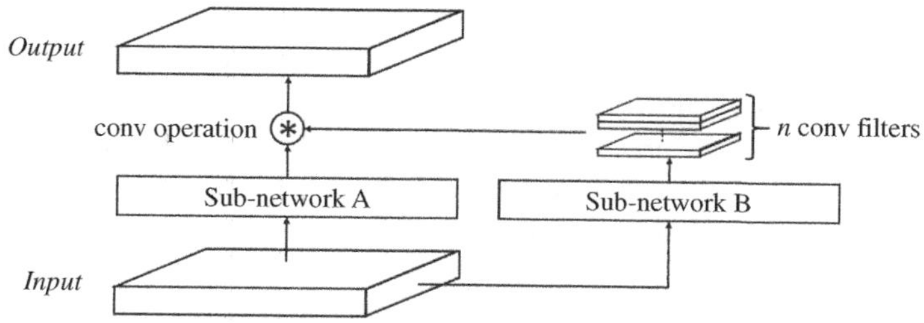

Figure 15.2 The dynamic convolutional layer. The input is fed into two sub-networks. The features are the result of sub-network A while the convolution filters are obtained from sub-network B. The final output of the dynamic convolution layer is computed by convolving the filters from sub-network B across the features from sub-network A.

convolution layer generates both the feature maps and the filters from the input and convolves the filters with the feature maps to get the output. The feature maps and the filters are obtained from the input with two sub-networks. Because the filters are dependent on the input, they will vary from one sample to another in the testing phase. The author concatenates four radar images as the input to predict the next radar image. Also, the author proposed a patch-by-patch synthesis technique which predicts a 10×10 patch in the output from a sequence of 70×70 patches in the input. Figure 15.2 illustrates the workflow of the dynamic convolution layer. Notice that this layer is different from the dynamic filter (Jia et al. 2016) layer that is introduce in section 15.4.2. In dynamic convolution layer, the filter is shared for all locations in the input, while they are adaptively selected in the dynamic filter layer.

Besides the radar images, satellite images are also commonly used as input in FNN-based models. In Lebedev et al. (2019b), satellite images and the observations from Global Forecast System (GFS) (Center 2003) are combined and used as the input. These two types of data are in different modalities and are misaligned with regard to spatial and temporal resolution. Thus, the author remapped them into the same spatial and temporal grid by interpolation. Lebedev et al. (2019b) also applied the U-Net architecture.

Similar to Lebedev et al. (2019b), Hernández et al. (2016) and Qiu et al. (2017) also deal with meteorological data from multiple modalities, including temperature, humidity, wind speed, barometric pressure, Dew point, etc. However, they do not consider the spatial dimension of these data. FNNs with 1D convolution layers and FC layers are built for the $1 \times D$ input data. The weather nowcasting problem is formulated as learning a deterministic mapping $\boldsymbol{Y}_{t+1} = f(\boldsymbol{X}_t)$ that maps the current meteorological observation $\boldsymbol{X}_t$ to the precipitation at next step $\boldsymbol{Y}_{t+1}$. Since the formulation has not fully utilized the spatiotemporal structure of the data, we will not go into the details here.

15.4.2 RNN-based Models

As mentioned in section 15.2, precipitation nowcasting can be formulated as a spatiotemporal sequence forecasting problem with the sequence of past radar maps as input and the sequence of future radar maps as output. In the advancement of DL, RNN-based architectures, such as *Gated Recurrent Unit* (GRU) (Cho et al. 2014) and LSTM (Hochreiter and Schmidhuber 1997), are proven to be effective for modeling sequential data (Sutskever et al.

2014; Karpathy and Fei-Fei 2015; Ranzato et al. 2014; Srivastava et al. 2015; Xu et al. 2015). Different from FNN-based models, which are designed for modeling inputs with static shapes, RNN-based models are designed for modeling dynamic systems. In this section, we introduce the RNN-based models for precipitation nowcasting. We first introduce the encoder-forecaster structure which is the common approach for constructing RNN-based models for spatiotemporal sequence forecasting. Then we introduce the *Convolutional LSTM* (ConvLSTM) network (Xingjian et al. 2015), which combines the advantage of CNN and RNN and is the first DL-based model for precipitation nowcasting. After that, we introduce other RNN-based models like the ConvLSTM with star-shaped bridge (Cao et al. 2019; Chen et al. 2019), *Predictive RNN* (PredRNN) (Wang et al. 2017d), *Memory In Memory* (MIM) Network (Wang et al. 2019b), and the *Trajectory GRU* (TrajGRU) (Shi et al. 2017), which improves upon ConvLSTM from different directions.

15.4.3 Encoder-forecaster Structure

The *Encoder-Forecaster* (EF) structure (Srivastava et al. 2015; Xingjian et al. 2015) is a widely-used neural network architecture for sequence forecasting. It first encodes the observations into a state with an encoder. The state can be a single vector, multiple vectors, or other mathematical objects. Based on the state, it generates the predictions with a forecaster. Following the same notation as in section 15.2, we can formulate the EF structure as follows:

$$\mathbf{H} = f(\mathbf{X}_{t-J+1:t}; \boldsymbol{\theta}_1), \quad \hat{\mathbf{X}}_{t+1:t+L} = g(\mathbf{H}; \boldsymbol{\theta}_2). \tag{15.6}$$

Here, $f(\cdot;\boldsymbol{\theta}_1)$ is the encoder parameterized by $\boldsymbol{\theta}_1$, $g(\cdot;\boldsymbol{\theta}_2)$ is the forecaster parameterized by $\boldsymbol{\theta}_2$, $\hat{\mathbf{X}}_{t+1:t+L}$ are the predictions.

15.4.4 Convolutional LSTM

In the DL community, *Fully-Connected LSTM* (FC-LSTM) is a type of RNN with gates and memory cells for dealing with the vanishing gradient problem (Goodfellow et al. 2016). The formula of FC-LSTM is given as follows:

$$\begin{aligned}
\boldsymbol{i}_t &= \sigma(\boldsymbol{W}_{xi}\boldsymbol{x}_t + \boldsymbol{W}_{hi}\boldsymbol{h}_{t-1} + \boldsymbol{w}_{ci} \odot \mathbf{c}_{t-1} + \boldsymbol{b}_i),\\
\boldsymbol{f}_t &= \sigma(\boldsymbol{W}_{xf}\boldsymbol{x}_t + \boldsymbol{W}_{hf}\boldsymbol{h}_{t-1} + \boldsymbol{w}_{cf} \odot \mathbf{c}_{t-1} + \boldsymbol{b}_f),\\
\mathbf{c}_t &= \boldsymbol{f}_t \odot \mathbf{c}_{t-1} + \boldsymbol{i}_t \odot \tau_h(\boldsymbol{W}_{xc}\boldsymbol{x}_t + \boldsymbol{W}_{hc}\boldsymbol{h}_{t-1} + \boldsymbol{b}_c),\\
\boldsymbol{o}_t &= \sigma(\boldsymbol{W}_{xo}\boldsymbol{x}_t + \boldsymbol{W}_{ho}\boldsymbol{h}_{t-1} + \boldsymbol{w}_{co} \odot \mathbf{c}_t + \boldsymbol{b}_o),\\
\boldsymbol{h}_t &= \boldsymbol{o}_t \odot \tau_o(\mathbf{c}_t),
\end{aligned} \tag{15.7}$$

in which $\boldsymbol{i}_t$, $\boldsymbol{f}_t$, $\boldsymbol{o}_t$ are correspondingly the input gate, forget gate, and output gate. The $\mathbf{c}_{t-1}$, $\mathbf{c}_t$ are the memory cells. $\tau_h(\cdot)$ and $\tau_o(\cdot)$ are the activations, e.g., the "tanh" function. $\boldsymbol{x}_t$s and $\boldsymbol{h}_t$s are input vectors and the hidden states.

However, FC-LSTM is not suitable for precipitation nowcasting in which the input and output are spatiotemporal sequences. The major drawback of FC-LSTM in handling spatiotemporal data is its usage of full-connections in input-state and state-state transitions that loses the spatial structure.

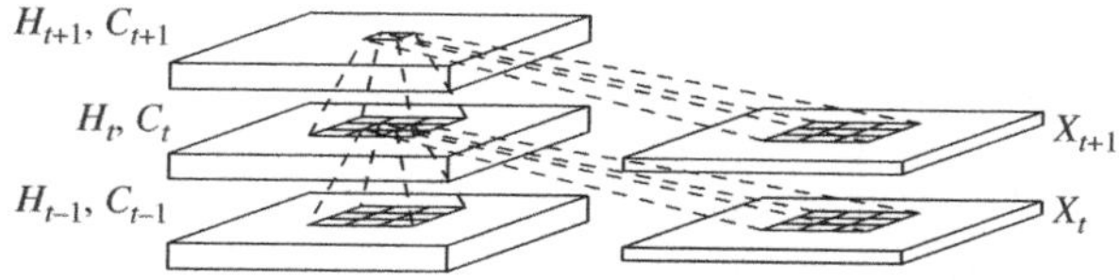

Figure 15.3 Inner structure of ConvLSTM. Source: (Xingjian et al. 2015).

To overcome the problem of FC-LSTM, ConvLSTM (Xingjian et al. 2015) extends FC-LSTM by having convolutional structures in both the input-state and state-state transitions. The key equations of ConvLSTM is given in Equation 15.8. A distinguishing feature of the design is that all the inputs $\mathbf{X}_t$s, cell states $\mathbf{C}_t$s, hidden states $\mathbf{H}_t$s, and gates $\mathbf{I}_t, \mathbf{F}_t, \mathbf{O}_t$ of the ConvLSTM are 3D tensors whose last two dimensions are spatial dimensions (rows and columns). To get a better picture of the inputs and states, we may imagine them as vectors standing on a spatial grid. The ConvLSTM determines the future state of a certain cell in the grid by the inputs and past states of its local neighbors. Figure 15.3 illustrates the connection structure of ConvLSTM.

$$
\begin{aligned}
\mathbf{I}_t &= \sigma(\mathbf{W}_{xi} * \mathbf{X}_t + \mathbf{W}_{hi} * \mathbf{H}_{t-1} + \mathbf{W}_{ci} \odot \mathbf{C}_{t-1} + \boldsymbol{b}_i) \\
\mathbf{F}_t &= \sigma(\mathbf{W}_{xf} * \mathbf{X}_t + \mathbf{W}_{hf} * \mathbf{H}_{t-1} + \mathbf{W}_{cf} \odot \mathbf{C}_{t-1} + \boldsymbol{b}_f) \\
\mathbf{C}_t &= \mathbf{F}_t \odot \mathbf{C}_{t-1} + \mathbf{I}_t \odot \tau_h(\mathbf{W}_{xc} * \mathbf{X}_t + \mathbf{W}_{hc} * \mathbf{H}_{t-1} + \boldsymbol{b}_c) \\
\mathbf{O}_t &= \sigma(\mathbf{W}_{xo} * \mathbf{X}_t + \mathbf{W}_{ho} * \mathbf{H}_{t-1} + \mathbf{W}_{co} \odot \mathbf{C}_t + \boldsymbol{b}_o) \\
\mathbf{H}_t &= \mathbf{O}_t \odot \tau_o(\mathbf{C}_t)
\end{aligned}
\tag{15.8}
$$

In order to ensure that each state has the same number of rows and columns, the author uses zero-padding in the convolution operator and views it as initializing the state of the outside world to be all zero. Also, the traditional FC-LSTM can be viewed as a special case of ConvLSTM with all features standing on a single cell.

The author adopts ConvLSTM as the building block for the EF architecture. The initial states and cell outputs of the forecasting network are copied from the last state of the encoding network. Both encoder and forecaster are formed by stacking several ConvLSTM layers. All states in the forecasting network are concatenated and fed into a 1×1 convolutional layer to generate the final prediction.

The nowcasting model based on ConvLSTM is compared with the OF-based ROVER (Woo and Wong 2017) algorithm operated in HKO and the FC-LSTM based model on a 97-day radar echo data in Hong Kong. Experiments show that ConvLSTM outperforms both two baselines. Also, the results showed that setting the kernel size of the state-state convolution to be larger than 1 is essential for the final performance.

15.4.5 ConvLSTM with Star-shaped Bridge

To make the feature flow in multi-layer ConvLSTM more robust, Cao et al. (2019) proposed the connection structure called star-shaped bridge. In this structure, the states of all ConvLSTM layers at timestamp t are concatenated and passed to a convolution layer with kernel size 1×1 to obtain a global state. The global state has residual connections to all ConvLSTM states at timestamp $t + 1$. The detailed structure is shown in Figure 15.4.

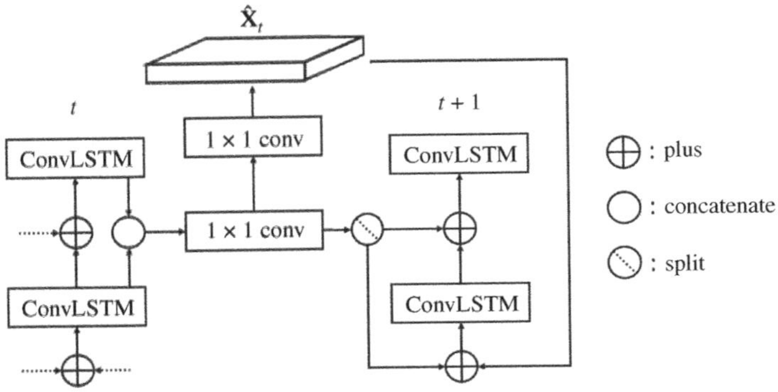

Figure 15.4 Connection structure of the star-shaped bridge.

Apart from the star-shaped bridge, the author also inserts *Group Normalization* (GN) (Wu and He 2018) between ConvLSTM layers. Ablation study shows that the best performance is obtained by combining ConvLSTM, star-shaped bridge, and GN. Also, experiments on 4-year radar echo data from Shanghai, China showed that the learning-based model outperforms the conventional COTREC method (Chen et al. 2019).

15.4.6 Predictive RNN

With memory cells being updated every time step inside the ConvLSTM block, the encoder-forecaster architecture is able to model the underlying temporal dynamics. However, the memory cells across different layers lack mutual communication, which are hence not powerful enough for capturing and memorizing spatial correlations. If we directly stack multiple ConvLSTM layers, the information goes only upwards and makes the features more and more abstract. However, since the network needs to predict a spatiotemporal sequence with fine details, information from the lower-level features, including the raw inputs, should be maintained. To solve the issue, Wang et al. (2017d) proposed *Spatiotemporal LSTM* (ST-LSTM) which keeps an extra memory cell $\mathbf{M}_t^l$ to enhance the memory capacity. The external memory is updated in a zigzag direction illustrated in Figure 15.5 and the author named the whole multi-layer architecture as *Predictive RNN* (PredRNN). The updating rule of ST-LSTM is formulated as follows:

$$\begin{aligned}
\mathbf{I}_t &= \sigma(\mathbf{W}_{xi} * \mathbf{X}_t + \mathbf{W}_{hi} * \mathbf{H}_{t-1}^l + \boldsymbol{b}_i),\\
\mathbf{F}_t &= \sigma(\mathbf{W}_{xf} * \mathbf{X}_t + \mathbf{W}_{hf} * \mathbf{H}_{t-1}^l + \boldsymbol{b}_f),\\
\mathbf{C}_t^l &= \mathbf{F}_t \odot \mathbf{C}_{t-1}^l + \mathbf{I}_t \odot \tanh(\mathbf{W}_{xc} * \mathbf{X}_t + \mathbf{W}_{hc} * \mathbf{H}_{t-1}^l + \boldsymbol{b}_c),\\
\mathbf{I}_t' &= \sigma(\mathbf{W}_{xi}' * \mathbf{X}_t + \mathbf{W}_{mi} * \mathbf{M}_t^{l-1} + \boldsymbol{b}_i'),\\
\mathbf{F}_t' &= \sigma(\mathbf{W}_{xf}' * \mathbf{X}_t + \mathbf{W}_{mf} * \mathbf{M}_t^{l-1} + \boldsymbol{b}_f'),\\
\mathbf{M}_t^l &= \mathbf{F}_t \odot \mathbf{M}_t^{l-1} + \mathbf{I}_t \odot \tanh(\mathbf{W}_{xm} * \mathbf{X}_t + \mathbf{W}_{hm} * \mathbf{M}_t^{l-1} + \boldsymbol{b}_m),\\
\mathbf{O}_t &= \sigma(\mathbf{W}_{xo} * \mathbf{X}_t + \mathbf{W}_{ho} * \mathbf{H}_{t-1}^l + \mathbf{W}_{co} \odot \mathbf{C}_t + \mathbf{W}_{mo} \odot \mathbf{M}_t^l + \boldsymbol{b}_o),\\
\mathbf{H}_t &= \mathbf{O}_t \odot \tanh(\mathbf{W}_{1\times 1} * [\mathbf{C}_t^l, \mathbf{M}_t^l]).
\end{aligned} \tag{15.9}$$

Here, $\mathbf{I}_t, \mathbf{F}_t, \mathbf{H}_t, \mathbf{O}_t$ have the same meaning as in Equation 15.8. $\mathbf{C}_t^l$ means the cell state at layer l at timestamp t, and $\mathbf{M}_t^l$ means the external memory cell that will be updated in a

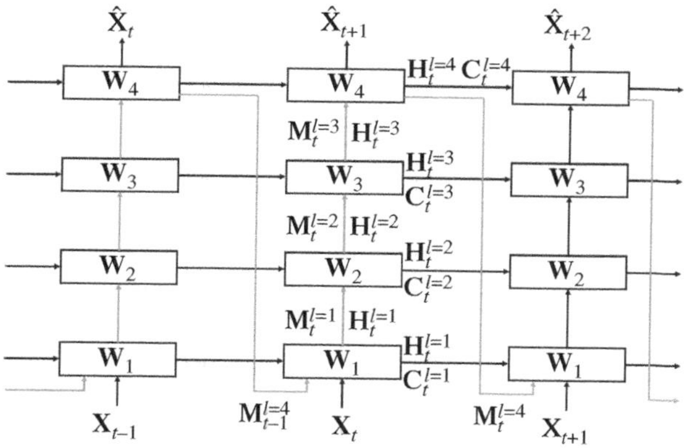

Figure 15.5 Connection structure of PredRNN. The orange arrows in PredRNN denote the flow of the spatiotemporal memory $\mathbf{M}_t^l$.

zigzag order. For the bottom ST-LSTM with $l = 1$, the memory cell from the previous layer is defined as $\mathbf{M}_t^{l-1} = \mathbf{M}_{t-1}^{L}$, which results in a zigzag update flow. Experiments show that PredRNN outperforms the ConvLSTM structure in precipitation nowcasting. The experiment is conducted on a dataset with 10,000 consecutive radar observations recorded every 6 minutes in Guangzhou, China. 10 frames are used as the input to predict the future 10 frames.

15.4.7 Memory in Memory Network

Temporal dynamics of spatiotemporal processes are usually non-stationary. Most RNN-based models approximate the non-stationary dynamics in a stationary manner. To learn a better representation of the underlying high-order non-stationary structure, *Memory In Memory* (MIM) (Wang et al. 2019b) extends the ST-LSTM by replacing the forget gate with another two embedded long short-term memories. It leverages the differential information between neighboring hidden states in the recurrent paths, and can gradually stationarize the spatiotemporal process by stacking multiple MIM blocks.

As shown in Figure 15.6, two cascaded temporal memory recurrent modules are designed to replace the temporal forget gate $\mathbf{F}_t$ in ST-LSTM. The first module additionally taking $\mathbf{H}_{t-1}^{l-1}$ as input is used to capture the non-stationary variations based on the difference $\mathbf{H}_t^{l-1} - \mathbf{H}_{t-1}^{l-1}$ between two consecutive hidden representations. Thus, it is named as the non-stationary module (shown as MIM-N in Figure 15.7). It generates differential features $\mathbf{D}_t^l$ based on the difference-stationary assumption. The other recurrent module takes the output $\mathbf{D}_t^l$ of the MIM-N module and the outer temporal memory $\mathbf{C}_{t-1}^l$ as inputs to capture the approximately stationary variations in spatiotemporal sequences. Thus, it is named as the stationary module (shown as MIM-S in Figure 15.7). By replacing the forget gate with the final output $\mathbf{T}_t^l$ of the cascaded non-stationary and stationary modules, the non-stationary dynamics can be captured more effectively. The complete formula of MIM is given as follows:

$$
\begin{aligned}
\mathbf{I}_t &= \sigma(\mathbf{W}_{xi} * \mathbf{X}_t + \mathbf{W}_{hi} * \mathbf{H}_{t-1}^l + \boldsymbol{b}_i),\\
\mathbf{F}_t &= \sigma(\mathbf{W}_{xf} * \mathbf{X}_t + \mathbf{W}_{hf} * \mathbf{H}_{t-1}^l + \boldsymbol{b}_f),\\
\mathbf{D}_t^l &= \text{MIM-N}(\mathbf{H}_t^{l-1}, \mathbf{H}_{t-1}^{l-1}, \mathbf{N}_{t-1}^l),\\
\mathbf{T}_t^l &= \text{MIM-S}(\mathbf{D}_t^l, \mathbf{C}_{t-1}^l, \mathbf{S}_{t-1}^l),
\end{aligned}
$$

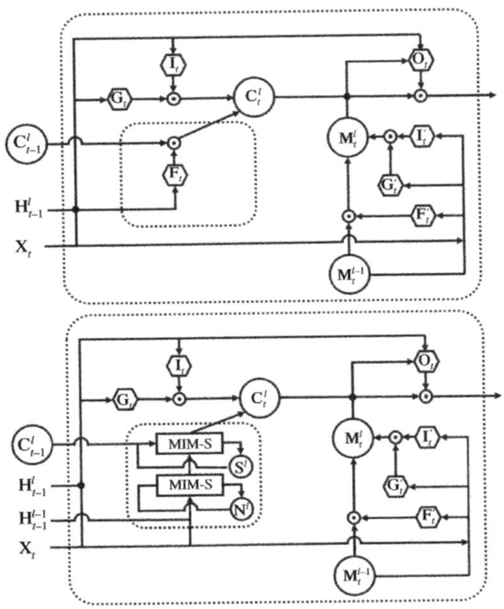

Figure 15.6 ST-LSTM block (top) and Memory In Memory block (bottom). For brevity, $\mathbf{G}_t = \tanh(\mathbf{W}_{xc} * \mathbf{X}_t + \mathbf{W}_{hc} * \mathbf{H}_{t-1}^l + \boldsymbol{b}_c)$, $\mathbf{G}_t' = \tanh(\mathbf{W}_{xm} * \mathbf{X}_t + \mathbf{W}_{hm} * \mathbf{M}_t^{l-1} + \boldsymbol{b}_m)$. MIM is designed to introduce two recurrent modules (yellow squares) to replace the forget gate (dashed box) in ST-LSTM. MIM-N is the non-stationary module and MIM-S is the stationary module.

Figure 15.7 The non-stationary module (MIM-N) and the stationary module (MIM-S), which are interlinked in a cascaded structure in the MIM block. Non-stationarity is modeled by differencing.

$$
\begin{aligned}
\mathbf{C}_t^l &= \mathbf{F}_t \odot \mathbf{T}_t^l + \mathbf{I}_t \odot \tanh(\mathbf{W}_{xc} * \mathbf{X}_t + \mathbf{W}_{hc} * \mathbf{H}_{t-1}^l + \boldsymbol{b}_c),\\
\mathbf{I}_t' &= \sigma(\mathbf{W}_{xi}' * \mathbf{X}_t + \mathbf{W}_{mi} * \mathbf{M}_t^{l-1} + \boldsymbol{b}_i'),\\
\mathbf{F}_t' &= \sigma(\mathbf{W}_{xf}' * \mathbf{X}_t + \mathbf{W}_{mf} * \mathbf{M}_t^{l-1} + \boldsymbol{b}_f'),\\
\mathbf{M}_t^l &= \mathbf{F}_t \odot \mathbf{M}_t^{l-1} + \mathbf{I}_t \odot \tanh(\mathbf{W}_{xm} * \mathbf{X}_t + \mathbf{W}_{hm} * \mathbf{M}_t^{l-1} + \boldsymbol{b}_m),\\
\mathbf{O}_t &= \sigma(\mathbf{W}_{xo} * \mathbf{H}_t^{l-1} + \mathbf{W}_{ho} * \mathbf{H}_{t-1}^l + \mathbf{W}_{co} \odot \mathbf{C}_t + \mathbf{W}_{mo} \odot \mathbf{M}_t^l + \boldsymbol{b}_o),\\
\mathbf{H}_t &= \mathbf{O}_t \odot \tanh(\mathbf{W}_{1\times 1} * [\mathbf{C}_t^l, \mathbf{M}_t^l]),
\end{aligned}
\tag{15.10}
$$

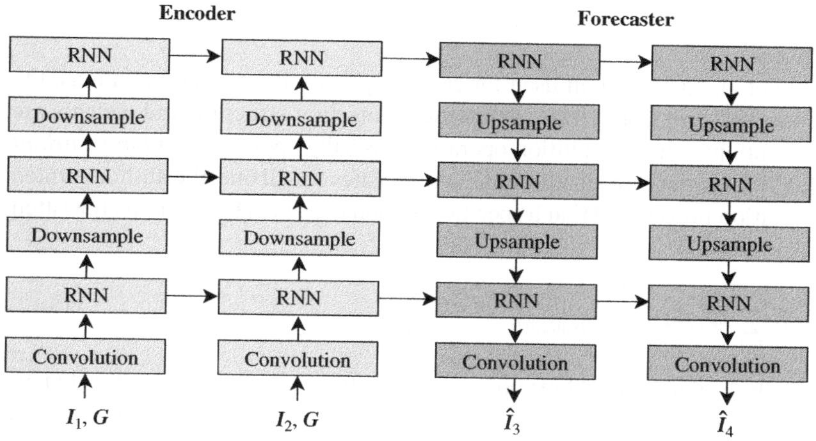

Figure 15.8 Encoder-forecaster architecture adopted in Shi et al. (2017). Source: Shi et al. (2017).

where **S** and **N** denote the horizontally-transited memory cells in the non-stationary module (MIM-N) and stationary module (MIM-S) respectively; $\mathbf{D}_t^l$s are the differential features learned by MIM-N; $\mathbf{T}_t^l$ is the memory passing the virtual "forget gate". MIM-N is a ConvLSTM with $\mathbf{H}_t^{l-1} - \mathbf{H}_{t-1}^{l-1}$ as the hidden state input. MIM-S is a ConvLSTM with $\mathbf{D}_t^l$ as the hidden state input. The detailed formula are omitted here and readers can refer to Wang et al. (2019b) for more details.

15.4.8 Trajectory GRU

Shi et al. (2017) proposed a U-Net-like modification to the EF architecture. As Figure 15.8 illustrates, the order of the forecaster network is reversed comparing to that in Xingjian et al. (2015). There are downsampling and upsampling layers between the RNNs, which are implemented by strided convolution and deconvolution. In this structure, the encoder adopts a local-to-global feature extraction process while the decoder adopts a coarse-to-fine generation process. There are "skip-connections" between the encoder and the forecaster to preserve the details from the raw inputs.

Along with the new EF structure, Shi et al. (2017) also pointed out a side-effect of the convolution operation in ConvLSTM. In essence, the location-invariant convolution filters are inefficient to capture location-variant spatiotemporal relationship. To overcome this problem, Shi et al. (2017) proposed the *Trajectory GRU* (TrajGRU) model which uses a sub-network to output the state-state connection structures before state transitions. TrajGRU extends upon *Convolutional GRU* (ConvGRU), which is a variant of ConvLSTM, and allows the state to be aggregated along some learned trajectories. The formula of ConvGRU is given as follows:

$$
\begin{aligned}
\mathbf{Z}_t &= \sigma(\mathbf{W}_{xz} * \mathbf{X}_t + \mathbf{W}_{hz} * \mathbf{H}_{t-1} + \boldsymbol{b}_z),\\
\mathbf{R}_t &= \sigma(\mathbf{W}_{xr} * \mathbf{X}_t + \mathbf{W}_{hr} * \mathbf{H}_{t-1} + \boldsymbol{b}_r),\\
\mathbf{H}_t' &= f(\mathbf{W}_{xh} * \mathbf{X}_t + \mathbf{R}_t \odot (\mathbf{W}_{hh} * \mathbf{H}_{t-1} + \boldsymbol{b}_h)),\\
\mathbf{H}_t &= (1 - \mathbf{Z}_t) \odot \mathbf{H}_t' + \mathbf{Z}_t \odot \mathbf{H}_{t-1}.
\end{aligned}
\tag{15.11}
$$

Here, $\mathbf{H}_t, \mathbf{R}_t, \mathbf{Z}_t, \mathbf{H}'_t \in \mathbb{R}^{C_h \times H \times W}$ are the memory state, reset gate, update gate, and new information, respectively.

As stated in Shi et al. (2017), when used for capturing spatiotemporal correlations, the deficiency of ConvGRU and ConvLSTM is that the connection structure and weights are fixed for all the locations. The convolution operation basically applies a *location-invariant* filter to the input. If the inputs are all zero and the reset gates are all one, the author pointed out that the calculation process of $\mathbf{H}'_t$ at a specific location (i,j), i.e, $\mathbf{H}'_{t,:,i,j}$, can be rewritten as follows:

$$\mathbf{H}'_{t,:,i,j} = f(\sum_{l=1}^{|\mathcal{N}^h_{i,j}|} \mathbf{W}^l_{hh} \mathbf{H}_{t-1,:,p(l,i,j),q(l,i,j)}), \tag{15.12}$$

in which $\mathcal{N}^h_{i,j}$ is the ordered neighborhood set at location (i,j) defined by the hyperparameters of the state-state convolution such as kernel size, dilation and padding (Yu and Koltun 2016). $(p(l,i,j), q(l,i,j))$ is the lth neighborhood location corresponding to position (i,j).

Based on this observation, TrajGRU uses the current input and previous state to generate the local neighborhood set for each location at each timestamp. The detailed formula is given in Equation 15.13. Here, L is the number of allowed links. $\mathbf{U}_t, \mathbf{V}_t \in \mathbb{R}^{L \times H \times W}$ are the flow fields that store the local connection structure generated by $\gamma(\mathbf{X}_t, \mathbf{H}_{t-1})$. The $\mathbf{W}^l_{hz}, \mathbf{W}^l_{hr}, \mathbf{W}^l_{hh}$ are the weights for projecting the channels and were chosen as 1×1 convolutions in the paper. The warp$(\mathbf{H}_{t-1}, \mathbf{U}_{t,l}, \mathbf{V}_{t,l})$ function selects the positions pointed out by $\mathbf{U}_{t,l}, \mathbf{V}_{t,l}$ from $\mathbf{H}_{t-1}$ via the bilinear sampling kernel (Jaderberg et al. 2015; Ilg et al. 2017; Shi et al. 2017).

$$\begin{aligned}
\mathbf{U}_t, \mathbf{V}_t &= \gamma(\mathbf{X}_t, \mathbf{H}_{t-1}),\\
\mathbf{Z}_t &= \sigma(\mathbf{W}_{xz} * \mathbf{X}_t + \sum_{l=1}^{L} \mathbf{W}^l_{hz} * \text{warp}(\mathbf{H}_{t-1}, \mathbf{U}_{t,l}, \mathbf{V}_{t,l})),\\
\mathbf{R}_t &= \sigma(\mathbf{W}_{xr} * \mathbf{X}_t + \sum_{l=1}^{L} \mathbf{W}^l_{hr} * \text{warp}(\mathbf{H}_{t-1}, \mathbf{U}_{t,l}, \mathbf{V}_{t,l})),\\
\mathbf{H}'_t &= f(\mathbf{W}_{xh} * \mathbf{X}_t + \mathbf{R}_t \odot (\sum_{l=1}^{L} \mathbf{W}^l_{hh} * \text{warp}(\mathbf{H}_{t-1}, \mathbf{U}_{t,l}, \mathbf{V}_{t,l}))),\\
\mathbf{H}_t &= (1 - \mathbf{Z}_t) \odot \mathbf{H}'_t + \mathbf{Z}_t \odot \mathbf{H}_{t-1}.
\end{aligned} \tag{15.13}$$

The advantage of such a structure is that it could learn the connection topology by learning the parameters of the subnetwork γ. γ has only a small number of parameters and thus adds nearly no cost to the overall computation. Compared to a ConvGRU with $K \times K$ state-state convolution, TrajGRU is able to learn a more efficient connection structure with $L < K^2$. For ConvGRU and TrajGRU, the number of model parameters is dominated by the size of the state-state weights, which is $O(L \times C_h^2)$ for TrajGRU and $O(K^2 \times C_h^2)$ for ConvGRU. If L is chosen to be smaller than K^2, the number of parameters of TrajGRU can also be smaller than the ConvGRU and the TrajGRU model is able to use the parameters more efficiently. Illustration of the recurrent connection structures of ConvGRU and TrajGRU is given in Figure 15.9.

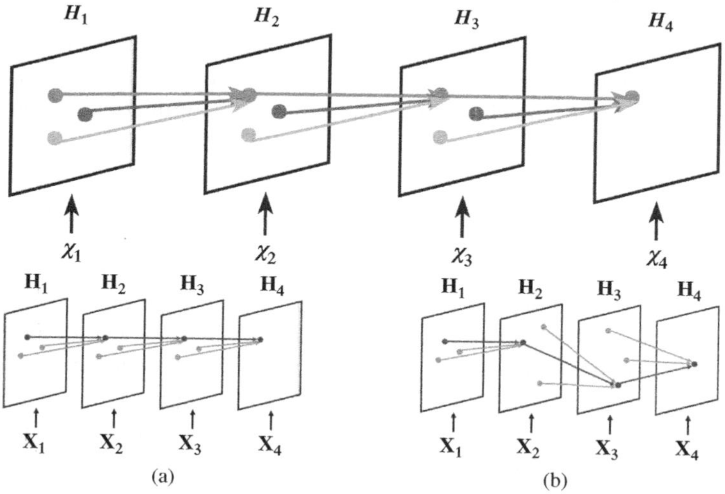

Figure 15.9 Top: For convolutional RNN, the recurrent connections are fixed over time. Bottom: For trajectory RNN, the recurrent connections are dynamically determined. Comparison of the connection structures of convolutional RNN and trajectory RNN. Links with the same color share the same transition weights. (Best viewed in color). Source of figure: Shi et al. (2017).

Experiments in the paper showed that TrajGRU outperforms ConvGRU, 2D CNN, 3D CNN, and the ROVER algorithm in precipitation nowcasting.

15.5 Benchmark

Despite the rapid development of DL models in solving this problem, the way to evaluate the models has some deficiencies. Firstly, the deep learning models are only evaluated on relatively small dataset containing limited data frames. Secondly, different models report evaluation results on different criteria. As the needs of real-world precipitation nowcasting system diverge from indicating raining or not to rainstorms alert, single criterion is not sufficient for demonstrating the algorithm's overall performance. Thirdly, in the real-world scenario, the meteorological data arrive in a stream and the nowcasting algorithm should be able to actively adapt to the new-coming sequences. Considering this online setting is not less important than considering offline setting with fixed-length input. In fact, as the area *deep learning for precipitation nowcasting* is still in its early stages, it is not clear how models should be evaluated to meet the needs of real-world applications.

Shi et al. (2017) proposed the large-scale HKO-7 benchmark for precipitation nowcasting to address this problem. HKO-7 benchmark is built on the HKO-7 dataset containing radar echo data from 2009 to 2015 near Hong Kong. Since the radar echo maps arrive in a stream in the real-world scenario, the nowcasting algorithms can adopt online learning to adapt to the newly emerging patterns dynamically. To take this setting into account, there are two testing protocols in this benchmark: the offline setting in which the algorithm can only use a fixed window of the previous radar echo maps and the *online setting* in which the algorithm is free to use all the historical data and any online learning algorithm. Another issue for the precipitation nowcasting task is that the proportions of rainfall events at different rain-rate thresholds are highly imbalanced. Heavier rainfall occurs less often but has a higher real-world impact. *Balanced Mean Squared Error* (B-MSE) and *Balanced Mean Absolute Error* (B-MAE) measures are thus introduced for training and evaluation, which assign more weights to heavier rainfalls in the calculation of MSE and MAE. Empirical study showed that the balanced variants of the loss functions are more consistent with the overall nowcasting performance at multiple rain-rate thresholds than the original loss functions. Moreover, training with the balanced loss functions is essential for deep learning models to achieve good performance at higher rain-rate thresholds.

Using the new dataset, testing protocols, and training loss, there are seven models being extensively evaluated, including a simple baseline model which always predicts the last frame, two OF-based models (ROVER and its nonlinear variant), and four representative deep learning models (TrajGRU, ConvGRU, 2D CNN, and 3D CNN). This large-scale benchmark for precipitation nowcasting is the first comprehensive benchmark of deep learning models for the precipitation nowcasting problem.

15.5.1 HKO-7 Dataset

The HKO-7 dataset used in the benchmark contains radar echo data from 2009 to 2015 collected by HKO. The radar CAPPI reflectivity images, which have resolution of 480 × 480 pixels, are taken from an altitude of 2 km and cover a 512 km×512 km area centered in Hong Kong. The data are recorded every 6 minutes and hence there are 240 frames per day. The raw logarithmic radar reflectivity factors are linearly transformed to pixel values via $\text{pixel} = \lfloor 255 \times \frac{\text{dBZ}+10}{70} + 0.5 \rfloor$ and are clipped to be between 0 and 255. The radar reflectivity values are converted to rainfall intensity values (mm/h) using the Z-R relationship: $\text{dBZ} = 10 \log a + 10b \log R$ where R is the rain-rate level, $a = 58.53$, and $b = 1.56$. As rainfall events occur sparsely, the rainy days are selected based on the rain barrel information to form the final dataset, which has 812 days for training, 50 days for validation and 131 days for testing.

The raw radar echo images generated by Doppler weather radar are noisy due to factors like ground clutter, sea clutter, anomalous propagation and electromagnetic interference (Lee and Kim 2017). To alleviate the impact of noise in training and evaluation, Noisy pixels are filtered out by generating the noise masks with a two-stage process.

15.5.2 Evaluation Methodology

As the radar echo maps arrive in a stream, nowcasting algorithms can apply online learning to adapt to the newly emerging spatiotemporal patterns. The evaluation protocol of HKO-7 benchmark consists of two settings: (i) the offline setting in which the algorithm always receives 5 frames as input and predicts 20 frames ahead, and (ii) the online setting in which the algorithm receives segments of length 5 sequentially and predicts 20 frames ahead for each new segment received. The testing environment guarantees that the same set of sequences is tested in both the offline and online settings for fair comparison.

For both settings, models are evaluated according to the skill scores for multiple thresholds that correspond to different rainfall levels to give an all-round evaluation of the algorithms' nowcasting performance. Table 15.1 shows the distribution of different

Table 15.1 Rain rate statistics in the HKO-7 benchmark.

Rain Rate (mm/h)			Proportion (%)	Rainfall Level
$0 \le$	x	< 0.5	90.25	No / Hardly noticeable
$0.5 \le$	x	< 2	4.38	Light
$2 \le$	x	< 5	2.46	Light to moderate
$5 \le$	x	< 10	1.35	Moderate
$10 \le$	x	< 30	1.14	Moderate to heavy
$30 \le$	x		0.42	Rainstorm warning

Source: (Shi et al. 2017).

rainfall levels in HKO-7 dataset. The thresholds 0.5, 2, 5, 10, 30 are selected to calculate the CSI and Heidke Skill Score (HSS) (Hogan et al. 2010). For calculating the skill score at a specific threshold τ, which is 0.5, 2, 5, 10 or 30, the pixel values in prediction and ground-truth are first converted to 0/1 by thresholding with τ. Then calculate the TP (prediction=1, truth=1), FN (prediction=0, truth=1), FP (prediction=1, truth=0), and TN (prediction=0, truth=0). The CSI score is calculated as $\frac{\text{TP}}{\text{TP+FN+FP}}$ and the HSS score is calculated as $\frac{\text{TP}\times\text{TN}-\text{FN}\times\text{FP}}{(\text{TP+FN})(\text{FN+TN})+(\text{TP+FP})(\text{FP+TN})}$. During the computation, the masked noisy points are ignored.

As shown in Table 15.1, the frequencies of different rainfall levels are highly imbalanced. Using weighted loss function helps solve this problem. Specifically, a weight $w(x)$ is assigned to each pixel according to its rainfall intensity x:

$$w(x) = \begin{cases} 1, & x < 2 \\ 2, & 2 \le x < 5 \\ 5, & 5 \le x < 10 \\ 10, & 10 \le x < 30 \\ 30, & x \ge 30 \end{cases}.$$

Also, the masked pixels have weight 0. The resulting B-MSE and B-MAE scores are computed as B-MSE $= \frac{1}{N}\sum_{n=1}^{N}\sum_{i=1}^{480}\sum_{j=1}^{480} w_{n,i,j}(x_{n,i,j} - \hat{x}_{n,i,j})^2$ and B-MAE $= \frac{1}{N}\sum_{n=1}^{N}\sum_{i=1}^{480}\sum_{j=1}^{480} w_{n,i,j}|x_{n,i,j} - \hat{x}_{n,i,j}|$, where N is the total number of frames and $w_{n,i,j}$ is the weight corresponding to the (i,j)th pixel in the nth frame. For the conventional MSE and MAE measures, all the weights are simply set to 1 except the masked points.

15.5.3 Evaluated Algorithms

There are seven nowcasting algorithms for evaluation in HKO-7 benchmark, including the simplest model which always predicts the last frame, two optical flow based methods (ROVER and its nonlinear variant), and four deep learning methods (TrajGRU, ConvGRU, 2D CNN, and 3D CNN). Specifically in the online setting, models are fine-tuned using AdaGrad (Duchi et al. 2011) with learning rate equal to 10^{-4}. The training objective during offline training and online fine-tuning is the sum of B-MSE and B-MAE. During the offline training process, all models are optimized by the Adam optimizer with learning rate equal to 10^{-4} and momentum equal to 0.5, with early-stopping on the sum of B-MSE and B-MAE. The ConvGRU model is also trained with the original MSE and MAE loss, which is named "ConvGRU-nobal" in the paper (Shi et al. 2017), to evaluate the improvement by training with the B-MSE and B-MAE loss.

15.5.4 Evaluation Results

The experiment results show that training with balanced loss functions is essential for good nowcasting performance of heavier rainfall. The ConvGRU model that is trained without balanced loss, which best represents the model in Xingjian et al. (2015), has a worse nowcasting score than the optical flow based methods at the 10 mm/h and 30 mm/h thresholds. Also, all the deep learning models that are trained with the balanced loss outperform the optical flow based models. Among the deep learning models, TrajGRU performs the best and 3D CNN outperforms 2D CNN, which shows that an appropriate network structure is crucial to achieving good performance. The improvement of TrajGRU over the other models is statistically significant because the differences in B-MSE and B-MAE are larger than three times their standard deviation. Moreover, the performance with online fine-tuning is consistently better than that without online fine-tuning, which verifies the effectiveness of online learning at least for this task.

The results of the Kendall's τ coefficients (Kendall 1938) between the MSE, MAE, B-MSE, B-MAE and the CSI, HSS at different thresholds show that B-MSE and B-MAE have stronger correlations with the CSI and HSS in most cases.

15.6 Discussion

In this chapter, we reviewed the DL-based methods for precipitation nowcasting. The architecture, building block, training objective function, metrics, and data source of the reviewed methods are summarized in Table 15.2. Precipitation nowcasting is formulated as a spatiotemporal sequence forecasting problem from the machine learning perspective. Thanks to the increased computational power and the amounts of data, the area is making rapid progress. Machine learning, specifically deep learning, facilitates the large amount of weather data and provides promising research directions for better modeling and understanding of precipitation nowcasting problem. Despite the success of DL-based methods achieved on precipitation nowcasting, this problem is still challenging. Below we list several major future research directions that are not solved or have not been explored:

- **Utilization of multi-source meteorological data**
  While existing DL-based models mainly focus on single-source data, typically radar echo maps or satellite images, multi-source meteorological data have become available thanks to rapidly developing sensing techniques as well as increasing data storage. Although DL-based models extract spatiotemporal features effectively, precipitation nowcasting using only single-source data is essentially ill-posed. The models are not offered complete knowledge on the dynamics of the meteorological system, hence they

Table 15.2 Summary of reviewed methods. The first half are FNN-based models and the second half are RNN-based models.

Method	Building Block	Architecture	Objectives	Metrics	Data Sources
Klein et al. (2015)	Dynamic CNN	Stacked CNN and dynamic CNN	MSE	MSE	Radar
Ayzel et al. (2019)	CNN	DozhdyaNet	MAE, MSE, Logcosh	MAE, CSI	Radar
Agrawal et al. (2019)	CNN	U-Net	Cross-entropy for three binary classifications	Precision, Recall	
Lebedev et al. (2019b)	CNN	U-Net	Cross-entropy for three binary classifications, Dice loss (Sudre et al. 2017)	Accuracy, Precision, Recall, F1 score	Satellite
Hernández et al. (2016)	MLP	AE, MLP	MSE	MSE	47 observational features
Qiu et al. (2017)	1D-CNN	Stacked 1D-CNN and FC layer	MSE with Frobenius norm	MSE, CSI, Correlation	Multi-site observational features
Xingjian et al. (2015)	ConvLSTM	Spatiotemporal encoder-forecaster	Cross-entropy	MSE, CSI, FAR, POD, Correlation	Radar
Shi et al. (2017)	TrajGRU	Encoder-forecaster with downsampling and upsampling	MSE, MAE, B-MSE, B-MAE	HKO-7 benchmark	Radar
Tran and Song (2019)	ConvRNN	encoder-forecaster with downsampling and upsampling	SSIM, MS-SSIM (Wang et al. 2003; Wang and Bovik 2009)	MSE, MAE, SSIM, MS-SSIM, PCC (Inamdar et al. 2018)	Radar
Cao et al. (2019)	ConvLSTM	Star-shaped Bridge	Multi-Sigmoid Loss	MSE, CSI	Radar
Wang et al. (2017d)	ST-LSTM	PredRNN	MSE, MAE	MSE	Radar
Wang et al. (2019b)	MIM block	MIM network	MSE	MSE, CSI	Radar

fail to accurately model it and infer its future evolution. Multi-source data, in contrast, provide multi-modal and multiscale meteorological information, giving the model a more holistic view of the system. Therefore exploring DL models that are able to jointly process complementary multi-source data can certainly help learn better representations of the observing systems.

- **Handling uncertainty**
  Precipitation nowcasting involves complex physics dynamics. According to chaos theory, chaotic behaviors in a meteorological system make it unpredictable due to high degree of uncertainty. Learning to capture the internal uncertainty is one of the major challenge in modeling and understanding the latent dynamics. However, most DL models address precipitation nowcasting in deterministic manner, which averages all possible futures into a single output, without describing its whole distribution. For some application scenarios such as rainstorm alert, not only the average and the most likely futures are concerned, but also some possible extreme cases should be considered. There are some recent works (Xue et al. 2016; Babaeizadeh et al. 2018; Denton and Fergus 2018; Lee et al. 2018a) that developed stochastic spatiotemporal models to predict different possible futures through variational inference. Though stochastic spatiotemporal models are not yet evaluated on precipitation nowcasting tasks, they are inspiring potential solutions for handling uncertainty in precipitation nowcasting.
- **Integration with numerical methods**
  Compared with theory-driven quantitative precipitation forecast (QPF) methods with clear physical meanings, deep learning models are data-driven and typically suffer from poor interpretability. Although theory-driven precipitation nowcasting models are derived from physical theories, they are essentially phenomenological models built by summarizing the empirical relationship of observations instead of deriving from first principles, which means theory-driven models are not entirely different from but in essence analogous to data-driven models. Theory-driven models consist of interpretable components to describe the observed data, while keeping consistent with physical laws including conservation of mass, momentum, energy, etc. They are determined by human experts and are hence hard to adjust according to different data from different distributions. On the contrary, data-driven models are equipped with high flexibility to adapt to different datasets since they directly learn parameters from data under few constraints. These two approaches are complementary in respect of interpretability and flexibility. Integrating theory-driven and data-driven approaches provides new opportunities in future precipitation nowcasting research, including but not limited to model calibration, recognizing unidentified observations, etc.

Appendix

The deep learning precipitation nowcast models introduced in this chapter have been utilized to support development of the operational rainfall nowcasting system, namely SWIRLS (Short-range Warning of Intense Rainstorms in Localized Systems) of the Hong Kong Observatory (HKO). In particular, TrajGRU has also been made available in the community version of SWIRLS (a.k.a. Com-SWIRLS) as part of core components under the Regional Specialized Meteorological Centre (RSMC) for Nowcasting of HKO, see https://rsmc.hko.gov.hk. The rainfall nowcasting models including TrajGRU are shared with the National Meteorological and Hydrological Services (NMHSs) of the World Meteorological

Organization to promote the development of nowcasting techniques. More information on Com-SWIRLS can be found at the website, see https://com-swirls.org/.

Acknowledgement

This research has been partially supported by General Research Fund 16207316 from the Research Grants Council of Hong Kong.

16

Deep Learning for High-dimensional Parameter Retrieval

David Malmgren-Hansen

16.1 Introduction

Various models in Earth Sciences e.g. related to climate, ecology, and meteorology, etc., are based on input parameters describing current and past states of different variables. The goal with Earth Science models is to understand the relationship between their parameters by describing the dynamic processes from which they vary. In this chapter we will refer to the parameters as bio-geo-physical parameters.

To be able to update our models' parameters we need to feed them with measurements from ground sensors (in-situ) or by retrieval from satellite observations, i.e. parameter retrieval. In-situ measurements often come with high cost, due to installment of sensors in remote places. Further, in-situ measurements provide often only sparse geographical coverage and the value of adding satellite observations, which provide frequent global coverage, is therefore high.

One example of a bio-geophysical parameter is atmospheric temperatures measured from sensors or satellites, which is often used as input to a Numerical Weather Prediction model. Many other examples of bio-geophysical parameters exist, and they can be grouped into:

- *biological*, e.g. *leaf area index* or other vegetation associated indices;
- *physical*, e.g. *soil moisture indices*, *temperature* or *humidity*, (see Figure 16.1);
- *chemical*, e.g. atmospheric trace gases, chlorophyll content in plants, hydrocarbon concentrations;
- *geographical*, *land cover*, *sea ice cover*, etc.

The task of retrieving parameters from observations is most commonly associated with finding functions that map observations to the parameter values by learning the inversion of a model that accounts for effects such as atmospheric distortion, geometry of the observation, and different noise sources. This is called the *inverse problem* because we map from effect (radiometric measurement) to cause (bio-geophysical parameter) rather than the opposite *forward problem* performed in e.g. radiative transfer models (RTMs). The inversion problems can be multi-modal or ill-posed, hence the mapping between observations and target parameters could have several or infinite solutions (Tsagkatakis et al. 2019). Further, the

Deep Learning for the Earth Sciences: A Comprehensive Approach to Remote Sensing, Climate Science, and Geosciences, First Edition. Edited by Gustau Camps-Valls, Devis Tuia, Xiao Xiang Zhu, and Markus Reichstein.

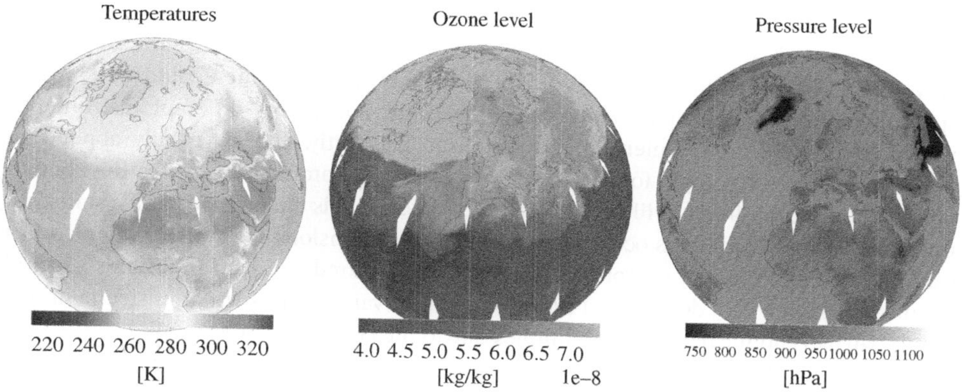

Figure 16.1 The plots are model forecasting parameters at the surface, extracted on measurement positions of the IASI sensor on board the MetOp satellite series, plotted over two days of orbits.

retrieval task can be based on high-dimensional sensor measurements or the target parameter can consist of several variables, which lead to statistically challenges when looking for significant mapping coefficients.

It is important to note that the difference between parameter retrieval and more generally prediction of a target variable from data, relates to how we use the predictions. For forecasting applications where satellite retrieved parameters are used we often consider dense predictions over a certain geographical area. Hence, predicting the crop types for a number of fields is not necessarily to be considered parameter retrieval. If, on the other hand, we apply our crop predictor's output in a model to understand agricultural parameters' effect on the surrounding environment, we would be solving a parameter retrieval problem.

Deep learning (DL) has the potential to play an increasing role in the future of bio-geophysical parameter retrieval from remote sensing data. The data can be high-dimensional and the relationships between measurements and the parameters often highly non-linear. These relationships can also have dependencies in time, space and spectral dimensions. DL offers a flexible framework to model such complex problems and the data we can use to train statistical models keeps growing. Despite the flexibility of deep learning, challenges must be overcome in order to extent the use further than today. These challenges include:

- how to efficiently perform statistical optimization in large model-parameter spaces;
- how to computationally handle large-scale problems on generally available computing platforms;
- how to interpret the findings to learn from these highly non-linear and deeply nested functions;
- how to perform cross domain and cross modality learning instead of redesigning and retraining algorithms for different sensors;
- how to exploit neighborhood relationships in all dimensions (for both input and output data); and
- how to ensure that the models generalize to new unseen data in operational scenarios.

The challenge of high dimensions is two-sided and is unlikely to be solved immediately. Firstly, we are bound by computational load when looking for modeling dependencies across different dimensions. Earth observations from satellites could be sampled spatially (2 or 3 dimensions), temporally, spectrally, and across different instruments, which could yield a 6+ dimensional data-tensor. Currently, DL is typically applied on relatively resolution sample for 2D- or 3D-problems. The problem related to the computational complexity in DL will be elaborated on later in this chapter. Secondly, statistical under-determined problems occur when any of our dimensions are densely sampled. An example is hyper-spectral sensors which can have hundreds or even thousands spectral channels. To get a well determined estimate of e.g. covariance in such an instrument will require an astronomical amount of samples due to the curse of dimensionality.

16.2 Deep Learning Parameter Retrieval Literature

Earth science related literature on bio-geophysical parameter retrieval spans a wide variety of applications across several scientific disciplines. We can group these problems in several ways, for example by,

- geography, e.g. land, ocean and cryosphere;
- science, e.g. biology, ecology, climatology, geophysics, meteorology, hydrology, etc.;
- applications, e.g. weather, agriculture, environment, etc.;
- EO-technology, e.g. optical imaging (visible), multi-/hyper-spectral instruments, passive microware radiometers, radar-based sensors, sounding instruments, etc.

Many parameters are used across several of these above categories, such as surface temperature, a versatile parameter used for weather prediction, agriculture, and environment, in ecology, hydrology, meteorology, and biology, with a component over land (land surface temperature, LST) over ocean (sea surface temperature, SST) or ice (ice surface temperature, IST). Several review papers relevant for DL parameter retrieval have already been published with each their focus area, e.g. with a focus on Earth Sciences (Reichstein et al. 2019), environmental applications (Yuan et al. 2020b), or remote sensing (Zhu et al. 2017).

For more in-depth literature reviews on the different subjects where bio-geophysical parameters are used, we refer the reader to other chapters in this book in Parts II and III.

16.2.1 Land

Land-based parameter retrieval often concerns bio-chemical parameters such as vegetation indices but can also include physical parameters such as LST. Deep Learning was used to retrieve LST from Microwave Radiometer data in Tan et al. (2019b), and the method was tested on reference data from both ground stations and other optical satellite data with good results.

Bio-chemical related retrieval often concerns indices such as Leaf-Area-Index (LAI) or Leaf-Chlorophyll-Content (LCC). Verrelst et al. (2015) provides an overview of bio-geophysical parameter retrieval and methods. Artificial Neural Networks (ANNs) are part of the standard toolbox for parameter retrieval, yet mostly confined to the use of shallower architectures (Verrelst et al. 2015). Simple ANNs with multiple layers (>1 hidden layer) can be considered as a part of the *deep learning* methods, but recent trends in *deep learning* research goes towards the use of neural networks variants such as Convolutional Neural Networks or Recurrent Neural Networks where depth is a common and powerful characteristic.

Research in farming applications also relates to biological parameter retrieval. Often though, the goal is not to use predictions as parameters in models, but as proxies for the health condition of the crops in so called smart farming applications. By monitoring and optimizing these vegetation indices the goal is to increase crop yield. The variables of interest, such as crop type, crop yield, soil moisture, and weather variables, can also be used to model and understand the ecosystems that farming effect (Kamilaris and Prenafeta-Boldú 2018). Most often though, it is applied to datasets covering only smaller regions of agricultural areas. As opposed to biological parameter retrieval applications, DL is frequently used in farming applications. Some country level work on agriculture has been done for e.g. corn crop yield (Kuwata and Shibasaki 2016), but little work exists on larger-scale studies where predictions could be used in models. Kim et al. (2017a) provides a comparison of several artificial intelligence methods on a case study in midwestern USA.

Forest cover and biomass are other types of biological parameters which is of high importance to understand and monitor the Earth. Martone et al. (2018) addresses forest cover with fuzzy clustering machine learning techniques to create global forest maps in 50×50 m grids. DL has also been applied to this problem although mostly on continental level scale, for example by Ye et al. (2019) where a Recurrent Neural Network (RNN), based on the Long-Short-Term-Memory (LSTM) architecture, outperformed the other Machine Learning methods in their benchmark. Khan et al. (2017) models forest dynamics over a 28-year period by stacking time-series and formulating the task as a change-classification problem. Zhang et al. (2019) predicts forest biomass from Lidar and Landsat 8 data, with a comparison between four different Machine Learning models. The best performing model is a DL model based on the concept of Stacked Sparse Autoencoders (SSAE). This model likely performs well due to the unsupervised pretraining.

16.2.2 Ocean

Sometimes deep learning can be beneficial in hybrid approaches where the model acts a post processing of physical-based retrieval. This is the case in Krasnopolsky et al. (2016) where a Neural Network works on physically retrieved ocean color chlorophyll to fill the gaps in data from NOAA's operational Visible Imaging Infrared Radiometer Suite (VIIRS). This approach can be seen as a general image post processing technique but is a clever way to exploit NNs' abilities to learn patterns, which is necessary to perform well in gap filling. Similar approach on microwave radiometer parameters were performed in Jo et al. (2018) to Chlorophyll-a content in oceans. Direct ocean parameter retrieval has also been done, e.g. from hyperspectral satellite images in Awad (2014), or from ground station measured water quality with temporal modeling by an LSTM in Cho et al. (2018). Hyperspectral data can also be modelled with a CNN, a more advanced deep learning technique that includes spatial information to extract chlorophyll (CHL), colored dissolved organic matter (CDOM), total suspended sediments (TSS) (Nock et al. 2019).

16.2.3 Cryosphere

Typical bio-geophysical parameters in cryospheric studies can be sea ice cover/concentrations (SIC), sea ice types (SIT), snow depth, snow water equivalent, and snow cover. Deep learning has been used for predicting SICs in Wang et al. (2016, 2017c); Karvonen (2017); Malmgren-Hansen et al. (2020) and for distinguishing SITs in Boulze et al. (2020). Both SIC and SIT are parameters used in climate models and models that forecast ice drift for marine users.

Snow cover applications of various kinds have been explored with ANN and Dl techniques e.g. in Tsai et al. (2019); Çiftçi et al. (2017); Nijhawan et al. (2019b); Gatebe et al. (2018). Çiftçi et al. (2017); Dobreva and Klein (2011) used physically derived parameters, NDSI and NDVI, as inputs to ANNs together with optical satellite data which performs well compared to model working only on optical satellite data alone. Snow depth and snow water equivalent has long been retrieved by ANNs, especially from microwave radiometer data (Davis et al. 1993; Tanikawa et al. 2015) and sometimes with auxiliary terrain information to correct for various effects (Bair et al. 2018).

16.2.4 Global Weather Models

A research field for which parameter retrieval long has been a very important task is meteorology (Isaacs et al. 1986). Here, global models with a large number of bio-geophysical parameters are necessary to predict the future weather. The parameters, in numerical weather prediction models (NWP) for example, are often physical-related and can be temperature, humidity, pressure, precipitation, wind-speed, moisture, clouds, and many more. Retrieval of these parameters, i.e. the regression task of mapping spectral bands intensities from RS sensors to the parameter values, enables us to update NWP models with measurements covering the Earth frequently. This lowers the uncertainty significantly and improves NWP models' predictions (Levizzani et al. 2007). Deep learning and artificial neural networks are having a growing impact on NWP parameter retrieval, e.g. for clouds in Mateo-García et al. (2020); Gómez-Chova et al. (2017) or temperatures in Aires et al. (2002); Malmgren-Hansen et al. (2019). Satellites' importance in meteorology is easily noticed by the establishment of dedicated satellite agencies for weather forecasts (NOAA and EUMETSAT). Compared to on-ground sensors, satellites' observations provide mean values over their ground resolution footprint and frequent global coverage.

16.3 The Challenge of High-dimensional Problems

Parameter retrieval problems can have spatial, temporal, and spectral dependencies, i.e. neighboring samples will be correlated in space, time, and frequency. The spatial and temporal dependencies can be present in both input data, as well as for the targets that we wish to retrieve. The spectral dependency often comes from the underlying process we observe which' emits or reflects a certain spectral signature. If we plot our parameters on a map we see the spatial dependencies as neighborhood correlation, i.e. patterns can be observed. An example can be seen on the surface temperatures in Figure 16.1 (a) where the poles exhibit colder temperature regions while equatorial locations are warmer. The spatio-spectral-temporal dependencies can be quantified by measures such as covariance/correlation, mutual-information, multi-information or the Rotation based iterative Gaussianization (RBIG) (Laparra et al. 2011). The RBIG method is used in Laparra and Santos-Rodríguez (2015); Johnson et al. (Submitted) to find the optimal balance between spatial and spectral neighborhood samples.

The temporal dependencies occur as the processes we observe change over time, e.g. seasonal vegetation changes. If we measure frequently enough compared to the time-constant of the temporal development we will be able to observe and potentially model this dependency.

Example: Let us consider an example of monitoring sea ice in the Arctic. If we take two images with days between them on the same location covering a large area with land-fast ice, it is most likely that the images overall will look the same, and we have time dependency in our samples. This is opposed to zooming in on a small area near the edge of land-fast ice where the ice moves fast due to sea currents and wind and the whole image would change in the course of a day.

While quantifying the dependencies is straightforward, building regression functions that capture them is a challenge. DL offers a flexible framework with the possibility do this due to:

- weight sharing in input nodes, i.e. convolutional filters;
- feature sharing between outputs in multi-output neural networks;
- loss functions that be formulated to incorporate neighborhood dependencies.

The variant of Neural Networks with built in convolutional filters, convolutional Neural Networks (CNN), can be extended to data samples that include all dependencies. The filters perform feature extraction and transform our data into a latent space where the problem is potentially easier to solve. For CNNs, data is typically structured in tensors (i.e. arrays) with two spatial dimensions, X, Y, e.g. (across-track, along-track) for satellite data. Further, one could add a spectral dimension, B, and a temporal component T, yielding a (X, Y, B, T) shaped tensor per sample of data. One can model the dependencies in this data tensor in several ways, and three ways are explained below and illustrated on Figure 16.2.

(i) (X, Y, B) could be considered a data cube and cubic convolutions with filter sizes of (H, W, D) (height, width, depth) could look for features across space and spectrum including T time-steps in each sequence. This would result in cubic filters with T channels.
(ii) One could perform 2D convolutions in spatial dimensions and stack spectral bands, B, from several time samples, T, as channels, with a tensor of $(X, Y, B \times T)$. In this way we perform 2D convolutions on $B \times T$ channels.
(iii) Another option is to build a hybrid model consisting of a CNN and an RNN, where the "visual" features in (X, Y, B) are extracted by the CNN one time-step at the time, and a feature vector from the CNN is parsed to the RNN that learns the temporal relationship with its sequential memory. The input CNN could be either with cubic convolutions or a 2D filters with spectral channels stacked.

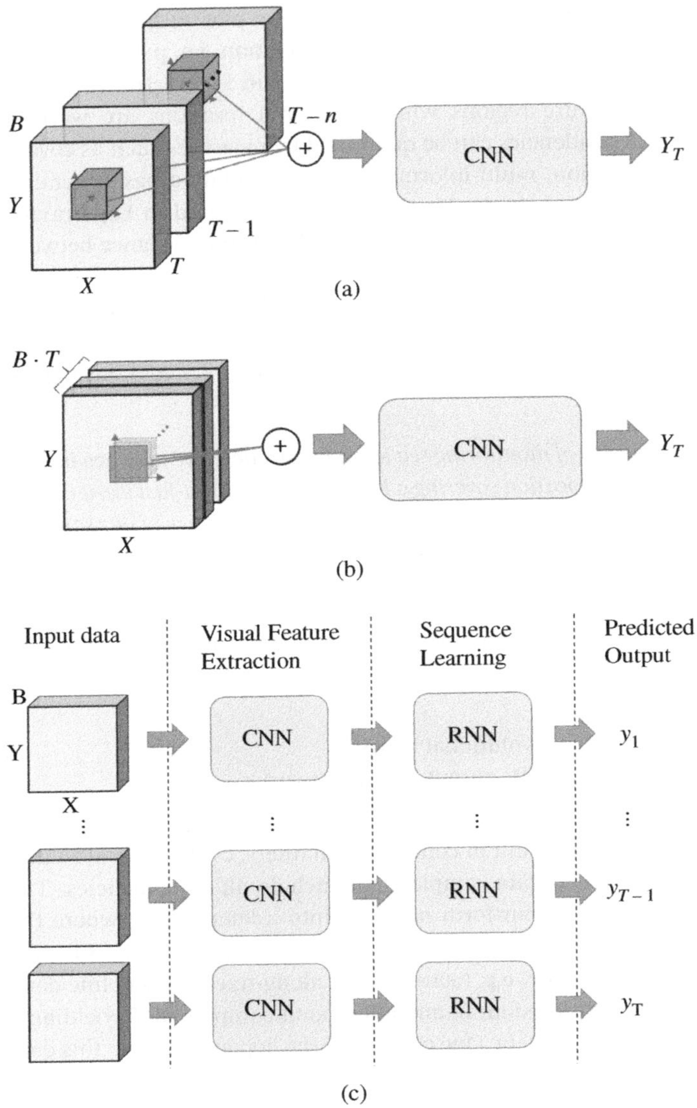

Figure 16.2 Three ways of modelling spatial, spectral and temporal relationships. (a) Cubic convolutions over space and frequencies and stacking the n time steps. (b) Stacking frequency, B, and time, T, and performing 2D convolutions over spatial dimensions (X, Y). (c) Hybrid approach of combining CNN to handle space and frequencies, (X, Y, B), with an RNN to handle time T. Originally, the concept was developed for visual recognition problems with time-varying inputs by Donahue et al. (2015).

It is important to note that when stacking channels, e.g. (B, T), the order of them is not taken into account by the model. Cubic convolutions or sequential learning with an RNN, on the other hand, assume the order to follow the natural order of the samples.

One could have input data with three spatial dimensions (X, Y, Z) although this is not commonly seen in RS research. Here, Z could be an altitude of the measurement.

This would require a cubic convolution over the three spatial dimensions and either stacking of time sequences, or handling time with an RNN hybrid approach.

A practical problem arises on most computing platforms with increasing dimensions. The memory requirements increase drastically and a trade-off between the sizes of each dimension per data sample, and number of dimensions must be made. Typically, DL research crops or resize 2D data into samples around 250×250 pixels and 3D data into $96 \times 96 \times 96$ voxels as described in Yang et al. (2009), for typical GPU-based computing platforms.

16.3.1 Computational Load of CNNs

In this section we look into the computational load of different configurations of CNN models. These properties can be relevant to consider when working with high-dimensional problems, e.g. when our input image is not a simple 3-channel RGB image, but a multi/hyper-spectral instruments with many spectral bands. Later in this chapter we will consider an experiment with retrieval of atmospheric temperatures from an infrared sounding interferometer for assimilation in numerical weather models using a CNN. The model for this atmospheric retrieval problem is presented in Table 16.1 and has 43.8M product-sum operations in total over its convolutional layers. Product-sum operations for convolutional layers can be calculated according to

$$PS_{ops} = (f_w \times f_h) \times N_{inp} \times (K_w \times K_h) \times N_{out}, \tag{16.1}$$

where f_w,f_h is the filter width and height, N_{inp}, the number of input channels, K_w, K_h the width and height of the output feature map and, N_{out} the number of output nodes/feature maps. It should be noted that it is occasionally found that convolutions with strides different from 1 are used. With e.g. stride=2 for a convolution, every second pixel position is skipped. This also reduces the number of product-sum operations as the output size, K_w, K_h, in Equation 16.1 is halved in width and height.

The model we use in the experiments later (IASI, section 16.4.1), works on spectral components from a dimensionality reduction algorithm. An extension to this approach could be to incorporate the dimensionality reduction into the very first layer of the network, but this comes with an increased computational cost. If we have 4699 spectral samples in a measurement and wish to decompose this into 125 components, this would be similar to what a principal component analysis (PCA) performs numerically. It would require a convolutional layer with 1x1 filter size and a 125 output features maps. For processing an input patch of 15x15 pixels it would require, $PS_{ops} = (1 \times 1) \times 4699 \times (15 \times 15) \times 125 = 132{,}159{,}375 \approx 132M$. This first layer alone, would more than triple the combined number of product-sum operations in our model. Now a combined computational load of <200M is not necessarily a problem for modern computing hardware like GPUs but another aspect of the computational burden is the increased amount of data per batch that needs to be transferred between storage and memory, and again between memory and processor (CPU or GPU). When using an orthogonal transformation like PCA or MNF we can perform the dimensionality reduction as pre-process and store a much smaller version of our data-set ($4699/125 = 37.6$ times smaller). Whereas when incorporating it into the CNN we need to load the full spectrum in spatio-spectral chunks for the network to learn the embedding alongside the learning objective. In the later example of atmospheric temperature retrieval we have chunks of sizes $BS * H * W * C * DS$ for Batch Size (BS), height (H), width (W), spectral channels (C), and

Table 16.1 Summary of CNN model used in the retrieval of atmospheric temperatures in Figure 16.4. The amount of Product-Sum Operations (PS_{ops}) is given per layer in the fourth column.

Layer-type	Output Shape	Param #	PS_{ops}
Conv2D	(None, 15, 15, 100)	112600	25,312,500
Conv2D	(None, 13, 13, 100)	90100	15,210,000
MaxPool2D	(None, 6, 6, 100)	0	0
BatchNorm	(None, 6, 6, 100)	400	
ReLU Actv.	(None, 6, 6, 100)	0	0
Dropout	(None, 6, 6, 100)	0	0
Conv2D	(None, 4, 4, 160)	144160	2,304,000
Conv2D	(None, 2, 2, 160)	230560	921,600
MaxPooling2D	(None, 1, 1, 160)	0	0
BatchNorm	(None, 1, 1, 160)	640	
ReLU Actv.	(None, 1, 1, 160)	0	0
Dropout	(None, 1, 1, 160)	0	0
Conv2D	(None, 1, 1, 240)	38640	38,400
BatchNorm	(None, 1, 1, 240)	960	
ReLU Actv.	(None, 1, 1, 240)	0	0
Dropout	(None, 1, 1, 240)	0	0
Conv2D	(None, 1, 1, 90)	21690	21600

Total params: 639,750

Trainable params: 638,750

Non-trainable params: 1,000

data type size (DS) in bytes. For the above example with batch size 32 and data stored in float32 (4 bytes per number) we have $32 * 15 * 15 * 125 * 4 = 3.6$MB per batch. Now with an increase of the spectral dimension to 4699 we have $32 * 15 * 15 * 4699 * 4 = 135.3MB$ per batch. A chunk size of 135.3MB can cause too large delays when transferring data to a GPU, resulting in non-optimal utilization of GPU cores. Whether in practice it is a problem, though, depends on the platform, the amount of training data, how many epochs are necessary, and how many weights the network has.

As previously discussed, one might consider spatial and spectral dimensions a data cube and perform cubic convolutions to exploit a combination of spatio-spectral feature extraction. This, however, comes at a higher computational cost. The extension of Equation 16.1 would be straightforward,

$$PS_{ops} = (f_w \times f_h \times f_d) \times N_{inp} \times (K_w \times K_h \times K_d) \times N_{out}, \tag{16.2}$$

with f_d being the filter-depth and the K_d the output cube depth. A consequence of considering the spectral dimension in the convolutions for the above IASI example is that N_{inp}

becomes 1, so for a layer with a $3 \times 3 \times 3$ filter kernel and 100 output nodes we have

$$PS_{ops} = (3 \times 3 \times 3) \times 1 \times (15 \times 15 \times 4699) \times 100 = 2{,}854{,}642{,}500 \approx 2.8G \tag{16.3}$$

for a single convolutional input layer. An alternative to cubic convolutions is the use of depth-wise separable convolutions. This method consist of first applying 2D spatial filter kernels to each input channel separately and then applying N_{out} number of $1 * 1 * f_d$ depth-wise filters to combine the spatially extracted information. This will split our product-sum formula in two terms and we see this will reduce Equation 16.3 to

$$PS_{ops} = (f_w \times f_h) \times N_{inp-chan} \times (K_w \times K_h) + f_d \times N_{inp-chan} \times N_{out-nod} \tag{16.4}$$

For our example, $PS_{ops} = (3 \times 3) \times 4699 \times (15 \times 15) + 3 \times 4699 \times 100 = 10{,}925{,}175 \approx 11M$, which is a drastic decrease in product-sum operations compared to the 2.8 billion for the cubic case.

16.3.2 Mean Square Error or Cross-entropy Optimization?

Retrieval is most often associated with least-square regression modeling. Traditionally, linear regressors optimized by least-square have been the preferred choice but are not always sufficient to capture the complexity of retrieval problems. A neural network optimized by the Mean-Square Error (MSE) loss function, is a one way of extending the least square linear regression with non-linearity. This approach is often chosen when the target is a sampled continuous function, like temperatures, sea-ice concentrations, pressure levels, humidity and moisture indices, etc. Alternatively, a probability distribution over possible outcomes can be modelled with Cross-Entropy based error functions. Cross-Entropy is generally associated with problems where we wish to label data, e.g. segmentation or classification. Retrieval of fractions/percentages can be easily be modelled by both approaches, though, and it is in the end up to the analyst to decide how to tackle a specific problem and test what gives the best result for the given dataset and problem at hand.

It can be shown (Bishop 2006) that the MSE (Equation 16.5) loss is the maximum likelihood solution to a problem where the target can take any real values ($t_n \in \mathrm{IR}$).

$$E(\mathbf{w}) = \sum_{n=1}^{N} (y(\mathbf{x}_n, \mathbf{w}) - t_n)^2 \tag{16.5}$$

In Equation 16.5 the n'th sample $\mathbf{x}_n$ can be modelled by our neural network output $y(\mathbf{x}_n, \mathbf{w})$ with the weights $\mathbf{w}$ and the probability t_n given an input sample is Gaussian distributed and has a $\mathbf{x}_n$ dependent mean. The output of $y(...)$ should be a linear projection, i.e. no output activation function.

When using a MSE loss function we assume Gaussian distributed prediction errors and that the optimal solution can be found by minimizing the conditional average of this error. This might not be true for all problems though. Problems of multi-modal character or ill-posed inversion problems will not follow the assumption of the optimum solution being the minimization of Eq. 16.5. For addressing this Bishop (2006) suggests a variant of neural networks called Mixture Density Networks (MDN). MDNs has an alternative formulation of the loss function over a space of continuous values by predicting a probability distribution parameterized by a Mixture Model, with e.g. Gaussian kernels. Instead of the model predicting a specific value as an MSE-trained network, one can sample from the probability distribution predicted by an MDN.

For Bernoulli distributed targets ($t_n \in [0, 1]$) such as how we would encode categorical problems, the maximum likelihood solution becomes the cross-entropy error function:

$$E(\mathbf{w}) = -\sum_{n=1}^{N}\sum_{k=1}^{K} t_n y_k(\mathbf{x}_n, \mathbf{w}). \tag{16.6}$$

For the output activation in this case, we need a canonical link that fulfills $y \in [0; 1]$ hence the activation should be mapped to probabilities by the logistic sigmoid for the binary case or the multi-class extension, softmax, when the output belongs to one of more classes.

It can be worth considering reformulating problems to one or the other loss function. Oord et al. (2016) saw improvements in building a Text-To-Speech model based on Neural Networks that predicted the wave signals directly by a probability distribution over its discretized values. In order not to have too many classes, K, and thereby making the problem intractable, the authors transformed the waveform with a μ-law algorithm to 255 discrete values rather than 65 536 values necessary to represent the full 16-bit. The authors in Oord et al. (2016) describe the advantages this modeling approach thus:

One of the reasons is that a categorical distribution is more flexible and can more easily model arbitrary distributions because it makes no assumptions about their shape.

In Wang et al. (2017c) the authors models concentrations of sea ice in the Gulf of St. Lawrence and Newfoundland Sea, Canada, with a convolutional neural network, feeding it Synthetic Aperture Radar images from the RADARSAT2 sensor. The authors choose to model the fractions of sea ice in square blocks of 18×18 km^2 with a least square error function, but a categorical error function could have been used as well. In the simplest case of modeling sea ice concentrations with cross-entropy error we could set the target equal to the percentage of sea ice in each block and have a model with one output that predicts this concentration.

16.4 Applications and Examples

This section describes two different retrieval problems. The first problem is retrieval of atmospheric temperatures from the Infrared Atmospheric Sounding Interferometer (IASI) instrument on board the MetOp-A/B satellites. Parameters retrieved from the IASI instrument are assimilated into NWP models and have greatly improved weather forecasts since its launch. The second problem is the task of predicting sea ice in the arctic from synthetic aperture radar images. This task when applied to a single Synthetic Aperture Radar (SAR) image is a simple prediction of a variable, but when applied to the whole Arctic for assimilation in sea ice forecast models becomes a parameter retrieval task.

16.4.1 Utilizing High-dimensional Spatio-spectral Information with CNNs

Retrieval from IASI measurements is inherently difficult due several properties, such as noise and complex relationship between observations and target variables. Another property, is the 8461 spectral variables, that makes it a high-dimensional retrieval problem.

Establishing the relationships between a measured IASI spectrum and e.g. the temperature at a certain altitude, is often a statistical under-determined problem. Due to this, dimensionality reduction is often applied as the first step (Pellet and Aires 2018).

IASI – Dataset The Infrared Atmospheric Sounding Interferometer (IASI) on board the MetOp satellite series measures the infrared spectrum with high resolution (Malmgren-Hansen et al. 2020). The ground footprint resolution of the instruments is 12 km at nadir, and a spectral resolution of 0.25 cm^{-1} in the spectrum between 645 cm^{-1} and 2760 cm^{-1}. This results in 8461 spectral samples covering 2200 km scan-swath with 60 points per line. IASI is an ideal instrument for monitoring different physical/chemical parameters in the atmosphere e.g. temperature, humidity, and trace gases such as ozone. Energy from different altitudes returns a different spectral shift. In this way atmospheric profiles can be obtained and these provides important information for e.g. NWP models. In the IASI dataset, channel selection has been performed reducing spectral components from its 8461 channels to 4699, before any further processing. For statistical modeling of atmospheric parameters, forecast models are used to provide dense target values for every point observed by IASI. The IASI dataset has been matched with forecasts by the Medium-Range Weather Forecasts (ECMWF) model. This includes temperatures, humidity, and ozone concentrations for 137 altitudes through the atmosphere. EUMETSAT which operates the MetOp satellites uses forecasting data together with IASI measurements to provide derived products. These retrievals are validated both on the forecasts and in-situ measurements from e.g. radiosondes. Temperatures can be derived down to 1K accuracy, humidity with 10% and ozone with 5%, all at 25 km ground resolution. For the following experiments 13 orbits from the 17-08-2013 were used, with the first 7 for training and the rest for test. In situ measurements were not used for validation as this is a relative comparison of the performance between a CNN and the traditionally used Ordinary Least Square (OLS) regression model.

Example of Atmospheric Parameter Retrieval An example of retrieving atmospheric temperatures in the 90 lowest altitudes from the IASI spectrum, can be done by applying a CNN on spatio-spectral chunks of the data. As a first step Minimum Noise Fractions (MNF) is applied to reduce the large spectral dimension (Green et al. 1988). MNF is an orthogonal dimensionality reduction method like Principal Component Analysis (PCA). Unlike PCA though, MNF does not maximize the signal variance but instead the signal-to-noise ratio and hence the noise must be known or estimated from data. Here, to get the full noise covariance matrix of the spectrum, we assume that the noise can be modelled as the residuals of a cubic polynomial fit over a 3×3 neighborhood of samples. This operation can be efficiently implemented as a image filter operation (Nielsen 1999). After the MNF decomposition, the IASI spectrum is represented by the 125 MNF components with highest Signal-to-Noise values.

The CNN used here is described in Table 16.1. The goal for the CNN is to learn spatial patterns that relates to the retrieval task. Hence, a spatial neighborhood of 15×15 samples is fed to the CNN and the atmospheric temperature profile over the center sample is predicted. A block of $120 \times 60 \times 260$ decomposed IASI data are shown in Figure 16.3 with 120 along track samples 60 across track samples and 260 spectral MNF components.

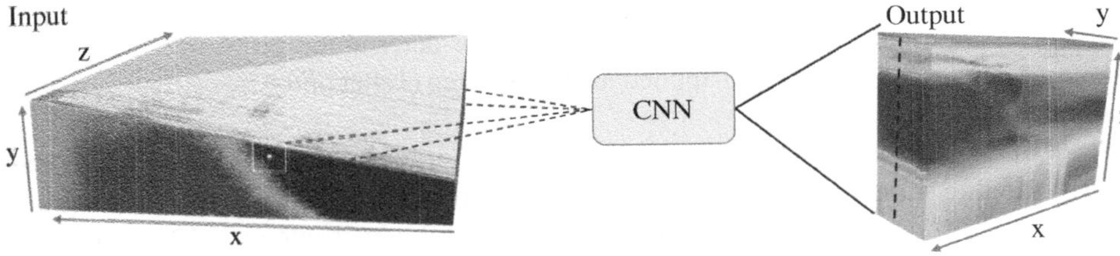

Figure 16.3 Input: Decomposed IASI spectrum using the MNF (260 components). x-axis is along track orbital direction and y-axis across track. z-axis represents the spectral MNF-components. The input cube is sliced in the corner to illustrate the rapid decreasing color intensity of the sorted MNF components as most of the information is compressed into the first components. Output: Atmospheric temperatures. x-axis is along track orbital direction and y-axis across track. z-axis represents altitudes in the atmosphere. The white square on the input depicts the 15 × 15 spatial neighborhood sample that is passed to the CNN.

While an OLS regression, would capture immediate correlation between spectral components and target, a neural network would be able to extend this to more complex and non-linear relationships. Further, an advantage of the CNN variant is that it shares all weights between all outputs, except for at the very last layer. On the contrary an OLS regression model will have 90 independent output predictions. This gives the CNN the advantage of smoother transitions between predictions as seen on the results in Figure 16.4.

On the test set the OLS regression model achieves a Root-Mean-Square-Error (RMSE) of 2.85 K while the CNN has a RMSE of 1.94 K. As opposed to many other studies, the RMSE is calculated over all measurements regardless whether they are marked as cloud-contaminated and contains samples over land as well as ocean. The main conclusion that can be drawn from this experiment is that the spatial dependencies are better modelled with the DL model compared to the traditional OLS regression. One of the advantages the CNN has over the OLS regression is the filtering operations that transforms the spatial dimensions into a feature space at a lower dimension than the

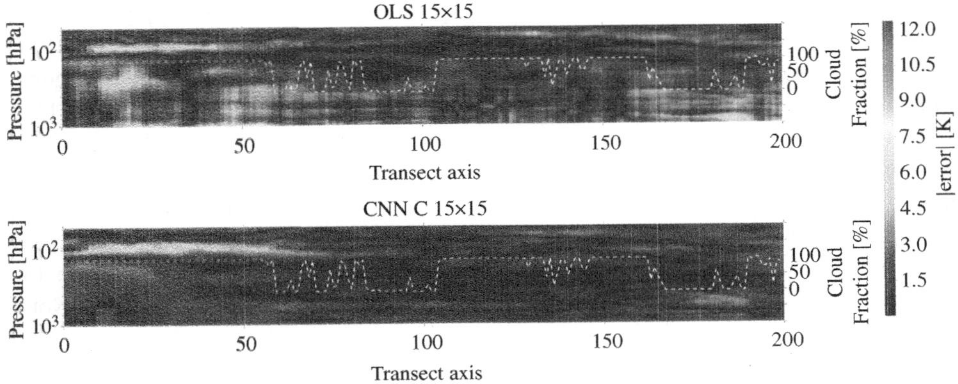

Figure 16.4 Transect profile of RMSE, Linear Regression (OLS), and CNN on cubes of IASI data 15 × 15 × 125 (height × width × MNF–components). The pressure on the y-axis corresponds to altitudes in the atmosphere and the x-axis shows distance along an arbitrary line on the surface of the Earth.

original input. This is both a good way to model spatial dependencies, but also a way to tackle the statistical under-determination which the problem suffers from. The OLS regression will estimate least square residuals from a regressor parameterized over input variables ($15 \times 15 \times 125 = 28{,}125$) for each target variable, here 90. Another advantage of the CNN filtering properties is that the filters can perform noise reduction (averaging), edge detection, or contrast enhancement, which might help the model tackle difficult predictions, e.g. around clouds, coastal areas, and between weather fronts.

16.4.2 The Effect of Loss Functions in Retrieval of Sea Ice Concentrations

Automating retrieval of sea ice information is inherently difficult due to a number of circumstances. Polar darkness and a dense cloud coverage in the polar regions prevents operational use of optical sensors, that would otherwise be ideal for separating the white ice from blue water. Microwave radiometers (MWR) measures the black body radiation of an object, or equivalent the brightness temperature. They are not as affected by clouds and have a high contrast between ice and water, but typically too low ground resolution for some sea ice use-cases. SAR sensors penetrate clouds and have a much higher resolution than MWR, but the measured intensities can be ambiguous, e.g. between open water in windy conditions and firm snow-covered sea ice. Further, ground truth on sea ice conditions is difficult to obtain due to the remote and vast areas that it covers. Relying on the expert labeled images is therefore the only possibility for collection of large datasets. This can be supported by smaller validation sets on occasional cloud-free optical imagery and in-situ validated conditions.

ASIP Sea Ice Dataset The ASIP Sea Ice Dataset version 1 (ASID-v1, publicly available at Malmgren-Hansen et al. (2020)) was collected for the Automated Sea Ice Products (ASIP) research project, that aimed at automating sea ice information retrieval in the arctic. Today, monitoring sea ice mainly consists of a time-consuming manual process where experts draw polygons, typically on Synthetic Aperture Radar (SAR) imagery, and assign information about ice conditions, referred to as ice charts or ice image analysis. ASID-v1 consists of Sentinel-1 SAR images matched with expert drawn ice charts containing polygons with sea ice concentrations. The concentrations are ranging from 0% to 100% in steps of 5%. The dataset covers the period from November 2014 to December 2017 and is gathered across 912 Sentinel-1 SAR scenes. All seasons are covered and all coastal areas of Greenland except for the north most region are represented. The polygons' geometry follow no strict definition and is based on the sea ice experts intuition on natural segments in the scenes. Further, the dataset also includes brightness temperatures from a Microwave Radiometer (MWR). MWR measurements from the Advanced Microwave Scanning Radiometer 2 (AMSR2) are recorded in 10×10 km grids at frequencies from 6.9 GHz to 36.5 GHz and in 5×5 km grid at 89 GHz, although the footprint resolution ranges from 35×62 km (6.9 GHz) to 3×5 km (89 GHz). All these frequencies, available in both horizontal and vertical polarization gives 14 brightness temperatures per measurement. Here, all AMSR2 measurements are resampled to a 2×2 km grid where each grid cell center align with every 50th Sentinel-1 pixel. The dataset is split in 90%/10% for training and test on scene level. In this way a independent test can be made as samples in training and test are separated in time or space, and it reflects the operational scenario of an automatic ice chart extraction algorithm.

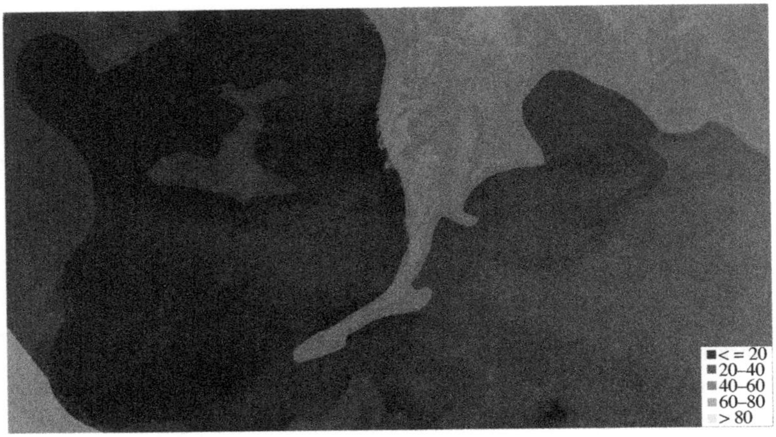

Figure 16.5 Polygon ice chart overlay on the HH polarization of a Sentinel-1 scene. After the ice experts have marked the points that outline the ice boundary, a spline curve fit is applied to make the polygon smooth. This results in ice occasionally not being encapsulated by the polygon along the edge.

The ice concentrations provided as averages over polygons pose a challenge when mapping them as target values with the SAR pixels. Assigning the average concentration to every pixel inside the polygon will lead to label errors as the distribution of ice within the polygon is unknown unless the sea ice concentration is 0% or 100%, see Figure 16.5. At the scale of the SAR image (40 × 40 m pixel-spacing), most pixels will be either wholly open water or wholly sea ice.

Example of Sea Ice Concentration Estimation In the following experiments we aim to apply a CNN to fuse the information in the SAR and MWR data and model the sea ice concentration values. Figure 16.6 shows a conceptual flow of the CNN fusion architecture that aims to combine SAR and MWR data to predict the pixel-wise presence of a sea ice concentration. The CNN takes in a patch of 300 × 300 SAR pixels with corresponding 6x6 MWR pixels and predicts ice-concentrations in 300 × 300 pixel prediction maps. The output activation

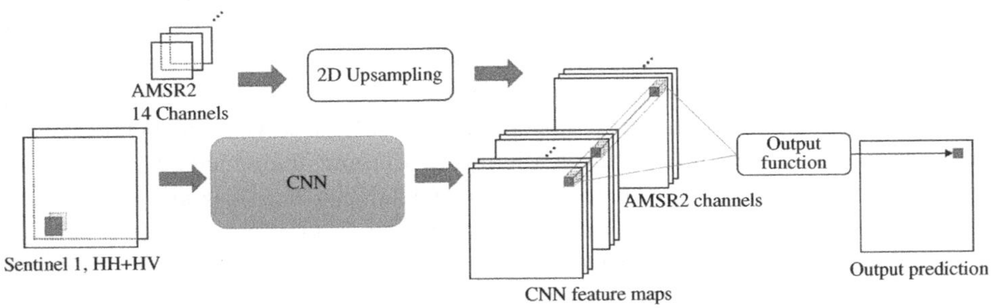

Figure 16.6 Conceptual flow of the prediction of sea ice maps with a CNN applied on SAR images for feature extraction that is concatenated with upsampled MWR measurements for fusion of satellite sources. The output function can either be linear with least-square loss or a sigmoid with binary cross entropy.

function is chosen according to the respective loss chosen (Bishop 2006), i.e. linear function with mean square loss or sigmoid function for cross-entropy loss.

One challenge here is the choice of loss function. If we model the sea ice concentrations as a continuous variable over pixels, a Mean Square Error (MSE) loss might be chosen, Equation 16.5. We know, though, that there are a lot of label errors associated with the pixels, and this might make it hard to minimize the error residuals. The problem could also be considered as modeling the probability of a pixel being ice. In this way we assume that a random sample from a polygon with a concentration of e.g. 40% has a 40% probability of being ice. Modeling this probability could be done with a Binary Cross-Entropy (BCE) loss function, where we replace the typically binary encoded categorical target with the discrete ice-probability values. From Equation 16.6 with $K = 1$ we have,

$$\mathbf{E}(\mathbf{w}) = -\sum_{i} \mathbf{t}_i \log(\mathbf{y}(\mathbf{w}, \mathbf{x}_i)), \tag{16.7}$$

where $\mathbf{t}_{n,i}$ is the pixel-wise sea ice concentration for the ith image patch $\mathbf{x}_i$. $\mathbf{y}(\mathbf{w}, \mathbf{x}_i)$ is the CNN output prediction map. A summary of the CNN architecture is given in Table 16.2. Both models were optimized with the Adam optimizer, with hyperparameters as given in Kingma and Ba (2014), for 80 epochs, i.e. runs over the training data.

Table 16.2 Summary of CNN model. *S* is the window size in the average pooling operation and *DR* is the Dilation-Rate of the convolutional filter, i.e. the pixel spacing between each filter coefficient.

Layer# - Type	Output Shape	Param	Input
1 - Conv2D	(None, 300, 300, 12)	228	—
2 - Conv2D	(None, 300, 300, 12)	1308	1
3 - BatchNorm	(None, 300, 300, 12)	48	2
4 - Dropout	(None, 300, 300, 12)	0	3
5 - Conv2D	(None, 300, 300, 18)	1962	4
6 - Conv2D	(None, 300, 300, 18)	2934	5
7 - BatchNorm	(None, 300, 300, 18)	72	6
8 - Dropout	(None, 300, 300, 18)	0	7
9 - BatchNorm	(None, 1, 1, 160)	640	8
10 - Conv2D	(None, 300, 300, 18)	2934	9
11 - Conv2D	(None, 300, 300, 18)	2934	10
12 - AveragePool $S = 2$	(None, 300, 300, 18)	0	11
13 - Conv2D $DR = 2$	(None, 300, 300, 24)	3912	12
14 - AveragePool $S = 4$	(None, 300, 300, 18)	0	11
15 - Conv2D $DR = 4$	(None, 300, 300, 24)	3912	14
16 - AveragePool $S = 8$	(None, 300, 300, 18)	0	11
17 - Conv2D $DR = 8$	(None, 300, 300, 24)	3912	16
18 - AveragePool $S = 16$	(None, 300, 300, 18)	0	11
19 - Conv2D $DR = 16$	(None, 300, 300, 24)	3912	18
20 - Concatenate	(None, 300, 300, 108)	0	2+12+14+16+18

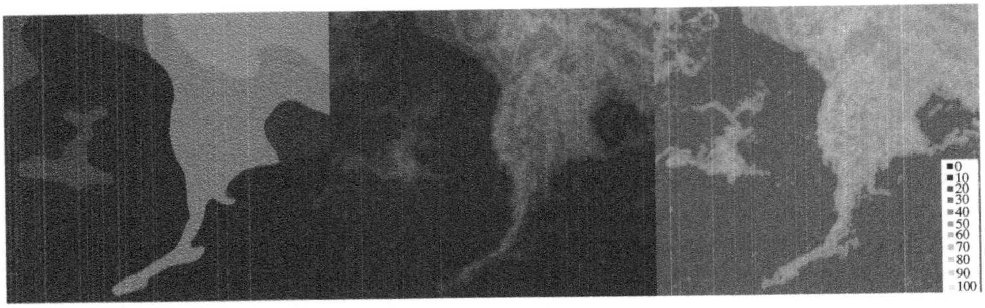

Figure 16.7 Results of Fusion-CNN. Left: ice chart from DMI experts, Mid: prediction from model with binary cross-entropy loss, Right: prediction from model optimized with mean square error.

Results from the two experiments can be seen in Figures 16.7 and 16.8. When comparing the results on Figure 16.7 it is natural that they do not match the label polygons exactly. The network learns to reflect input SAR backscatter values with ice concentrations and contain more details than the polygons. For validation it is necessary to compare predictions at the same scale as the polygons, by comparing the average prediction within each.

This can be done by comparing the mean concentration of a polygon with the mean of the pixel prediction within that polygon. Figure 16.8 shows such a comparison for the test set data. As each scene contains several polygons a mean (red dot) and a std error (black vertical lines) is estimated for the prediction of each unique ice concentration value.

The resolution of Sentinel-1 (EW-mode: 93×87 m) and AMSR2 (at 6.9 GHz: 35×62 km) are magnitudes apart but still, in the sea ice case, it makes sense to fuse these due to the different advantages of each sensor. The model presented in Table 16.2 by-passes the AMSR2 input around the CNN layers to merge information from all 14 channels at the end layer. Other approaches could have been chosen but since there only

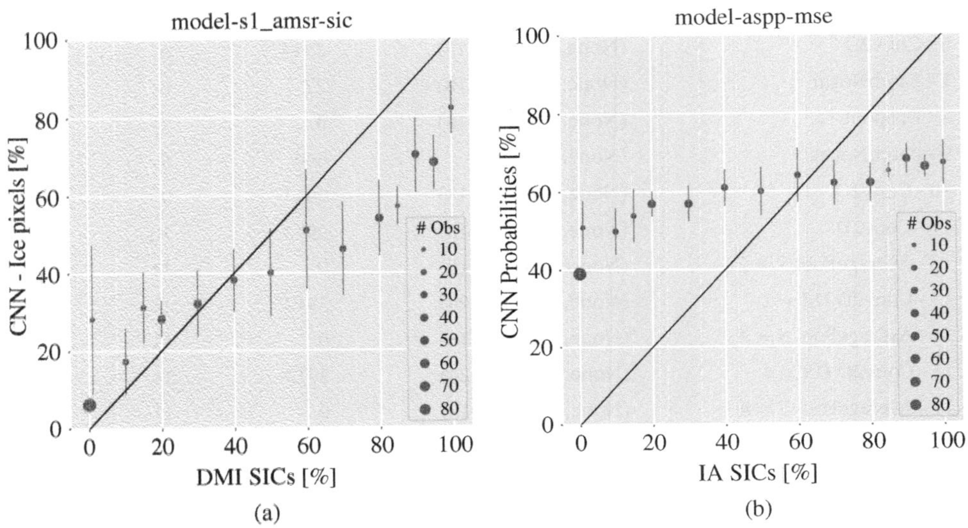

Figure 16.8 (a): BCE loss. (b): MSE loss

are 6x6 pixels of AMSR2 data for every patch, there are much lesser spatial features and applying the same amount of filters to this data leads to redundant convolutional operations. A typical approach with CNNs for sensor fusion is to resample all data to same pixel spacing and stack it as extended "color" channels. If the 2 channels of the Sentinel-1 image where stacked with the 14 channels upsampled AMSR2 the amount of PS_{ops} in the first layer would, according to Equation 16.1, rise from $(3 \times 3) \times 2 \times (300 \times 300) \times 12 = 19.44e6$ to $(3 \times 3) \times 16 \times (300 \times 300) \times 12 = 155.52e6$. This is a large increase of computational load to introduce and therefore is the stacking method not optimal when resolutions in fusion datasets differs so much.

Generally, we can conclude that the BCE loss does significantly better in optimizing the weights for sea ice predictions at 40×40 m scale. The BCE trained model aligns much better with the expert annotated mean sea ice concentrations, Figure 16.8(a), as oppose to the MSE trained model that struggles to properly catch the full range of values, Figure 16.8(b). Other approaches to loss functions could have been taken as well. Following the previously discussed approach in Oord et al. (2016), with a multi class categorical cross-entropy (CCE) loss, we would assign each unique ice concentration a class. Another approach could be to use the MDN networks proposed by Bishop (2006). Both a CCE-trained network and the MDN network uses several output nodes to model the probability distribution over the full range of possible outputs, as opposed to the single output approaches shown in these experiments.

16.5 Conclusion

Many challenges exist for deep learning in bio-geophysical parameter retrieval problems. Since we are modeling the Earth's state, we need to apply algorithms on large amounts of data. Further, we have many sources of variance in our observations caused by e.g. seasonal, yearly and geographical variation. There is also many different types of sensors, some measuring very small signal values resulting in noise problems, and sometime we need to fuse data from sensors to optimize predictions. Deep learning has several advantages though. The end-to-end learning concept makes it possible to map between highly non-linear relationships and the latent feature space inside a neural networks impose some sparsity when working with high-dimensional problems. Further, the architectures of DL models are flexible and can be tailored to the many sources of Earth observation data. Once trained, DL models are typically not a large computational burden, which often makes it possible to incorporate in operational pipelines. It therefore very likely that the Earth parameter retrieval will see increasing use of the DL framework in the future.

17

A Review of Deep Learning for Cryospheric Studies

Lin Liu

17.1 Introduction

The cryosphere refers to the Earth's surface where water is frozen. Its major components include snow cover, glaciers, ice sheets, permafrost and seasonally frozen ground, sea ice, lake ice, river ice, ice shelves, and icebergs. Storing about 75% of the world's fresh water in frozen state, the cryosphere plays an important role in the global water cycles. As an integrated part of the Earth system, the cryosphere modulates the surface energy and gas exchange with the biosphere, atmosphere, and ocean. We refer interested readers to Barry and Gan (2011) and Marshall (2011) for a comprehensive and detailed description of the cryosphere.

In recent decades, the cryosphere has been undergoing strong warming and area reduction. The Special Report on the Ocean and Cryosphere in a Changing Climate, released in September 2019 by the United Nations' Intergovernmental Panel on Climate Change (IPCC), provides an up-to-date and comprehensive summary of the past, ongoing, and future changes of the cryosphere (IPCC 2019). For instance, according to the most recent Ice Sheet Mass Balance Inter-comparison Exercises (IMBIE), both the Greenland and Antarctic Ice Sheets have been losing ice mass at accelerated rates in the past two decades (IMBIE 2018, 2019). According to space-borne passive microwave measurements, the Arctic sea ice extent in September has decreased by $\sim 12.8\%$ per decade between 1979 and 2018 (Onarheim et al. 2018). Globally, the ground temperature at the depth of zero amplitude in the continuous permafrost zone rose by $\sim 0.39°C$ from 2007 to 2016 (Biskaborn et al. 2019).

The rapid changes of the cryosphere have numerous profound implications for human society such as opening of new shipping routes in the Arctic, inundation and land loss associated with rising sea level, glacial lake outburst flood, and slope instability and infrastructure damage from permafrost degradation (IPCC 2019). Cryospheric changes also affect the global climate system through feedbacks associated with the decrease of

Deep Learning for the Earth Sciences: A Comprehensive Approach to Remote Sensing, Climate Science, and Geosciences, First Edition. Edited by Gustau Camps-Valls, Devis Tuia, Xiao Xiang Zhu, and Markus Reichstein.

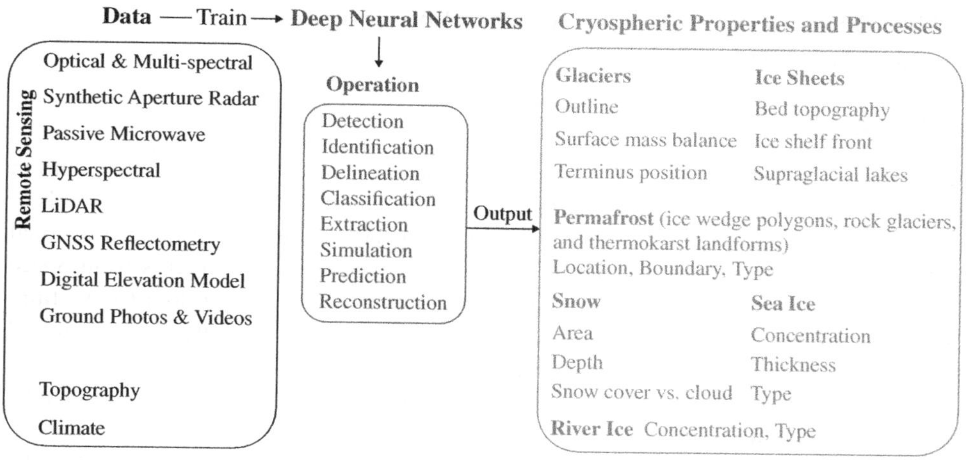

Figure 17.1 Deep-learning-based studies of the cryosphere.

Earth's albedo (Flanner et al. 2011), modifications to ocean circulation caused by the influx of fresh meltwater (Böning et al. 2016), and the release of carbon from thawing permafrost (Schuur et al. 2015).

This chapter reviews the use of deep learning for cryospheric studies, aiming to showcase the diverse applications of deep learning for tackling research tasks that are becoming more challenging to conventional approaches. Even though such applications are still in an early stage, deep learning has been utilized to characterize nearly all the cryospheric components in a diverse manner (see Figure 17.1 for a graph summary). In terms of datasets, a majority of deep learning studies have been applied to remote sensing observations, largely due to the effectiveness and versatility of remote sensing tools to observe cryospheric systems in remote and often inaccessible places (Tedesco 2014). Moreover, many deep learning algorithms that have been well-developed for computer vision can be directly used to remote sensing observations (also see chapters in Part I). Some modeling studies make use of deep learning for parameterization and prediction. Some deep learning studies on Arctic vegetation (e.g., Langford et al. 2019), Arctic wetland (e.g., Jiang et al. 2019a), and land surface temperature (e.g., Tan et al. 2019), all related to the cryospheric variables and processes, but are beyond the scope of this review.

Because the applications are diverse with no single common dataset, methodology, evaluation metrics, we are not going to compare various works. Instead, we will highlight a few studies that represent the growing literature on DL applications, summarize some innovative and unique use of DL, and offer some thoughts on common strengths and limitations as well as future directions.

Section 17.2 summarizes DL-based remote sensing studies of various components of the cryosphere, including glaciers, ice sheets, snow cover, permafrost, sea ice, and freshwater ice. Section 17.3 describes the use of deep learning for modeling the cryosphere. Section 17.4 concludes this review by highlighting the key achievements and future directions of deep learning, followed by a list of public data and codes in the appendix.

17.2 Deep-learning-based Remote Sensing Studies of the Cryosphere

17.2.1 Glaciers

Remote sensing has been widely used to study glaciers: their location, extent, changes in thickness, mass, volume, etc. (see Raup et al. 2014). One of the most intuitive tasks is to delineate glacial boundaries from remote sensing imagery, which has produced regional and global glacier inventories such as the Global Land Ice Measurements from Space (GLIMS, Raup et al. (2007)) and the Randolph Glacier Inventory (The RGI Consortium 2017). In addition to such a static view, it is of great interest to quantify glaciers' frontal changes over time, especially in areas where changes are fast.

With the continuous accumulation in the past decades and by many recent space missions, the data volume of remote sensing imagery over glaciers has increased dramatically. Automated mapping is therefore superior to manual methods because of greater productivity and reliability and a lower cost. There were only a few feature-extracting methods developed in the past (e.g., Sohn and Jezek 1999; Liu and Jezek 2004; Seale et al. 2011) but they were not widely adopted due to the need for extensive prior knowledge and experience. Moreover, case-specific modifications to these methods are needed when applying to different glaciers or remote sensing images because of changes in glacier textures due to variations in snow cover, wetness, roughness, grain size, and internal structure.

In 2019 alone, there were three independent studies utilizing DL towards automated delineation of glacial terminus (Baumhoer et al. 2019; Mohajerani et al. 2019; Zhang et al. 2019). All these studies used U-Net, a robust semantic segmentation architecture originally developed for biomedical applications (Ronneberger et al. 2015). Yet, they have significant differences in their source images and input data, preparation of training data, and pre- and post-processing strategies. For instance, Mohajerani et al. (2019) used Landsat-5 (green band only), -7 and -8 (panchromatic band only) images and reprojected and resampled all the input images into a common grid that is aligned to the glacier flow direction, to reduce the diversity brought by four major outlet glaciers in Greenland. They then directly applied U-Net to isolate buffered calving fronts and extracted the ice front by finding the most possible path based on the network's output. Their results are affected by similarly-looking features such as icebergs and crevasses. Zhang et al. (2019) used synthetic aperture radar (SAR) images taken by the TerraSAR-X satellite over one outlet glacier, i.e., Jakobshavn Isbræ in western Greenland. They classified the surface into ice-mélange and non-ice-mélange regions, and their boundary was extracted as the calving front (Figure 17.2). For extracting glacier and ice shelf fronts at 9 locations across Antarctica, Baumhoer et al. (2019)'s input data have four channels: the first three are HH, HV, and HH to HV ratio, all from Sentinel-1 SAR and the fourth is the TanDEM DEM. The inclusion of the DEM, which is static, was to provide approximate elevation information for distinguishing water and multi-year sea ice versus ice shelves at higher elevations. Zhang et al. (2019) and Baumhoer et al. (2019) kept the orientation and spatial resolution of the original SAR images but needed to subdivide into small patches to accommodate the required input size of U-Net. These two works classified the surface into two classes and extracted the boundary between them, which is different from the strategy adopted by Mohajerani et al. (2019) that isolated buffered calving

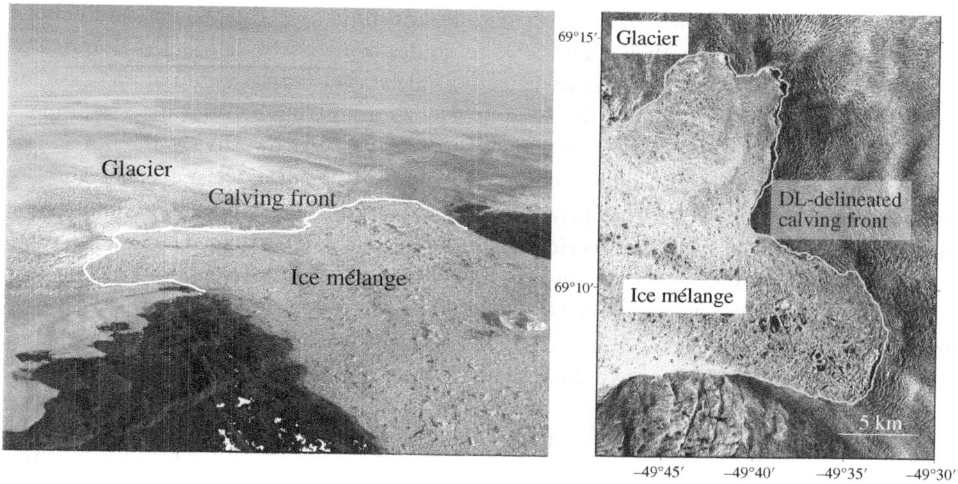

Figure 17.2 Jakobshavn Isbræ in western Greenland. Left: aerial photo (oblique view) of the glacier. Photo credit: NASA/OIB/John Sonntag (https://www.jpl.nasa.gov/news/news.php?feature=7356). Its calving front, manually delineated, separates the glacier and ice mélange (a mixture of calved icebergs and sea ice). Right: TerraSAR-X image taken on August 28, 2013. The calving front was delineated by Zhang et al. (2019) using DL. This figure was modified from Figure 4a in Zhang et al. (2019) with the authors' permission.

fronts directly. Zhang et al. (2019) only mapped one glacier, and data diversity originates from hundreds of SAR images taken in both summer and winter seasons over several years. Mohajerani et al. (2019) and Baumhoer et al. (2019) tested the networks' performance on glaciers/regions not used in training, showing good transferability of their DL networks.

In mountain areas, the ablation zones of many valley glaciers are covered by debris at the surface. The supraglacial debris presents similar spectral properties as the surrounding landscapes on remote sensing imagery, making it challenging to map the boundary of debris-covered glaciers. Xie et al. (2020) took the first step towards a DL-based mapping of debris-covered glaciers. Their input data consist of 17 layers, including all 11 Landsat-8 bands, 1 DEM, and 5 DEM-derived layers containing topo-geomorphic parameters (slope angle, slope-azimuth divergence index, etc.). They used the manually-delineated boundaries from the GLIMS dataset to train a feed-forward neural network. Testing in two regions in central Karakoram and Nepal Himalaya, they evaluated the network's performance when using different portions of splitting the ground truth data for training and demonstrated an overall high accuracy despite in complex cases where lakes and proglacial moraine are present. They also conducted a transfer-learning experiment that used the pre-trained network with the Karakoram debris-covered glaciers as the base model to train the Nepal data. This strategy helped to reduce the training time and slightly improve the mapping accuracy.

17.2.2 Ice Sheet

Leong and Horgan (2020) presented a new and interesting use of DL to improve the spatial resolution of the bed elevation map of the Antarctica Ice Sheet. Their model, dubbed DeepBedMap, built on a Generative Adversarial Network (GAN). Because the basis input, a preexisting 1-km-resolution bed elevation map, alone cannot meet their goal of producing a higher-resolution bed map, they added three remote-sensing-based datasets, including the surface elevation (100 m resolution), ice velocity (500 m), and snow accumulation (1 km) as conditional input grids. Trained with 250-m-gridded ground-truth bed elevation resampled from ice-penetrating radar surveys at five locations in western Antarctica, the GAN generated a higher 250-m-resolution bed elevation map over the entire ice sheet. The authors demonstrated that this new map is generally realistic but still exaggerates the roughness or introduces obvious artifacts such as ridges and speckles in some places. The major limitations are that the training dataset only covers an extremely small fraction of the ice sheet and the high-resolution details in the GAN-based bed elevation rely on the conditional data from the surface of the ice sheet, especially the surface elevation. More work is needed to enhance the quantity of training data and incorporate glacial flow mechanics into the DL model.

Yuan et al. (2020a) extracted supraglacial lakes in central west Greenland from Landsat 8 Operational Land Imager (OLI) data and further documented their changes during the melt seasons from 2014 to 2018. Their input data are the mean of Bands 1 to 8, Normalized Difference Water Index calculated from the green and near-infrared bands, and Normalized Difference Vegetation Index calculated from the red and near-infrared bands. Their training data are manual-delineated supraglacial lake outlines from Landsat 8 RGB images. Comparing with an unsupervised image thresholding method (Otsu), and two supervised methods (Random Forests and Support Vector Machine), they demonstrated that the CNN outputs contain the least noise and omission errors.

17.2.3 Snow

Due to the high albedo of snow, its presence can often be easily identified from visible and passive microwave images. Numerous hand-crafted retrieval algorithms, tailed towards specific sensors, haven been developed towards routine products of snow extent, melt, albedo, depth, and water equivalent (see Tedesco 2014, Chapters 3, 4, 5, 6).

Recently, a few studies explored the use of DL to extract snow cover from remote sensing imagery and evaluated its potentials against conventional methods. For instance, Xia et al. (2019) used a multi-dimensional deep residual network to classify snow and cloud cover images on multi-spectral HuanJing-1 satellite imagery over Tibet. Nijhawan et al. (2019a) developed a hybrid method that integrates an AlexNet-based DL network with Sentinel-2 optical images as the input and a Random Forest classifier with hand-crafted features based on Sentinel-1 SAR and SRTM DEM to extract snow cover in northern India. Validating against ground truth based on field observations, they showed that their hybrid method gave the best accuracy (98.1%) and highest Kappa coefficient (0.98) compared with conventional machine-learning methods (accuracies ranging from 77% to 95%). Guo et al. (2020) firstly trained DeepLabv3+ (Chen et al. 2018a) with a 30-m snow-cover product based on Landsat-8-based Normalized Difference Snow Index and then fine-tuned the DL network using a smaller amount of training data based on 3.2-m-resolution Gaofen-2 imagery. Their initial experiments demonstrated that such a DL-based model can differentiate snow from clouds and even recognize snow in image shadows.

In addition to snow cover, DL has also been used to retrieve snow depth. Braakmann-Folgmann and Donlon (2019) proposed a DL-based retrieval algorithm for estimating snow depth on Arctic sea ice from passive microwave radiometer measurements. Their input data included three AMSR2-based brightness temperature (Tb) ratios (vertically polarized Tb at 18.7 vs. at 36.5 GHz, vertically polarized Tb at 6.9 vs. 18.7 GHz, and vertically vs. horizontally polarized Tb at 36.5 GHz) plus one SMOS-based Tb ratio (vertically vs. horizontally polarized Tb at 1.4 GHz). They used snow depth measured from an airborne radar by NASA's airborne Operation IceBridge campaigns to train their DL network that consists of five fully connected hidden layers. Comparing with three empirical snow depth algorithms, the authors showed that the DL-based one gives the highest accuracy. They further demonstrated that the DL-estimated snow depth could improve the retrieval of sea ice thickness when converting altimeter measurements of ice freeboard to sea ice thickness.

17.2.4 Permafrost

Permafrost refers to the ground that remains at or below 0 °C for at least two consecutive years (French 2017). Because permafrost is an underground thermal phenomenon, it is challenging to observe directly using remote sensing. The use of DL has thus far been limited to mapping ice wedge polygons and hillslope thermokarst landforms from remote sensing data.

In a pioneer study, Zhang et al. (2018) applied DL to detect, delineate, and classify ice wedge polygons from aerial orthophotos with spatial resolution ranging from 0.15 m to 1 m) taken in northern Alaska. They manually annotated the training samples as high-centered and low-centered polygons (a high/low-centered polygon is slightly higher/lower at the center than at its rim). Then they carried out object instance segmentation using Mask R-CNN and outputs binary mask (ice wedges or non-ice-wedges) with classification. They separately evaluated their accuracies of detection, delineation, and classification and reported that the DL-based method could detect about 79% of ice wedge polygons across a 134-km^2 area. Demonstrating the degree of transferability of DL by applying the network trained to coarser-resolution images (0.5 m to 1 m) taken in a new area, they showed that the DL can still achieve a 72% detection accuracy.

Abolt et al. (2019) applied a simpler, 10-layer CNN to 50-cm-resolution DEMs constructed by using airborne LiDAR data over two areas on the Arctic coast in northern Alaska. The training data are manually-delineated ice wedge boundaries and non-ice-wedge boundaries. After the CNN-based operation, they applied a watershed analysis based on DEM and measured microtopography, classify them into high- and low-centered polygons. They detected up to 5000 ice wedge polygons per square kilometer and more than 1,000,000 over an area of 1200 km^2 near Prudhoe Bay (Abolt and Young 2020). It is arguably only feasible to use DL to generate such kind of high-resolution, high-density, and extensive maps.

Thermokarst is a generic term that refers to "the process by which characteristic landforms result from the thawing of ice-rich permafrost or the melting of massive ice" (Harris et al. 1988). Thermokarst landforms are important surface expressions and visual indicators of permafrost degradation. According to their geomorphological and hydrological characteristics, thermokarst landforms are classified into more than 20 types, such as thermokarst lakes, thermo-erosion gullies, active layer detachments, and retrogressive

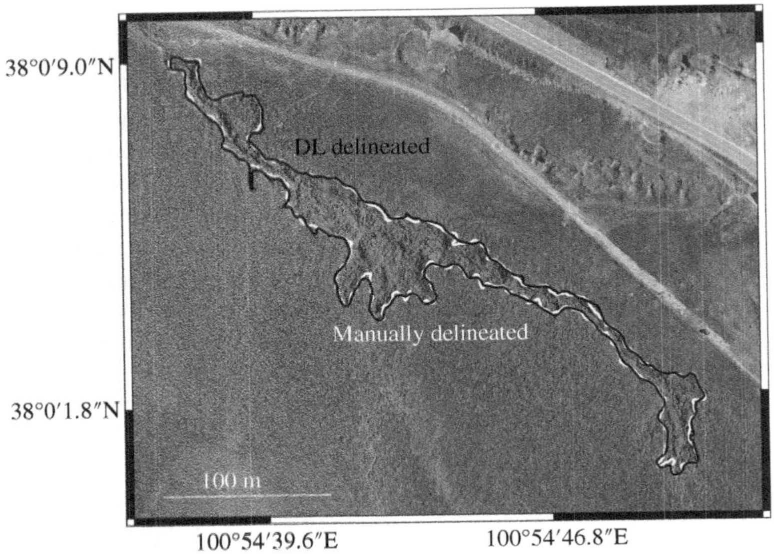

Figure 17.3 Deep-learning-based delineation of thermokarst landforms. This example shows one of 16 landforms that Huang et al. (2018) identified from a high-resolution UAV image (background) using DL. This figure was modified from Figure 8c in Huang et al. (2018) with the authors' permission.

thaw slumps (Jorgenson 2013). Thermokarst landforms are common on the Qinghai-Tibet Plateau and high mountains of China, but their locations and surface dynamics are still poorly quantified or understood, especially compared with their counterparts in the Arctic.

Huang et al. (2018) identified thermo-erosion gullies in a small watershed (6 km^2) in northeastern Tibet. They applied DeepLab v2 (Chen et al. 2016), to a 0.15-m-resolution digital orthophoto map constructed aerial photographs taken by an unmanned aerial vehicle (UAV). Validating the results against field-mapped boundaries, they showed that the DL method successfully mapped all the 16 thermokarst landforms in the watershed (see one of them in Figure 17.3). They also showed the drastic improvement of DL over a conventional object-based image analysis method, the latter of which has many challenges in identifying thermo-erosion gullies with complex geometric and geomorphic features. Applying a newer and improved version DeepLabv3+ to 3-m-resolution CubeSat images taken by the Planet constellation, Huang et al. (2020) successfully delineated 220 retrogressive thaw slumps within an area of 5200 km^2 in central Tibet. They also proved the robustness of their results based on more than 100 experiments with different data augmentation strategies and portions of ground truth data used for training.

17.2.5 Sea Ice

Several studies have used DL to retrieve sea ice concentration, namely the fractional amount of sea ice coverage within an area, from various types of remote sensing data (Wang et al. 2016; Wang et al. 2017b; Yan et al. 2017; Yan and Huang 2018; Cooke and Scott 2019; Chi et al. 2019; Xu and Scott 2019). In one of the first of such efforts, Wang et al. (2016) applied a simple CNN (two convolutional layers, two max-pooling layers, and a fully-connected layer) to dual-polarized (HH and HV) RADARSAT-2 ScanSAR images taken over the Beaufort Sea (Arctic). The training data they used were ice concentration charts manually produced by experts from visual interpretation of SAR images. Validating against with AMSR-E ice concentration products, they showed the robustness of their DL-based method even in cases of significant SAR speckle noise, varying incidence angle, and in areas of low ice concentration. Cooke and Scott (2019) presented an innovative use of DL that is trained using passive microwave sea ice concentration products (from AMSR-E) and inferring on higher-resolution RADARSAT-2 SAR images. Yan et al. (2017); Yan and Huang (2018) used DL to detect the presence of sea ice and estimate its concentration from TechDemoSat-1 Delay-Doppler maps obtained using Global Navigation Satellite System Reflectometry.

Mei et al. (2019) estimated sea ice thickness in the Ross Sea (Antarctica) from profiles of snow surface acquired from terrestrial laser scanning. The highlight of this work is that the input does not include any snow depth or surface densities. Instead, the DL, which consists of three convolutional layers and two fully connected layers, learns 3D geomorphic features in the laser scanning and builds a non-linear link with sea ice thickness. Additionally, DL has been used to classify sea ice types from Earth Observing-1 (EO-1) hyperspectral imagery (Han et al. 2019) and detect sea ice changes from multi-temporal SAR images (Gao et al. 2019).

17.2.6 River Ice

Singh et al. (2019) used DL networks for river ice segmentation from video data captured by UAV and fixed cameras from two Alberta rivers during the winters of 2016 and 2017. They classified the into water and two types of sea ice: randomly oriented needlelike frazil ice and sediment-carrying anchor ice. Subsequently, the concentration of each type of river ice could be easily calculated. Comparing the performance of four networks: U-Net, SegNet, DeepLabv3+, DenseNet, they found that DenseNet shows a high-level of generalization; DeepLab gave the highest accuracy; whereas U-Net casts a good balance between accuracy and generalization.

17.3 Deep-learning-based Modeling of the Cryosphere

As of early 2020, only a limited number of studies have explored the use of DL to model cryospheric processes. Three studies have used DL in long-term (one year or longer) and short-term (one month) predictions of sea ice concentration (Chi and Kim 2017; Kim et al. 2018, 2020). In one of the most recent works, Kim et al. (2020) used DL to build a non-linear link between passive microwave sea ice concentration with daily sea surface temperature and monthly reanalysis of eight predictors (such as sea ice concentration one-year and one-month before, sea surface and air temperature one-month before) over the Arctic ocean for 30 years (1988–2017). Their CNN model consists of three convolutional layers and one fully-connected layer. Comparing with the results of Random Forest and an anomaly persistence prediction model, they showed that the CNN model gives the best predictions in both space and time domains according to five accuracy metrics (anomaly correlation coefficient,

Nash-Sutcliffe efficiency, mean absolute error, root-mean-square error, and its normalization). They also demonstrated the superior performance of their CNN model for predicting sea ice concentration in extreme cases such as the significant sea ice loss in the summers of 2007 and 2012.

Bolibar et al. (2020) used DL for simulating and reconstructing the surface mass balance (SMB) of glaciers in the French Alps. In contrary to empirical or physics-based models, their data-driven method used DL to parameterize the non-linear link between annual glacier-wide SMB with topographic variables such as mean and maximum altitude and meteorological/climatic variables such as cumulative positive degree days, snow precipitation and temperature anomalies. Because of the small size of the annual SMB data, covering 32 glaciers spanning 31 years, they designed a simple, 6-layer feed-forward fully-connected DL and implemented it in Keras. Comparing with two classic linear regression methods, they showed that the DL-based model gives improved explained variance (by up to 64% in space and 100% in time) and better accuracy (by up to 47% in space and 58% in time). In particular, the DL model captures about one-third of non-linear variabilities in the temporal changes. This DL-based SMB reconstruction is now included in the open-source ALpine Parameterized Glacier Model (ALPGM, https://github.com/JordiBolibar/ALPGM). However, because the DL model was trained using data from the French Alps, it needs to be retrained if being applied to other regions.

17.4 Summary and Prospect

As summarized in this review, deep learning has emerged as an innovative and important tool in cryosphere remote sensing and modeling and has been applied to nearly all cryospheric components in the Arctic, Antarctica, and high mountain regions. Even though all of the published works are demonstrative in nature, they have proved the feasibility and potential of DL. A few modeling-focused studies utilized DL for establishing non-linear links between cryospheric processes and various topo-climatic variables. Many studies showed its superior performance over conventional hand-crafted methods in learning non-linear patterns of spatial, temporal, and spectral signatures of the cryosphere in remote sensing data. DL is also able to automate detection, segmentation, and classification operations on multi-sensor, multi-temporal remote sensing datasets for highly dynamics cryosphere systems.

However, DL applications in the cryosphere are not in full-fledged status or ready for routine operation or used by multiple research groups. One of the major bottlenecks is the lack of labeled data that can be directly used to train DL networks. Most of the studies mentioned above prepared their own labeled data, which tend to be limited in quantity and may only be applicable to specific cases. Moreover, it is logistically challenging and expensive to collect ground truth. Due to the diverse nature of cryospheric systems and the datasets used, it is challenging to compile a few benchmark datasets similar to those in computer vision. The datasets published by individual research groups, such as those listed in the appendix, could serve the first step. It is also important to establish standard protocols and guidelines for choosing DL architectures, generating label data, preparing training data, conducting cross-validation and testing, assessing model performance. Close coordination among international communities in cryospheric sciences, data sciences, space agencies, and data centers is needed.

Diverse datasets pose challenges, but also opportunities in maximizing/taking advantage of DL's transferability and generalization to process multiple datasets in one single network, or using one set of data to train and inference with another data, or combining them into input nodes, or using independent data in testing and accuracy evaluation.

It is worth noting that the cryosphere research community has shown a strong and ever-growing interest in deep learning. For instance, two of the world's largest earth-science conferences, the American Geophysical Union (AGU) Fall Meeting and European Geosciences Union (EGU) General Assembly, launched inaugural "AI for cryosphere" sessions in 2019 and 2020, respectively. The 2019 AGU session alone received more than 20 papers that used DL. Some of these new studies tackled cryospheric systems such as icebergs, crevasses, ice shelf hydro-fractures, rock glaciers, lake ice, as well as important processes such as melt on ice sheets and glacial flow, none of which is included in this soon-to-be-outdated review. DL will continue to offer new and exciting opportunities for better and more comprehensive quantification and modeling of the cryosphere. We foresee that DL will be developed into full-fledged status be used to tackle more complex cryospheric problems and to investigate the interactions of the cryosphere with the other earth systems.

Appendix: List of Data and Codes

Here are the major data centers, repositories, and providers for cryospheric studies:

- US National Snow and Ice Data Center (https://nsidc.org)
- US National Science Foundation Arctic Data Center (https://arcticdata.io)
- US Antarctic Program Data Center (http://www.usap-dc.org)
- European Space Agency Climate Change Initiative (http://cci.esa.int)
  - Antarctic Ice Sheet (http://esa-icesheets-antarctica-cci.org)
  - Greenland Ice Sheet (http://esa-icesheets-greenland-cci.org)
  - Glaciers (http://www.esa-glaciers-cci.org)
  - Permafrost (http://cci.esa.int/Permafrost)
  - Sea ice (http://cci.esa.int/seaice)
  - Snow (https://climate.esa.int/en/projects/snow)
- Canadian Cryospheric Information Network (https://www.ccin.ca)
- China National Tibetan Plateau Data Center (https://data.tpdc.ac.cn/en/)

Below we list the data and codes published in the cryospheric studies reviewed in this chapter, grouped by the cryospheric components.

1. Glaciers
   - Detection of glacier calving margins with convolutional neural networks (Mohajerani et al. 2019)
     Code and data: https://github.com/yaramohajerani/FrontLearning

- Automatically delineating the calving front of Jakobshavn Isbræ from multitemporal TerraSAR-X images (Zhang et al. 2019)
  Code: https://github.com/enzezhang/Front_DL3
  Training and test data: https://doi.org/10.1594/PANGAEA.897066
- ALpine Parameterized Glacier Model (ALPGM) (Bolibar et al. 2020)
  Code and sample data: https://github.com/JordiBolibar/ALPGM

2. Ice sheet
   - DeepBedMap: Antarctica Ice Sheet bed elevation using a super resolution deep neural network (Leong and Horgan 2020)
     Code: https://github.com/weiji14/deepbedmap
     Training experiments: https://www.comet.ml/weiji14/deepbedmap
     Digital bed elevation model: https://doi.org/10.17605/OSF.IO/96APW
3. Snow
   - Estimating snow depth on Arctic sea ice using satellite microwave radiometry and a neural network (Braakmann-Folgmann and Donlon 2019)
     Sample code and data: https://github.com/AnneBF/snownet
4. Permafrost
   - Automatic mapping of thermokarst landforms from remote sensing images using deep learning (Huang et al. 2018)
     Code: https://github.com/yghlc/DeeplabforRS
   - Using deep learning to map retrogressive thaw slumps from CubeSat images (Huang et al. 2020)
     Code: https://github.com/yghlc/Landuse_DL
     Training and test data: https://doi.pangaea.de/10.1594/PANGAEA.908909
   - High-resolution mapping of spatial heterogeneity in ice wedge polygon geomorphology near Prudhoe Bay, Alaska (Abolt and Young 2020)
     Code and data: https://doi.org/10.1594/PANGAEA.910178
5. River ice
   - River ice segmentation with deep learning (Singh et al. 2019)
     Code: https://github.com/abhineet123/river_ice_segmentation
     Data: https://ieee-dataport.org/open-access/alberta-river-ice-segmentation-dataset

18

Emulating Ecological Memory with Recurrent Neural Networks

Basil Kraft, Simon Besnard, and Sujan Koirala

18.1 Ecological Memory Effects: Concepts and Relevance

Ecological memory can be broadly defined as the encoding of past environmental conditions in the current ecosystem state that affects its future trajectory. The consequent effects, known as *memory effects*, are the direct influence of ecological memory on the current ecosystem functions (Peterson 2002; Ogle et al. 2015). Such memory effects are prevalent across several spatial and temporal scales. For example, at the seasonal scale, the variability of spring temperature affects ecosystem productivity over the subsequent summer and autumn (Buermann et al. 2018). Inter-annually, moisture availability over the previous year is linked to contemporary ecosystem carbon uptake (Aubinet et al. 2018; Barron-Gafford et al. 2011; Ryan et al. 2015). Furthermore, less frequent and large extreme events (e.g., heat waves, frost, fires, or insect outbreaks) can lead to short-term phenological changes (Marino et al. 2011) or long-term damage to the ecosystem with diverse effects on present and future ecosystem dynamics (Larcher 2003; Lobell et al. 2012; Niu et al. 2014). This evidence highlights the relevance of short to long-term temporal dependencies on past environmental conditions in terrestrial ecosystems. However, due to the large spectrum of the environmental conditions and their consequent effects on the ecosystem, quantifying and understanding the strength and persistence of memory effects is often challenging.

Ecological memory effects may comprise *direct* and *indirect* influences of *external* and *internal* factors (Ogle et al. 2015) that are either *concurrent* or *lagged* in time. For instance, a drought may directly decrease ecosystem productivity, with indirect concurrent effects on loss of biomass due to the drought-induced fire (t_3 in Figure 18.1). Additionally, ecosystems may not only be responding to concurrent factors, but also the lagged effects of past environmental conditions. A drought event can further impact the ecosystem productivity for months to years through a direct but lagged effect. On the other hand, indirect lagged effects involve external factors that affect the ecosystem productivity during a drought, e.g., disturbances like tree mortality and deadwood accumulation (t_4 in Figure 18.1), which may lead to insect outbreaks with further influences on the ecosystem (t_5 and t_6 in Figure 18.1).

Deep Learning for the Earth Sciences: A Comprehensive Approach to Remote Sensing, Climate Science, and Geosciences,
First Edition. Edited by Gustau Camps-Valls, Devis Tuia, Xiao Xiang Zhu, and Markus Reichstein.

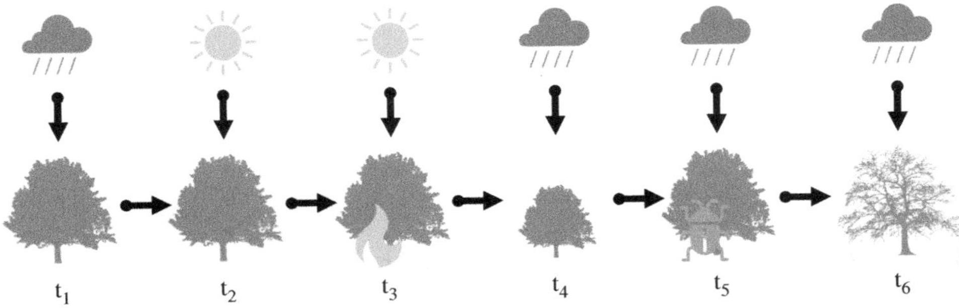

Figure 18.1 Schematic diagram illustrating the temporal forest dynamics during and post-disturbance: drought occurring in t_2 and t_3 conditions, fire event in t_2, and insect outbreaks in t_5.

Memory effects are not exclusive to ecosystem productivity, but encompass a large number of Earth system processes of carbon (Green et al. 2019) and water cycles (Humphrey et al. 2017). A key variable that encodes memory effects in the Earth system is soil moisture. Soil moisture is controlled by instantaneous and long-term climate regimes, vegetation properties, soil hydraulic properties, topography, and geology. As such, soil moisture exhibits complex variabilities in space, time, and along with soil depth. Owing to its central role but large complexity, most physical models are built around the parameterization of moisture storage, which in turn affects the responses of land surface to environmental conditions. Nevertheless, physical models have inherent uncertainties due to differences in structure and complexity and input data as well as unconstrained model parameters.

Several data-driven methods have therefore been developed to address the shortcomings of physical models for understanding Earth system processes as observed in the data. But, the data-driven methods may also be limited by data quality and availability. For example, the vegetation state over the land surface can be observed with satellite remote sensing. Yet, state variables such as soil moisture, which have imprints of ecological memory, are difficult to measure to meaningful soil depths and across larger scales. This poses a key challenge in capturing the memory effects using conventional data-driven methods. As such, dynamic statistical methods (cf. Chapter 8), such as Recurrent Neural Networks (RNNs, LeCun et al., 2015), may address these shortcomings, as they do not necessarily require measurements or observations of state variables. In this context, RNNs have a large potential for bringing the data-driven estimates on par with Earth system models with regards to capturing the ecological memory effects on land surface responses. This chapter focuses on this aspect and demonstrates the capabilities of RNNs to quantify memory effects with and without the use of state variables.

18.2 Data-driven Approaches for Ecological Memory Effects

18.2.1 A Brief Overview of Memory Effects

Conceptually, the memory effects on a system response Y_t, at time t, encompass the influences of forcing X_{t-k} in previous $k \geq 1$ time steps. As such, the memory effects propagate through time via the system state S_t, at every time step, which can be expressed as

$$S_t = f(S_{t-1}, X_t). \tag{18.1}$$

The response Y_t is, in turn, a function of S_t as

$$Y_t = g(S_t). \tag{18.2}$$

The S_t encodes all memory effects needed to compute Y_t, and it can be interpreted as the ecological memory. From a data-driven perspective, the memory S_t emerges solely from the effects of "unobserved" previous states that are not directly encoded in any given observations (Jung et al. 2019). For example, if instantaneous vegetation state (e.g., vegetation greenness) and current climatic conditions (e.g., air temperature or rainfall) are included in the observed state O_t, their effects are not necessarily encoded in S_t. Therefore, S_t can be mathematically expressed as

$$S_t = f(S_{t-1}, X_t, O_t). \tag{18.3}$$

18.2.2 Data-driven Methods for Memory Effects

Following Equation 18.3, several data-driven statistical methods have been employed to account for ecological memory and quantify their effects on ecohydrological responses. Given the lack of observed state variables, a common practice is to use hand-designed features (Tramontana et al. 2016; Papagiannopoulou et al. 2017), such as lag or cumulative variables of past time-steps, in sequence-agnostic machine learning methods (e.g., random forest, feed-forward networks) (Tramontana et al. 2016; Papagiannopoulou et al. 2017). Although these methods generally work well, they do not capture the long-term dependencies of ecohydrological processes on past environmental conditions and interactions among different variables, as well as their complex temporal dynamics (Lipton et al. 2015). Alternatively, Bayesian non-linear mixed-effects methods that consider joint probability distributions of different variables have shown promising avenues to represent interactions and understand environmental and biological memory (Ogle et al. 2015; Liu et al. 2019).

Lastly, dynamic deep learning methods, such as RNNs, are capable of extracting temporal features. As such, they can represent ecosystem responses to past environmental conditions and capture ecological memory effects. In RNNs, analogous to Equation 18.1, a hidden state S_t is updated from the past state S_{t-1} and concurrent observations X_t (cf. Chapter 8). Owing to that, dynamic methods have been successfully applied in sequence learning (e.g., speech recognition) and land cover classification (Rußwurm and Körner 2017a).

With the increasing availability of remote sensing and climate data that span several decades, new avenues to employ temporally dynamic statistical methods like RNNs have opened for exploring and understanding the known and unknown temporal dynamics of Earth system processes. Such methods have already been applied to dynamically incorporate the effects of recent and past vegetation and climate dynamics on, for instance, ecosystem productivity (Reichstein et al. 2018), and the memory effects therein (Kraft et al. 2019). Compared to static methods, the dynamic methods improve the prediction of seasonal dynamics of net carbon dioxide fluxes, with varying degrees of memory effects across different climate and ecosystem types (Besnard et al. 2019).

18.3 Case Study: Emulating a Physical Model Using Recurrent Neural Networks

As shown in previous studies, RNNs can potentially learn ecological memory (Reichstein et al. 2018; Besnard et al. 2019; Kraft et al. 2019). It is, however, unclear under what conditions the RNNs can emulate the ecosystem responses, and to what extent the ecological memory plays a role in defining these responses. Using RNNs for such questions in the real-world is often challenging due to the data availability (e.g., gaps in the remote sensing data), data uncertainty, and data inconsistency. Despite the limitations in data quality, the RNNs provide useful insights on ecosystem responses to past environmental conditions, albeit with inherent uncertainties. The validation of RNNs prediction would require more data including those from natural control and factorial experiments, but such data are hardly available.

To address this, we implement a series of experiments on a complete set of simulated data, i.e., a simulation from a physical land surface model, to test whether—and to what extent—an RNN can learn ecological memory and simulate its effects on ecohydrological processes. The physical model simulation circumvent known limitations in measured Earth observation data, such as noise and biases, limited length of the time-series with potentially limited representation of the full range of environmental conditions, or incomplete set of variables. It should be noted that the physical model simulations are not the observed reality, but they provide a viable test bed for evaluating RNNs. Given the same input data, RNNs should be able to replicate the underlying processes included in the physical model. Such an exercise provides a robust assessment of the usefulness of dynamic statistical models for Earth system science. More specifically, in the upcoming sections, we demonstrate the capabilities of RNNs to:

1. emulate global spatio-temporal distributions of daily Evapotranspiration (ET) simulations from a physical land surface model;
2. quantify the effect of land surface states (e.g., soil moisture state) that are not directly provided as input to RNN; and
3. evaluate the capability of RNNs to capture the seasonal dynamics of ET under normal and extreme climatic conditions.

18.3.1 Physical Model Simulation Data

The test dataset for the RNN experiments was obtained from the simulations of a physically-based global land surface model, the MATSIRO (Takata et al. 2003; Koirala et al. 2014). The MATSIRO is a land surface scheme of an Earth system model that simulates the water and energy budget over the land surface using physically-based representations of hydrological fluxes such as runoff, ET, and a cascade of storage components including snow, soil and groundwater. In the MATSIRO model, the hydrological fluxes are diagnosed based on the prognostic variation of hydrological storages. As such, memory effects of past climatic and environmental conditions on current fluxes are explicitly considered through their dependence on storage. In essence, the temporal variations of storage variables are

Table 18.1 Datasets used in MATSIRO model simulation.

	Variables	Native resolution	
		spatial	temporal
Spatial	Plant functional types, soil texture	0.5 degree	—
Spatial, seasonal and interannual	Rainfall, snowfall, air temperature, snowfall, downward short-wave and long-wave radiation, wind speed, specific humidity, surface pressure, cloud cover, leaf area index	0.5 degree	3-hourly

constrained by physical mass balance equations and can be represented as

$$S_t = f(S_{t-1}, X_t, Z_t), \tag{18.4}$$

where X_t represents the input drivers controlling the soil moisture S_t, such as precipitation, vegetation activity, and soil characteristics, and Z_t represents the output fluxes such as runoff and ET.

The output variables, Z_t, at any time, are non-linear and complex functions of climatic conditions and moisture storage, and thus include the memory effects of past conditions. Due to the physical constraints of the mass balance equations, the model responses are mathematically tangible and depend exclusively on the input data and physical processes encoded in the model. A brief overview of the input variables and their features is provided in Table 18.1.

18.3.2 Experimental Design

To assess the capability of an RNN to emulate the MATSIRO model, we implemented experiments to predict ET and its dependence on ecological memory provided through soil moisture. To do so, we also use the exact set of input variables (Table 18.1) from Matsiro simulations aggregated at daily scale.

Different RNN model setups were contrasted in a 2×2 factorial experiment design (Table 18.2). The RNN setups use at least the meteorological drivers and the static variables as inputs. We used the Long Short-Term Memory (LSTM) architecture (Hochreiter and Schmidhuber 1997), capable of learning long-term dependencies and therefore accounting for ecological memory effects (cf. Chapter 8). If the temporal model without soil moisture ($LSTM_{\neg SM}$) is capable of learning the memory effects implicitly, its performance should be on par with a temporal model with soil moisture as an additional input ($LSTM_{SM}$), as ET is only dependent on soil moisture state in the MATSIRO model (cf. section 18.3.1). In addition, two non-temporal models based on multiple Fully Connected (FC) layers were trained, one without soil moisture ($FC_{\neg SM}$), and one with soil moisture as input (FC_{SM}). While both models do not have access to past observations conceptually, the latter can use the concurrent soil moisture state. Contrasting the FC models allows the local importance of soil moisture to be assessed.

Table 18.2 Factorial experimental design: the four models are trained individually to assess the capability of an LSTM to learn ecological memory ($LSTM_{SM}$, with soil moisture *vs.* $LSTM_{\neg SM}$, without soil moisture as input) and to quantify the local relevance of soil moisture for ET (FC_{SM} *vs.* $FC_{\neg SM}$). The temporal models learn a mapping from the concurrent and past features $X_{\leq t}$ to the target Y_t, while the non-temporal models have access to the concurrent features X_t only. S_t is the ecosystem state, i.e., soil moisture.

		model type			
		temporal		non-temporal	
model input	w/ SM	$LSTM_{SM}$	$Y_t = f(X_{\leq t}, S_t)$	FC_{SM}	$Y_t = f(X_t, S_t)$
	w/o SM	$LSTM_{\neg SM}$	$Y_t = f(X_{\leq t})$	$FC_{\neg SM}$	$Y_t = f(X_t)$

The predictions from four model setups were evaluated against the MATSIRO simulation at global and regional scales. At the grid-scale, the overall performances were evaluated using the Nash–Sutcliffe model efficiency coefficient (NSE) (Nash and Sutcliffe 1970) and the Root Mean Square Error (RMSE) (Omlin and Reichert 1999). Globally, the performances are also summarized across different temporal (daily, daily anomalies, daily seasonal cycle, interannual variation) scales. At the regional scale, our evaluation focused on the capability of LSTM to simulate temporal ET dynamics in two focus regions: the humid Amazon and semi-arid Australia. In these two example cases, the mean seasonal cycle for the period 2001–2013 and seasonal anomalies observed during climate extreme events (2005 drought in the Amazon (Phillips et al. 2009) and the 2010 La Niña in Australia (Boening et al. 2012)) were evaluated. Table 18.3 summarizes the main features of the evaluations.

18.3.3 RNN Setup and Training

As described in the previous section, two different models were used: a temporal model (LSTM) and a non-temporal model (FC), i.e., with stacked fully connected layers. All setups had the same input features as the MATSIRO model, and optionally soil moisture as added as an input variable (see Table 18.1). The models were trained on the MATSIRO ET simulations, with Mean Square Error (MSE) as a loss function.

Table 18.3 Summary of the scope of the experiments.

	Objective	Regions assessed	Period assessed	Input used
Analysis 1	Use of RNNs for emulating physical models	global	2001–2013	Original input+ soil moisture + physical model outputs
Analysis 2	Simulating seasonal dynamics under normal and extreme conditions	Amazon region and Australia	2001–2013, 2005 and 2010	Original input+ soil moisture + physical model outputs

The LSTM takes the multivariate time-series and static variables as input, which is followed by a hyperbolic tangent activation and a linear layer that maps the LSTM output at each time step to a single value: the predicted ET. The FC models consist of several fully connected layers, each followed by a non-linear activation function, except for the output layer, where no activation function is used. The FC model takes the static variables and only a single time-step of the time-series as input.

The final model architectures (Table 18.4) were selected using a hyper-parameter optimization approach: the Bayesian optimization hyper-band algorithm (Falkner et al. 2018). The state-of-the-art optimization algorithm efficiently finds optimal hyper-parameters by combining an early stopping mechanism (dropping non-promising runs early) and a Bayesian sampling of promising hyper-parameters, with a surrogate loss model for the existing samples. To prevent over-fitting of the hyper-parameters, we used only every 6th latitude/longitude grid-cell (approximately 3% of the data) during hyper-parameter optimization. To avoid over-fitting of the residuals caused by temporal

Table 18.4 The model and training parameters from hyper-parameter optimization and their ranges searched. Both LSTM models (SM vs. ¬SM) consist of several LSTM layers, followed by multiple fully connected layers. The non-temporal FC models consist of several stacked fully connected layers. In all setups, dropout was enabled for the input data and between all layers. Note that a dropout of 0.0 means that no dropout is applied.

Parameter	Search range	SM	¬SM
LSTM			
dropout (input)	(0.0, 0.5)	0.0	0.0
LSTM number of layers	(1, 3)	2	1
LSTM hidden size	(50, 300)	300	200
LSTM dropout	(0.0, 0.5)	0.4	0.3
FC number of layers	(2, 6)	3	5
FC hidden size	(50, 300)	300	300
FC activation	{ReLU, softplus, tanh}	ReLU	ReLU
FC dropout	(0.0, 0.5)	0.3	0.1
learning rate	(0.001, 0.0001)	0.001	0.001
weight decay	(0.01, 0.0001)	0.01	0.01
FC			
dropout (input)	(0.0, 0.5)	0.0	0.0
FC number of layers	(2, 6)	6	4
FC hidden size	(50, 600)	200	200
FC activation	{ReLU, softplus, tanh}	ReLU	ReLU
FC dropout	(0.0, 0.5)	0.0	0.0
learning rate	(0.001, 0.0001)	0.01	0.01
weight decay	(0.01, 0.0001)	0.001	0.001

auto-correlation and to test how the model generalizes, the data were split into two sets: training data from 1981 to 1999 inclusive and test data from 2000 to 2013 inclusive. For both sets, an additional period of 5 years was used for model warm-up. For all four setups, the hyper-parameter optimization and model training were carried out independently.

18.4 Results and Discussion

18.4.1 The Predictive Capability Across Scales

In this section, we evaluate the performances of the different RNN setups against the MATSIRO simulations. In general, the LSTM model setups perform considerably better than the FC models (Figure 18.2). In fact, outside the tropical humid regions, the LSTM models achieve a systematically higher predictive capacity than the FC models. The LSTM models have a higher median NSE ($LSTM_{SM}$: 0.98, $LSTM_{\neg SM}$: 0.97) and lower RMSE ($LSTM_{SM}$: 0.15, $LSTM_{\neg SM}$: 0.19) than the FC models (NSE of FC_{SM}: 0.93, $FC_{\neg SM}$: 0.89, and RMSE of FC_{SM}: 0.28, $FC_{\neg SM}$: 0.33). However, within the tropical humid regions, all setups have lower performance than in other regions (median NSE of 0.78, 0.75, 0.69, 0.57 and median RMSE of 0.45, 0.48, 0.55, 0.61 for $LSTM_{SM}$, $LSTM_{\neg SM}$, FC_{SM}, and $FC_{\neg SM}$, respectively). This may be possibly associated with larger variability in the water fluxes leading to a low signal-to-noise ratio in this region.

It can be hypothesized that an LSTM model can learn the ecological memory effect of soil moisture, even when soil moisture is not included as an input variable. Along this line, we find that the $LSTM_{SM}$ and $LSTM_{\neg SM}$ setups perform better compared to FC setups. This provides evidence that the two LSTM model architectures, with or even without soil moisture, are suitable for learning information content related to unseen state variable, such as soil moisture.

Yet, differentiating $LSTM_{SM}$ and $LSTM_{\neg SM}$ setups does not provide information on where the ecological memory of soil moisture is the strongest. We, therefore, plot the differences in term of predictive capacity between the model setups with and without soil moisture as an input variable (Figure 18.3). As expected, contrasting $LSTM_{\neg SM}$ with $LSTM_{SM}$ shows no substantial differences across the globe (first row of Figure 18.3), The comparison of the FC models (second row of Figure 18.3) suggested that the performances of the model with and without soil moisture can vary significantly in space. This is also reflected in the global model performance (NSE): While the 75th percentile of the temporal ($LSTM_{\neg SM}$: 0.98) versus the non-temporal ($FC_{\neg SM}$: 0.95) models are similar, the 25th percentile differs largely ($LSTM_{\neg SM}$: 0.94, $FC_{\neg SM}$: 0.86). This shows that the $LSTM_{\neg SM}$ model is capable of learning heterogeneous global dynamics, while the $FC_{\neg SM}$ model struggles in particular regions, which are, as we argue here, the ecosystems exhibiting strong memory effects. The differences in $FC_{\neg SM}$ and the FC_{SM} model setups were mostly apparent in water-limited regions. In these semi-arid regions, the memory effects through soil moisture that are present, and influential (Koirala et al. 2014), in the MATSIRO simulations cannot be well reproduced by FC models, especially when soil moisture is not provided as an input variable.

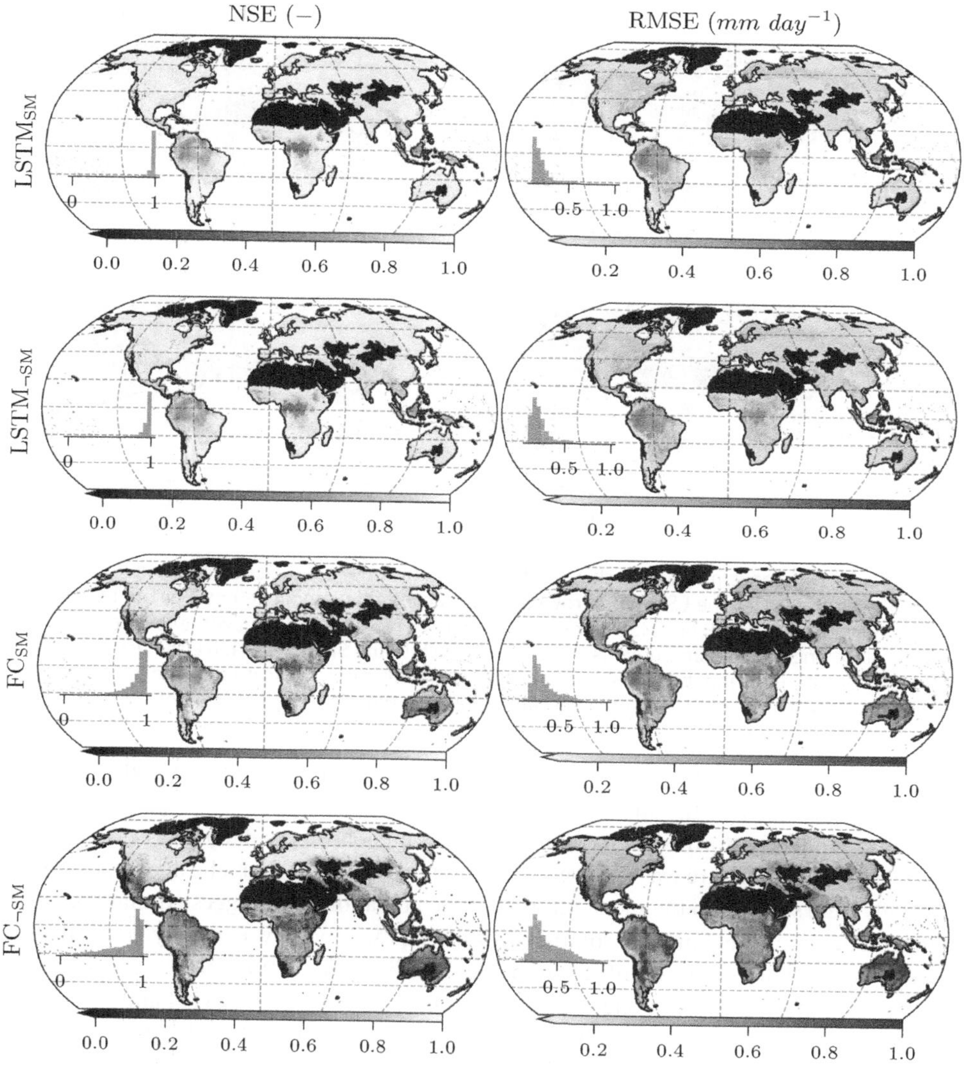

Figure 18.2 Global distributions of performances of different model setups based on daily model predictions from the test dataset. Nash-Sutcliffe model efficiency coefficient (NSE) is shown in the left and Root Mean Square Error (RMSE) in the right column for the temporal LSTM and non-temporal FC models with (SM) and without (¬SM) soil moisture input, respectively. The inset histogram represents the global distribution of the metrics.

We further investigate the performances of the model experiments across different temporal scales in training and test sets (Figure 18.4). As shown in Figure 18.2, the two LSTM models (shown in the box and whisker plots) were able to learn the spatio-temporal daily patterns with NSE values close to 1 and a low variation across different grid-cells. We further found that the performances of LSTM models are relatively weaker for the predictions of daily and annual anomalies than that for the mean daily seasonal cycle. The

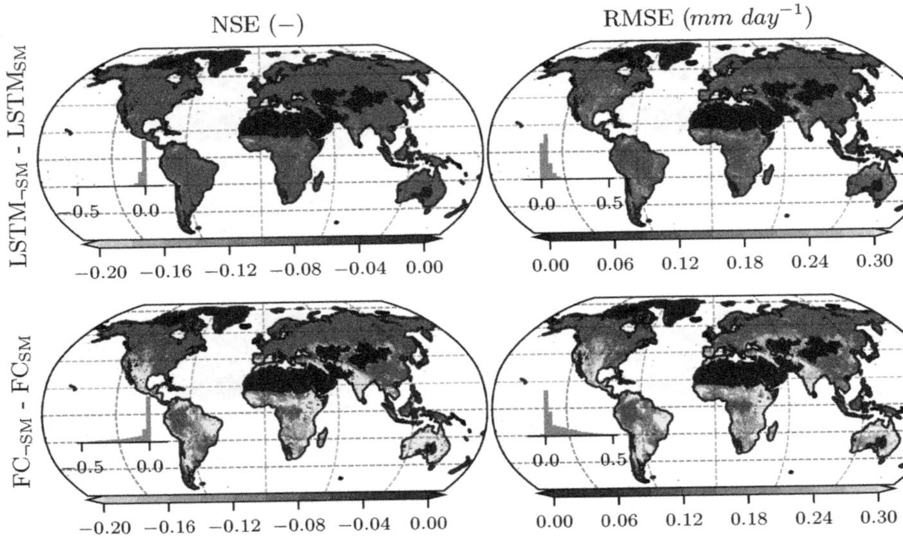

Figure 18.3 Difference maps of Nash-Sutcliffe model efficiency coefficient (NSE) and Root Mean Square Error (RMSE) for the LSTM (first row) and FC (second row) model setups. For the LSTM models, differences in NSE or RMSE were computed as $LSTM_{\neg SM}$ − $LSTM_{SM}$, while for the FC models, differences were computed as $FC_{\neg SM}$ − FC_{SM}. While the SM models have the ecosystem state (soil moisture) as input, the ¬SM do not have it. Darker colors indicate that the SM model performs better than ¬SM. The inset histogram represents the global distribution of the differences.

performances of the LSTM models were still good with a median NSE of 0.91 ($LSTM_{SM}$) and 0.88 ($LSTM_{\neg SM}$) for the anomalies.

The FC models performed worse than the LSTM models on the daily time series, particularly when soil moisture was not used as an input variable ($FC_{\neg SM}$). The decomposition of the daily time series into the mean seasonal cycle and anomalies suggested that the lower performance of the FC models compared to the LSTM model, was mostly controlled by weaker performance with regards to anomalies (median NSE of 0.75 for FC_{SM} and 0.63 for $FC_{\neg SM}$). The mean seasonal cycle was captured similarly well in the LSTM and FC models (median NSE from 0.97 to 1.00, where lowest is $FC_{\neg SM}$ and highest is $LSTM_{SM}$), although with a larger variability across grid-cells, with a 25th to 75th percentile of 0.95 to 1.00 (FC_{SM}) and 0.82 to 0.99 ($FC_{\neg SM}$) versus 1.00 to 1.00 ($LSTM_{SM}$) and 0.97 to 1.00 ($LSTM_{\neg SM}$). The model performances for anomalies were substantially lower for FC models compared to the LSTM models. These results suggest that ecological memory effects appear to be especially relevant for improving the model performance of capturing the daily and annual anomalies.

Surprisingly, the FC_{SM} model performed worse than the LSTM models, particularly for the anomalies, even though the only relevant state variable for a given time step, SM_t, was known to the model. This contradiction may be associated with several factors. First, in MATSIRO simulation, the ET is based on the transient soil moisture with losses and gains within a day between the SM_{t-1} and SM_t. In the experiment here, SM_{t-1} was used as an input for the FC_{SM} model, and as such, one would expect some minor differences. Additionally,

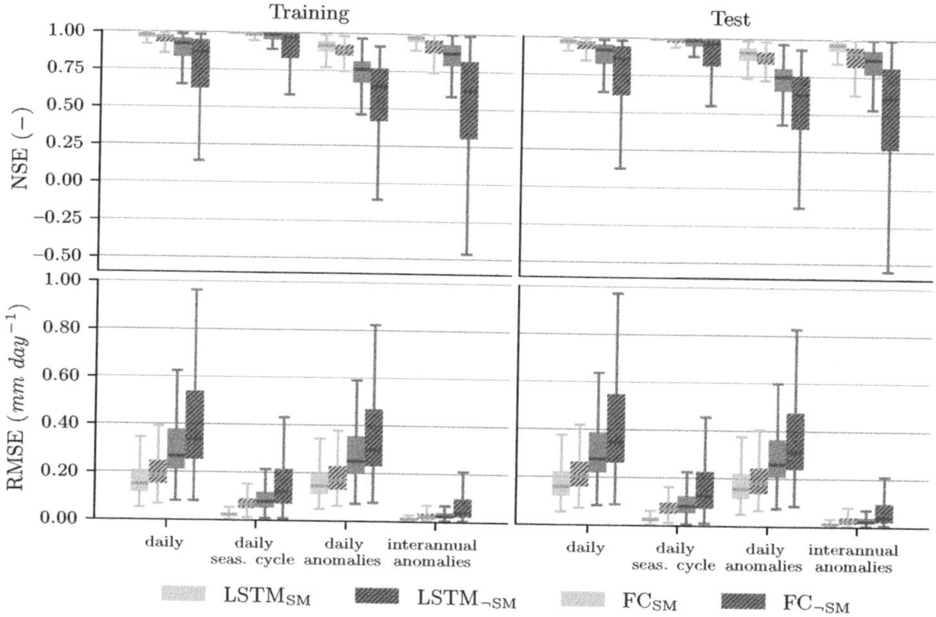

Figure 18.4 Box and whisker plots showing grid-level model performances across timescales (i.e., daily, daily seasonal cycle, daily anomalies, and annual anomalies) for the training and test sets. Daily seasonal cycle are calculated as the mean of each day across different years, daily anomalies are calculated as the difference between daily raw estimates and the mean of each day and annual anomalies are calculated as the difference between mean annual and mean estimates within each grid-cell. Nash-Sutcliffe model effiiency coefficient (NSE) and Root Mean Square Error (RMSE) are shown. The whiskers represent the 1.5 · inter-quartile range (IQR) of the spatial variability of the model performances.

albeit hypothetical, the FC may not have enough capacity to extract high-level features for an instantaneous mapping from the concurrent time step of the input data, while the LSTM models can learn complex representations from a series of past time steps. Therefore, the LSTM can learn part of the ecological memory effects through temporal dynamics of soil moisture in addition to instantaneous soil moisture, compared to information used by the FC_{SM}. This also extends to the potential utilization of the distribution of input data by the LSTM model, which has access to the full global distribution of all the input data.

18.4.2 Prediction of Seasonal Dynamics

We have shown evidences of the capabilities of RNN models in emulating a land surface model globally across different temporal scales. It is also worthwhile to analyze whether the model setups can emulate temporal dynamics under *normal* and *extreme/anomalous* climatic condition. This is an important factor, as extreme conditions are rare and only represent a fraction of the full data, that RNNs use to learn about the dynamics.

In general, for the mean seasonal cycles of 2001–2013, the FC model is farther from the MATSIRO simulations in both the Amazon and Australian regions (Figure 18.5, top

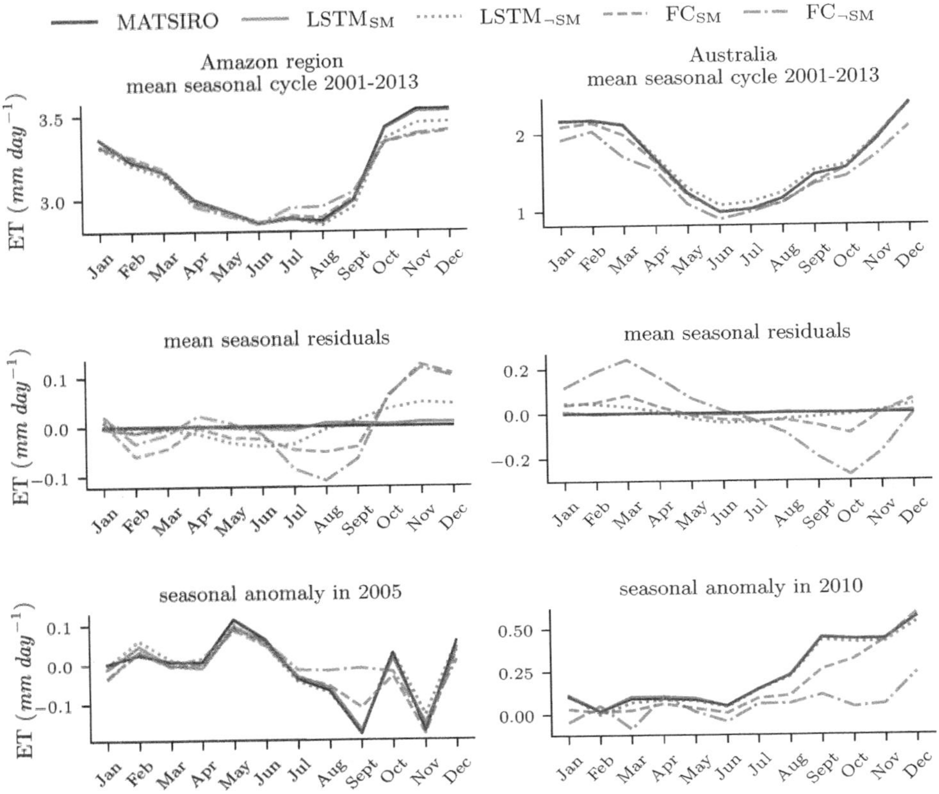

Figure 18.5 Seasonal cycle (first row), seasonal variation of the residuals (second row) and seasonal anomaly (third row) in the Amazon region (first column) and Australia (second column). Seasonal residuals were computed as ET residuals$_i$ = [ET MATSIRO$_i$ - mean(ET MATSIRO)] – [ET predicted$_i$ – mean(ET predicted)], where i is a month. Seasonal anomalies are shown for the years 2005 and 2010 for the Amazon region and Australia, respectively.

row). But, not all the models perform well under all conditions. For example, in humid Amazon, the LSTM$_{SM}$ performs the best across all months, while other models perform relatively worse in drier condition (July-December). The mean seasonal variations of the residuals (second row) show that the LSTM models can better learn temporal dynamics of ET than FC models, as the residuals for these models (blue lines) is closer to zero over the entire year. The FC models have larger residuals, with particularly high values for FC$_{\neg SM}$ model, especially during the dry season in the Amazon region and over the growing season in Australia (August to May). The high values in the seasonal patterns of residuals in Australia for the FC$_{\neg SM}$ experiment but not in the FC$_{SM}$ model suggested apparent importance of soil moisture in controlling ET in this region.

We further investigate the performance of LSTM models under two extreme climatic conditions: the 2005 drought in the Amazon, and the 2010 La Niña in Australia (Figure 18.5, bottom row). The LSTM$_{SM}$ (dashed blue line) and LSTM$_{\neg SM}$ (solid blue line) models can reproduce the MATSIRO simulation of strong seasonal anomalies even under the extreme conditions (second row). As also shown in the previous sections, the FC_{SM} model cannot reproduce the seasonal anomalies as well as the LSTM models do.

18.5 Conclusions

This chapter provided an overview of ecological memory effects in the Earth system, along with a case of the application of a deep learning method, the RNNs, for representing ecological memory effects. The case study used the simulations of a physical model as a pseudo-observation to evaluate the capabilities of RNNs models to predict ET and ecological memory effects therein.

The LSTM model was able to capture the ecological memory effects inherent in the physical model. Moreover, the difference in the performances of the LSTM model with and without soil moisture state was found to be negligible. This appeared to be consistent from daily to annual temporal scales, and over most regions globally. This finding demonstrated that the LSTM, through its hidden states, is indeed able to learn the memory effects that are explicitly encoded in the state variables of a physical model.

We further found that the LSTM was able to predict the soil moisture-ET dynamics even during anomalous climatic conditions demonstrating that the predictions of the LSTM are general and applicable under a wide range of environmental conditions. This was true for seasonal responses of ET to the 2005 dry spell in the Amazon, and the 2010 La Niña event in Australia. The non-temporal FC models generally performed worse, especially with regards to anomalies when soil moisture was not given as input ($FC_{\neg SM}$). Under the assumption that the physical model is analogous to reality, the poorer performance of the model can be interpreted as the importance of memory effects of soil moisture on ET. The relatively weaker performance of the FC model, which has access to soil moisture (FC_{SM}), compared to the LSTM architectures could not be explained conceptually. We hypothesize that access to the distribution of the past climate observations in the LSTM models and the LSTMs being able to compensate for biases emerging from temporal aggregation may be associated with its better performance.

In summary, our results compared with the simulations of a physical model demonstrated the usefulness of LSTM model architecture for learning the dynamics and the ecological memory of unobserved state variables, such as soil moisture. This justifies the need, and provides confidence, for use of a dynamic statistical model, such as LSTM, when investigating temporally dependent ecohydrological processes using (often limited) observation-based dataset. The coupling of dynamic data-driven methods either with physically-based models (i.e., hybrid modeling, cf. Chapter 22) or with complementary machine learning approaches (e.g., convolutional neural networks, cf. Chapter 2) will pave the way for a better understanding of the known as well as unknown Earth system processes.

Part III

Linking Physics and Deep Learning Models

19

Applications of Deep Learning in Hydrology

Chaopeng Shen and Kathryn Lawson

19.1 Introduction

Hydrologists have had a long history working with neural networks in myriad applications including rainfall runoff modeling, groundwater management, water quality, stream salinity, and precipitation forecasting (Gupta et al. 2000; Govindaraju 2000). As one of the largest fields in geoscience by population, hydrology was also one of the early geoscientific fields to adopt deep learning (DL) (Shen et al. 2018; Shen 2018a). Following several pioneering applications (Tao et al. 2016; Fang et al. 2017; Kratzert et al. 2018), DL has gradually taken hold in hydrology. As a 2018 open-access review paper has already summarized some of the applications of hydrologic DL (Shen 2018a), the main purpose of this chapter is to account for the recent trends from late 2017 to early 2020 and provide some outlooks into the next stage.

The 2017-early 2020 era marked a proof-of-capability phase for DL in hydrology as well as a period of fast researcher onboarding and radically increasing hydrologic applications for many topics (section 19.2). DL has been evolving from a niche tool to a method of choice for some prediction tasks, while a wide range of approaches have been attempted to offer the full suite of services commonly provided by traditional hydrologic models (e.g., dynamical modeling, forecasting, data collection). Nevertheless, at the same time, DL is still a skill wielded by a minority in the field. This may be mainly because the educational background required for DL is fundamentally different from the traditional hydrology curriculum (Shen et al. 2018). However, with the current growth rate, it is possible that DL will one day be an integral component of the hydrologic discipline (Shen 2018b).

DL was developed primarily as a tool to learn from data and extract information. It is no surprise that hydrologists first used DL to learn from the most prevalent hydrologic datasets, including both satellite-based and gage-based observations. Within this realm, the applications can be grouped into big-data DL (section 19.2.1.1), small-data DL (section 19.2.1.2), and information retrieval (section 19.2.3) (Figure 19.1). However, applied mathematicians and modelers have come to realize that the fundamental approach of DL, including the tracking of the derivative chain and the back-propagation of error, provides new ways to support scientific computing and new ways to ask questions (section 19.2.2).

Deep Learning for the Earth Sciences: A Comprehensive Approach to Remote Sensing, Climate Science, and Geosciences, First Edition. Edited by Gustau Camps-Valls, Devis Tuia, Xiao Xiang Zhu, and Markus Reichstein.

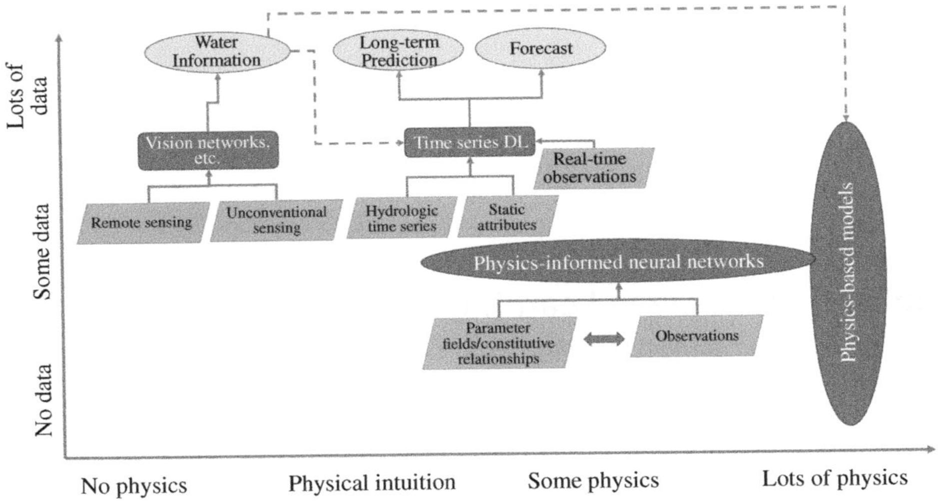

Figure 19.1 A summary of recent progress on deep learning applications in hydrology. DL has been used to process raw data and extract information, which can then be consumed by other models (dashed line). With physical intuition and physics-inspired problem setup, data-driven time series models have been created to successfully model a range of hydrologic variables, in both small-data and large-data domains, for both forward model runs and short-term forecasting. When data is limited but the physics is relatively straightforward, physically-informed neural networks can be incorporated to allow forward modeling or inference of parameters or constitutive relationships under a limited-data regime. This figure has been inspired by Figure 1 in Tartakovsky et al. (2018). Dark grey-colored components represent the models.

19.2 Deep Learning Applications in Hydrology

19.2.1 Dynamical System Modeling

Perhaps one of the biggest findings in the 2017–2019 period for hydrologic DL was that time series DL is a highly competent dynamical modeling tool for tasks addressed by traditional hydrologic models. In particular, the long short-term memory (LSTM) (Hochreiter et al. 2001) network has emerged from being a niche tool to a mainstream modeling method. LSTM is a self-trained memory system with storage units that can mimic system storage and fluxes. It was originally designed for sequence modeling, and has been widely used in machine translation and handwriting recognition (Graves and Schmidhuber 2009). The memory capability makes it effective for modeling hydrologic systems.

19.2.1.1 Large-scale Hydrologic Modeling with Big Data

In this type of application, atmospheric forcings such as precipitation, temperature, radiation, etc., along with certain attributes of the landscape, are passed into a LSTM unit which outputs the target variable, which is then compared to the observations. This problem is posed in a fundamental way such that we attempt to include all physical factors that exert influence on the target. We then train the model using many instances with different factor combinations, and ask the model to learn the underlying mathematical relations between the inputs and outputs. If the network can find a good relationship for a complex variety of instances, then it is likely to have learned the fundamental relationship. The use of big data is the differentiating factor from previous neural network applications, which fitted curves to a single site/basin. Optionally, one can also provide simulated results from another model as additional input items.

Time series deep learning models have demonstrated unrivaled predictive accuracy: they directly learn from the target data and excel at reproducing dynamics of the observations. Fang et al. (2017) demonstrated that LSTM could learn from soil moisture dynamics observed by the Soil Moisture Active Passive (SMAP) satellite with high fidelity. Inputs include atmospheric forcings and land surface characteristics such as soil texture and slope. The problem is posed in a fundamental and universal way so the trained LSTM is essentially a hydrologic model that predicts surface soil moisture, and could replace the corresponding component in land surface models. The test error of the conterminous United States (CONUS)-scale LSTM model achieved 0.027, significantly smaller than SMAP's design accuracy. In addition, even if only trained for 3 years, the model could capture multi-year trends in surface and root-zone soil moisture for an unseen period (Fang et al. 2018), and is thus applicable in the face of mild climate non-stationarity. However, this evaluation has only been demonstrated on mild trends, as soil moisture is bounded and has limited length of memory. The effectiveness of LSTM in predicting stronger trends has not been assessed yet. In rainfall-runoff modeling, Kratzert et al. (2018) showed that a regional-scale LSTM model trained without basin attributes can produce high mean Nash-Sutcliffe model efficiency coefficients (NSE). Once basin attributes were included, a CONUS-scale model gave the highest performance metrics compared to other conceptual rainfall-runoff models when evaluated on hundreds of basins, with a median NSE of around 0.75 for their ensemble mean streamflow (Kratzert et al. 2019). Similar metrics were reported in Feng et al. (2020a), with the forcing data introducing some minor differences. Rainfall-runoff modeling is the most classic modeling task for hydrologists, and arguably attracts the most attention. Countless rainfall-runoff models have been developed in the past with different mechanisms and complexities. The demonstration that a DL model was able to outperform many different hydrologic models should have been shocking, but similar feats have already been accomplished in many domains such as chemistry (Goh et al. 2017) and physics (Baldi et al. 2014).

Yang et al. (2019b) used basin-average climate forcings along with simulated outputs from a global hydrologic model to predict floods for global basins. The authors showed that several global hydrologic models performed reasonably well in terms of amplitude of peak discharge, but poorly in terms of their timing. Utilizing an LSTM model that received inputs from the global hydrologic model simulations, the authors pushed the median NSE from −0.54 to 0.49 on a global scale. Although not mentioned by the authors, the results also suggested that the LSTM model was learning principles of routing from the mismatch between model and observations.

For forecasting, all available information including the most recent (or lagged) observations should be employed with any model to maximally improve the forecast accuracy. Traditionally, the primary way of achieving such a forecast was either autoregression for statistical models or data assimilation for process-based models. For autoregression models, the choice of formulation was limited for describing the coupling between lagged observations and environmental variables. For data assimilation with ensemble-Kalman-filtering approaches (Evensen 1994), expert knowledge is needed to make complex choices regarding the assimilation scheme, bias correction (De Lannoy et al. 2007), what variables to include in the covariance matrix, and how to solve the matrix. Such choices are non-trivial, often arbitrarily made, and could involve a great deal of time for testing and experimenting.

Fang and Shen (2020) added an adaptive, "closed-loop" data integration (DI) kernel to assimilate the most recent SMAP observations in order to improve 1-day, 2-day, and 3-day soil moisture forecasts. This kernel offers a generic solution to the common situation where part of the input stream is irregularly missing, which could otherwise cause the algorithm to crash. The kernel first uses LSTM to make a prediction of the soil moisture for a day, but if an actual observation is available for the day, it replaces the model prediction with that observation. The output from this kernel is then used as an input to make a prediction for the next day. The use of this kernel requires an architectural change from the code that directly uses lagged observations as an input. Lagged observations are typically supplied directly, meaning that all training data can be prepared before training starts. However, this cannot be done for the closed-loop kernel. Instead, this kernel is called at each time step at runtime (both in training and testing), so that each observation point may influence the training data of the next time steps. In essence, the LSTM model itself serves as a forward extrapolator. The end result is a very high forecast accuracy of soil moisture over the CONUS, with a median 1-day forecast error of less than 0.021, which also serves as an upper bound estimate for the random component of the SMAP satellite. Utilizing this model, the authors showed processes that are difficult to capture with integrating observations, including irrigation and lake and riverine inundation.

Applying a similar DI technique to streamflow modeling, Feng et al. (2020a) showed that the CONUS-scale median daily NSE with the Catchment Attributes and Meteorology for Large Sample Studies (CAMELS) dataset could be boosted to an unprecedented value of 0.86 (Figure 19.2). In addition, the data integration method flexibly accepted observations of various forms and latency, including daily, weekly, or monthly averages, running averages, or snapshot types. Moreover, because the study was on the CONUS scale and included physiographic attributes in the inputs, it was able to offer explanations for the performance of LSTM and LSTM with data integration based on geographic patterns. For example, the Prairie Potholes Region in the northern-central US and states with extensive lakes are difficult to simulate directly because the extent of surface inundation is challenging to model, but these simulations can be improved greatly by DI as the connected water bodies lead to high autocorrelation in flow. Strong performance was also noted with LSTM for short-term (hourly-scale) flood forecasting (Xiang and Demir 2020), and when upstream gage data was available (Xiang et al. 2020). Similar models were applied with success to water quality indicators including stream temperature (Rahmani et al., 2021) and dissolved oxygen (Zhi et al. 2021).

Time series DL is not limited to LSTM. Just like convolutional neural networks (CNNs) have been used in machine translation, they are also applicable to time series modeling. Sun et al. (2019) used a CNN to learn the mismatch between the terrestrial water storage anomaly simulated by global hydrologic models and that measured by the Gravity Recovery And Climate Experiment (GRACE) satellite mission. The GRACE data is monthly, with 10 years of records, making it arguably too little data for an LSTM model to learn from. In

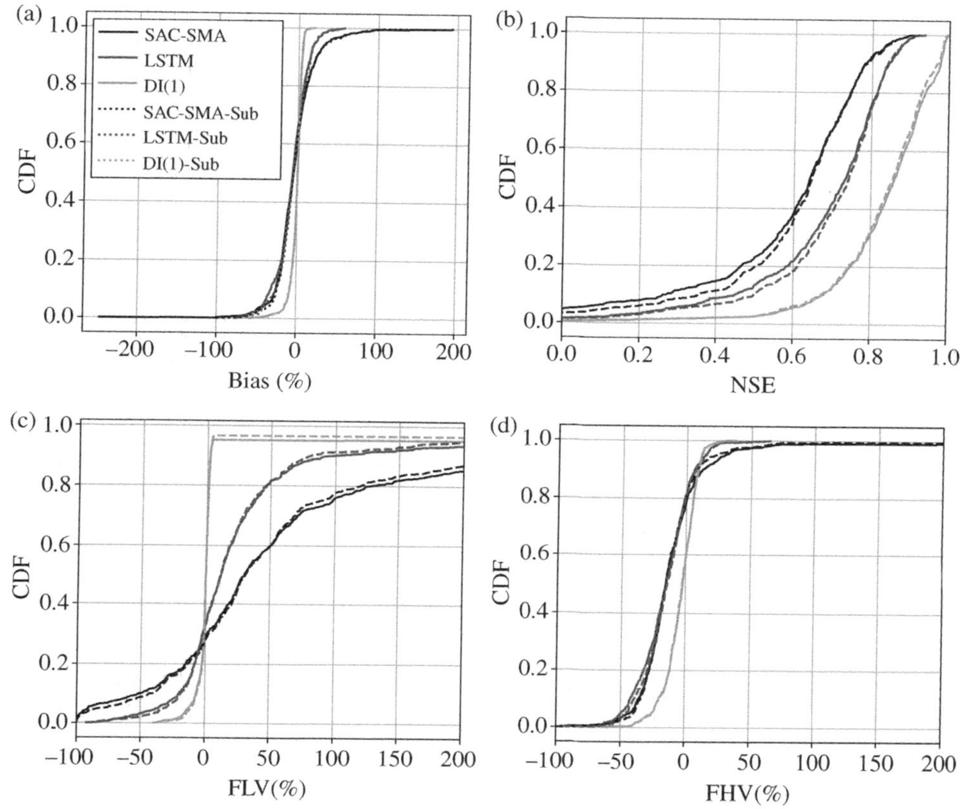

Figure 19.2 Performance of the LSTM forecast model for the CAMELS data, in comparison to the SAC-SMA model, a well-established operational hydrologic model. Figure is from Feng et al. (2020a) with permission from the authors. DI(1) is the forecast model with data integration of the 1-day-lag streamflow observations. The "-Sub" suffix refers to the 531-basin subset used in previous studies (a) FLV: The percent bias of low flow regime (bottom 30%); FHV: The percent bias of high flow regime (top 2%).

this case, the CNN was able to predict the mismatch, and greatly reduce the error with the simulated water storage.

Rather than directly learning from observations and building a DL model, machine learning can be employed to estimate parameter sets for process-based models. Krapu et al. (2019) compared automatic differentiation variational inference, a Bayesian gradient-based optimization method scheme, to several Markov Chain Monte Carlo schemes for estimating parameter distributions for a hydrologic model. They coded a hydrologic model in Theano, a deep learning platform, which allowed the tracking of derivatives through the hydrologic model, and hence, gradient-based parameter adjustment. The approach was reported to be a highly effective parameter estimator. At the core, this scheme still solves an inverse problem. Tsai et al., 2020 proposed a parameter learning scheme that linked deep learning with a hydrologic model. They turned the parameter estimation problem into a big data machine learning problem, and demonstrated substantial advantages as the method scales with more data.

19.2.1.2 Data-limited LSTM Applications

Since the publication of the early hydrologic DL papers in 2017 and 2018, LSTM applications have sprung up like mushrooms in data-limited settings, where data were collected from only one or a few geographical sites, but with sufficient historical length. These applications often have very targeted use cases with direct social impact. Or, stated another way, for many practical problems, there is not a large collection of instances with varied input attributes, forcings, and observations to learn their implications.

For groundwater flow problems, Huang et al. (2018a) used an LSTM model to simulate monthly groundwater recharge, which was estimated using groundwater level data and the water table fluctuation method. Zhang et al. (2018b) used monthly water diversion, evaporation, precipitation, temperature, and time as inputs to predict region-averaged water table depth in different sub-areas of an irrigation district in China. Following the study above, Bowes et al. (2019) used a LSTM model to forecast groundwater level in observed wells with forecasted rainfall and sea level data as inputs along with lagged groundwater levels in a coastal city. The model was able to provide accurate groundwater forecasts for up to 18 hours of lead time. They also found that, for this limited dataset, using data from only periods with storm events produced a better model. Because groundwater well responses are largely influenced by the position of the well and its connection to the seawater, it would have been interesting if they had used the trained model to investigate the impact of sea level, and how it was related to the location of the wells.

LSTM is applicable to connected surface water systems. In a complex river system, a LSTM model was trained to predict the water level in Dongting Lake in China, using lagged lake level information and discharge data from five upstream flow gages (Liang et al. 2018). The prediction errors were on the centimeter level. The authors further explored the impacts of the Three Gorges Dam's outflow on the lake level. Producing the same predictions with a hydrodynamic model would have required a substantial amount of additional input data, model calibration, and computational resources. In another study, de la Fuente et al. (2019) supplied near-term weather forecasts, real-time measured flow conditions, and geomorphological variables (area, length of mainstream, average slope, maximum and minimum elevation) to an LSTM model to obtain a flood forecast model for nine stations in Chile. The authors presented hourly NSE of higher than 0.97 (presumably because the previous-hour discharge, which is included in the inputs, is highly correlated to the next hour), and the LSTM model performed significantly better than a simple artificial neural network (ANN). LSTM was also shown to be stronger than ANN and the Soil & Water Assessment Tool (SWAT) model in another modeling effort (Fan et al. 2020). Mouatadid et al. (2019) used LSTM and a discrete wavelet transform in a model to forecast irrigation flow (water demand) in an irrigation district in Spain. The wavelet-LSTM model was demonstrated to have the best results among the models tested, although the necessity of the wavelet transform was not explored.

Reservoir operations depend on a large number of complex rules, weather forecasts, and economic and political considerations, and can be challenging to fully describe in a process-based or rule-based approach. For a limited number of applications, LSTM has been shown to successfully model reservoir operation at large scales, especially for the many smaller reservoirs in a basin (Ouyang et al. 2020). Yang et al. (2019a) showed that an LSTM model can capture the operation of a reservoir. The inputs include the inflow of the previous two days, forecasts of inflow for the current and next two days based on a short-term weather forecast, and reservoir characteristics (storage volume). The model achieved NSE values of 0.85, 0.93, and 0.66 for three reservoirs in Southeast Asia with catchment sizes of 26,386 km^2, 13,130 km^2, and 4,254 km^2 respectively, which seems to suggest that smaller dams are more difficult to predict. The outflow in the dry season of 2012 was underestimated because of an abrupt change in the operation rules during that period, which had never been seen in the training data. For the Gezhouba reservoir in China (catchment area $>1\times106$ km^2), Zhang et al. (2018a) used current inflow rates, lagged inflow and outflow rates, water levels in the downstream region, and month of the year as inputs to an LSTM model. They showed that their LSTM model outperformed other data-driven methods, such as a support vector machine and a simpler neural network. While these results showed the promise of LSTM models, it was less clear what features LSTM constructed to model the operation, and is still less clear whether longer-range forecasts can succeed with using lagged outflows, as this variable may not be attainable for longer-term simulations.

Apart from LSTM, Yen et al. (2019) used deep echo state network (also called reservoir computing), a different form of recurrent neural network, to forecast rainfall using hourly meteorological variables as inputs, including air pressure, temperature, humidity, wind speed, wind direction, current precipitation, and sea level. Perhaps because the analysis scope was local (no spatial dependence or large-scale dataset was considered), the forecast had limited capability. The authors also did not compare the results with LSTM. However, this work is mentioned here because reservoir computing has shown great success in modeling chaos (Pathak et al. 2017), and should gain more attention from hydrologists.

We are currently witnessing a surge in time series DL applications in a data-limited setting. Despite the solid progress and widely reported superior performance, there are nonetheless some potential concerns regarding such DL applications. The most significant one is that because the data from a small number of sites are employed for both training and test, there are limited opportunities to train or evaluate the model for extreme conditions. These pitfalls are typically mitigated by exposing the model to myriad situations in a big data setting, but are more problematic for data-limited scenarios. This situation can lead to reduced reliability for rare events, as witnessed by Yang et al. (2019a) described above. The implications of such limitations should be carefully considered for mission-critical applications. Also, as such a model is entrained in the specific dynamics at a site and therefore does not need to understand the implications of control factors, these models do not capture the fundamental processes, e.g., rainfall-runoff responses or reservoir operation policies, and thus cannot be migrated to other regions.

It is well-known that DL, and especially reinforcement learning, can be used for decision making. AlphaGo (Silver et al. 2016) has rocked the world with an AI (a combination of Monte Carlo tree search and CNNs that assess the strength of the player's position in the game and make proposals of next moves) that is capable of very long-term, superior-than-human decision-making for an extremely complex game. Such decision-making applications are still rare in hydrology, but the reservoir management problem seems like an obvious target. Matheussen et al. (2019) combined direct policy search methods with DL to optimize the reservoir management in a Norwegian system of two reservoirs. The authors used a direct policy search to optimize reservoir operations and obtain maximum power generation profits given variables such as hydropower price, given hydropower price, inflows, constraints in minimum reservoir levels, and start-stop costs of machinery. Then, they ran an ensemble of simulations to produce inflow and price scenarios and obtain their respective best policies, which were used as a training dataset for a multilayer perceptron (MLP) network. This network could then directly predict the optimum policy, and was used in a sequential simulation to determine the overall system performance in terms of profits.

Urban water systems are highly complex systems to model, yet data-driven approaches can prove to be a cost-effective option to enable rapid deployment and fast response to infrastructure management. Karimi et al. (2019) simulated wastewater flow rate in a sewage network based on hour index, rainfall rate, and groundwater level, and were able to obtain an R^2 value of around 0.81 for different periods. They showed that including groundwater data improved the model, suggesting a connection between groundwater and the sewage network. However, more evidence in different scenarios and locations would be needed to confirm that the model is not overfitting to this signal. Liu et al. (2019) forecasted water quality parameters for city drinking water intakes based on LSTM and lagged water quality data. However, it was not entirely clear if weather forcing attributes were included in the inputs, and if not, what inputs were driving the model.

With the help of DL and more instrumentation, the operation of the urban water system can be automated, providing more lead time and allowing for more monitoring. On the other side, due to the uniqueness of each water system, most of the applications belong in the data-limited setting, meaning that models built in one city cannot be directly employed in other cities, and so only places with sufficient history of monitoring could tap into this prediction potential.

The features of DL, e.g., high accuracy, high efficiency, low cost, and low barriers, are bound to increase the public's access to hydrologic predictions. If better prediction is the sole purpose, DL offers not only highly accurate results but orders-of-magnitude lower cost in terms of both model preparation/validation and run-time computation efforts. If understanding the relationships or causes and effects is the priority, then we need to employ more interpretable AI techniques, which are so far rare in geosciences.

19.2.2 Physics-constrained Hydrologic Machine Learning

While most of the above-mentioned models directly learn from data, a handful of applications have sought to combine data-driven models with physics. They sought to demonstrate that when physics are included in machine learning models, they tend to perform better than pure machine learning models when extrapolating to instances unseen in the training database.

Zhao et al. (2019) compared two MLP models that estimate parameters associated with evapotranspiration by learning from latent heat dynamics, along with support inputs such as photosynthetically active radiation and carbon dioxide concentration. The first model was a purely data-driven model that output latent heat fluxes. The second model used the inputs and LSTM to estimate stomatal resistance, which was then coupled to an explicitly coded Penman–Monteith evapotranspiration equation inside the LSTM. The model was then trained to predict the latent heat flux. Thus, the second model was said to conserve the surface energy balance, which the purely data-driven model was not capable of doing. We need to trust the Penmon-Monteith equation for giving the correct assumptions. Although the two models performed similarly, the second model extrapolated much better to sites not in the training dataset. However, this model required many supporting variables that were only available at the monitoring sites.

Read et al. (2019) created a process-guided deep learning framework to simulate lake temperature. This model was based on LSTM, but modified to impose a soft penalty for violating the conservation of energy. The authors also employed model pre-training to initialize the network weights, using outputs from a process-based model. They reported a superior performance of the hybrid model as compared to either the process-based model or a data-driven model.

It is worth mentioning that the principles that can be integrated into DL models so far are few, and it is more difficult for more complicated models. The engineering complexity will undoubtedly increase with more complex systems. It also requires delicate decisions with respect to which physical laws are to be retained. However, these hybrid models seem to be an important direction for improvement in the accuracy of future DL-based hydrological solutions.

19.2.3 Information Retrieval for Hydrology

For information retrieval from remotely-sensed data, DL has such an enormous practical utility that it is essentially redefining the textbooks of remote sensing (Zhu et al. 2017d). Compared to multispectral methods, which are based on expert-defined signatures, DL has proven to be stronger at finding the signatures and capturing fine-grained spatial texture. The variety of variables of interest and the methods employed are growing in diversity very rapidly. Transfer learning, the technique to migrate trained model components from one task to another, is a common tool for such applications as it allows the borrowing of useful features trained from much larger datasets (Ma et al. 2021).

Based on Landsat 8 images, Fang et al. (2019) sought to identify global man-made reservoirs and separate them from natural lakes. They tested transferring several pretrained networks, and found that a ResNet-50 model best suited their tasks. In line with previous research on camera rainfall gages, Jiang et al. (2019a) proposed a convex optimization algorithm that separates rainfall streaks from the background of surveillance camera motion images, and estimates rainfall intensities using predefined equations and geometrical optics estimated from the images. Haurum et al. (2019) showed rainfall intensities can also be estimated directly using a 3D CNN network.

Several efforts have proposed to extract rain or snow information from remotely-sensed data. Tao et al. (2016) already reported superior performance of a DL model (stacked denoising autoencoder) compared to older-generation ANNs at retrieving precipitation from satellite images. The authors later added additional input channels (water vapor) and even further improved the methodology (Tao et al. 2017, 2018). Tang et al. (2018a) trained an MLP model to retrieve rain and snow in high latitudes, with inputs including passive microwave data from the Global Precipitation Measurement (GPM) Microwave Imager, infrared data from MODerate resolution Imaging Spectroradiometer, and environmental data from European Centre for Medium–Range Weather Forecasts. The target was GPM precipitation, and the authors showed superior performance compared to an operational algorithm. One would expect a CNN-type model to further improve the accuracy because it is better at extracting spatial structures. Nonetheless, the authors have already demonstrated the potential for precipitation retrieval from a mixture of data sources including microwave data.

The current level of AI makes it possible to retrieve information from massive and unconventional datasets, and leverage help from citizen scientists. For example, the PhenoCam network consists of nearly 500 cameras in North America for vegetation monitoring (Seyednasrollah et al. 2019; Richardson et al. 2018). Kosmala et al. (2018) enlisted crowdsourced labeling on whether snow exists on the images, which they first verified against expert labels and then used to further tune a pre-trained CNN. They replaced the classification layer of the model by a Support Vector Machine (SVM) and trained the SVM, which obtained better results than classifying the scene with Places365-VGG and determining if one of the top five categories for an image contained snow. Despite the heterogeneity in camera model, camera view/configuration, and background vegetation, the model with SVM with Places365-VGG features has produced an accuracy of higher than 97%.

Jiang et al. (2018a) used transfer learning and Inception V3, a pre-trained CNN model, to extract urban water-logging depths from video images. They showed an R^2 of $0.75 - 0.98$ and a root-mean-square error of $0.031 - 0.033$ m for their two test datasets. Pan et al. (2018) proposed a low-cost system to read water level information from unmanned surveillance videos.

Pan et al. (2019) trained a CNN to estimate precipitation from geopotential height and precipitable water data from the National Centers for Environmental Prediction (NCEP) North American Regional Reanalysis (NARR) dataset, which was obtained by regionally downscaling NCEP global reanalysis. For the western and eastern coasts of the United States, where there is more rainfall, the CNN model was stronger than conventional stochastic downscaling products and the reference NARR precipitation data, a high-quality baseline that has already assimilated precipitation observations. It is noteworthy that this CNN model is closer to downscaling weather model than to information retrieval. Since it has to perform forecasts, the model needs to capture how precipitation evolves over time from given initial conditions. The authors argued that this model can be seamlessly transferred to numerical weather modeling, suggesting the model could be an improvement over our present precipitation prediction methods.

19.2.4 Physically-informed Machine Learning for Subsurface Flow and Reactive Transport Modeling

Another major finding of this era is that DL offers a brand new way of data-driven scientific computing that offers fresh new capabilities. Flow and reactive transport modeling in porous media is a research area with a long history and many practical applications in hydrology, pollution management, resource development, etc. The field is not a data-rich field, as the subsurface is challenging to instrument, and substantial heterogeneity across scales make it difficult to adequately sample the variations. There are many mature process-based numerical codes for modeling flow and reactive transport, some with high-performance computing capability (Molins et al. 2012). Yet, in such a mature field, DL allows us to pose questions in a novel format that was not possible before.

The first new use for DL is to identify the inverse mapping from observed model states (e.g. hydraulic head) to parameters (e.g. hydraulic conductivity). Typically, numerical models are used in a forward mode, i.e., solving for model states when given parameters. The inverse problem is solved via various algorithms that run the costly forward simulations many times to maximize the fit between the simulated states and the observations, or to estimate the distribution for the parameter sets that make the observations possible. However, DL provides a novel and efficient way to obtain such inverse mapping. Of course, the inverse mappings are not unique, so the uncertainty needs to be estimated. Sun (2018) demonstrated that it was possible to use the Generative Adversarial Network (GAN) method to generate conductivity (K) fields using hydraulic head (H) fields, and vice versa. The GAN, composed of two CNNs, was trained using 400 pairs of K fields (generated by a geostatistical method) and their resulting H fields (obtained using a groundwater flow solver after the solutions were evolved for a fixed time step. Both log-normal K fields with a correlation length and K fields with bimodal distribution were tested, and in both scenarios the GAN-estimated fields were similar to the original ones. Although there could be some concerns regarding the sample size and the model performance in real-world situations, the study showed the possibility for DL models to directly learn the inverse problem. Compared to multi-point statistical methods, the CNNs can capture highly complex spatial structure that was not limited to the first two statistical moments (Laloy et al. 2018). Similarly, Mo et al. (2019a, b) turned the multi-phase flow and reactive transport prediction problem into an image-to-image regression problem with a densely convolutional neural network (DenseNet), where the time step was used as an input argument whose effects are learned. DenseNet was essentially used as a surrogate model, but compared to previous surrogate models, it can reproduce the full 3D dynamics governed by the partial differential equation (PDE) to enable fast simulations and uncertainty estimates.

The networks used in Sun (2018) and Mo et al. (2019b) were trained entirely on numerical solutions. If the problem is complex with varied boundary and forcing conditions, it could require a large number of expensive numerical solutions to train, which weakens the motivation for using a data-driven approach. The branch of physics-informed neural networks (PINNs) has developed rapidly to address this issue. It should be noted that PINN has a different scope than previously advocated theory-guided data science (TGDS) (Karpatne et al. 2017a). TGDS is an overall concept to bring physical principles such as mass conservation into neural networks. PINN is more targeted toward data-driven scientific computing, and seeks to encode PDEs into the formulation of the neural network.

Raissi et al. (2019) proposed a form of PINN by supervising the derivatives of a network with the PDE, instead of learning all the physics from numerical solutions. Such supervision is possible because modern machine learning infrastructure allows one to calculate the derivation of the output with respect to its inputs. If a neural network can predict $u = f(x, t)$, then its derivatives $\partial u/\partial x$ and $\partial u/\partial t$ can be extracted by automatic differentiation. These derivatives can be put together as dictated by the PDE, and thus a network can be trained to respect a PDE. They demonstrated that this approach can infer solutions with multiple governing equations, and can also identify system equations, with a limited amount of training data. Tartakovsky et al. (2018) extended this framework for saturated groundwater flow problems to estimate (i) a heterogeneous K field using scattered observations of H; and (ii) a nonlinear K as a function of H. Problem (i) is somewhat similar to Sun (2018) and Tartakovsky et al. (2018) only trained the network on steady-state solutions and scattered observations, which is closer to reality, but the inclusion of physics allowed the model to be trained with one K field. Because of its unique way to model $u = f(x, t)$, PINN is also useful for data assimilation (He et al. 2020). It is noteworthy that the PINN framework can be cast in either a discretized time-step version or a continuous version, and it was reported that just learning the constitutive relationships (the parameter depending on the system states) produced more reliable results than learning the whole dynamics (Tipireddy et al. 2019). Furthermore, this framework has been scaled to very high performance on the Summit supercomputer (Yang et al. 2019). A different flavor of PINN proposed to transform the PDE into a minimization problem (Zhu and Zabaras 2018; Zhu et al. 2019c). However, this kind of transformation requires substantial customized adaptation of the method to each physical equation.

PINN is still nascent and has its limitations, e.g., for every combination of initial and boundary conditions, the PINN needs to be retrained. Thus it is more suitable for solving inverse problems than being used as a replacement for traditional numerical modeling. Ultimately, the advantages of DL include allowing us to ask questions in a novel manner (mapping relationships that could be not pursued using traditional modeling approaches) and the ability to continue to learn from data beyond known physics. While research on data-driven scientific computing is advancing rapidly, there are currently still some limitations with respect to flexibly handling boundaries, time stepping, and full 3D simulations.

19.2.5 Additional Observations

Some of the founding work with DL in hydrology arose from groups based in the United States (e.g. Fang et al. (2017); Mo et al. (2019b); Raissi et al. (2019); Read et al. (2019)). However, it is noteworthy to mention that a large amount of ensuing efforts have arisen from China (section 19.2.1.2). In particular, numerous applications for data-limited settings have emerged from the country, which has held Artificial Intelligence (AI) to a position of national strategic importance (Allen 2019) and has been consistently making investments. In comparison, work from the US and Europe often tend to utilize large datasets. Apart from an elevated investment level, the rise of AI in China could also be partially attributed to the psychological shock inflicted by the victory of AlphaGo in the oriental game of Go. While the policies and management of data in China are still imposing barriers to big-data geoscientific research, which can be witnessed through the lack of genuinely big-data geoscientific DL work, we can expect rapid advances in applications and physics-informed machine learning from there.

19.3 Current Limitations and Outlook

While the growth of DL in hydrology has been impressive, there are still some visible challenges. First, except for limited visualization efforts (Pan et al. 2019), studies on the interpretation of the DL model are still weak or largely missing. We anticipate that more progress in this regard will likely come from hydrologists rather than AI generalists. The methods introduced in general interpretable AI may be of value, but not directly applicable to the field in terms of the explanations needed by hydrologists. The domain scientists are responsible for customizing DL for knowledge discovery.

Presently most DL models focus on a single task and are tailored to their respective applications and domains. In the numerical modeling domain, multiple components for different tasks can be coupled together as integrated land surface hydrologic models (Maxwell et al. 2014; Ji et al. 2019; Lawrence et al. 2018). For them, the interfacing requires specific handling and skills (Kollet and Maxwell 2006; Shen et al. 2016; Camporese et al. 2010). Interface handling may similarly be needed when networks are put together to enable multi-physics large-domain simulations, for example, if a network predicting groundwater level fluctuation is coupled to a network predicting streamflow.

Uncertainty quantification remains challenging for varied model architectures. A central question is if we know a new instance is close to the training dataset. There has been some initial investigation, e.g., Fang et al. (2020) tested the Monte Carlo dropout scheme, which argued that running the Monte Carlo version of the network through a randomized dropout mask is similar to running an approximate variational Bayesian inference (Gal and Ghahramani 2015). However, much more testing and algorithm improvement is required for different models under different application scenarios.

For physics-guided machine learning for scientific computing, many of the demonstrated cases are for steady-state solutions, 1D or 2D. Learning to reproduce 3D transient solutions remains difficult, as it can require too much training data or too large a computational demand. In addition, numerical solutions can accommodate various boundary and initial conditions, where these different configurations would need to be covered in the training dataset and thus could entail significant effort in training data preparation. Future efforts will need to develop ways to more easily teach DL models the meaning of different boundary and source conditions.

Present applications of time series deep learning models have been mostly learning directly from data, and are thus limited to variables that can be directly observed. I anticipate future uses will feature a deep meshing and integration between DL and physically-based models, to overcome multiple issues facing purely data-driven models and to use DL as a knowledge discovery tool. Process-based models could be used to assess causal controls and distinguish between competing factors in an adversarial fashion Fang et al. (2020). The next stage may see in-depth modification of DL algorithms to fit the needs of hydrology and to offer a full suite of services to fit the tasks that society asks of hydrologists.

Acknowledgments

This work was supported by National Science Foundation Award OAC #1940190 and the Office of Biological and Environmental Research of the U.S. Department of Energy under contract DE-SC0016605.

20

Deep Learning of Unresolved Turbulent Ocean Processes in Climate Models

Laure Zanna and Thomas Bolton

20.1 Introduction

Current climate models do not resolve many nonlinear turbulent processes, which occur on scales smaller than 100 km, and are key in setting the large-scale ocean circulation and the transport of heat, carbon and oxygen in the ocean. The spatial-resolution of the ocean component of climate models, in the most recent phases of the Coupled Model Intercomparison Project, CMIP5 and CMIP6, ranges from 0.1° to 1° (Taylor et al. 2012; Eyring et al. 2016b). For example, at such resolution, mesoscale eddies, which have characteristic horizontal scales of 10–100 km, are only partially resolved – or not resolved at all – in most regions of the ocean (Hallberg 2013). While numerical models contribute to our understanding of the future of our climate, they do not fully capture the physical effects of processes such as mesoscale eddies. The lack of a resolved mesoscale eddy field leads to biases in ocean currents (e.g., the Gulf Stream or the Kuroshio Extension), stratification, and ocean heat and carbon uptake (Griffies et al. 2015).

To resolve turbulent processes, we can increase the spatial resolution of climate models. However, we are limited by the computational costs of an increase in resolution (Fox-Kemper et al. 2014). We must instead approximate the effects of turbulent processes, which cannot be resolved in climate models. This problem is known as the parameterization (or closure) problem. For the past several decades, parameterizations have conventionally been derived from semi-empirical physical principles, and when implemented in coarse resolution climate models, they can lead to improvements in the mean state of the climate (Danabasoglu et al. 1994). However, these parameterizations remain imperfect and can lead to large biases in ocean currents, ocean heat and carbon uptake.

The amount – and availability – of data from observations and high-resolution simulations has been increasing. These data contain spatio-temporal information that can complement or surpass our theoretical understanding of the effects of unresolved (subgrid) processes on the large-scale, such as mesoscale eddies. Efficient and accurate deep learning algorithms can now be used to leverage information within this data, exploiting subtle patterns previously inaccessible to former data-driven techniques. The ability of deep learning to extract complex spatio-temporal patterns can be used to improve the parameterizations of subgrid scale processes, and ultimately improve coarse resolution climate models.

Deep Learning for the Earth Sciences: A Comprehensive Approach to Remote Sensing, Climate Science, and Geosciences, First Edition. Edited by Gustau Camps-Valls, Devis Tuia, Xiao Xiang Zhu, and Markus Reichstein.

20.2 The Parameterization Problem

The generic nonlinear equations for the evolution of a variable Y (e.g., velocity component, temperature, and salinity) in the ocean are given by

$$\frac{\partial Y}{\partial t} = -(\mathbf{u} \cdot \nabla)Y + F, \tag{20.1}$$

where $\mathbf{u} = (u, v, w)$ is the three-dimensional velocity (momentum), $\frac{\partial}{\partial t}$ the Eulerian acceleration, $\nabla = \left(\frac{\partial}{\partial x}, \frac{\partial}{\partial y}, \frac{\partial}{\partial z}\right)$ is the 3D gradient operator, and F is a set of forces or sources/sinks of the oceanic quantity Y. If, for example, the variable Y represents ocean momentum, then the force F includes all individual forces, such as those from the pressure, Coriolis effect, viscosity, and all external influences (e.g., wind forcing).

In ocean models, Equation 20.1 are numerically integrated at a finite spatial and temporal resolution. This leads to equations for the resolved fields, denoted by $\overline{()}$, as illustrated below:

$$\frac{\delta \overline{Y}}{\delta t} = -\left(\overline{u}\frac{\delta}{\delta x} + \overline{v}\frac{\delta}{\delta y} + \overline{w}\frac{\delta}{\delta z}\right)\overline{Y} + \overline{F}. \tag{20.2}$$

The discretization, and the truncation at a finite resolution, eliminates processes occurring at scales below the model resolution. Therefore, the ocean model described by Equation 20.2 is not an accurate representation of the true system described by Equation 20.1. To improve the fidelity of the climate model, the effects of unresolved (subgrid) scales on the large-scale flow (i.e., $\overline{Y}, \overline{\mathbf{u}}$) must be approximated – this is the parameterization problem.

The crux of parameterizing subgrid physical processes within models is done by introducing using some function P which only depends on $\overline{Y}$ on the right hand side of Equation 20.2. The challenge lies in constructing $P(\overline{Y})$, which: (i) accurately captures the physical processes being parameterized; (ii) respects physical principles such as conservation laws; (iii)) is numerically stable when implemented into a climate model; and (iv) generalizes to new dynamical regimes. There are two main ways of constructing parameterizations: physics-driven and data-driven.

The physics-driven approach has been the prevailing and conventional approach for the past few decades. This approach starts with some physical principle, mechanism, or theory, related to the process to be parameterized. From this physical knowledge, a bulk formula is derived to approximate the effect of that process on the resolved flow. The main advantage of physics-driven parameterizations is their interpretability. However, physics-driven parameterizations only include the bulk effect of a process and parameters are often not observable and/or not constrained by observations. As a result of these caveats, ocean models, which mainly implement physics-driven parameterizations, still exhibit large biases in both the mean and variance of the climate state. Hence, alternative and complementary approaches are necessary to address these shortcomings. Data-driven approaches can help overcome these issues.

The data-driven approach makes no physical assumptions about the process to be parameterized. Here, data – from high-resolution models, observations, or a combination of both – guides the construction of the parameterization. An algorithm (e.g., simple linear regression, or machine learning techniques) is employed to directly learn the parameterization from the data. In general, the functional form and parameters are empirically estimated from the data. However, the choice of algorithm will implicitly make assumptions regarding the parameterization and its functional form. Some early attempts of data-driven ocean eddy (subgrid) parameterizations showed interesting results in idealized model setups (e.g., Berloff 2005; Mana and Zanna 2014; Zanna et al. 2017). More generally, data-driven modeling of turbulence has advanced in recent years (Duraisamy et al. 2019). The advent of new tools from machine learning can improve the computational efficiency and generalization of data-driven parameterizations. Combined with the increasing wealth of data from observations and high-resolution simulations, it is now possible to begin investigating more thoroughly how data-driven parameterizations can improve the representation of unresolved processes and reduce model biases in long-range climate simulations.

20.3 Deep Learning Parameterizations of Subgrid Ocean Processes

20.3.1 Why DL for Subgrid Parameterizations?

There are a plethora of machine learning algorithms which are well suited for the supervised regression problem needed for parameterization (linear regressions, random forests, support vector machines, etc.). However, one particular form of algorithm has stood out in recent years: neural networks (NNs).

Computational resources have now reached the levels required to train large NNs containing thousands or millions of parameters, across a range of architectures such as deep fully-connected networks, deep belief networks, recurrent neural networks, and convolutional neural networks (CNNs) (Goodfellow et al. 2016). For example, the power and success of CNNs comes from the fact that the convolution layers – which typically extract the most vital information from 2D spatial fields – are learned from data. A disadvantage of convolution layers is their computational cost when forming a prediction during implementation, compared to simpler physics-driven parameterizations - however, parameterizing with CNNs is still computationally cheaper than running high-resolution models.

20.3.2 Recent Advances in DL for Subgrid Parameterizations

This chapter is concerned with ocean subgrid parameterizations, but there have been many developments for the parameterization of atmospheric processes which are described in Chapter 20.

Data-driven parameterizations of turbulence, that use deep learning, initially appeared in direct numerical simulation studies (Tracey et al. 2015; Ling et al. 2016a, b), which concern spatial scales that are orders of magnitude smaller than the spatial scales of climate models. For example, Ling et al. (2016b) used a deep fully-connected NN to learn an eddy momentum parameterization of the anisotropic stress tensor. They introduced a physical constraint into their NN using the Galilean-invariant tensor basis of Pope (1975), $\mathbf{T}^{(n)}$, which ensures the data-driven parameterization has particular symmetries. The predicted momentum parameterization ($\hat{\mathbf{S}}_{\mathbf{u}}$) of Ling et al. (2016b) takes the form

$$\hat{\mathbf{S}}_{\mathbf{u}} = \begin{pmatrix} \hat{S}_x \\ \hat{S}_y \end{pmatrix} = \overline{\nabla} \cdot \left(\sum_n g_n \mathbf{T}^{(n)} \right), \tag{20.3}$$

where the coefficients g_n are predicted by the NN using only Galilean-invariant inputs. The deep NN outperform linear regression models, only after applying these physical constraints. Integrating physical principles into data-driven algorithms is important for fidelity, but can also boost the predictive skill of the resulting parameterization.

Additional studies have used NNs to parameterize eddy momentum fluxes in models of freely-decaying 2D turbulence (Maulik and San 2017; San and Maulik 2018; Maulik et al. 2019; Cruz et al. 2019) or in large-eddy simulations (Zhou et al. 2019). For example, Maulik et al. (2019) used NNs to parameterize eddy vorticity fluxes, which were then implemented back into the same model. They used the conventional Smagorinsky and Leith eddy viscosity functions (Smagorinsky 1963; Leith 1968) as input features to one of the NNs: this did not improve the predictive skill of the NN but did improve the numerical stability of the turbulence model once the NN is implemented. By doing so, they removed upgradient momentum fluxes to stabilize the numerical simulations but therefore altered the physics of turbulence processes. Nonetheless, incorporating physical and mathematical properties into DL algorithms is an important step for making parameterizations physically-consistent and potentially improving their performance when implemented into a climate model.

For parameterizations of ocean turbulence, a handful of studies using CNNs have emerged. Bolton and Zanna (2019) and Zanna and Bolton (2020) used CNNs to parameterize ocean mesoscale eddies in idealized models. They showed that CNNs can be extremely skillful for eddy momentum parameterizations, with predictions that generalize very well to different dynamical regimes (e.g., different ocean conditions and turbulence regimes). Another idealized study by Salehipour and Peltier (2018) showed the potential of CNNs to parameterize ocean vertical mixing rates. The DL algorithm could predict the mixing efficiency well beyond the range of the training data, producing a more universal parameterization compared to previous studies.

20.4 Physics-aware Deep Learning

CNNs, and NNs in general, are good candidates to capture the spatio-temporal variability of the subgrid eddy momentum forcing and potentially other subgrid ocean processes which are not resolved in current climate models. However, one of the main criticism of NNs and DL approaches is that they do not include physical constraints. Physics-based parameterizations, on the other hand, are often developed using physical constraints, together with conservation and symmetry laws. We have presented several examples highlighting the improvements of DL parameterizations when physical constraints are included. The constraints ensure that the parameterizations remain faithful to the physics of the underlying process which we are trying to capture, as well as improving the numerical stability of the global climate model when the new parameterizations are implemented. There are several routes to embed physics-constraints in machine learning-based parameterizations. For example, pre-processing the input data or the post-process the output (Bolton and Zanna 2019) could be used, but will likely introduce some biases. However, there are other avenues that might show more promise.

The first avenue is the most conventional and entails using DL to optimally learn the unknown coefficients used in physics-based parameterizations (Schneider et al. 2017a).

However, the underlying assumptions behind this approach is that the structural form of the parameterization is a correct representation of a given process, and no other parameterizations than the ones already in use are needed. Neither of these assumptions are valid in ocean models (Zanna et al. 2018), since not all parameterizations included in ocean models are correct or encompass all the missing processes.

A second avenue is a change in the loss function used during optimizations. The loss function can be adjusted to include additional constraints, such as global conservation of mass, momentum, salt, or energy. This simple approach helps ensure that the system tends toward such conservation principles (Beucler et al. 2019). However, the conservation laws may not be strictly enforced, only approximately, unless hard constraints are used.

The third avenue is to modify the architecture of the NNs (Ling et al. 2016a; Zanna and Bolton 2020). For example, Zanna and Bolton (2020) used maps of resolved velocity components as input to the CNN in order to predict both components of the subgrid eddy momentum forcing $\hat{S}_x$ and $\hat{S}_y$. To physically-constrain the architecture, they used a specifically-constructed final convolutional layer with fixed parameters (Figure 20.1). The activation maps of the second-to-last convolution layer represent the elements of an eddy stress tensor **T**. The final convolution layer then takes the spatial derivatives of the activation maps of the second-to-last convolutional layer (i.e., the eddy tensor elements) using fixed filters, representing central-difference stencils to form the two outputs $\hat{S}_x$, and $\hat{S}_y$ for eddy momentum forcing. This ensures that the final prediction originates from taking the divergence of a symmetric eddy stress tensor, achieving global momentum and

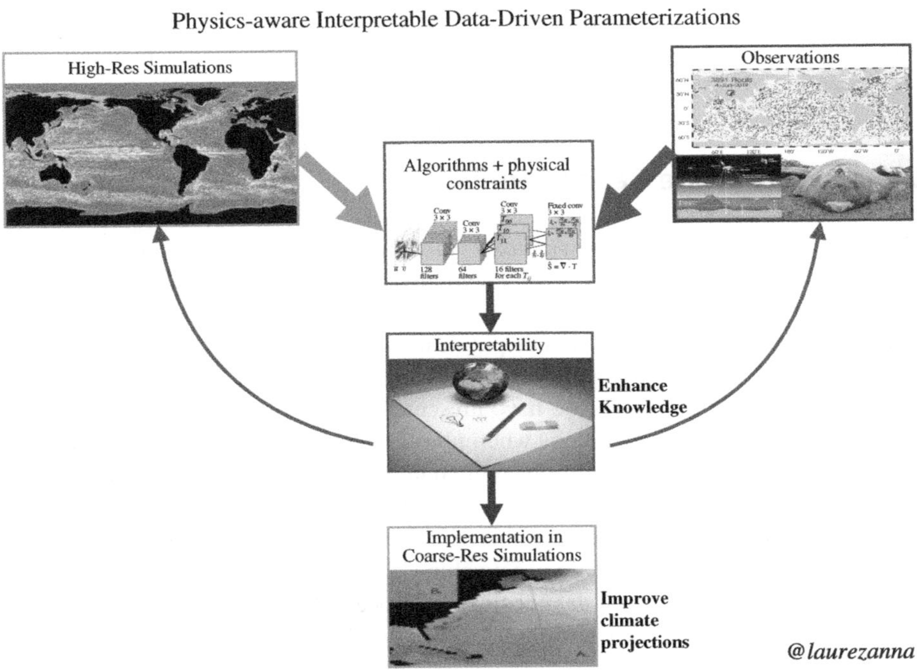

Figure 20.1 Schematic of physics-aware deep learning parameterizations for implementation in coarse-resolution models.

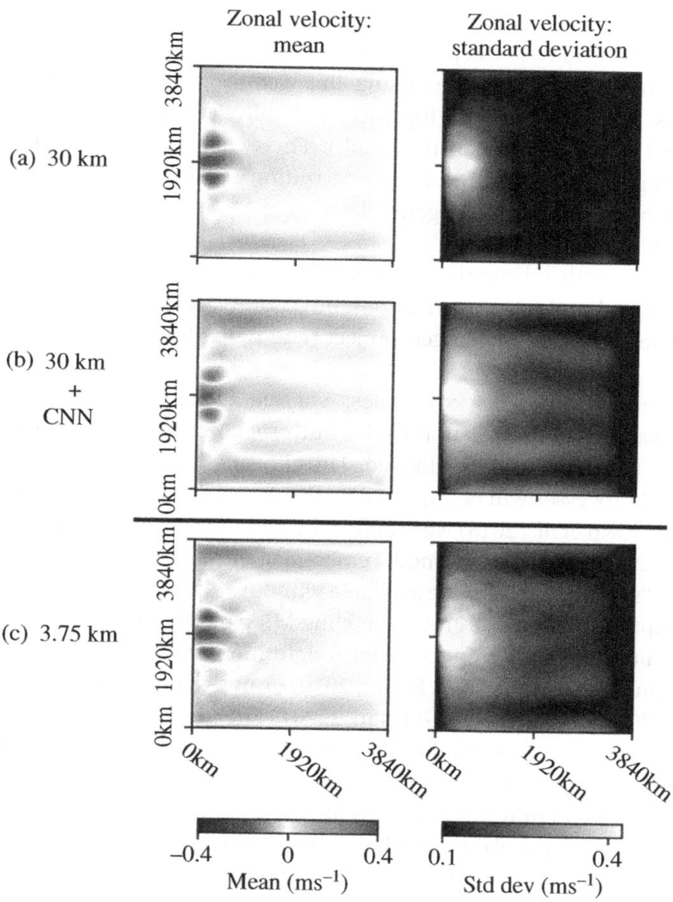

Figure 20.2 Evaluation in an idealized model, zonal velocity (time-mean, left and standard deviation, right): a) Coarse-resolution (30 km), b) Coarse-resolution (30 km) with physics-aware CNN parameterization implemented, c) High-resolution (3.75 km).

vorticity conservation (via the divergence theorem) by prescribing appropriate boundary conditions (the latter might be difficult in climate simulations with complex geometry). This approach, directly integrating physical-principles with data-driven algorithms, leads to more physically-robust ML parameterizations and vastly superior results compared to purely physics-driven parameterizations (Figure 20.2) (Zanna and Bolton 2020). However, it requires the largest knowledge of the practitioner: expertise in both deep learning and physics is necessary.

20.5 Further Challenges ahead for Deep Learning Parameterizations

Parameterizations of unresolved ocean processes will be in demand in climate models for many decades to come. The traditional approach of physics-driven parameterizations, while showing some success, remain sub-optimal as many processes remain poorly represented or are missing from models. Deep learning can help bridge the gap and improve the representation of missing processes using the wealth of new data from high-resolutions simulations and observations, together with physical constraints (as described in section 20.4 and other chapters of this book). There are, however, several challenges ahead in developing physics-aware ML parameterizations, which relate to: how and what to learn from data; how to improve the generalization of ML parameterizations; the interpretability of the resulting algorithm.

Learning from data. Dealing with substantial amounts of data to train ML algorithms remains an obstacle in deriving subgrid parameterizations, but coordinated efforts are well underway to break this barrier, such as the Pangeo project (e.g. Eynard-Bontemps et al. 2019).

However, defining "subgrid" (or unresolved) scales, via an averaging procedure, from either model or observational data is a non-trivial but crucial component of any data-driven parameterization which is often overlooked. The choice of subgrid definition directly impacts what physical processes will be captured by the data-driven parameterization. Gentine et al. (2018) and Rasp et al. (2018) were able to by-pass this problem by using data directly extracted from a 2D high-resolution model embedded into a coarse-resolution climate model; therefore, the "subgrid" scales were available without additional processing. However, this case is an exception. Most other groups tackling ML parameterizations have so far used spatial coarse-graining, which produces a local definition of eddy forcing, on uniform grid (Bolton and Zanna 2019; Zanna and Bolton 2020) or non-spherical geometry (Brenowitz and Bretherton 2018; Yuval and O'Gorman 2020). The choice of averaging procedure has a significant impact on the nature of the resulting subgrid forcing and the separation of scales (as illustrated in Figure 20.3 for the subgrid momentum forcing). The choice of how to separate resolved and unresolved scales can lead to artifacts in the evaluation of nonlinear subgrid forcing (e.g., panel d in which a simple coarse-graining procedure is used), or can produce different patterns and magnitudes of subgrid forcings (e.g., panels d–f, which show the effects of using course-graining, a low-pass filter, or a combination of a low pass filter with coarse-graining). If using a (low-pass) filter, the spatial scale of the filter should also be carefully considered when dealing with spherical coordinates as the subgrid forcing will change in spatial scale as well; e.g., the Rossby deformation scale at which mesoscale eddies are resolved varies with latitudes. Whether these definitions (panels d–f) are truly representative of the missing forcing in a coarse-resolution model remains to be determined.

Generalization of ML parameterizations. Another obstacle to accurate ML parameterizations is their ability to generalize to different regimes or conditions (i.e., to extrapolate outside the range on which they were trained). While CNNs for ocean eddy momentum parameterizations have shown great success in generalizing to different turbulent regimes (Bolton and Zanna 2019), when implemented into ocean models, even if physical constraints imposed, they can lead to unphysical behaviors without ad-hoc tuning (Zanna and Bolton 2020). There are several ways to improve generalizations of ML-parameterizations, which include: (i) learning from a range of high-resolution simulation under different regimes (O'Gorman and Dwyer 2018) and optimally combining the resulting DL parameterizations as suggested by Bolton and Zanna (2019) while imposing physical constraints; (ii) the use of causal inference to target physical relationships in the training data to be used

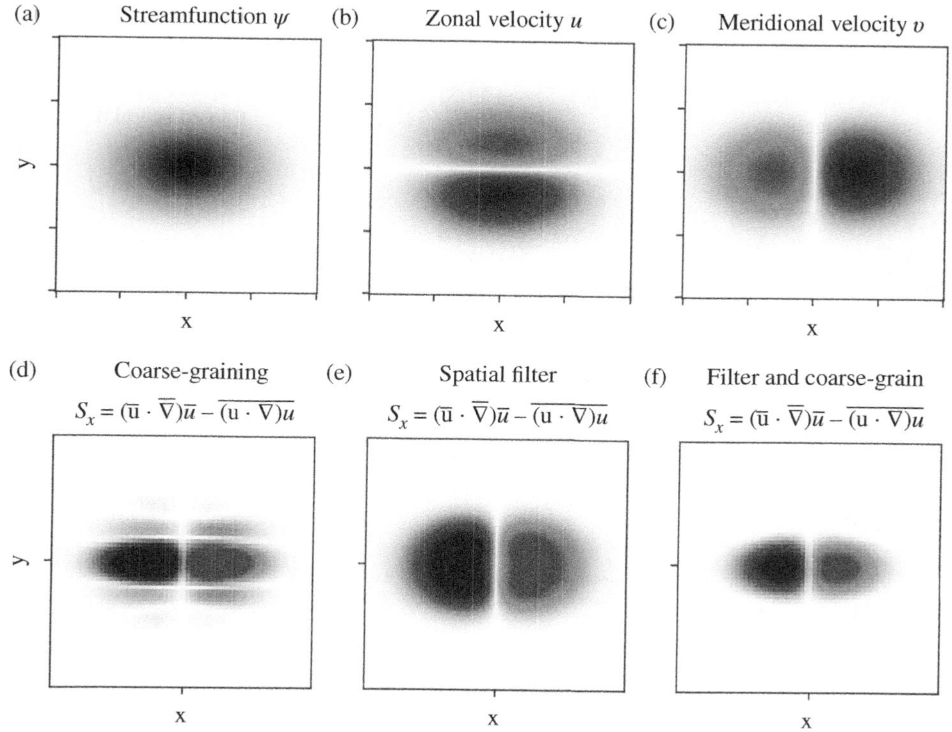

Figure 20.3 Illustrative example considering the effects of averaging procedure on the corresponding zonal eddy momentum forcing (S_x): Assume a Gaussian ellipse streamfunction $\psi \propto e^{-(ax^2+by^2)}$, which emulates a coherent vortex (panel a), and associated velocity components $u = -\frac{\partial \psi}{\partial y}$ (panel b) and $v = \frac{\partial \psi}{\partial x}$ (panel c). Panel (d) shows coarse-graining, Panel (e) a Gaussian spatial filter, and Panel (f) a Gaussian spatial filtering followed by coarse-graining.

as input in DL algorithm; this has the potential to select variables which co-vary according to physical laws and therefore constrain the algorithm to reproduce that relationship, even in unseen conditions. The training data, whether from high-resolution numerical models or observations, possess some biases which may limit the performance or accuracy of the ML parameterizations. A potential way forward is to use transfer learning: one trains ML algorithms with abundant model data and re-tune the ML parameterizations with observations Chattopadhyay et al. (2020), which have less biases. Transfer learning could also potentially improve the generalization of these deep learning models.

Interpretability. Finally, for deep learning models in general, predictive skill is valued above other factors such as interpretability. In general, it is difficult to understand how deep learning methods transform an input into the target variable. The final prediction of a CNN is a culmination of the information extracted from the previous convolutional layers of the network. We can talk broadly about how convolution layers automate feature extraction, and then attempt to dissect the feature maps of the intermediate layers, but identifying exactly what features are being extracted by the many learnt filters can be cumbersome or sometimes completely unfeasible. For example, in Figure 20.4, the first layers extract

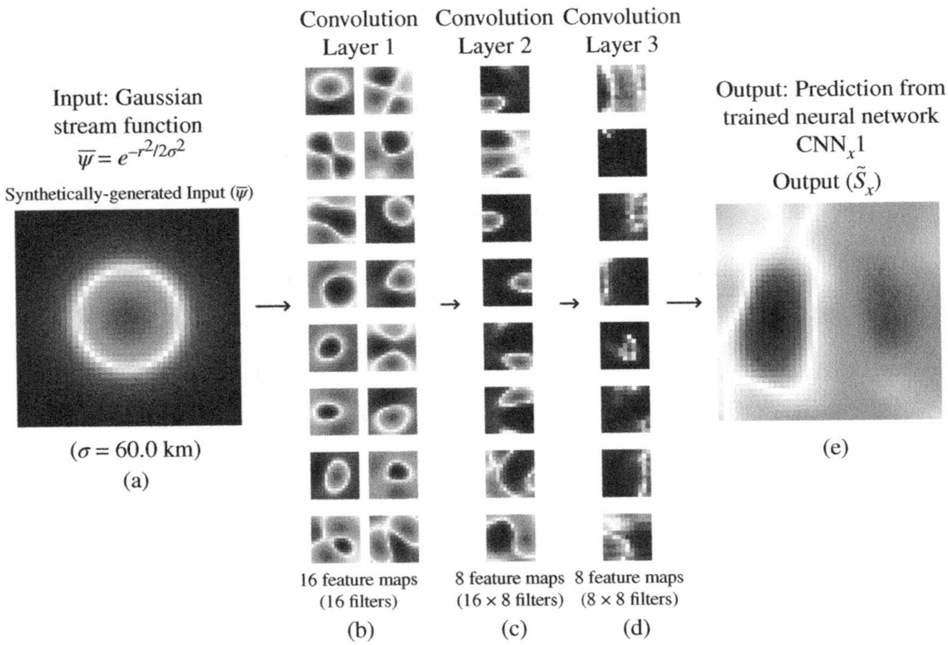

Figure 20.4 Interpretability: Activation maps are the result of the convolution acting on the previous layers output, and then passing it through the activation function. Here, a radially-symmetric Gaussian function to generate an eddy is fed into the already-trained NN for an ocean subgrid parameterization by Bolton and Zanna (2019). The activation maps for each convolutional layers are shown. The activation maps for the first convolution layer are collection of first- and second-order derivatives. Therefore, without a-priori knowledge, the neural network learns to take derivatives of the input streamfunction, which corresponds to velocities and velocity shears. This is a robust feature across all of the NNs trained to predict the eddy momentum forcing.

derivatives but subsequent features are harder to interpret. Interpretability is particularly hard to extract in large networks with 10^5–10^6 parameters. For simpler NNs architecture, different techniques are currently being developed (Toms et al. 2020), and it would be interesting though challenging to applying them to more complex NN architectures. Finally, other machine learning methods, such as data-driven equation-discovery, can lead to interpretable parameterizations of ocean eddy forcing (Zanna and Bolton 2020); this approach aims to construct a closed-form equation from data, harnessing the power of data-driven algorithms while retaining interpretability. Data-driven equation discovery could be used in conjecture with deep learning methods, to extract information from the complex NNs to go beyond the traditional parameterizations already in use.

We are at the very beginning of what deep learning can bring to development of ocean parameterizations, with many challenges ahead, but with the exciting potential to discover new physics, further our understanding and representation of ocean processes, and improve the fidelity and reliability of climate models.

21

Deep Learning for the Parametrization of Subgrid Processes in Climate Models

Pierre Gentine, Veronika Eyring, and Tom Beucler

21.1 Introduction

Earth system models simulate the physical climate and biogeochemical cycles under a wide range of forcings (e.g., greenhouse gases, land use, and land cover changes). Given their complexity and number of processes represented, there is persistent inter-model spread in their projections even for a given prescribed carbon dioxide (CO_2) concentration pathway (IPCC 2013; Schneider et al. 2017b). Despite significant progress in climate modeling over the last decades (Taylor et al. 2012; Eyring et al. 2016c), the simulated range for effective climate sensitivity (ECS), i.e. the change in global mean surface temperature for a doubling of atmospheric CO_2 concentration, has not decreased since the 1970s. It still ranges between 2.1 and 4.7 °C and is even increasing in the newest generation of climate models participating in the World Climate Research Programme (WCRP) Coupled Model Intercomparison Project Phase 6 (CMIP6, Eyring et al. (2016c)) (see Figure 21.1).

One of the largest contributions to this uncertainty stems from differences in the representation of clouds and convection (i.e, deep clouds) occurring at scales smaller than the model grid resolution (Schneider et al. 2019; Stevens and Bony 2013; Bony et al. 2015; Stevens et al. 2016; Sherwood et al. 2014). These processes need to be approximated in global models using so-called parametrizations, i.e. an empirical representation of the process at play, because the typical horizontal resolution of today's global Earth system and climate models is around 100 km or more. This limits the models' ability to accurately project global and regional climate changes, as well as climate variability, extremes and their impacts on ecosystems and biogeochemical cycles. Yet, accurate projections are essential for efficient adaptation and mitigation strategies (IPCC 2018) and for assessing targets to limit global mean temperature increase below 1.5 °C above pre-industrial levels, as defined in the Paris Agreement (UNFCCC 2015). Reducing uncertainties in climate projections in the next decade can also significantly reduce associated economic costs (Hope 2015).

The long-standing deficiencies in cloud parametrizations (Randall et al. 2003; Boucher et al. 2013; Flato et al. 2013) have motivated the developments of high-resolution global cloud-resolving climate models with the ultimate goal to explicitly resolve clouds and convection (Schneider et al. 2017b; Stevens et al. 2019), as well as shorter duration large-eddy simulations (LES), resolving most of the energy-containing atmospheric turbulence and

Deep Learning for the Earth Sciences: A Comprehensive Approach to Remote Sensing, Climate Science, and Geosciences, First Edition. Edited by Gustau Camps-Valls, Devis Tuia, Xiao Xiang Zhu, and Markus Reichstein.

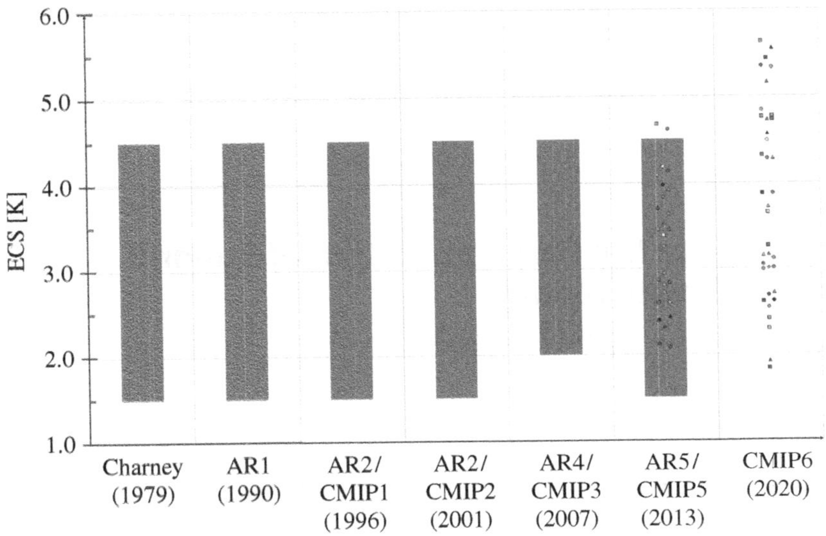

Figure 21.1 Effective Climate Sensitivity. Assessed range of effective climate sensitivity in IPCC Reports over the years (gray bars). ECS values from individual models participating in CMIP5 and CMIP6 are shown in addition (symbols). Source: Modified from Meehl et al. (2020).

covering up to a few hundreds of kilometers (Tonttila et al. 2017). Yet, these simulations are extremely computationally demanding so they can only be run for a few days to months and cannot be used for long-term climate projections in a foreseeable future. Coarse-scale model simulations, in particular those from Earth system models that include additional Earth system components beyond the physical climate such as the carbon cycle, will therefore continue to be required. These additional processes are needed to represent key feedbacks that affect climate change, such as the biogeochemistry cycle, but are also likely to increase the spread of climate projections across the multi-model ensemble even further. For instance, future terrestrial carbon uptake remains one of the most uncertain processes in the Earth system, as even its mere sign in the future is unknown (Friedlingstein et al. 2014).

Yet, as many cloud and convection processes are explicitly resolved in high-resolution simulations (Figure 21.2), these simulations can serve as important sources of information to constrain small-scale representation (parametrizations) in coarse-resolution Earth system and climate models. With the recent developments in machine learning, in particular deep learning, this provides unique new opportunities for the development of improved Earth system and climate models.

In this chapter, we present pioneering results and progress on the use of machine learning for cloud parametrizations that can replace typical parametrizations in coarse-scale Earth system and climate models (section 21.2) and discuss studies that particularly address generalization and the implementation of physical constraints into machine learning algorithms (section 21.3). Section 21.4 closes with an outlook of remaining challenges for this new and exciting interdisciplinary research field with what we argue has huge potential to improve understanding and modeling of the Earth system, in particular if guided both by data and by physical knowledge.

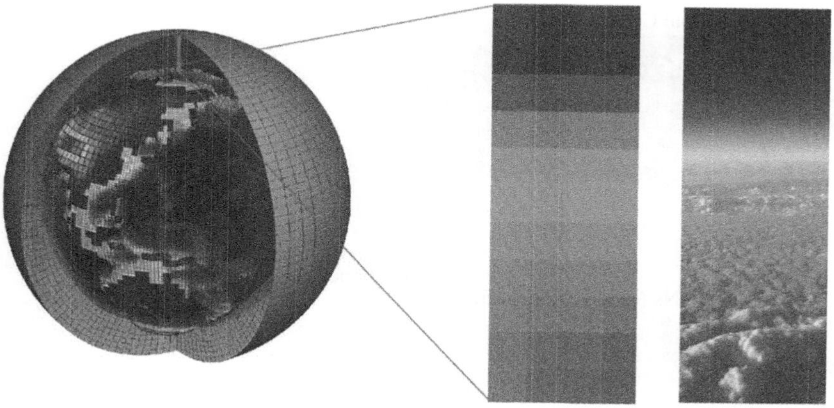

Figure 21.2 Schematic representation of clouds in current climate models and the objective to represent them similarly to very fine resolution models. In coarse-scale climate models (left) small-scale physical processes need to be empirically represented as a function of the coarse-scale resolved variables such as mean temperature or humidity over the grid, at a given level (level). These small-scale processes can be explicitly resolved in high-resolution cloud resolving models (right).

21.2 Deep Neural Networks for Moist Convection (Deep Clouds) Parametrization

Gentine et al. (2018) and Rasp et al. (2018) demonstrated that deep convection simulated by a cloud-resolving model (CRM) could be correctly emulated by a deep neural network, at a scale comparable to a coarse global climate model (GCM). The authors used a super-parametrization (SP) of convection, where 2-dimensional (2D) CRMs (in the y and z directions) were embedded in a coarse GCM. They further idealized the setup by prescribing oceanic surface conditions with a steady latitudinal temperature gradient, and without continents, topography, and sea ice (aquaplanet setup as in Stevens and Bony (2013)). This strategy allowed the authors to bypass the coarse-graining of CRMs typically required for ML subgrid-scale parametrizations (Figure 21.3), a notoriously difficult step.

The NN maps the vertical temperature $\overline{T}(z)$, specific humidity $\overline{q}(z)$, surface pressure p_s, solar insolation S_0, surface sensible flux H, and latent heat flux LE (inputs) to the heating and moistening (tendencies) of the coarse-grained SP model (outputs). In more technical terms, this means that the predicted heating $\frac{\partial \overline{T}}{\partial t}$ and moistening $\frac{\partial \overline{q}}{\partial t}$ physics tendencies (rate of change of temperature and moisture due to the physical components of the model unlinked to advection) of the coarse-pixels (Figure 21.3) can be written as:

$$\frac{\partial \overline{X}}{\partial t} = \mathrm{NN}(\overline{T}, \overline{q}, p_s, S_0, H, LE), \tag{21.1}$$

with $\overline{X}$ the grid mean value of either temperature $X = T$ or specific humidity q.

Rasp et al. (2018)'s NN was trained on labeled data from the simulation's first year and validated on data from the second year. The training required more than 6 months of data

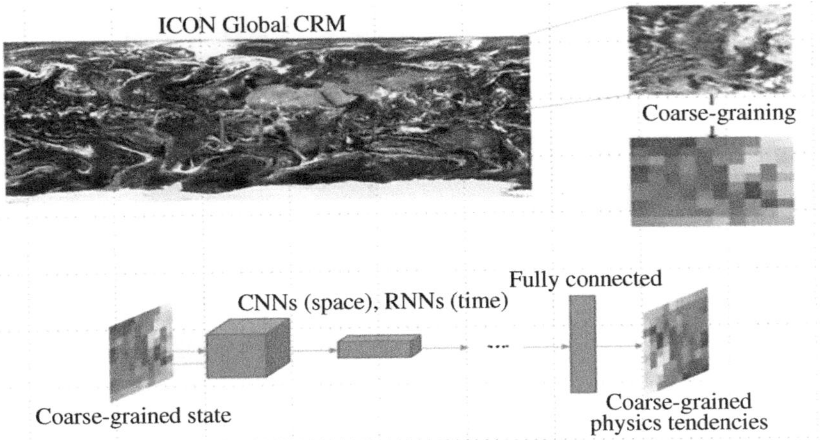

Figure 21.3 Schematic diagram for ML-based cloud parametrizations for climate models. High-resolution cloud-resolving model simulations are coarse-grained to the scale of the climate modes (~100 km) with the help of convolutional or recurrent NNs to learn the impact of convection on the resolved coarse-scale variables.

(approximately 140 million training samples) to reach final convergence. Alternatively, the NN can be trained on less samples (e.g., 40 million), and computational resources can be invested to tune its hyperparameters (e.g., number of layers, number of nodes per layer, learning rate, etc.) instead to guarantee optimal performance on the validation dataset. The NN used for this chapter's figures was trained using this second strategy (see Beucler et al. (2019) for details). Note that both NNs did not include any temporal nor spatial covariations of the coarse resolution pixels, similar to the embedded 2D CRM.

The NN reproduced not only the heating and moistening due to deep convection, but also due to all other subgrid processes such as turbulence, radiation, waves, and shallow convection. The NN was able to correctly reproduce the CRM, as seen in uncoupled mode (i.e., by prescribing at each time step the input features $\overline{T}, \overline{q}, p_s, S_0, H, LE$ from the CRM model) with mostly the correct spatial structure (Figure 21.4), even if less stochasticity (i.e., random noise) was present. The heating $\frac{\partial \overline{T}}{\partial t}$ and moistening $\frac{\partial \overline{q}}{\partial t}$ profiles were also well reproduced in terms of both the mean and standard deviation for the total moistening and heating as well as for the radiative components only in the longwave and shortwave (Figure 21.5).

Similarly, Brenowitz and Bretherton (2018) showed that a NN could reproduce a full 3D CRM in offline mode, but this 3D model was more difficult to use in coupled mode and generated coupled model instabilities. These instabilities led the model to blow up. To solve this issue, they removed the upper-atmospheric temperature and moisture inputs to guarantee a bottom-up only effect of convection (Brenowitz and Bretherton 2019). In addition they developed a general diagnostic tool that identified regimes during which NNs were creating unrealistic convection in two different climate models (Brenowitz et al. 2020). Based on a linear stability analysis of the NN convection scheme coupled to a simplified wave dynamics model, this tool was used to diagnose and prevent the instability of NN convection schemes. Brenowitz et al. (2020) additionally showed that regardless of the climate model they were trained on, NNs exhibited physically-consistent behavior, such as increased convective

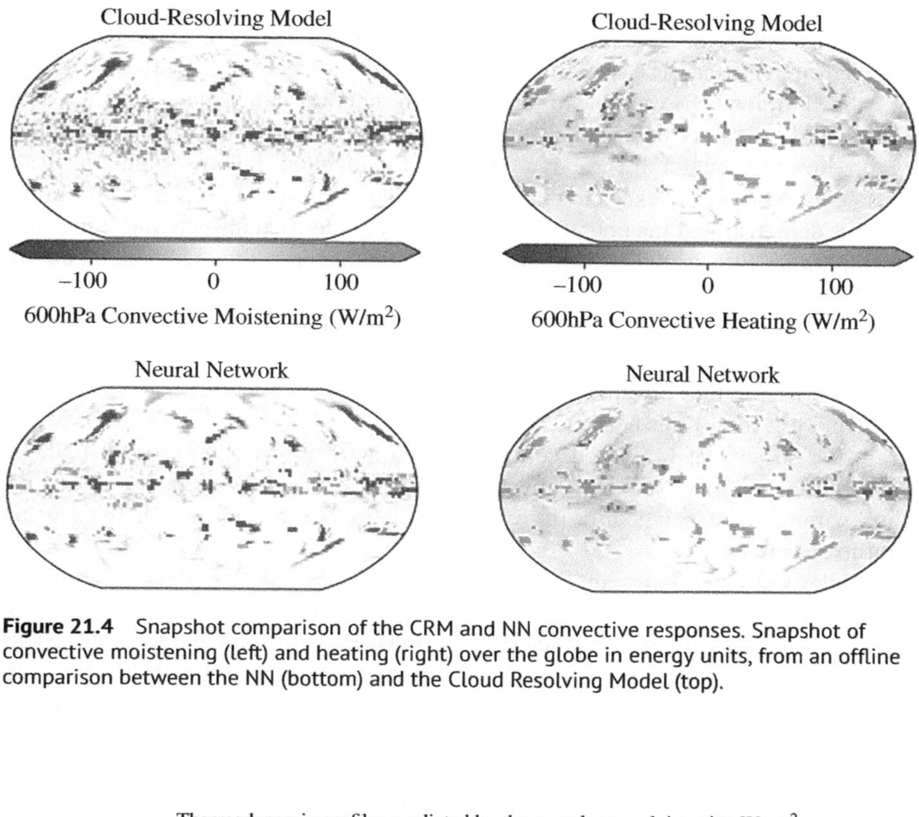

Figure 21.4 Snapshot comparison of the CRM and NN convective responses. Snapshot of convective moistening (left) and heating (right) over the globe in energy units, from an offline comparison between the NN (bottom) and the Cloud Resolving Model (top).

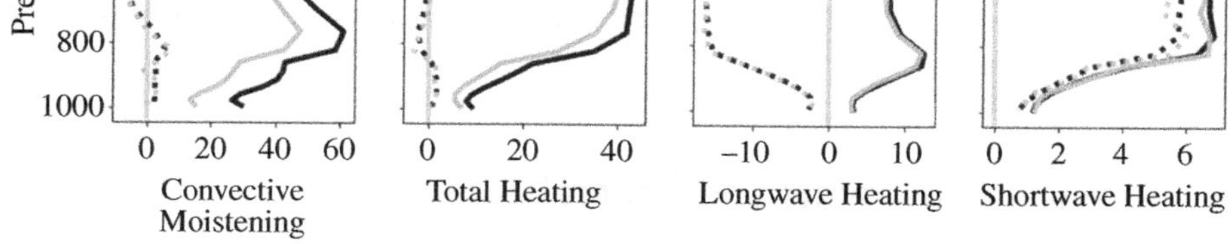

Figure 21.5 Comparison of the thermodynamic profiles predicted by the CRM and NN. Ensemble mean (dotted lines) and standard deviation (full lines) of total subgrid moistening (left), total subgrid heating (second to the left), subgrid longwave heating (second to the right), and subgrid shortwave heating (right) in energy units, from an offline comparison between the NN (gray) and the Cloud Resolving Model (black).

heating and moistening in response to increased lower-tropospheric moisture. This emphasizes that there are important features of convection that are resolved in a variety of cloud resolving approaches, that are nicely learnt and summarized by the NN algorithms.

In coupled online mode (i.e., once coupled to the dynamical core i.e. the advection scheme), the SP-trained NN was able to correctly represent many important characteristics of the original SP model. In particular, the NN version of the model physics was able to better reproduce the equatorial wave dynamics and the Madden–Julian Oscillation (Rasp et al. 2018), as well as a Walker circulation that was not present in the initial training dataset. This demonstrated the potential of the NN physics to partially generalize in some unseen conditions. Finally, the NN was able to better characterize the precipitation distribution: while traditional convective parametrizations tend to "drizzle" too much (Wang et al. 2016), Rasp et al. (2018)'s NN was able to drastically improve the representation of precipitation extremes, and Yuval and O'Gorman (2020) showed that it was possible to almost perfectly reproduce precipitation extremes by training a random forest using a very fine time step (less than 1 min) instead of a deep NN.

Overall, the success of ML-based convective parametrizations demonstrates that, in coarse-resolution models, subgrid convective processes can be accurately estimated and thus parametrized using only the coarse-resolution state information (e.g. mean temperature, humidity).

Importantly, we note that standard neural networks or random forest algorithms (O'Gorman and Dwyer 2018) are deterministic functions of the coarse-grain variables only. Stochasticity is an inherent characteristic of convection (Teixeira and Reynolds 2008; Plant and Craig 2008) that is not accounted for in this approach, although it is possible to use ensembles of NNs to account for stochasticity (Krasnopolsky 2013). In addition, memory effects are not included in this framework but could potentially be relevant as convection exhibits memory especially in the lower part of the atmosphere (the boundary layer) (Coppin and Bony 2017). Additionally, this NN structure does not include spatial covariations in the larger-scale features $\overline{X}$. In the SP-CAM setup, there is a reason to do so: the SP setup assumes a 1D periodicity in the y direction of the subgrid 2D (y-z) CRM so that there is a clear scale separation between the inner CRM scales and the outer GCM scale. This is not the case in a regular full 3D CRM such as CRMs used in the DYnamics of the Atmospheric general circulation MOdeled on Non-hydrostatic Domains (DYAMOND) intercomparison (Stevens et al. 2019) and spatial covariations could be included such as recently done for turbulence (Cheng et al. 2019) to account for vertical and horizontal variations in convection.

21.3 Physical Constraints and Generalization

Even though the ML techniques discussed in section 21.2 were able to improve upon traditional parametrizations, there were several issues using those techniques.

First, the deep NN was not conserving mass (water) and energy (Figure 21.6b). Even though well-tuned networks have relatively small sources and sinks of mass and energy (Krasnopolsky et al. 2010) (e.g., 1 W m^{-2}), mass and energy conservation are strict requirements for climate modeling. This issue was recently addressed by Beucler et al.

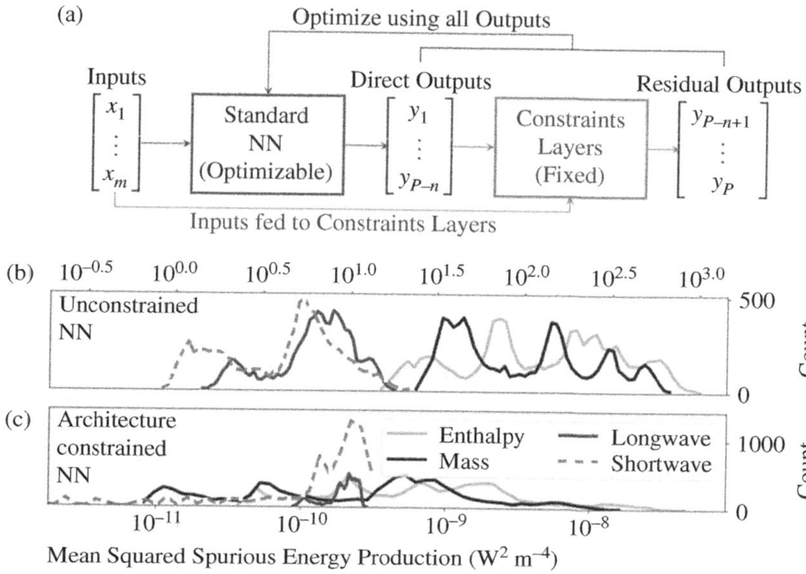

Figure 21.6 Architecture-constrained NNs can enforce conservation laws to within machine precision ($\sim 10^{-8}\mathrm{W}^2\ \mathrm{m}^{-4}$ compared to $\sim 10^2\mathrm{W}^2\ \mathrm{m}^{-4}$). (a) Schematic of the architecture-constrained network from Beucler et al. (2019). Histogram of the mean squared spurious energy production associated to enthalpy (light gray), mass (black), longwave (dark gray), and shortwave (stripe line) conservation for (b) a standard NN and (c) a constrained NN.

(2019). The authors demonstrated how strict physical constraints can be imposed within an NN architecture. The conservation of mass and energy can be strictly imposed through the addition of so-called "constraints layers" that combine inputs and outputs in order to impose strict equalities (Figure 21.6a). In the example of moist convection, these equalities are enthalpy, mass, as well as longwave and shortwave radiation conservation, which the constrained-NN enforces to within machine precision (Figure 21.6c). This goes beyond traditional way of imposing constraints in a soft way, using a regularization of the loss function with Lagrange multiplier (Márquez-Neila et al. 2017; Karpatne et al. 2017d). Indeed, in this latter approach the physical constraint is only approximately true. However, for climate modeling energy and mass conservation need to be exactly satisfied at every time step. Note that this framework also goes beyond parametrizations that only enforce *linear* constraints by default, such as random forests (O'Gorman and Dwyer 2018; Yuval and O'Gorman 2020), as it can enforce non-linear constraints as long as they are analytic.

A second challenge relates to generalization way outside of the regime of training. For instance, Rasp et al. (2018) tested the capacity of the NN trained on a given climate (0 Kelvin experiment) to generalize to a much warmer world (+4 Kelvin). The algorithm failed and exhibited the typical double intertropical convergence zone bias (i.e. two tropical rain bands) similarly to many global climate models (Oueslati and Bellon 2015; Flato et al. 2013). A similar experiment showed that the model was also unable to generalize to a colder climate (-4 Kelvin), but this time mostly at the poles, again emphasizing the difficulty to generalize. Krasnopolsky et al. (2008) suggested training a NN to anticipate generalization errors made by the ML-parametrization and automatically switch back to the traditional parametrization when needed, while Beucler et al. (2020) proposed rescaling the NN inputs and outputs to transform extrapolation into interpolation, thus avoiding the NN generalization issue. However, these two methods of circumventing generalization problems have only been tested on prototype models and most current ML subgrid parametrization schemes fail to generalize when tested outside of their training set.

21.4 Future Challenges

As discussed, there are inherent challenges in using machine learning to represent physical climate processes. The first one related to the strict conservation of physical laws can be fixed as demonstrated by Beucler et al. (2019). The second, more challenging one, is related to the fact that climate change by definition implies a shift in the distribution – not only in the mean but also in the tails of the distributions and in the extremes (Schär and Jendritzky 2004). We consider this as the primary roadblock for the more widespread use of machine learning. In addition to the challenges on physical constraints and generalization, there are other open issues with this approach that need to be further addressed.

This includes the use of three-dimensional (3D) simulations, and related inclusion of momentum in addition to temperature and moisture, which is essential for organized convection and for precipitation extremes (Houze Jr 2004), the inclusion of topography and land surfaces, as well as the decomposition of waves. High-resolution CRMs (few km in horizontal scale) alleviate many of the biases observed in coarse-resolution models. Specifically, CRMs dramatically improve the representation of deep clouds and convection, organized convection such as mesoscale convective systems, their atmospheric heating/moistening, wave propagation, as well as precipitation (Stevens et al. 2019; Kooperman et al. 2016). However, even cloud-resolving simulations still use a parametrization to represent shallow clouds, which are crucial for climate sensitivity though (Schneider et al. 2017b). Further work in that direction is therefore needed.

22

Using Deep Learning to Correct Theoretically-derived Models

Peter A. G. Watson

Earth system simulators are dynamical models based on the laws of physics, chemistry, and biology as far as we understand them. However, the laws cannot be used directly as this would be too computationally costly. Instead, approximate equations are used, leading to substantial errors in the output. Examples of particular difficult problems are predicting climate change feedbacks due to clouds (Vial et al. 2013) and representing tropical rainfall variability (e.g. Westra et al. 2014; Watson et al. 2017). Reducing these errors would be very valuable for giving better warnings of severe climatic events (also see Chapter 21).

Some attempts to apply deep learning to produce better simulators have proposed learning all of the equations from data (e.g. in the case of the atmosphere by Dueben and Bauer 2018; Weyn et al. 2019), but as of the time of writing, these have not come very close to matching the skill of state-of-the-art weather forecasts.

It is potentially more promising to combine deep learning with the theoretically-derived models we currently have. Karpatne et al. (2017a) present a set of such approaches they call "theory-guided data science", which includes:

- using theory to restrict statistical models to have physically-consistent structures (for example, ensuring that quantities like rainfall cannot become negative);
- guiding these models to learn physically-consistent solutions, such as by including known scientific laws in objective functions (for example, penalizing energy non-conservation);
- using theory to refine models' outputs (for example, using a data-driven model to produce possible solutions to a problem and then validating these with a theoretically-derived model);
- creating hybrids of models based on theory and statistical learning (for example, using a statistical model to predict the error term of a theoretically-derived model);
- enhancing theory-based models using statistics (for example, by finding optimal parameter settings).

Reichstein et al. (2019) reviewed approaches to integrate deep learning with theory-based modeling, including by learning parameterizations for processes that are particularly hard to derive from theory and emulating parts of physical models to enable them to be run more

Deep Learning for the Earth Sciences: A Comprehensive Approach to Remote Sensing, Climate Science, and Geosciences, First Edition. Edited by Gustau Camps-Valls, Devis Tuia, Xiao Xiang Zhu, and Markus Reichstein.

efficiently, potentially allowing more complex and realistic simulators to be deployed.

This chapter will focus on one particular method to use deep artificial neural networks (ANNs) together with theoretically-derived models, which is to use deep learning algorithms to correct these models' errors. The total form of the dynamical model is then

$$\frac{\Delta x}{\Delta t} = \mathcal{M}(x) + \epsilon(x), \tag{22.1}$$

where x is the vector of state variables, $\Delta x/\Delta t$ is the tendency over one discretized time step Δt, $\mathcal{M}(x)$ is the prediction of the theoretically-derived model and $\epsilon(x)$ is the correction produced by the algorithm (each presumed not to have explicit time-dependence). $\epsilon(x)$ is trained to maximize some cost function based on differences between the overall predictions and data from the target system. $\epsilon(x)$ affects predictions made for subsequent time steps. In this set-up, the theoretically-derived model retains a critical role and keeps a link between the predictions and our theoretical understanding, whilst the deep learning component allows potential skill increases. This potentially hugely reduces the complexity of the necessary machine learning algorithms compared with learning to represent $\Delta x/\Delta t$ entirely (as in Dueben and Bauer 2018; Weyn et al. 2019), since achieving a skill improvement does not require learning all the knowledge encoded in the existing simulators. $\epsilon(x)$ does not need to be as complex as the theoretically-derived model – for example, not all input and output variables need to be included, as long as what is included is enough to give a skill gain. This is a particular advantage for subject areas that use very complex simulators, such as climate modeling. This may make it a more practical approach for a research program, since improvements can be made using simple algorithms and then built on from there, rather than waiting many years for an algorithm that outperforms current models. Additionally, keeping the theoretically-derived model components will help to maintain interpretability of the simulator's behavior. This approach may also work better in novel physical situations than replacing the full equations because the theoretically-derived model provides a prediction based on theory and it is not all left to the algorithm, as long as $|\epsilon(x)|$ is not generally much larger than $|\mathcal{M}(x)|$ (though it would be a challenge to firmly guarantee reasonable behavior of $\epsilon(x)$).

Using machine learning algorithms to correct output from dynamical models in "offline" mode (without using the correction at one time step in calculating the tendency at the next time step) has been applied in several contexts. Xu and Valocchi (2015) found that support vector machines and random forests could improve the skill of predictions of groundwater flow. Their approach reduced the bias in the mean predicted flow from 18% to nearly zero and reduced the mean absolute error by 50%. Karpatne et al. (2017d) trained a neural network to correct predictions from a dynamic model of lake temperatures. As well as minimizing the root mean square error (RMSE) of prediction errors of standardized temperatures at each time step, they included a "physics-based" term in the loss function to penalize the model if it predicted that the water density would increase with height, and attained a 30% reduction in the RMSE. Rasp et al. (2018) used ANNs to postprocess weather forecasts of temperature in Germany and found that this gave better skill than some other frequently-used statistical methods. Together, these results indicate that machine learning has the potential to considerably reduce errors in predictions from dynamical models.

There do not seem to be many examples in the literature of applying this approach to Earth science applications in "online" mode, where the corrected output of dynamical models is used as input for predicting the next time step. Early work on this method was done by Forssell and Lindskog (1997), who addressed a laboratory-scale problem of predicting the water level in a tank that was being filled by a pump driven by a time-varying voltage. Combining a theoretically-derived model with a neural network to account for phenomena that the theoretically-derived model did not represent (for example, eddies in the tank) reduced prediction RMSE by a factor of two. It also did not predict unphysical situations like the water level becoming negative, which did happen when a neural network alone was used to try to solve the problem.

Cooper and Zanna (2015) used a stochastic linear algorithm added to a low-resolution model to improve simulations of an idealized two-dimensional ocean with a double gyre. "Truth" simulations were produced using a horizontal resolution of 7.5 km and it was attempted to reproduce the long-term statistics of this run using models with a resolution of 30 km. The targeted statistics were the mean, variance, and autocorrelation in time at each grid point. An iterative approach was used to learn parameters of the algorithm that produced the best results, requiring ~100 integrations of the low-resolution model. The resulting simulations have mean squared errors in the bias and variance of the horizontal flow velocities that are reduced by more than a factor of 10, and also substantially reduced errors of their 5-day lag covariance. The response to a change in wind forcing of the system, whilst still far from perfect, had a mean squared error that was 50% and 60% of that for the uncorrected low-resolution model for the eastward and northward components of the flow respectively. However, it is unclear whether their linear approach could give substantial improvements of nonlinear processes in the atmosphere that are difficult to model, such as convection (Arakawa 2004).

Pathak et al. (2018) applied an algorithmic correction approach with a non-linear machine learning method to simulate low-dimensional chaotic systems, whose variability was coming entirely from internal dynamics rather than being forced. They tested "knowledge-based" predictors of these systems that differed from the true equations through a parameter change, a "reservoir computer", and a hybrid of the two. A reservoir computer is similar to a neural network in that it consists internally of a large-number of non-linear functions relating inputs to outputs, but only parameters relating the outputs of these functions to the predictions are learnt. So, unlike in deep learning, the non-linear functions between the inputs and outputs are not optimized for the prediction task. For both the Lorenz (1963) system of equations (that which produces the famous "butterfly attractor") and the one-dimensional Kuramoto–Sivashinsky equations, the hybrid system made predictions that remained close to those from the true equations for typically ~2–3 times longer than predictions made by the separate knowledge-based and reservoir-based models.

22.1 Experiments with the Lorenz '96 System

The aforementioned work did not show whether deep learning models could be used with this method to give solutions that simultaneously have better skill in short initialized forecasts and also run stably for a long time and produce simulations with long-term statistics that are closer to the truth. This is a key aim in Earth system modeling in order to have models to use both for making short-range predictions and simulating long-term effects of climate change. As well as it being more practical to have one set of models for both tasks rather than maintaining them separately, showing that models used for climate change projections can make good short-range forecasts can also increase our confidence in their ability to correctly simulate the dynamics of phenomena like extreme weather events (discussed more in section 22.1.3).

To test whether this is possible, and also to give a more detailed explanation of how the error-correction approach might be applied, an example based on the chaotic Lorenz '96 dynamical system (Lorenz 1996, sometimes also referred to as the Lorenz '95 system) will now be presented in detail. This follows the approach of Watson (2019), but focuses more on how varying the ANN structure affects the prediction skill and demonstrates how it can be used to test ideas like seamless prediction of weather and climate (Palmer et al. 2008).

The Lorenz '96 system is instructive for testing new concepts for approaches to improve Earth system models and has been used numerous times before in this way (e.g. Wilks 2005; Arnold et al. 2013; Schneider et al. 2017a; Dueben and Bauer 2018), although it is a lot simpler than a model of the Earth's climate. The set up is to have a "Truth" system with complex, fine scale behavior and try to simulate its behavior using a system that only has coarse-scale information. This is analogous to trying to simulate the Earth, with all of its important small-scale phenomena such as tropical thunderstorms and oceanic eddies, using weather and climate models with ~10–100 km scale resolution, which is the limit of what can currently be afforded computationally. The experimental method is described below, with more details given by Watson (2019).

22.1.1 The Lorenz'96 Equations and Coarse-scale Models

The Lorenz '96 system equations are

$$\begin{aligned}\frac{dX_k}{dt} &= -X_{k-1}(X_{k-2} - X_{k+1}) - X_k + F - (hc/b)\sum_{j=1}^{J} Y_{j,k},\\ \frac{dY_{j,k}}{dt} &= -cbY_{j+1,k}(Y_{j+2,k} - Y_{j-1,k}) - cY_{j,k} + (hc/b)X_k,\end{aligned} \tag{22.2}$$

with cyclic boundary conditions $X_k = X_{k+K}$ and $Y_{j,k} = Y_{j,k+K}$. This work uses parameter values $K = 8, J = 32, h = 1, F = 20, b = 10$ and $c = 4$, following the work of Arnold et al. (2013). This ensures that X_k, defined for $k = 1, \dots, K$, are slowly-varying variables and $Y_{j,k}$, defined additionally for $j = 0, \dots, J + 2$, are quickly-varying. The Y variables are connected in a ring with $Y_{0,k} = Y_{J,k-1}$, $Y_{J+1,k} = Y_{1,k+1}$, and $Y_{J+2,k} = Y_{2,k+1}$. This means there are J unique $Y_{j,k}$ variables associated with each X_k variable. Lorenz (1996) suggested that the $Y_{j,k}$ be considered analogous to a convective-scale quantity in the real atmosphere and X_k analogous to an environmental variable that favors convective activity. These equations are integrated in time using a fourth-order Runge–Kutta time-stepping scheme with a time step of 0.001 time units. Equation 22.2 is hereafter referred to as the "Truth" system.

22.1.1.1 Theoretically-derived Coarse-scale Model

In order to obtain a set of equations to simulate equation 22.2 much more computationally cheaply, following Wilks (2005) and Arnold et al. (2013), we can try not to explicitly simulate the Y variables but parameterize their effect on the X variables. This is conceptually similar to how the effect of unresolved phenomena on larger scales is treated in Earth system models. Thus we obtain the system of equations

$$\frac{\mathrm{d}X_k^*}{\mathrm{d}t} = -X_{k-1}^*(X_{k-2}^* - X_{k+1}^*) - X_k^* + F - U(X_k^*), \tag{22.3}$$

with $X^*k = X^*k + K$. $U(X_k^*)$ is a function that parameterizes the effect of the Y variables on the X variables. The time step is lengthened to 0.005 time units, which is analogous to how Earth system models do not properly simulate processes that occur on very short timescales, as well as those on short length scales.

$U(X_k^*)$ is assumed to be a cubic polynomial, following Arnold et al. (2013), such that

$$U(X) = \sum_{n=0}^{3} a_n X^n.$$

Its parameters are derived using a coarse-graining approach, and have values $a_0 = -0.207$, $a_1 = 0.577$, $a_2 = -0.00553$ and $a_3 = -0.000220$, following Watson (2019).

The model given by equation 22.3 will be referred to as the "No-ANN model" from here on.

22.1.1.2 Models with ANNs

Error-correcting ANNs to be used with equation 22.3 were chosen to have a multilayer perceptron architecture (Nielsen 2015; Goodfellow et al. 2016). They were trained to predict the true system tendency over one 0.005 time unit time step minus the prediction of equation 22.3 given the X values at the start of that time step, namely

$$\epsilon_k = \frac{\mathrm{d}X_k}{\mathrm{d}t} - \frac{\mathrm{d}X_k^*}{\mathrm{d}t}.$$

The ANNs predict a tendency for one grid point, being applied K times per time step, and taking as inputs standardized X variables up to a distance of two grid points away. This means the prediction at a grid point only depends on information in the region of the point. The inputs include one X value that is not used in the No-ANN model, illustrating how ANNs can use inputs that are not easy to include in theoretically-derived models – Watson (2019) found that this improved the skill of initialized forecasts of the system but not simulation of its long-range statistics, and qualitatively similar results are obtained if the extra input is excluded.

The ANNs used had between 1 to 3 hidden layers (the "depth") and 2, 4, 8, 16, 32, or 64 neurons in each layer (the "width"). They use rectified linear unit activation functions for the hidden neurons and a linear output activation function.

The ANNs were trained using 1000 time units of data from the truth system (200,000 time steps). Results are not sensitive to the amount of training data used, and Watson (2019) reported that using as little as 2 time units of data for training gave good results. The ANNs' cost function was the sum of the squared prediction error and an L_2 regularization term for the weights with coefficient 10^{-4}. Training was done using stochastic gradient descent with the Adam algorithm (Kingma and Welling 2014). Minibatches of size 200 sets of inputs and output were used together with a learning rate of 0.001. Training stopped when the squared prediction error failed to decrease by at least 10^{-4} twice consecutively after iterating over the whole training dataset.

A validation dataset for evaluating the models with ANNs was produced by separately creating 3000 time units of data from equation 22.2.

22.1.2 Results

Here, performance of the models with ANNs is presented, and the difference made by choosing different ANN structures is explored. Note that sampling variability is not generally large compared to the differences between coarse models with and without ANNs, with very similar results being found using the first and second halves of the datasets only. (The exception is the biases in the time-mean of the X variables, which differed between the first and second halves of the datasets, and were not found to be statistically significantly different between the models in most cases (not shown), so there is substantial uncertainty in the patterns in the results for this statistic, but it is not important for the conclusions.)

22.1.2.1 Single-timestep Tendency Prediction Errors

Figure 22.1 shows the RMSE of tendency predictions over a single coarse time step (0.005 time units) for coarse-resolution models with error-correcting ANNs, plotted as a

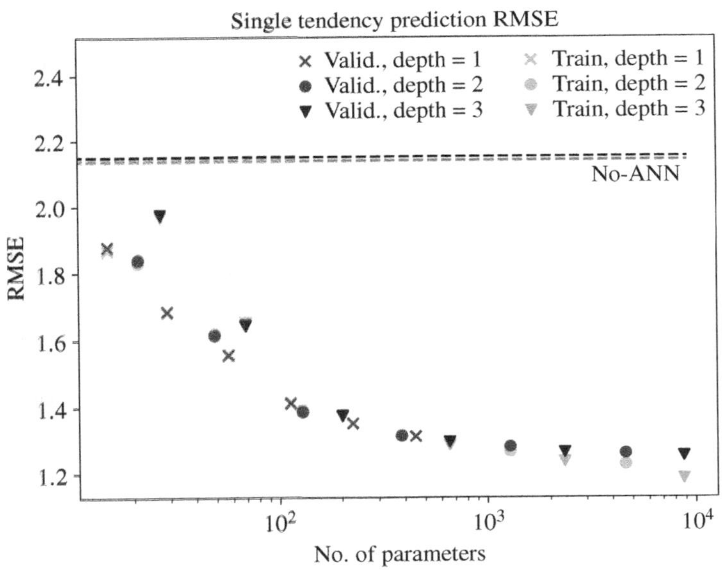

Figure 22.1 RMSEs of single-time step tendency predictions by coarse-scale models with error-correcting ANNs as a function of the number of ANN parameters. Different symbols are used for models with ANNs of different depths (number of hidden layers). Opaque symbols show RMSEs on validation data and partially transparent symbols those on training data. The black and grey horizontal dashed lines show the RMSE for the No-ANN model in the validation and training datasets respectively. Note the horizontal axis is logarithmic. The models with ANNs robustly outperform the No-ANN model, with increases in the number of parameters giving smaller gains as the number of parameters increases.

function of the number of ANN parameters, which indicates the ANN complexity and is approximately proportional to the number of floating point operations required to make a prediction. ANNs are grouped according to the number of hidden layers. The RMSE is calculated using 10,000 randomly chosen time steps in the training and validation datasets separately, with the same time steps selected for each ANN structure.

Every ANN tested reduces the RMSE compared to the No-ANN model (whose RMSE is shown by the dashed line). This demonstrates that even ANNs of low complexity can give a better performance, including an ANN with just two neurons in one hidden layer. The RMSE decreases as the number of parameters increases, by up to 42% for the validation data, and has not saturated for the ANNs tested here – further error reductions are probably possible, but this is not tested here. The relationship between the RMSE and the number of parameters seems quite independent of the number of hidden layers when the number of parameters is more than about 100. The RMSEs on the training data are not very much lower than those on the validation data, indicating that overfitting is not occurring to a large extent.

Watson (2019) showed that ANNs in this set up could also improve predictions for extreme positive and negative tendencies in the validation dataset. This suggests the ANNs have actually learnt to better represent the dynamics, rather than learning how to reproduce examples seen in training. This is very important for applications like Earth system models, for which predictions of extreme situations are a large part of the total value they produce.

22.1.2.2 Forecast and Climate Prediction Skill

Figure 22.2 shows the anomaly correlation coefficient (ACC) and RMSE of forecasts made by coarse-resolution models tasked with predicting the trajectory of the Truth system, at a lead time of one time unit (approximately analogous to a "medium-range" weather forecast a few days to a week ahead, based on the autocorrelation timescale of the system). The forecasts were made by taking initial conditions for the X variables from the Truth validation run. Small perturbations were applied to these initial conditions, sampled from a Gaussian distribution, in order to create 10-member ensemble forecasts. The skill of the ensemble-mean's predictions is what is quantified in Figure 22.2. More complete details of the method to create the ensemble forecasts are given by Watson (2019), where the skill as a function of lead time is also shown.

All but two of the models with ANN error-correctors achieved better skill than the No-ANN model – the two ANNs that failed to give a skill increase were very simple, having only two neurons per layer. This shows that in this system, it is possible to train the ANNs to perform well on very short timescales and give improved skill on much longer timescales (with the lead time of these forecasts being 200 times the length of the time step used in training).

The improvements are quite subtle, however, with the ACC only increasing from ~0.46 to ~0.49 and the RMSE decreasing from 5.89 to 5.73 at the most. Part of the reason is likely to be that the No-ANN model is already close to attaining the maximum possible skill (as found by making the same forecasts using the Truth system with the same initial condition perturbations, which gives an ACC of 0.52 and RMSE of 5.59). Therefore the improvements produced by the ANNs are about half of the maximum possible improvement. So it is

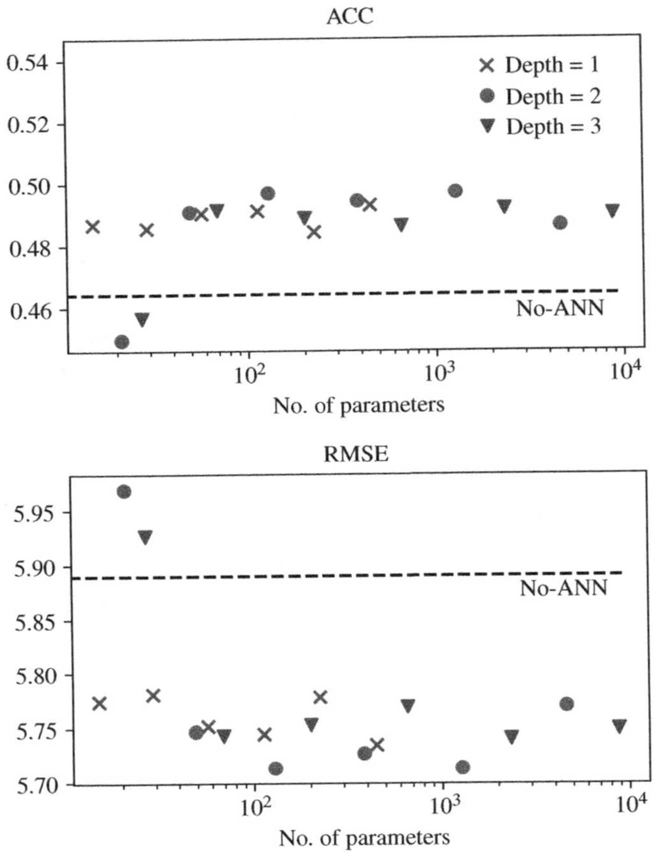

Figure 22.2 The ACC and RMSE of ensemble forecasts of the Truth validation run trajectory made by models with error-correcting ANNs as a function of the number of ANN parameters, plotted as in Figure 22.1 but not showing results for evaluation using the training dataset. It is better to have a larger ACC and a smaller RMSE. All but two models with ANNs outperform the No-ANN model, but there is no large gain with increasing model complexity.

possible that when predicting more complex systems, where the theoretically-derived models have substantially lower skill compared to the maximum attainable, the improvements made by using ANNs in this way would be relatively much greater.

Increasing the ANN's complexity does not appear to give higher forecast skill improvements beyond having ~50 parameters, despite improvements being found for predicting single-time step tendencies (Figure 22.1).

Figure 22.3 shows diagnostics of the quality of long-term statistics diagnosed from free-running 3000 time unit simulations with each coarse-resolution model – these are analogous to "climate" statistics for the Earth system. The diagnosed biases in the time-mean of the X variables for the models with ANNs lie both above and below the value for the No-ANN model, and are not generally statistically significantly different from it, apart from in the cases of the two models with the highest biases (not shown). However, the two-sample Kolmogorov–Smirnov statistic is improved by nearly all ANN structures.

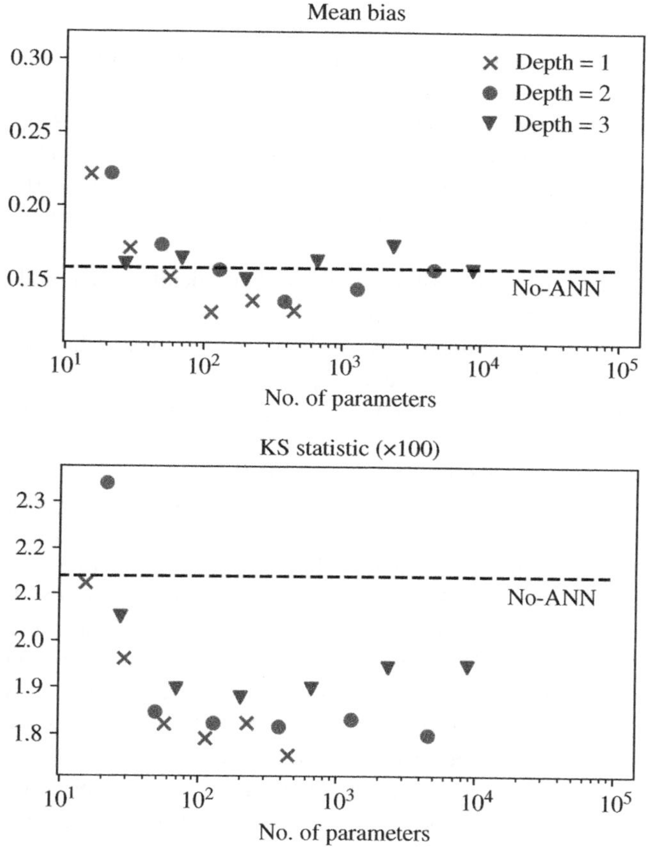

Figure 22.3 The bias in the climate mean and Kolmogorov–Smirnov (KS) statistic of long time series produced by models with error-correcting ANNs as a function of the number of ANN parameters, plotted as in Figure 22.2. It is better for both diagnostics to be as small as possible. The ANNs do not give a clear improvement in the mean bias, but the KS statistic is improved in all but one case, with greater model complexity giving improved results up to having ~100 parameters.

This is the maximum difference between the cumulative distribution functions of X in the Truth system and in each respective coarse-resolution model, and therefore depends on the shape of the distributions of X variables as well as their means. Watson (2019) shows that this reflects the fact that the ANNs reduce the excessive occurrence of X values near the central peak of the distribution and reduce the deficit in its negative flank. However, the ANNs do not improve the low frequency bias of extreme X values. Subsequent work by Chattopadhyay et al. (2019) found that algorithms that can incorporate memory over sequences of time steps can simulate the frequency of extreme values well in a system very similar to the one being considered here. Therefore, a promising direction is to modify error-correcting algorithms to also be able to do this.

For the KS statistic, there is some evidence of improving skill with increasing model complexity up to ~100 parameters. Using a 3-layer ANN gives a slightly worse performance than a 2-layer ANN with a comparable number of parameters.

22.1.3 Testing Seamless Prediction

One open question in Earth system modeling is to what extent does improving simulation skill on short timescales give improved skill on longer timescales. If short-range prediction skill is very related to longer range skill, then weather forecast skill metrics would be useful for judging how good particular climate models are likely to be at simulating climate statistics, and may be a lot cheaper to compute. Palmer et al. (2008) and Matsueda et al. (2016) have made the case for using relatively short-range forecast diagnostics to evaluate climate models.

Using an ensemble of models that is composed of one or more theoretically-based models combined with differing machine learning algorithms, with systematic sampling of the algorithms' hyperparameters, can help to quantify relationships such as these. This can complement methods of creating model ensembles such as perturbing the parameters of a single model structure (e.g. Sexton et al. 2019), which can struggle to give skill improvements in contrast to the results shown here, or using a set of models with varying structures (Tebaldi and Knutti 2007), which is very difficult to do in a systematic way.

Figure 22.4 shows relationships between diagnostics of relatively short-range and long-range prediction skill of the models with ANNs tested here. Panels (a) and (b) show the skill of forecasts 1 time unit into the future against the RMSE of single-time step tendency predictions, and (c) and (d) show similar relationships for the statistics of the X variables in long runs. (e) and (f) show the performance at simulating long-term statistics against 1 time unit forecasts. These relationships do show evidence of prediction skill on short timescales being indicative of skill on long timescales, with correlations between the variables having magnitudes between 0.51–0.81. However, if ANNs of width 2 are excluded, which includes all the models that perform worse than the No-ANN model on at least one diagnostic, then the correlation magnitudes fall substantially to be between 0.15–0.38. The signs of the correlations are always consistent with improved skill on short timescales being associated with improved skill on long timescales, though, and they are the same when computed for separate halves of the validation dataset. Therefore this does provide evidence that diagnostics of short-range prediction skill can help to select models that are better at longer-range predictions for this system, but the relationships are quite weak. Also, the magnitude of the improvements made to long range skill are much less than what might be expected from the improvement found for single time step tendency predictions (Figure 22.1). Stronger relationships may be found in Earth system models, given the links that have been found between biases on short and long timescales in simulations (Ma et al. 2013; Sexton et al. 2019), and this would be interesting to test.

22.2 Discussion and Outlook

The abovementioned studies and results show there is much promise for using machine learning together with theoretically-derived models to get better simulators of complex dynamical systems, including the Earth system. The error-correction approach is one potential method to do this. However, the research on this approach to date has focused on reproducing the statistics of fairly simple systems, and much needs to be done to show whether this will also give good results for much more complex systems and perform well

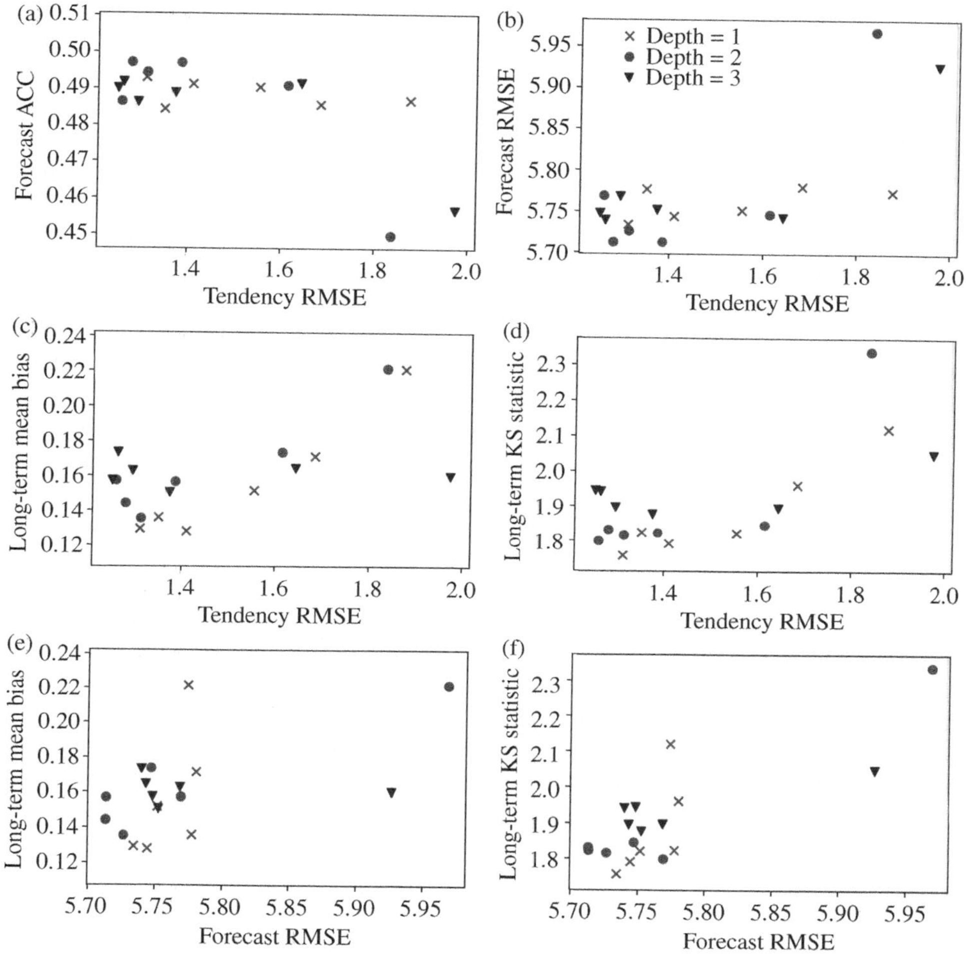

Figure 22.4 Simulation quality diagnostics plotted against each other: (a) and (b) show forecast ACC and RMSE at lead time 1 time unit against the single-time step prediction error respectively; (c) and (d) show similar results for the long-term mean bias and KS statistic; and (e) and (f) show the long-term mean bias and KS statistic plotted against the forecast RMSE at lead time 1 time unit. Different symbols are used for ANNs with different numbers of hidden layers. There are substantial correlations between these diagnostics, signaling that skill at simulating shorter timescales is indicative of skill on longer timescales, but these are sensitive to the exclusion of outliers (see text).

at tasks like predicting how such systems will respond to external forcing, which is very relevant for getting better climate change projections. The final part of this chapter discusses particular challenges that deserve attention.

22.2.1 Towards Earth System Modeling

The approach demonstrated above for the Lorenz '96 system, where an algorithm was used to learn corrections to single time step tendencies, could not be applied in the same way for the Earth system because observations in a given place are generally separated by six hours or more. The time steps in state-of-the-art dynamical Earth system models are a lot shorter, which is likely to be needed so the models are numerically stable, and is desirable in order to better represent the true continuous-time equations. The method therefore needs to be extended to allow an algorithm to be learnt when there is only data many time steps apart. In a free-running system, the impact of perturbations to the algorithm's parameters on predictions over multiple time steps needs to be taken into account. One possibility is to use the "backpropagation through time" algorithm (Werbos 1990a), as used in recurrent ANNs (Funahashi and Nakamura 1993). This would require the tangent linear approximation of the theoretically-derived model. This becomes more complicated if the algorithm is "local", so predictions at a point only depend on inputs from nearby points, which is likely to be beneficial for parallel computing and also incorporates spatial invariance of the prediction equations if the same algorithm is used at all grid points, as in the above work. Then the effect of perturbing parameters on predictions at those nearby points likely also needs to be taken into account, so that backpropagation needs to be done "backwards through time and sideways through space".

Once appropriate algorithms have been shown to work in simple cases, improving Earth system simulations would require them to be trained on data either from high-quality simulators that we cannot afford to use in all of the experiments we would like to, or on data based on observations of the real system. Reanalysis data is a possible choice for the latter (e.g. Dee et al. 2011). Although it is not perfect, it is likely to be closer to the behavior of the real system than existing simulators, so learning to simulate it would yield modeling improvements. Using the improved models to create a better reanalysis could then give a self-improvement cycle, where improved reanalysis is used to produce improved simulators, which are used to make better a reanalysis, and so on. Alternatively, Bocquet et al. (2020) have shown that the model and system state may be learnt together.

22.2.2 Application to Climate Change Studies

Using machine learning approaches to improve future climate predictions is a difficult challenge because it means understanding states that we have not observed. The approaches described above are all designed for situations where we have detailed observations of the system we want to simulate. So can deep learning add anything of value to this?

It is firstly worth noting that not all climate change-related problems involve predicting the far future. For example, it is greatly valuable to improve estimates of present day severe climatic risks. Climate change means we cannot just use past observations to estimate present risks, since they have changed, and simulations of large numbers of weather events in the present-day climate can be valuable for this task (e.g. Thompson et al. 2017). It may be possible to make significant improvements to simulators using machine learning based only on meteorological observations from recent decades, which include information about the effect of varying climate change forcings, and the simulators could then predict current risks much better.

Another key problem is attributing recent observed extreme weather events to climate change (Allen 2003; National Academies of Sciences Engineering and Medicine 2016). This involves simulating the risk of the events using external forcing values at the time that they occur and counterfactual values corresponding to a world with less anthropogenic climate change. If the counterfactual values are taken from an earlier observed period, then it may be possible to train systems to predict weather risk for both cases based on observations, and so derive the difference between them, without requiring extrapolation into unseen climates.

In order to predict the Earth system's behavior in future climatic states, it is clearly necessary to represent the effect of changing forcings, for example the increasing concentrations of carbon dioxide. Our current simulators rely on data based on laboratory experiments and calculations from the laws of physics for this, back to the work of Tyndall (1861). I do not know a way to integrate this data into simulators trained to reproduce observed Earth system behavior with no theoretically-derived structure. Combining machine learning components with theoretically-derived modules has the advantage that the theoretically-derived part can incorporate such knowledge. Therefore, a theoretically-derived simulator with an error correction algorithm could reproduce effects such as the warming produced by carbon dioxide, whilst also giving a more realistic simulation of things like weather variability. The algorithm would become increasingly less trustworthy as the climate changed more, but may still add value. For example, Scher and Messori (2019) found some skill for neural networks simulating simple dynamical systems with external forcings for some way outside the range seen in training.

Tests of whether combining theoretically-derived and machine learnt components perform better at predicting the effect on chaotic systems of changes in external forcings than using either approach in isolation would be very valuable.

22.3 Conclusion

This chapter has discussed using machine learning algorithms to correct errors in simulated tendencies of theoretically-derived models of dynamical systems. This is a promising approach to produce improved simulators, including for the Earth system. It has been shown here that this can yield better quality simulations in the chaotic Lorenz '96 system and produce insights into theories like seamless prediction. The main challenges for future development are applying this in more complex models in situations where observational data is sparse in time and making it reliable at predicting the impact of changing external boundary conditions, such as greenhouse gas emissions.

23

Outlook

Markus Reichstein, Gustau Camps-Valls, Devis Tuia, and Xiao Xiang Zhu

Deep learning has in the past decade surpassed the boundaries of pure machine learning and computer vision research, and became a state-of-the-art tool in almost every scientific discipline and is exponentially growing. On Google Scholar, more than half of the total literature on the terms "`Earth Science` and `deep learning`" is recorded since 2019 (except for pre-2016 articles, where "deep learning" is found with an educational meaning). More importantly, while the success of deep learning has started with "black-box" classification tasks, deep learning is contributing more and more in diverse ways to the scientific process of knowledge generation, with some latest examples reported in the chapters of this book. As an example, remote sensing (Chapters 2–11) exploited the ability of deep learning to align and fuse heterogeneous sources of data (Chapters 9 and 10) before extracting spatial, often geometrical features (Chapters 5 and 6), as well as temporal features in observational multitemporal data (Chapters 8 and 18), as indicated in Figure 23.1, arrow A. Both discriminative and generative deep learning models, as well as the whole continuum from fully supervised to unsupervised approaches is applicable. In particular, to deal with domain adaptation (Chapter 7) and sparse labels, smart combinations of supervised and unsupervised approaches need to be researched further: purely unsupervised (Chapter 2), semi-supervised, and self-supervised (Chapter 4) are strong candidates in this direction. Another strand of deep learning for geosciences has been emerging later, which relates to exploiting synergies with system modeling (Figure 23.1B). System modeling is often termed "physical modeling", which is too narrow, because chemical, biological, ecophysiological, and physical processes may be modeled here. System modeling (also called process-based modeling or mechanistic modeling) refers to an approach which attempts to create the behavior of a system from the behavior and interactions of its components, ideally derived from fundamental laws or at least ample empirically established knowledge. With respect to system modeling, deep learning can be used as an accelerator of the calibration processes: the examples in this book include emulation of dynamical spatio-temporal systems (Chapter 18), parameterizations of processes which (mostly for computational reasons) cannot be explicitly resolved at the required resolution (Chapters 20 and 21), and bias corrections of system models (Chapter 14). These examples mostly deal indeed

Deep Learning for the Earth Sciences: A Comprehensive Approach to Remote Sensing, Climate Science, and Geosciences, First Edition. Edited by Gustau Camps-Valls, Devis Tuia, Xiao Xiang Zhu, and Markus Reichstein.

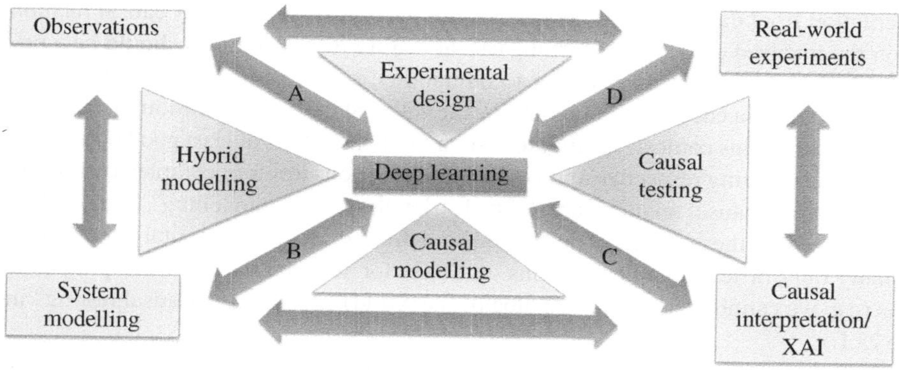

Figure 23.1 Future challenges of deep learning in linking to observations, experiments, causal interpretation and system modeling.

with physical systems, but the approaches have high potential also for geo-biological and eco-physiological processes in the future.

There are a number of future challenges ahead, which relate to the integration of deep learning with other approaches, most notably four main pillars: experimental design, hybrid modeling and causal testing and causal modeling. In Figure 23.1, they are found as the triangles. For instance, bringing together Deep learning, System modeling and observations in a hybrid modeling framework, which complements physics-guided machine learning (Reichstein et al. 2019), see triangle "Hybrid modelling" in Figure 23.1. While technically this sort of hybrid modeling can be addressed seamlessly within a differentiable programming framework (e.g. `PyTorch`, `JuliaFlux`), and first examples exist (de Bezenac et al. 2019; Kraft et al. 2020), there are a lot of conceptual questions still to be addressed. One of the advantages is that in such a framework, physically interpretable parameters or states can be estimated as in a data assimilation system, but the identifiability of the parameters and states is a big challenge in particular if several of these are estimates, together with all the weights of the DNN. Regularization techniques have to be explored, for instance.

DL models are often intransparent and overparameterized, and the learned relations are difficult to understand and visualize. One would need to decompose models responses in interpretable building blocks, in order to understand the drivers of the different (climatic, perceptive, demographic, ...) processes being modeled by the network. Explainable deep learning is a rising trend (Lapuschkin et al. 2019; Samek et al. 2019), which is also picking up in spatial studies (Marcos et al. 2019; Roscher et al. 2020). Making the black boxes grayer will definitely help not only in gaining confidence and trust in machine decisions, but also would certainly make a decisive step forward understanding through data analysis.

The other big challenge lies in the links between deep learning and XAI/Causal inference. By construction DNNs are not causal, but simply find the "best" associations/correlations between data patterns, given a defined cost function. There are two ways of making linkages: (i) using causal theory to explain the functioning (e.g. feature relevance) of a DNN (e.g. Tibau et al. (2020)) and (ii) using DNN for causal hypothesis generation or causal inference in general (e.g. Bengio et al. (2019)). DL models learn a useful discriminative feature representation from data, but the network has not necessarily learned any causal relation between the involved covariates. Learning causal feature representations should be a priority, particularly in current times where accountability, explainability, and interpretability (see last point) are needed in general, and in the relevant case of attribution of causes to climate change. In this context, the link to system modeling offers at least good opportunities for ground-truthing respective approaches across various levels of complexity, since in system models the causal relations are defined (triangle "Causal modelling", Figure 23.1). In the real world, these kinds of tests can be achieved via experimental approaches, yet usually only in less complex systems and only for selected variables for pragmatic reasons (e.g. we cannot yet build an analog of the world), cf. triangle "Causal testing" in Figure 23.1.

Last but not least, real-world experiments are important to test the hypotheses we generate from observations or from theoretical reasoning (Cuttler et al. 2020). Optimal experimental design attempts to design experiments that can best constrain parameters of a model, or distinguish between different model structures, but also for non-parametric estimation (Winer 1962). A geo-scientific non-parametric example strongly related to machine learning is an estimation of a spatio-temporal stochastic process where optimal experimental design tells where in space observations should be placed for maximal information. Certainly, deep generative models can play an interesting role in the "Experimental design" triangle (Figure 23.1). For model parameter estimation and model selection, it will be interesting to link this to hybrid modeling as well, for instance asking which are most informative experiments or spatio-temporal sampling strategies to constrain hybrid models and/or the physically interpretable parameters and latent states. While Bayesian nonparametric models has been very active in this regard, it appears that the deep learning literature is lacking on this topic, hence there are very good prospects to make impactful first contributions.

The future of the interface between DL are Earth and Climate sciences is bright and exciting. We have now access to operational tools that allow optimizing arbitrary networks, losses, and physics-aware architectures. Besides, current methods are able to make sense of the learned latent network representations: interpretability is just the first step; eXplainable AI (XAI) and causal inference have to guide network training. Our long-term vision is tied to these open frontiers and foster research towards algorithms capable of discovering knowledge from Earth data, a stepping stone before the more ambitious final goal of machine reasoning of anthropogenic climate change.

However, while the field of machine/DL has traditionally progressed very rapidly, we observe that this is not the case in tackling such grand challenges. Cognitive barriers are still on our pathway: domain knowledge is elusive and difficult to encode, interaction between computer scientists and physicists is still complicated, and education in these synergistic concepts will be a hard task to achieve in the upcoming years. The ways forward we have promoted, based on experimental design, hybrid DL, interpretability, and causal discovery, definitely call for an active and continuous interaction between domain knowledge experts and computer scientists. The new era for AI in geoscience is knocking at the door and shouting out: "collaborate, collaborate!"

Bibliography

C.J. Abolt, M.H. Young, A.L. Atchley, and C.J. Wilson. Brief communication: Rapid machine-learning-based extraction and measurement of ice wedge polygons in high-resolution digital elevation models. *The Cryosphere*, 13(1):237–245, 2019. doi: 10.5194/tc-13-237-2019.

C.J. Abolt and M.H. Young. High-resolution mapping of spatial heterogeneity in ice wedge polygon geomorphology near Prudhoe Bay, Alaska. *Scientific Data*, 7(1):87, 2020. doi: 10.1038/s41597-020-0423-9.

D.H. Ackley, G.E. Hinton, and T.J. Sejnowski. A learning algorithm for Boltzmann machines. *Cognitive Science*, 9(1):147–169, 1985.

S.V. Adams, R.W. Ford, M. Hambley, J.M. Hobson, I. Kavčič, C.M. Maynard, T. Melvin, E.H. Müller, S. Mullerworth, A.R. Porter, M. Rezny, B.J. Shipway, and R. Wong. Lfric: Meeting the challenges of scalability and performance portability in weather and climate models. *Journal of Parallel and Distributed Computing*, 132:383–396, 2019. ISSN 0743-7315. doi: https://doi.org/10.1016/j.jpdc.2019.02.007. URL http://www.sciencedirect.com/science/article/pii/S0743731518305306.

S. Agarwal, Y. Furukawa, N. Snavely, I. Simon, B. Curless, S.M. Seitz, and R. Szeliski. Building Rome in a day. *Communications of the ACM*, 54(10):105–112, 2011.

S. Agrawal, L. Barrington, C. Bromberg, J. Burge, C. Gazen, and J. Hickey. Machine learning for precipitation nowcasting from radar images. *arXiv preprint arXiv:1912.12132*, 2019.

M. Aharon, M. Elad, and A. Bruckstein. K -SVD: An algorithm for designing overcomplete dictionaries for sparse representation. *IEEE Transactions on Signal Processing*, 54(11): 4311–4322, Nov 2006. ISSN 1053-587X. doi: 10.1109/TSP.2006.881199.

F. Aires, W.B. Rossow, N.A. Scott, and A. Chédin. Remote sensing from the infrared atmospheric sounding interferometer instrument 2. Simultaneous retrieval of temperature, water vapor, and ozone atmospheric profiles. *Journal of Geophysical Research: Atmospheres*, 107(D22), 2002.

G.C. Allen. Understanding China's AI strategy: Clues to Chinese strategic thinking on artificial intelligence and national security. Technical report, Center for a New American Security, February 2019. URL https://www.cnas.org/publications/reports/understanding-chinas-ai-strategy.

M. Allen. Liability for climate change. *Nature*, 421:891–892, 2003. ISSN 02624079. doi: 10.1016/S0262-4079(10)62047-7.

Deep Learning for the Earth Sciences: A Comprehensive Approach to Remote Sensing, Climate Science, and Geosciences, First Edition. Edited by Gustau Camps-Valls, Devis Tuia, Xiao Xiang Zhu, and Markus Reichstein.

L. Alparone, S. Baronti, A. Garzelli, and F. Nencini. A global quality measurement of pan-sharpened multispectral imagery. *IEEE Geoscience and Remote Sensing Letters*, 1(4):313–317, 2004.

L. Alparone, B. Aiazzi, S. Baronti, A. Garzelli, F. Nencini, and M. Selva. Multispectral and panchromatic data fusion assessment without reference. *Photogrammetric Engineering and Remote Sensing*, 74(2):193–200, 2008.

H. Altwaijry, E. Trulls, J. Hays, P. Fua, and S. Belongie. Learning to match aerial images with deep attentive architectures. In *2016 IEEE Conference on Computer Vision and Pattern Recognition (CVPR)*, June 2016. doi: 10.1109/CVPR.2016.385.

American Meteorological Society. *Atmospheric river*. Glossary of Meteorology, cited 2017a. URL http://glossary.ametsoc.org/wiki/Atmospheric_river.

American Meteorological Society. *Blocking*. Glossary of Meteorology, cited 2017b. URL http://glossary.ametsoc.org/wiki/Blocking.

American Meteorological Society. *Front*. Glossary of Meteorology, cited 2017c. URL http://glossary.ametsoc.org/wiki/Front.

M. Anthimopoulos, S. Christodoulidis, L. Ebner, T. Geiser, A. Christe, and S. Mougiakakou. Semantic segmentation of pathological lung tissue with dilated fully convolutional networks. *IEEE Journal of Biomedical and Health Informatics*, 2018. doi: 10.1109/JBHI.2018.2818620.

E. Aptoula. Remote sensing image retrieval with global morphological texture descriptors. *IEEE Transactions on Geoscience and Remote Sensing*, 52(5):3023–3034, May 2014.

A. Arakawa. The cumulus parameterization problem: Past, present, and future. *J. Clim.*, 17:2493–2525, 2004.

J. Arenas-García, K.B. Petersen, G. Camps-Valls, and L.K. Hansen. Kernel multivariate analysis framework for supervised subspace learning: A tutorial on linear and kernel multivariate methods. *IEEE Signal Processing Magazine*, 30(4):16–29, 2013.

H.M. Arnold, I.M. Moroz, and T.N. Palmer. Stochastic parametrizations and model uncertainty in the Lorenz '96 system. *Philosophical Transactions of the Royal Society A*, 371, 2013. doi: 10.1098/rsta.2011.0479. URL http://rsta.royalsocietypublishing.org/content/371/1991/20110479.short.

M. Aubinet, Q. Hurdebise, H. Chopin, A. Debacq, A. De Ligne, B. Heinesch, T. Manise, and C. Vincke. Inter-annual variability of Net Ecosystem Productivity for a temperate mixed forest: A predominance of carry-over effects? *Agricultural and Forest Meteorology*, 262:340–353, 2018. doi: 10.1016/j.agrformet.2018.07.024.

N. Audebert, B. Le Saux, and S. Lefèvre. Semantic segmentation of earth observation data using multimodal and multi-scale deep networks. In *Asian Conference on Computer Vision (ACCV)*, 2016.

N. Audebert, B. Le Saux, and S. Lefèvre. Beyond RGB: Very high resolution urban remote sensing with multimodal deep networks. *ISPRS Journal of Photogrammetry and Remote Sensing*, 140:20–32, 2018.

N. Audebert, A. Boulch, B. Le Saux, and S. Lefèvre. Distance transform regression for spatially-aware deep semantic segmentation. *Computer Vision and Image Understanding*, 189:102809, 2019a.

N. Audebert, B. Le Saux, and S. Lefèvre. Deep learning for classification of hyperspectral data: A comparative review. *IEEE Geoscience and Remote Sensing Magazine*, 7(2):159–173, 2019b.

N. Audebert, B. Le Saux, and S. Lefevre. Deep learning for classification of hyperspectral data: A comparative review. *IEEE Geoscience and Remote Sensing Magazine*, 7(2):159–173, 2019.

M. Awad. Sea water chlorophyll-a estimation using hyperspectral images and supervised artificial neural network. *Ecological informatics*, 24:60–68, 2014.

G. Ayzel, M. Heistermann, A. Sorokin, O. Nikitin, and O. Lukyanova. All convolutional neural networks for radar-based precipitation nowcasting. *Procedia Computer Science*, 150:186–192, 2019.

A. Azarang, H.E. Manoochehri, and N. Kehtarnavaz. Convolutional autoencoder-based multispectral image fusion. *IEEE access*, 7:35673–35683, 2019.

S.M. Azimi, E. Vig, R. Bahmanyar, M. Körner, and P. Reinartz. Towards multi-class object detection in unconstrained remote sensing imagery. In *ACCV*, pages 150–165, 2018. doi: 10.1007/978-3-030-20893-6_10.

S.M. Azimi, C. Henry, L. Sommer, A. Schumann, and E. Vig. Skyscapes – fine-grained semantic understanding of aerial scenes. In *2019 IEEE/CVF International Conference on Computer Vision (ICCV)*, pages 7392–7402, 2019.

M. Babaeizadeh, C. Finn, D. Erhan, R.H. Campbell, and S. Levine. Stochastic variational video prediction. *arXiv preprint arXiv:1710.11252*, 2017.

M. Babaeizadeh, C. Finn, D. Erhan, R.H. Campbell, and S. Levine. Stochastic variational video prediction. In *6th International Conference on Learning Representations, ICLR 2018*, 2018.

V. Badrinarayanan, A. Kendall, and R. Cipolla. Segnet: A deep convolutional encoder-decoder architecture for image segmentation. *IEEE Transactions on Pattern Analysis and Machine Intelligence*, 39(12):2481–2495, 2017.

E.H. Bair, A.A. Calfa, K. Rittger, and J. Dozier. Using machine learning for real-time estimates of snow water equivalent in the watersheds of Afghanistan. *Cryosphere*, 12(5), 2018.

M. Baktashmotlagh, M. Harandi, B. Lovell, and M. Salzmann. Unsupervised domain adaptation by domain invariant projection. In *International Conference on Computer Vision*, pages 769–776, 2013.

G. Balakrishnan, A. Zhao, M.R. Sabuncu, J. Guttag, and A.V. Dalca. Voxelmorph: A learning framework for deformable medical image registration. *IEEE Transactions on Medical Imaging*, 38(8):1788–1800, Aug 2019. ISSN 1558-254X. doi: 10.1109/TMI.2019.2897538.

P. Baldi and K. Hornik. Neural networks and principal component analysis: Learning from examples without local minima. *Neural Networks*, 2(1):53–58, 1989.

P. Baldi, P. Sadowski, and D. Whiteson. Searching for exotic particles in high-energy physics with deep learning. *Nature Communications*, 5, Jul 2014. ISSN 2041-1723. doi: 10.1038/ncomms5308. URL http://www.nature.com/doifinder/10.1038/ncomms5308.

E.A. Barnes, J. Slingo, and T. Woollings. A methodology for the comparison of blocking climatologies across indices, models and climate scenarios. *Climate Dynamics*, 38(11):2467–2481, Jun 2012. ISSN 1432-0894. doi: 10.1007/s00382-011-1243-6. URL https://doi.org/10.1007/s00382-011-1243-6.

G.A. Barron-Gafford, R.L. Scott, G.D. Jenerette, and T.E. Huxman. The relative controls of temperature, soil moisture, and plant functional group on soil CO_2 efflux at diel, seasonal, and annual scales. *Journal of Geophysical Research: Biogeosciences*, 116, 2011. doi: 10.1029/2010JG001442.

R. Barry and T.Y. Gan. *The Global Cryosphere: Past, Present and Future*. Cambridge University Press, 2011.

P. Bauer, A. Thorpe, and G. Brunet. The quiet revolution of numerical weather prediction. *Nature*, 525, 2015. URL https://doi.org/10.1038/nature14956.

L.E. Baum and T. Petrie. *Statistical Inference for Probabilistic Functions of Finite State Markov Chains.* volume 37, pages 1554–1563. *Institute of Mathematical Statistics*, 1966.

C.A. Baumhoer, A.J. Dietz, C. Kneisel, and C. Kuenzer. Automated extraction of antarctic glacier and ice shelf fronts from Sentinel-1 imagery using deep learning. *Remote Sensing*, 11(21), 2019. doi: 10.3390/rs11212529.

H. Bay, A. Ess, T. Tuytelaars, and L. Van Gool. Speeded-up robust features (SURF). *Computer Vision and Image Understanding*, 110(3):346–359, June 2008. ISSN 1077-3142. doi: 10.1016/j.cviu.2007.09.014. URL https://doi.org/10.1016/j.cviu.2007.09.014.

M. Belkin and P. Niyogi. Laplacian eigenmaps for dimensionality reduction and data representation. *Neural computation*, 15(6):1373–1396, 2003.

J.A. Benediktsson, M. Pesaresi, and K. Arnason. Classification and feature extraction for remote sensing images from urban areas based on morphological transformations. *IEEE Transactions in Geoscience and Remote Sensing*, 41(9):1940–1949, 2003.

S. Bengio, O. Vinyals, N. Jaitly, and N. Shazeer. Scheduled sampling for sequence prediction with recurrent neural networks. In *NIPS*, 2015.

Y. Bengio, T. Deleu, N. Rahaman, R. Ke, S. Lachapelle, O. Bilaniuk, A. Goyal, and C. Pal. A meta-transfer objective for learning to disentangle causal mechanisms. 01 2019.

Y. Bengio, P. Simard, and P. Frasconi. Learning long-term dependencies with gradient descent is difficult. *IEEE Transactions on Neural Networks*, 5(2):157–166, 1994. doi: 10.1109/72.279181.

Y. Bengio, J. Louradour, R. Collobert, and J. Weston. Curriculum learning. In *ICML*, 2009.

Y. Bengio, A.C. Courville, and P. Vincent. Representation learning: A review and new perspectives. *IEEE TPAMI*, 35(8):1798–1828, 2013.

S.G. Benjamin, S.S. Weygandt, J.M. Brown, M. Hu, C.R. Alexander, T.G. Smirnova, J.B. Olson, E.P. James, D.C. Dowell, G.A. Grell, et al. A North American hourly assimilation and model forecast cycle: The rapid refresh. *Monthly Weather Review*, 144(4): 1669–1694, 2016.

C. Bentes, D. Velotto, and S. Lehner. Target classification in oceanographic SAR images with deep neural networks: Architecture and initial results. In *2015 IEEE International Geoscience and Remote Sensing Symposium (IGARSS)*, pages 3703–3706, 2015.

K.J. Bergen, P.A. Johnson, V. Maarten, and G.C. Beroza. Machine learning for data-driven discovery in solid earth geoscience. *Science*, 363(6433):eaau0323, 2019.

P.S. Berloff. Random-forcing model of the mesoscale oceanic eddies. *Journal of Fluid Mechanics*, 529:71–95, 2005.

J.D. Bermudez, P.N. Happ, R.Q. Feitosa, and D.A.B. Oliveira. Synthesis of multispectral optical images from SAR/optical multitemporal data using conditional generative adversarial networks. *IEEE Geoscience and Remote Sensing Letters*, 16(8):1220–1224, Aug 2019.

J. Berner, U. Achatz, L. Batté, L. Bengtsson, A. de la Cámara, Hannah M. Christensen, M. Colangeli, D.R.B. Coleman, D. Crommelin, S.I. Dolaptchiev, C.L.E. Franzke, P. Friederichs, P. Imkeller, H. Järvinen, S. Juricke, V. Kitsios, F. Lott, V. Lucarini, S. Mahajan, T.N. Palmer, C. Penland, M. Sakradzija, J.-S. von Storch, A. Weisheimer, M. Weniger, P.D. Williams, and J.-I. Yano. Stochastic parameterization: Toward a new view of weather and climate models. *Bulletin of the American Meteorological Society*, 98(3):565–588, 2017. doi: 10.1175/BAMS-D-15-00268.1.

S. Besnard, N. Carvalhais, M.A. Arain, A. Black, B. Brede, N. Buchmann, J. Chen, J.G.P.W. Clevers, L.P. Dutrieux, F. Gans, M. Herold, M. Jung, Y. Kosugi, A. Knohl, Beverly E. Law, E. Paul-Limoges, A. Lohila, L. Merbold, O. Roupsard, R. Valentini, S. Wolf, X. Zhang, and M. Reichstein. Memory effects of climate and vegetation affecting net ecosystem CO_2 fluxes in global forests. *PLoS ONE*, 14:e0211510, 2019. doi: 10.1371/journal.pone.0211510.

A. Bettge, R. Roscher, and S. Wenzel. Deep self-taught learning for remote sensing image classification. In *ESA Big Data from Space*, 2017. accepted.

T. Beucler, M. Pritchard, S. Rasp, P. Gentine, J. Ott, and P. Baldi. Enforcing analytic constraints in neural-networks emulating physical systems. Sep 2019. URL http://arxiv.org/abs/1909.00912.

T. Beucler, M. Pritchard, P. Gentine, and S. Rasp. Towards physically-consistent, data-driven models of convection. Feb 2020. URL http://arxiv.org/abs/2002.08525.

J. Biercamp, P. Bauer, P. Dueben, and B. Lawrence. A roadmap to the implementation of 1km earth system model ensembles. *ESiWACE Deliverable*, D1.2, 2019.

J.M. Bioucas-Dias, A. Plaza, N. Dobigeon, M. Parente, Q. Du, P. Gader, and J. Chanussot. Hyperspectral unmixing overview: Geometrical, statistical, and sparse regression-based approaches. *IEEE JSTARS*, 5(2):354–379, 2012.

C.M. Bishop. Pattern recognition. *Machine Learning*, 128, 2006.

B.K. Biskaborn, S.L. Smith, J. Noetzli, H. Matthes, G. Vieira, D.A. Streletskiy, P. Schoeneich, V.E. Romanovsky, A.G. Lewkowicz, A. Abramov, et al. Permafrost is warming at a global scale. *Nature Communications*, 10(1):264, 2019.

M. Bocquet, J. Brajard, A. Carrassi, and L. Bertino. Data assimilation as a learning tool to infer ordinary differential equation representations of dynamical models. *Nonlinear Processes in Geophysics*, 26(3):143–162, 2019. doi: 10.5194/npg-26-143-2019. URL https://www.nonlin-processes-geophys.net/26/143/2019/.

M. Bocquet, J. Brajard, A. Carrassi, and L. Bertino. Bayesian inference of chaotic dynamics by merging data assimilation, machine learning and expectation-maximization. *Foundations of Data Science*, 2(1):55–80, 2020. doi: 10.3934/fods.2020004.

C. Boening, J.K. Willis, F.W. Landerer, R.S. Nerem, and J. Fasullo. The 2011 la niña: So strong, the oceans fell. *Geophysical Research Letters*, 39, 2012. doi: 10.1029/2012GL053055.

J. Bolibar, A. Rabatel, I. Gouttevin, C. Galiez, T. Condom, and E. Sauquet. Deep learning applied to glacier evolution modelling. *The Cryosphere*, 14(2):565–584, 2020. doi: 10.5194/tc-14-565-2020.

T. Bolton and L. Zanna. Applications of deep learning to ocean data inference and subgrid parameterization. *Journal of Advances in Modeling Earth Systems*, 11(1):376–399, 2019.

D. Bonafilia, B. Tellman, T. Anderson, and E. Issenberg. Sen1floods11: A georeferenced dataset to train and test deep learning flood algorithms for Sentinel-1. In *The IEEE/CVF Conference on Computer Vision and Pattern Recognition (CVPR)Workshops*, Jun 2020.

C.W. Böning, E. Behrens, A. Biastoch, K. Getzlaff, and J.L. Bamber. Emerging impact of Greenland meltwater on deepwater formation in the North Atlantic Ocean. *Nature Geoscience*, 9(7):523–527, 2016.

S. Bony, B. Stevens, D.M.W. Frierson, C. Jakob, M. Kageyama, R. Pincus, T.G. Shepherd, S.C. Sherwood, A.P. Siebesma, A.H. Sobel, et al. Clouds, circulation and climate sensitivity. *Nature Geoscience*, 8(4):261–268, 2015.

Y. Boualleg and M. Farah. Enhanced interactive remote sensing image retrieval with scene classification convolutional neural networks model. In *IEEE International Geoscience and Remote Sensing Symposium*, pages 4748–4751, July 2018.

O. Boucher, D. Randall, P. Artaxo, C. Bretherton, G. Feingold, P. Forster, V.-M. Kerminen, Y. Kondo, H. Liao, U. Lohmann, P. Rasch, S.K. Satheesh, S. Sherwood, B. Stevens, and X.Y. Zhang. Clouds and aerosols. In *Climate Change 2013: The Physical Science Basis. Contribution of Working Group I to the Fifth Assessment Report of the Intergovernmental Panel on Climate Change* [T.F. Stocker, D. Qin, G.-K. Plattner, M.M.B. Tignor, S.K. Allen, J. Boschung, A. Nauels, Y. Xia, V. Bex and P.M. Midgley (eds.)]. Cambridge University Press, Cambridge, United Kingdom and New York, NY, USA, 2013.

A. Boulch. Generalizing discrete convolutions for unstructured point clouds. In *Eurographics 3DOR*, April 2019.

A. Boulch and R. Marlet. Deep learning for robust normal estimation in unstructured point clouds. *Computer Graphics Forum*, 2016.

A. Boulch, J. Guerry, Be. Le Saux, and N. Audebert. SnapNet: 3d point cloud semantic labeling with 2d deep segmentation networks. *Computers & Graphics*, 71:189–198, 2018. ISSN 00978493.

H. Boulze, A. Korosov, and J. Brajard. Classification of sea ice types in Sentinel-1 SAR data using convolutional neural networks. *Remote Sensing*, 12(13):2165, 2020.

H. Bourlard and Y. Kamp. Auto-association by multilayer perceptrons and singular value decomposition. *Biological Cybernetics*, 59(4–5):291–294, 1988.

B.D. Bowes, J.M. Sadler, M.M. Morsy, M. Behl, and J.L. Goodall. Forecasting groundwater table in a flood prone coastal city with long short-term memory and recurrent neural networks. *Water*, 11 (5):1098, May 2019. ISSN 2073-4441. doi: 10.3390/w11051098. URL http://dx.doi .org/10.3390/w11051098.

A. Braakmann-Folgmann and C. Donlon. Estimating snow depth on arctic sea ice using satellite microwave radiometry and a neural network. *The Cryosphere*, 13(9):2421–2438, 2019. doi: 10.5194/tc-13-2421-2019.

J. Brajard, A. Carrassi, M. Bocquet, and L. Bertino. Combining data assimilation and machine learning to emulate a dynamical model from sparse and noisy observations: a case study with the Lorenz 96 model. *Geoscientific Model Development Discussions*, pages 1–21, May 2019. ISSN 1991-962X. doi: 10.5194/gmd-2019-136. URL https://www.geosci-model-dev-discuss.net/gmd-2019-136/.

N.D. Brenowitz and C.S. Bretherton. Prognostic validation of a neural network unified physics parameterization. *Geophysical Research Letters*, 45(12):6289–6298, 2018.

N.D. Brenowitz and C.S. Bretherton. Spatially extended tests of a neural network parametrization trained by coarse-graining. Apr 2019. URL http://arxiv.org/abs/1904.03327.

N.D. Brenowitz, T. Beucler, M. Pritchard, and C.S. Bretherton. Interpreting and stabilizing machine-learning parametrizations of convection. *arXiv preprint arXiv:2003.06549*, 2020.

H. Bristow, A. Eriksson, and S. Lucey. Fast convolutional sparse coding. In *Proceedings of CVPR*, pages 391–398, 2013.

G. Buchsbaum. A spatial processor model for object colour perception. *Journal of the Franklin Institute*, 310(1):1–26, 1980.

M. Buchwitz, M. Reuter, O. Schneising, W. Hewson, R.G. Detmers, H. Boesch, O.P. Hasekamp, I. Aben, H. Bovensmann, J.P. Burrows, et al. Global satellite observations of column-averaged carbon dioxide and methane: The GHG-CCI XCO_2 and XCH_4 CRDP3 data set. *Remote Sensing of Environment*, 203:276–295, 2017.

W. Buermann, M. Forkel, M. O'Sullivan, S. Sitch, P. Friedlingstein, V. Haverd, A.K. Jain, E. Kato, M. Kautz, S. Lienert, D. Lombardozzi, J.E.M.S. Nabel, H. Tian, A.J. Wiltshire, D. Zhu, W.K. Smith, and A.D. Richardson. Widespread seasonal compensation effects of spring warming on northern plant productivity. *Nature*, 562:110, 2018. doi: 10.1038/s41586-018-0555-7.

M. Bujisic, V. Bogicevic, H.G. Parsa, V. Jovanovic, and A. Sukhu. It's raining complaints! How weather factors drive consumer comments and word-of-mouth. *Journal of Hospitality & Tourism Research*, 43(5): 656–681, 2019.

W. Burger and M.J. Burge. *Image Matching and Registration*, pages 565–585. Springer London, London, 2016. ISBN 978-1-4471-6684-9. doi: 10.1007/978-1-4471-6684-9_23. URL https://doi.org/10.1007/978-1-4471-6684-9_23.

A. Buslaev, A. Parinov, E. Khvedchenya, V.I. Iglovikov, and A.A. Kalinin. Albumentations: Fast and flexible image augmentations. *ArXiv e-prints*, 2018.

T. Bürgmann, W. Koppe, and M. Schmitt. Matching of terrasar-x derived ground control points to optical image patches using deep learning. *ISPRS Journal of Photogrammetry and Remote Sensing*, 158:241–248, 2019. ISSN 0924-2716. doi: https://doi.org/10.1016/j.isprsjprs.2019.09.010.

M. Calonder, V. Lepetit, M. Ozuysal, T. Trzcinski, C. Strecha, and P. Fua. Brief: Computing a local binary descriptor very fast. *IEEE Transactions on Pattern Analysis and Machine Intelligence*, 34(7):1281–1298, July 2012. ISSN 1939-3539. doi: 10.1109/TPAMI.2011.222.

M. Camporese, C. Paniconi, M. Putti, and S. Orlandini. Surface-subsurface flow modeling with path-based runoff routing, boundary condition-based coupling, and assimilation of multisource observation data. *Water Resources Research*, 46(2):W02512, Feb 2010. ISSN 0043-1397. doi: 10.1029/2008WR007536. URL http://www.agu.org/pubs/crossref/2010/2008WR007536.shtml.

M. Campos-Taberner, A. Romero-Soriano, C. Gatta, G. Camps-Valls, A. Lagrange, B. Le Saux, A. Beaupère, A. Boulch, A. Chan-Hon-Tong, S. Herbin, H. Randrianarivo, M. Ferecatu, M. Shimoni, G. Moser, and D. Tuia. Processing of extremely high resolution LiDAR and RGB data: Outcome of the 2015 IEEE GRSS Data Fusion Contest. Part A: 2D contest. *IEEE Journal of Selected Topics in Applied Earth Observations and Remote Sensing*, 9(12): 5547–5559, 2016.

G. Camps-Valls, L. Gómez-Chova, J. Muñoz-Marí, J. Vila-Francés, and J. Calpe-Maravilla. Composite kernels for hyperspectral image classification. *IEEE Geoscience and Remote Sensing Letters*, 3(1):93–97, 2006.

G. Camps-Valls, D. Svendsen, L. Martino, J. Muñoz-Marí, V. Laparra, M. Campos-Taberner, and D. Luengo. Physics-aware Gaussian processes in remote sensing. *Applied Soft Computing*, 68:69–82, Jul 2018a. doi: https://doi.org/10.1016/j.asoc.2018.03.021.

G. Camps-Valls, J. Verrelst, J. Munoz-Mari, V. Laparra, F. Mateo-Jimenez, and J. Gomez-Dans. A survey on Gaussian processes for earth-observation data analysis: A comprehensive investigation. *IEEE Geoscience and Remote Sensing Magazine*, 4(2):58–78, 2016.

G. Camps-Valls, L. Martino, D.H. Svendsen, M. Campos-Taberner, J. Muñoz-Marí, V. Laparra, D. Luengo, and F.J. García-Haro. Physics-aware Gaussian processes in remote sensing. *Applied Soft Computing*, 68:69–82, 2018b.

G. Camps-Valls, D. Sejdinovic, J. Runge, and M. Reichstein. A perspective on Gaussian processes for earth observation. *National Science Review*, 2019.

R. Cao, Q. Zhang, J. Zhu, Q. Li, Q. Li, B. Liu, and G. Qiu. Enhancing remote sensing image retrieval using a triplet deep metric learning network. *International Journal of Remote Sensing*, 41(2):740–751, January 2020.

Y. Cao, B. Liu, M. Long, and J. Wang. Hashgan: Deep learning to hash with pair conditional wasserstein gan. In *IEEE Conference on Computer Vision and Pattern Recognition*, pages 1287–1296, June 2018.

Y. Cao, Q. Li, L. Chen, J. Zhang, and L. Ma. Video prediction for precipitation nowcasting. *arXiv preprint arXiv:1907.08069*, 2019.

M. Castelluccio, G. Poggi, C. Sansone, and L. Verdoliva. Land use classification in remote sensing images by convolutional neural networks. *ArXiv e-prints*, 2015.

J. Castillo-Navarro, N. Audebert, A. Boulch, B. Le Saux, and S. Lefèvre. Semi-supervised semantic segmentation in earth observation: the minifrance suite, dataset analysis and multi-task network study. *To appear*, 2020. URL http://dx.doi.org/10.21227/b9pt-8x03.

W.E. Chapman, A.C. Subramanian, S.P. Delle Monache, L. and Xie, and F. M. Ralph. Improving atmospheric river forecasts with machine learning. *Geophysical Research Letters*, 46(17-18):10627–10635, 2019.

A. Chattopadhyay, P. Hassanzadeh, K. Palem, and D. Subramanian. Data-driven prediction of a multi-scale Lorenz 96 chaotic system using a hierarchy of deep learning methods: Reservoir computing, ANN, and RNN-LSTM. *EarthArxiv*, 2019. doi: 10.31223/osf.io/fbxns.

A. Chattopadhyay, A. Subel, and P. Hassanzadeh. Data-driven super-parameterization using deep learning: Experimentation with multi-scale Lorenz 96 systems and transfer-learning, 2020.

B. Chaudhuri, B. Demir, L. Bruzzone, and S. Chaudhuri. Region-based retrieval of remote sensing images using an unsupervised graph-theoretic approach. *IEEE Geoscience and Remote Sensing Letters*, 13(7):987–991, July 2016.

B. Chaudhuri, B. Demir, S. Chaudhuri, and L. Bruzzone. Multilabel remote sensing image retrieval using a semisupervised graph-theoretic method. *IEEE Transactions on Geoscience and Remote Sensing*, 56(2):1144–1158, February 2018.

U. Chaudhuri, B. Banerjee, and A. Bhattacharya. Siamese graph convolutional network for content based remote sensing image retrieval. *Computer Vision and Image Understanding*, 184:22–30, July 2019.

C. Chen, Y. Li, W. Liu, and Z. Huang. Sirf: Simultaneous satellite image registration and fusion in a unified framework. *IEEE Transactions on Image Processing*, 24(11):4213–4224, 2015.

J. Chen and A. Zipf. Deepvgi: Deep learning with volunteered geographic information. In *Proceedings of the 26th International Conference on World Wide Web Companion*, pages 771–772, 2017.

L.-C. Chen, G. Papandreou, I. Kokkinos, K. Murphy, and A.L. Yuille. Deeplab: Semantic image segmentation with deep convolutional nets, atrous convolution, and fully connected crfs. *IEEE Transactions on Pattern Analysis and Machine Intelligence*, 40(4):834–848, 2017a.

L.-C. Chen, Y. Zhu, G. Papandreou, F. Schroff, and H. Adam. Encoder-decoder with atrous separable convolution for semantic image segmentation. In *European Conference on Computer Vision*, 2018a.

L.-C Chen, Y. Cao, L. Ma, and J. Zhang. A deep learning based methodology for precipitation nowcasting with radar. *Earth and Space Science*, page e2019EA000812, 2019.

S. Chen and D. Zhang. Semisupervised dimensionality reduction with pairwise constraints for hyperspectral image classification. *IEEE Geoscience and Remote Sensing Letters*, 8(2):369–373, Mar 2011.

S. Chen, H. Wang, F. Xu, and Y. Jin. Target classification using the deep convolutional networks for SAR images. *IEEE Transactions on Geoscience and Remote Sensing*, 54(8):4806–4817, 2016.

S. Chen, X. Li, Y. Zhang, R. Feng, and Ch. Zhang. Local deep hashing matching of aerial images based on relative distance and absolute distance constraints. *Remote Sensing*, 9(12), 2017b. ISSN 2072-4292. doi: 10.3390/rs9121244.

S. Chen, X. Yuan, W. Yuan, J. Niu, F. Xu, and Y. Zhang. Matching multi-sensor remote sensing images via an affinity tensor. *Remote Sensing*, 10(7), 2018c. ISSN 2072-4292. doi: 10.3390/rs10071104. URL https://www.mdpi.com/2072-4292/10/7/1104.

T. Chen, S. Kornblith, M. Norouzi, and G. Hinton. A simple framework for contrastive learning of visual representations, 2020. URL http://arxiv.org/abs/2002.05709.

X. Chen, Y. Duan, R. Houthooft, J. Schulman, I. Sutskever, and P. Abbeel. Infogan: Interpretable representation learning by information maximizing generative adversarial nets. In D.D. Lee, M. Sugiyama, U.V. Luxburg, I. Guyon, and R. Garnett, editors, *Advances in Neural Information Processing Systems 29*, pages 2172–2180. Curran Associates, Inc., 2016.

Y. Chen, Z. Lin, X. Zhao, G. Wang, and Y. Gu. Deep learning-based classification of hyperspectral data. *IEEE Journal of Selected Topics in Applied Earth Observations and Remote Sensing*, 7(6):2094–2107, 2014.

Y. Chen, J. Mairal, Z. Harchaoui, et al. Fast and robust archetypal analysis for representation learning. In *CVPR 2014-IEEE Conference on Computer Vision & Pattern Recognition*, 2014.

G. Cheng, J. Han, and X. Lu. Remote sensing image scene classification: Benchmark and state of the art. *Proceedings of the IEEE*, 105(10):1865–1883, 2017.

G. Cheng, P. Zhou, and J. Han. Learning rotation-invariant convolutional neural networks for object detection in VHR optical remote sensing images. *IEEE Transactions in Geoscience and Remote Sensing*, 54(12):7405–7415, 2016a.

G. Cheng, P. Zhou, and J. Han. RIFD-CNN: Rotation-invariant and fisher discriminative convolutional neural networks for object detection. In *CVPR*, pages 2884–2893, 2016b.

Y. Cheng, M. Giometto, P. Kauffmann, L. Lin, C. Cao, C. Zupnick, H. Li, Q. Li, R. Abernathey, and P. Gentine. Deep learning for subgrid-scale turbulence modeling in large-eddy simulations of the atmospheric boundary layer. *arXiv preprint arXiv:1910.12125*, 2019.

A.M. Cheriyadat. Unsupervised feature learning for aerial scene classification. *IEEE Transactions in Geoscience and Remote Sensing*, 52(1):439–451, Jan 2014. ISSN 0196-2892. doi: 10.1109/TGRS.2013.2241444.

F. Chevallier, F. Chéruy, N.A. Scott, and A. Chédin. A neural network approach for a fast and accurate computation of a longwave radiative budget. *Journal of Applied Meteorology*, 37(11):1385–1397, 1998. doi: 10.1175/1520-0450(1998)037⟨1385:ANNAFA⟩2.0.CO;2.

G. Chevillon. Direct multi-step estimation and forecasting. *Journal of Economic Surveys*, 21(4):746–785, 2007.

J. Chi and H. Kim. Prediction of Arctic sea ice concentration using a fully data driven deep neural network. *Remote Sensing*, 9(12), 2017. doi: 10.3390/rs9121305.

J. Chi, H. Kim, S. Lee, and M.M. Crawford. Deep learning based retrieval algorithm for arctic sea ice concentration from AMSR2 passive microwave and modis optical data. *Remote Sensing of Environment*, 231:111204, 2019. doi: https://doi.org/10.1016/j.rse.2019.05.023.

M.T. Chiu, X. Xu, Y. Wei, Z. Huang, A. Schwing, R. Brunner, H. Khachatrian, H. Karapetyan, I. Dozier, G. Rose, D. Wilson, A. Tudor, N. Hovakimyan, T.S. Huang, and H. Shi. *Agriculture-vision: A large aerial image database for agricultural pattern analysis*, 2020.

H. Cho, U. Choi, and H. Park. Deep learning application to time-series prediction of daily chlorophyll-a concentration. *WIT Transactions on Ecology and the Environment*, 215:157–63, 2018.

K. Cho, B. Van Merriënboer, C. Gulcehre, D. Bahdanau, F. Bougares, H. Schwenk, and Y. Bengio. Learning phrase representations using rnn Encoder-Decoder for Statistical machine translation. *arXiv preprint arXiv:1406.1078*, 2014.

F. Chollet. Xception: Deep learning with depthwise separable convolutions. In *IEEE Conference on Computer Vision and Pattern Recognition*, 2017.

S. Christodoulidis, M. Sahasrabudhe, M. Vakalopoulou, G. Chassagnon, M-P Revel, S. Mougiakakou, and N. Paragios. Linear and deformable image registration with 3D convolutional neural networks. In *Image Analysis for Moving Organ, Breast, and Thoracic Images*, 2018.

J. Chung, C. Gulcehre, K. Cho, and Y. Bengio. Empirical evaluation of gated recurrent neural networks on sequence modeling. In *Conference on Neural Information Processing Systems (NIPS) Workshop on Deep Learning*, 2014.

B.B. Çiftçi, S. Kuter, Z. Akyürek, and G.W. Weber. Fractional snow cover mapping by artificial neural networks and support vector machines. *ISPRS Annals of the Photogrammetry, Remote Sensing and Spatial Information Sciences*, 4:179, 2017.

A. Coates and A. Ng. The importance of encoding versus training with sparse coding and vector quantization. In *ICML*, pages 921–928, 2011.

J. Collins, J. Sohl-Dickstein, and D. Sussillo. Capacity and trainability in recurrent neural networks. In *International Conference on Learning Representations (ICLR)*, 2017.

C. L.V. Cooke and K.A. Scott. Estimating sea ice concentration from SAR: Training convolutional neural networks with passive microwave data. *IEEE Transactions on Geoscience and Remote Sensing*, 57(7):4735–4747, Jul 2019. doi: 10.1109/TGRS.2019.2892723.

F.C. Cooper and L. Zanna. Optimisation of an idealized ocean model, stochastic parameterisation of sub-grid eddies. *Ocean Modelling*, 88: 38–53, 2015. ISSN 14635003. doi: 10.1016/j.ocemod.2014.12.014. URL http://dx.doi.org/10.1016/j.ocemod.2014.12.014.

D. Coppin and S. Bony. Internal variability in a coupled general circulation model in radiative-convective equilibrium. *Geophysical Research Letters*, 44(10):5142–5149, 2017.

G.W. Cottrell and J.D. Willen. Image compression within visual system constraints. *Neural Networks*, 1:487, 1988.

N. Courty, R. Flamary, D. Tuia, and A. Rakotomamonjy. Optimal transport for domain adaptation. *IEEE Transactions on Pattern Analysis and Machine Intelligence*, 39(9): 1853–1865, 2017.

M.A. Cruz, R.L. Thompson, L.E.B. Sampaio, and R.D.A. Bacchi. The use of the Reynolds force vector in a physics informed machine learning approach for predictive turbulence modeling. *Computers & Fluids*, page 104258, 2019.

B.C. Csáji et al. Approximation with artificial neural networks. *Faculty of Sciences*, Etvs Lornd University, Hungary, 24(48):7, 2001.

G. Csurka. *Domain Adaptation in Computer Vision Applications*. Springer, 2017.

A. Cutler and L. Breiman. Archetypal analysis. *Technometrics*, 36:338–347, 1994.

C. Cuttler, R.S. Jhangiani, and D.C. Leighton. Research methods in psychology-open textbook library. 2020.

D. Dai and W. Yang. Satellite image classification via two-layer sparse coding with biased image representation. *IEEE Geoscience and Remote Sensing Letters*, 8(1):173–176, 2010.

J. Dai, Y. Li, K. He, and J. Sun. R-FCN: object detection via region-based fully convolutional networks. In *NIPS*, pages 379–387, 2016.

J. Dai, H. Qi, Y. Xiong, Y. Li, G. Zhang, H. Hu, and Y. Wei. Deformable convolutional networks. *CoRR, abs/1703.06211*, 1 (2):3, 2017.

O.E. Dai, B. Demir, B. Sankur, and L. Bruzzone. A novel system for content-based retrieval of single and multi-label high-dimensional remote sensing images. *IEEE Journal of Selected Topics in Applied Earth Observations and Remote Sensing*, 11(7):2473–2490, July 2018.

N. Dalal and B. Triggs. Histograms of oriented gradients for human detection. In *2005 IEEE Computer Society Conference on Computer Vision and Pattern Recognition (CVPR'05)*, volume 1, pages 886–893. IEEE, 2005.

B.B. Damodaran, B. Kellenberger, R. Flamary, D. Tuia, and N. Courty. DeepJDOT: Deep joint distribution optimal transport for unsupervised domain adaptation. In *European Conference on Computer Vision*, pages 467–483. Springer International Publishing, 2018.

G. Danabasoglu, J.C. McWilliams, and P.R. Gent. The role of mesoscale tracer transports in the global ocean circulation. *Science*, 264(5162): 1123–1126, 1994.

R.C. Daudt, B. Le Saux, A. Boulch, and Y. Gosseau. Multitask learning for large-scale semantic change detection. *Computer Vision and Image Understanding*, 187:102783, 2019.

R. Caye Daudt, B. Le Saux, A. Boulch, and Y. Gousseau. Urban change detection for multispectral earth observation using convolutional neural networks. In *IEEE International Geoscience and Remote Sensing Symposium (IGARSS)*, Valencia, Spain, 2018.

D.T. Davis, Z. Chen, L. Tsang, J.-N. Hwang, and A.T.C. Chang. Retrieval of snow parameters by iterative inversion of a neural network. *IEEE Transactions on Geoscience and Remote Sensing*, 31(4):842–852, 1993.

E. De Bézenac, A. Pajot, and P. Gallinari. Towards a hybrid approach to physical process modeling. Technical report, 2017.

E. de Bezenac, A. Pajot, and P. Gallinari. Deep learning for physical processes: Incorporating prior scientific knowledge. *Journal of Statistical Mechanics: Theory and Experiment*, 2019(12):124009, 2019.

A. de la Fuente, V. Meruane, and C. Meruane. Hydrological early warning system based on a deep learning runoff model coupled with a meteorological forecast. *Water*, 11(9):1808, Aug 2019. ISSN 2073-4441. doi: 10.3390/w11091808. URL http://dx.doi.org/10.3390/w11091808.

G.J.M. De Lannoy, R.H. Reichle, P.R. Houser, V.R.N. Pauwels, and N.E.C. Verhoest. Correcting for forecast bias in soil moisture assimilation with the ensemble Kalman filter. *Water*

Resources Research, 43(9), Sep 2007. ISSN 00431397. doi: 10.1029/2006WR005449. URL http://doi.wiley.com/10.1029/2006WR005449.

D.P. Dee, S.M. Uppala, A.J. Simmons, P. Berrisford, P. Poli, S. Kobayashi, U. Andrae, M.A. Balmaseda, G. Balsamo, P. Bauer, P. Bechtold, A.C.M. Beljaars, L. van de Berg, J. Bidlot, N. Bormann, C. Delsol, R. Dragani, M. Fuentes, A.J. Geer, L. Haimberger, S.B. Healy, H. Hersbach, E. V. Hólm, L. Isaksen, P. Kållberg, M. Köhler, M. Matricardi, A.P. McNally, B.M. Monge-Sanz, J.-J. Morcrette, B.-K. Park, C. Peubey, P. de Rosnay, C. Tavolato, J.-N. Thépaut, and F. Vitart. The ERA-Interim reanalysis: configuration and performance of the data assimilation system. *Quarterly Journal of the Royal Meteorological Society*, 137(656): 553–597, Apr 2011. ISSN 00359009. doi: 10.1002/qj.828. URL http://doi.wiley.com/10.1002/qj.828.

B. Demir and L. Bruzzone. A novel active learning method in relevance feedback for content-based remote sensing image retrieval. *IEEE Transactions on Geoscience and Remote Sensing*, 53(5):2323–2334, May 2015.

B. Demir and L. Bruzzone. Hashing-based scalable remote sensing image search and retrieval in large archives. *IEEE Transactions on Geoscience and Remote Sensing*, 54(2):892–904, February 2016.

I. Demir, K. Koperski, D. Lindenbaum, G. Pang, J. Huang, S. Basu, F. Hughes, D. Tuia, and R. Raskar. Deepglobe 2018: A challenge to parse the earth through satellite images. *2018 IEEE/CVF Conference on Computer Vision and Pattern Recognition Workshops (CVPRW)*, Jun 2018. doi: 10.1109/cvprw.2018.00031. URL http://dx.doi.org/10.1109/CVPRW.2018.00031.

J. Deng, W. Dong, R. Socher, L.J. Li, K. Li, and L. Fei-Fei. ImageNet: A large-scale hierarchical image database. In *IEEE Conference on Computer Vision and Pattern Recognition*, pages 248–255. IEEE, 2009.

J. Deng, Z. Zhang, E. Marchi, and B. Schuller. Sparse autoencoder-based feature transfer learning for speech emotion recognition. In *2013 Humaine Association Conference on Affective Computing and Intelligent Interaction*, pages 511–516. IEEE, 2013.

Z. Deng, H. Sun, S. Zhou, J. Zhao, and H. Zou. Toward fast and accurate vehicle detection in aerial images using coupled region-based convolutional neural networks. *J-STARS*, 10(8):3652–3664, 2017.

Z. Deng, H. Sun, S. Zhou, and J. Zhao. Learning deep ship detector in SAR images from scratch. *IEEE Transactions on Geoscience and Remote Sensing*, 57(6):4021–4039, 2019.

E. Denton and R. Fergus. Stochastic video generation with a learned prior. In *International Conference on Machine Learning*, pages 1174–1183, 2018.

R. Dian, S. Li, A. Guo, and L. Fang. Deep hyperspectral image sharpening. *IEEE transactions on neural networks and learning systems*, 29(11):5345–5355, 2018.

J. Ding, N. Xue, Y. Long, G.-S. Xia, and Q. Lu. Learning roi transformer for oriented object detection in aerial images. In *CVPR*, pages 2849–2858, 2019.

I.D. Dobreva and A.G. Klein. Fractional snow cover mapping through artificial neural network analysis of modis surface reflectance. *Remote Sensing of Environment*, 115(12):3355–3366, 2011.

C. Doersch. Tutorial on variational autoencoders, 2016. URL http://arxiv.org/abs/1606.05908. cite arxiv:1606.05908.

C. Doersch, A. Gupta, and A.A. Efros. Unsupervised visual representation learning by context prediction. In *International Conference on Computer Vision (ICCV)*, 2015.

J. Donahue, L.A. Hendricks, S. Guadarrama, M. Rohrbach, S. Venugopalan, K. Saenko, and T. Darrell. Long-term recurrent convolutional networks for visual recognition and description. In *Proceedings of the IEEE Conference on Computer Vision and Pattern Recognition*, pages 2625–2634, 2015.

G. Dong, W. Huang, W.A.P. Smith, and P. Ren. A shadow constrained conditional generative adversarial net for srtm data restoration. *Remote Sensing of Environment*, 237:111602, 2020.

J. Dong, R. Yin, X. Sun, Q. Li, Y. Yang, and X. Qin. Inpainting of remote sensing sst images with deep convolutional generative adversarial network. *IEEE Geoscience and Remote Sensing Letters*, 16(2):173–177, Feb 2019.

Y. Dong, W. Jiao, T. Long, L. Liu, G. He, Ch. Gong, and Y. Guo. Local deep descriptor for remote sensing image feature matching. *Remote Sensing*, 11(4), 2019. ISSN 2072-4292. doi: 10.3390/rs11040430.

R.H. Douglas. The stormy weather group (Canada). In *Radar in Meteorology*, pages 61–68. Springer, 1990.

J. Duchi, E. Hazan, and Y. Singer. Adaptive subgradient methods for online learning and stochastic optimization. *Journal of Machine Learning Research*, 12(Jul):2121–2159, 2011.

P. Dueben, P. Bauer, J.-N. Thepaut, V.-H. Peuch, A. Geer, and S. English. Machine learning at ECMWF. *ECMWF Memorandum*, 2019.

P.D. Dueben and P. Bauer. Challenges and design choices for global weather and climate models based on machine learning. *Geoscientific Model Development*, 11(10):3999–4009, Oct 2018. ISSN 19919603. doi: 10.5194/gmd-11-3999-2018.

V. Dumoulin and F. Visin. A guide to convolution arithmetic for deep learning. *arXiv preprint arXiv:1603.07285*, 2016.

K. Duraisamy, G. Iaccarino, and H. Xiao. Turbulence modeling in the age of data. *Annual Review of Fluid Mechanics*, 51(1):357–377, 2019. doi: 10.1146/annurev-fluid-010518-040547.

M. Elad. *Sparse and Redundant Representations: From Theory to Applications in Signal and Image Processing*. Springer Science & Business Media, 2010.

M. Elad and M. Aharon. Image denoising via sparse and redundant representations over learned dictionaries. *IEEE Transactions on Image processing*, 15(12):3736–3745, 2006.

J.L. Elman. Finding structure in time. *Cognitive Science*, 14(2): 179–211, 1990. doi: 10.1016/0364-0213(90)90002-E.

A. Elshamli, G.W. Taylor, A. Berg, and S. Areibi. Domain adaptation using representation learning for the classification of remote sensing images. *IEEE Journal of Selected Topics in Applied Earth Observations and Remote Sensing*, 10(9):4198–4209, Sep. 2017.

S. En, A. Lechervy, and Fr. Jurie. Ts-net: Combining modality specific and common features for multimodal patch matching. In *2018 25th IEEE International Conference on Image Processing (ICIP)*, pages 3024–3028, 2018. doi: 10.1109/icip.2018.8451804.

G. Evensen. Sequential data assimilation with a nonlinear quasi-geostrophic model using Monte Carlo methods to forecast error statistics. *Journal of Geophysical Research*, 99(C5):10143, 1994. ISSN 0148-0227. doi: 10.1029/94JC00572. URL http://doi.wiley.com/10.1029/94JC00572.

M. Everingham, L. Van Gool, C. Williams, J. Winn, and A. Zisserman. The pascal visual object classes (VOC) challenge. *International Journal of Computer Vision*, 88(2):303–338, 2010.

G. Eynard-Bontemps, R. Abernathey, J. Hamman, A. Ponte, and W. Rath. The pangeo big data ecosystem and its use at cnes. In *Big Data from Space (BiDS'19). … Turning Data into insights…19-21 February 2019, Munich, Germany*, 2019.

V. Eyring, S. Bony, G.A. Meehl, C.A. Senior, B. Stevens, R.J. Stouffer, and K.E. Taylor. Overview of the coupled model intercomparison project phase 6 (cmip6) experimental design and organization. *Geoscientific Model Development*, 9(5):1937–1958, 2016a. doi: 10.5194/gmd-9-1937-2016. URL https://www.geosci-model-dev.net/9/1937/2016/.

V. Eyring, S. Bony, G.A. Meehl, C.A. Senior, B. Stevens, R.J. Stouffer, and K.E. Taylor. Overview of the coupled model intercomparison project phase 6 (cmip6) experimental design and organization. *Geoscientific Model Development (Online)*, 9 (LLNL-JRNL-736881), 2016b.

V. Eyring, M. Righi, A. Lauer, M. Evaldsson, S. Wenzel, C. Jones, Alessandro Anav, O. Andrews, I. Cionni, and E.L. Davin. ESMValTool (v1. 0) – a community diagnostic and performance metrics tool for routine evaluation of Earth system models in CMIP. *Geoscientific Model Development*, 9:1747–1802, 2016c.

S. Falkner, A. Klein, and F. Hutter. Bohb: Robust and efficient hyperparameter optimization at scale. *arXiv preprint arXiv:1807.01774*, 2018.

H. Fan, M. Jiang, L. Xu, H. Zhu, J. Cheng, and J. Jiang. Comparison of long short term memory networks and the hydrological model in runoff simulation. *Water*, 12(1):175, Jan 2020. ISSN 2073-4441. doi: 10.3390/w12010175. URL http://dx.doi.org/10.3390/w12010175.

K. Fang and C. Shen. Near-real-time forecast of satellite-based soil moisture using long short-term memory with an adaptive data integration kernel. *Journal of Hydrometeorology*, pages JHM–D–19–0169.1, Jan 2020. ISSN 1525-755X. doi: 10.1175/JHM-D-19-0169.1. URL http://journals.ametsoc.org/doi/10.1175/JHM-D-19-0169.1.

K. Fang, C. Shen, D. Kifer, and X. Yang. Prolongation of SMAP to spatio-temporally seamless coverage of continental US using a deep learning neural network. *Geophysical Research Letters*, 44:11030–11039, 2017. doi: 10.1002/2017GL075619. URL https://arxiv.org/abs/1707.06611.

K. Fang, M. Pan, and C. Shen. The value of SMAP for long-term soil moisture estimation with the help of deep learning. *IEEE Transactions on Geoscience and Remote Sensing, pages* 1–13, 2018. ISSN 0196-2892. doi: 10.1109/TGRS.2018.2872131. URL https://ieeexplore.ieee.org/document/8497052/.

K. Fang, D. Kifer, K. Lawson, and C. Shen. Evaluating the potential and challenges of an uncertainty quantification method for long short-term memory models for soil moisture predictions, *Water Resources Research*, 2020. doi: 10.1029/2020WR028095.

K. Fang, W.-P. Tsai, X. Ji, K. Lawson, C. Shen, Revealing causal controls of storage-streamflow relationships with a data-centric Bayesian framework combining machine learning and process-based modeling. *Frontiers in Water-Water and Hydrocomplexity*, 2020. doi:10.3389/frwa.2020.583000.

W. Fang, C. Wang, X. Chen, W. Wan, H. Li, S. Zhu, Y. Fang, B. Liu, and Y. Hong. Recognizing global reservoirs from Landsat 8 images: A deep learning approach. *IEEE Journal of Selected Topics in Applied Earth Observations and Remote Sensing*, 12(9):3168–3177, Sep. 2019. ISSN 2151-1535. doi: 10.1109/JSTARS.2019.2929601.

M. Fauvel, Y. Tarabalka, J.A. Benediktsson, J. Chanussot, and J.C. Tilton. Advances in spectral-spatial classification of hyperspectral images. *Proceedings of the IEEE*, 101(3):652–675, 2013.

D. Feng, K. Fang, and C. Shen. Enhancing streamflow forecast and extracting insights using long-short term memory networks with data integration at continental scales. *Water Resources Research*, 2020a. doi: 10.1029/2019WR026793.

S. Feng, H. Yu, and M.F. Duarte. Autoencoder based sample selection for self-taught learning. *Knowledge-Based Systems*, 192:105343, 2020b.

S. Fernández, A. Graves, and J. Schmidhuber. Sequence labelling in structured domains with hierarchical recurrent neural networks. In *International Joint Conference on Artificial Intelligence (IJCAI)*, pages 774–779, 2007.

R. Fernandez-Beltran, B. Demir, F. Pla, and A. Plaza. Unsupervised remote sensing image retrieval using probabilistic latent semantic hashing. *IEEE Geoscience and Remote Sensing Letters*, February 2020. doi: 10.1109/LGRS.2020.2969491.

D.J. Field. What is the goal of sensory coding? *Neural Computation*, 6(4): 559–601, 1994.

T. Finn, G. Geppert, and F. Ament. Deep assimilation: Adversarial variational inference for implicit and non-linear data assimilation How to combine data assimilation and deep learning? Simple linear Gaussian model.

M.A. Fischler and R.C. Bolles. Random sample consensus: A paradigm for model fitting with applications to image analysis and automated cartography. *Communications of the ACM*, 24(6):381–395, June 1981. ISSN 0001-0782. doi: 10.1145/358669.358692.

R. Flamary, D. Tuia, B. Labbé, G. Camps-Valls, and A. Rakotomamonjy. Large margin filtering. *IEEE Transactions on Signal Processing*, 60(2):648–659, 2012.

M.G. Flanner, K.M. Shell, M. Barlage, D.K. Perovich, and M.A. Tschudi. Radiative forcing and albedo feedback from the Northern Hemisphere cryosphere between 1979 and 2008. *Nature Geoscience*, 4(3):151–155, 2011.

G. Flato, J. Marotzke, B. Abiodun, P. Braconnot, S.C. Chou, W. Collins, P. Cox, F. Driouech, S. Emori, V. Eyring, C. Forest, P. Gleckler, E. Guilyardi, C. Jakob, V. Kattsov, C. Reason, and M. Rummukainen. *Evaluation of Climate Models.* In: *Climate Change 2013: The Physical Science Basis. Contribution of Working Group I to the Fifth Assessment Report of the Intergovernmental Panel on Climate Change* [T.F. Stocker, D. Qin, G.-K. Plattner, M.M.B. Tignor, S.K. Allen, J. Boschung, A. Nauels, Y. Xia, V. Bex, and P.M. Midgley (eds.)]. Cambridge University Press, Cambridge, United Kingdom and New York, NY, USA, 2013.

L. Foresti, I.V. Sideris, D. Nerini, L. Beusch, and U. Germann. Using a 10-year radar archive for nowcasting precipitation growth and decay – a probabilistic machine learning approach. *Weather and Forecasting*, pages WAF–D–18–0206.1, Jul 2019. ISSN 0882-8156. doi: 10.1175/WAF-D-18-0206.1. URL http://journals.ametsoc.org/doi/10.1175/WAF-D-18-0206.1.

U. Forssell and P. Lindskog. Combining Semi-Physical and Neural Network Modeling: An Example of its Usefulness. *IFAC Proceedings* 30(11):767–770, 1997. ISSN 14746670. doi: 10.1016/s1474-6670(17)42938-7. URL http://dx.doi.org/10.1016/S1474-6670(17)42938-7.

B. Fox-Kemper, S. Bachman, B. Pearson, and S. Reckinger. Principles and advances in subgrid modeling for eddy-rich simulations. *CLIVAR Exchanges*, 19(2):42–46, Jul 2014.

H.M. French. *The Periglacial Environment*. John Wiley & Sons, 2017.

P. Friedlingstein, M. Meinshausen, V.K. Arora, C.D. Jones, A. Anav, S.K. Liddicoat, and R. Knutti. Uncertainties in cmip5 climate projections due to carbon cycle feedbacks. *Journal of Climate*, 27(2):511–526, 2014. doi: 10.1175/JCLI-D-12-00579.1. URL https://doi.org/10.1175/JCLI-D-12-00579.1.

G. Fu, C. Liu, R. Zhou, T. Sun, and Q. Zhang. Classification for high resolution remote sensing imagery using a fully convolutional network. *Remote Sensing*, 9(5):498, 2017.

Y. Fu, T. Zhang, Y. Zheng, D. Zhang, and H. Huang. Hyperspectral image super-resolution with optimized RGB guidance. In *Proceedings of the IEEE Conference on Computer Vision and Pattern Recognition*, pages 11661–11670, 2019.

O. Fuhrer, T. Chadha, T. Hoefler, G. Kwasniewski, X. Lapillonne, D. Leutwyler, D. Lüthi, C. Osuna, C. Schär, T.C. Schulthess, and H. Vogt. Near-global climate simulation at 1km resolution: establishing a performance baseline on 4888 GPUs with Cosmo 5.0. *Geoscientific Model Development*, 11 (4):1665–1681, 2018. doi: 10.5194/gmd-11-1665-2018. URL https://www.geosci-model-dev.net/11/1665/2018/.

K. Fukushima. Neocognitron: A self-organizing neural network model for a mechanism of pattern recognition unaffected by shift in position. *Biological Cybernetics*, 36(4):193–202, 1980.

K. Funahashi and Y. Nakamura. Approximation of dynamical systems by continuous time recurrent neural networks. *Neural Networks*, 6(6):801–806, 1993. ISSN 0893-6080. doi: https://doi.org/10.1016/S0893-6080(05)80125-X. URL http://www.sciencedirect.com/science/article/pii/S089360800580125X.

G.B. Goldstein False-alarm regulation in log-normal and weibull clutter. *IEEE Transactions on Aerospace and Electronic Systems*, 9(1):84–92, 1973.

Y. Gal and Z. Ghahramani. A theoretically grounded application of dropout in recurrent neural networks. *arxiv preprint*, Dec 2015. URL http://arxiv.org/abs/1512.05287.

Y. Ganin, E. Ustinova, H. Ajakan, P. Germain, H. Larochelle, F. Laviolette, M. Marchand, and V. Lempitsky. Domain-adversarial training of neural networks. *Journal of Machine Learning Research*, 17(1):2096–2030, 2016.

Y. Gao, F. Gao, J. Dong, and S. Wang. Transferred deep learning for sea ice change detection from synthetic-aperture radar images. *IEEE Geoscience and Remote Sensing Letters*, 16(10):1655–1659, Oct 2019. doi: 10.1109/LGRS.2019.2906279.

A. Garzelli, F. Nencini, and L. Capobianco. Optimal mmse pan sharpening of very high resolution multispectral images. *IEEE Transactions on Geoscience and Remote Sensing*, 46(1):228–236, 2007.

C. Gatebe, W. Li, N. Chen, Y. Fan, R. Poudyal, L. Brucker, and K. Stamnes. Snow-covered area using machine learning techniques. In *IGARSS 2018-2018 IEEE International Geoscience and Remote Sensing Symposium*, pages 6291–6293. IEEE, 2018.

P. Gentine, M. Pritchard, S. Rasp, G. Reinaudi, and G. Yacalis. Could machine learning break the convection parameterization deadlock? *Geophysical Research Letters*, 45(11):5742–5751, Jun 2018. ISSN 19448007. doi: 10.1029/2018GL078202.

F.A. Gers and J. Schmidhuber. Recurrent nets that time and count. In *IEEE-INNS-ENNS International Joint Conference on Neural Networks (IJCNN)*, volume 3, pages 189–194, 2000. doi: 10.1109/IJCNN.2000.861302.

F.A. Gers and J. Schmidhuber. LSTM recurrent networks learn simple context-free and context-sensitive languages. *IEEE Transactions on Neural Networks*, 12(6):1333–1340, 2001. ISSN 1941-0093. doi: 10.1109/72.963769.

F.A. Gers, J. Schmidhuber, and F. Cummins. Learning to forget: Continual prediction with LSTM. *Neural Computation*, 12(10):2451–2471, 2000. doi: 10.1162/089976600300015015.

F.A. Gers, N. Schraudolph, and J. Schmidhuber. Learning precise timing with LSTM recurrent networks. *Journal of Machine Learning Research*, 3:115–143, 2003. doi: 10.1162/153244303768966139.

P. Ghamisi and N. Yokoya. IMG2DSM: Height simulation from single imagery using conditional generative adversarial net. *IEEE Geoscience and Remote Sensing Letters*, 15(5):794–798, May 2018.

P. Ghamisi, B. Rasti, N. Yokoya, Q. Wang, B. Hofle, L. Bruzzone, F. Bovolo, M. Chi, K. Anders, R. Gloaguen, et al. Multisource and multitemporal data fusion in remote sensing: A comprehensive review of the state of the art. *IEEE Geoscience and Remote Sensing Magazine*, 7(1):6–39, 2019.

P.B. Gibson, S.E. Perkins-Kirkpatrick, P. Uotila, A.S. Pepler, and L.V. Alexander. On the use of self-organizing maps for studying climate extremes. *Journal of Geophysical Research: Atmospheres*, 122(7):3891–3903, 2017.

S. Gidaris, P. Singh, and N. Komodakis. Unsupervised representation learning by predicting image rotations. In *International Conference on Learning Representations (ICLR)*, 2018.

R.C. Gilbert, M.B. Richman, T.B. Trafalis, and L.M. Leslie. Machine learning methods for data assimilation. In *Intelligent Engineering Systems through Artificial Neural Networks, Volume 20*, pages 105–112. ASME Press, Nov 2010. doi: 10.1115/1.859599.paper14.

N. Girard, G. Charpiat, and Y. Tarabalka. Aligning and updating cadaster maps with aerial images by multi-task, multi-resolution deep learning. In C.V. Jawahar, Hongdong Li, Greg Mori, and Konrad Schindler, editors, *Computer Vision – ACCV 2018*, pages 675–690, Cham, 2019. Springer International Publishing. ISBN 978-3-030-20873-8.

R. Girshick. Fast R-CNN. In *CVPR*, pages 1440–1448, 2015.

R. Girshick, J. Donahue, T. Darrell, and J. Malik. Rich feature hierarchies for accurate object detection and semantic segmentation. In *CVPR*, pages 580–587, 2014.

X. Glorot and Y. Bengio. Understanding the difficulty of training deep feedforward neural networks. In *International Conference on Artificial Intelligence and Statistics*, pages 249–256, 2010.

G.B. Goh, N.O. Hodas, and A. Vishnu. Deep learning for computational chemistry. *Journal of Computational Chemistry*, 38(16):1291–1307, Jun 2017. ISSN 01928651. doi: 10.1002/jcc.24764. URL http://doi.wiley.com/10.1002/jcc.24764.

C. Goller and A. Küchler. Learning task-dependent distributed representations by backpropagation through structure. In *International Conference on Neural Networks (ICNN)*, volume 1, pages 347–352, Jun 1996. doi: 10.1109/ICNN.1996.548916.

L. Gómez-Chova, D. Tuia, G. Moser, and G. Camps-Valls. Multimodal classification of remote sensing images: A review and future directions. *Proceedings of the IEEE*, 103(9):1560–1584, Sep 2015. ISSN 0018-9219. doi: 10.1109/JPROC.2015.2449668.

L. Gómez-Chova, G. Mateo-García, J. Muñoz-Marí, and G. Camps-Valls. Cloud detection machine learning algorithms for proba-v. In *2017 IEEE International Geoscience and Remote Sensing Symposium (IGARSS)*, pages 2251–2254. IEEE, 2017.

M. Gong, X. Niu, P. Zhang, and Z. Li. Generative adversarial networks for change detection in multispectral imagery. *IEEE Geoscience and Remote Sensing Letters*, 14(12):2310–2314, Dec 2017.

L. Gonog and Y. Zhou. A review: Generative adversarial networks. In *2019 14th IEEE Conference on Industrial Electronics and Applications (ICIEA)*, pages 505–510, 2019.

R.C. Gonzalez and R.E. Woods. *Digital Image Processing (3rd Edition)*. Prentice-Hall, Inc., Upper Saddle River, NJ, USA, 2006. ISBN 013168728X.

I. Goodfellow, J. Pouget-Abadie, M. Mirza, B. Xu, D. Warde-Farley, S. Ozair, A. Courville, and Y. Bengio. Generative adversarial nets. In *Advances in Neural Information Processing Systems*, pages 2672–2680, 2014a.

I. Goodfellow, Y. Bengio, and A. Courville. *Deep Learning*. Book in preparation for MIT press, 2016. URL http://www.deeplearningbook.org.

I.J. Goodfellow, J. Pouget-Abadie, M. Mirza, B. Xu, D. Warde-Farley, S. Ozair, A.C. Courville, and Y. Bengio. Generative adversarial networks. *CoRR*, abs/1406. 2661, 2014b.

R.S. Govindaraju. Artificial neural networks in hydrology. ii: Hydrologic applications. *Journal of Hydrologic Engineering*, 5(2):124–137, 2000. doi: 10.1061/(ASCE)1084-0699(2000)5:2(124).

A. Graves, A. Mohamed, and G. Hinton. Speech recognition with deep recurrent neural networks. In *Acoustics, Speech and Signal Processing (ICASSP), 2013 IEEE International Conference on*, pages 6645–6649. IEEE, 2013.

A. Graves and J. Schmidhuber. Framewise phoneme classification with bidirectional LSTM and other neural network architectures. *Neural Networks*, 18(5):602–610, 2005. doi: 10.1016/j.neunet.2005.06.042.

A. Graves and J. Schmidhuber. Offline handwriting recognition with multidimensional recurrent neural networks. In D. Koller, D. Schuurmans, Y. Bengio, and L. Bottou, editors, *Advances in Neural Information Processing Systems 21*, pages 545–552. Curran Associates, Inc., 2009. URL http://papers.nips.cc/paper/3449-offline-handwriting-recognition-with-multidimensional-recurrent-neural-networks.pdf.

A.A. Green, M. Berman, P. Switzer, and M.D. Craig. A transformation for ordering multispectral data in terms of image quality with implications for noise removal. *IEEE Transactions on Geoscience and Remote Sensing*, 26(1):65–74, 1988.

J.K. Green, S.I. Seneviratne, A.M. Berg, K.L. Findell, S. Hagemann, D.M. Lawrence, and P. Gentine. Large influence of soil moisture on long-term terrestrial carbon uptake. *Nature*, 565:476–479, 2019. doi: 10.1038/s41586-018-0848-x.

K. Greff, R. Kumar Srivastava, J. Koutník, B.R. Steunebrink, and J. Schmidhuber. LSTM: A search space odyssey. *IEEE Transactions on Neural Networks and Learning Systems (TNNLS)*, 28(10):2222–2232, 2017. doi: 10.1109/TNNLS.2016.2582924.

K. Gregor, I. Danihelka, A. Graves, D. Rezende, and D. Wierstra. Draw: A recurrent neural network for image generation. In F. Bach and D. Blei, editors, *Proceedings of the 32nd International Conference on Machine Learning, volume 37 of Proceedings of Machine Learning Research*, pages 1462–1471, Lille, France, 07–09 Jul 2015. PMLR.

S.M. Griffies, M. Winton, W.G. Anderson, R. Benson, T.L. Delworth, C.O. Dufour, J.P. Dunne, P. Goddard, A.K. Morrison, A. Rosati, et al. Impacts on ocean heat from transient mesoscale eddies in a hierarchy of climate models. *Journal of Climate*, 28(3): 952–977, 2015.

F. Groh, P. Wieschollek, and H.P.A. Lensch. Flex-convolution (million-scale point-cloud learning beyond grid-worlds). In *Asian Conference on Computer Vision (ACCV)*, 2018.

C. Grohnfeldt, M. Schmitt, and X. Zhu. A conditional generative adversarial network to fuse SAR and multispectral optical data for cloud removal from Sentinel-2 images. In *IGARSS 2018-2018 IEEE International Geoscience and Remote Sensing Symposium*, pages 1726–1729. IEEE, 2018.

P. Grönquist, T. Ben-Nun, N. Dryden, P. Dueben, L. Lavarini, S. Li, and T. Hoefler. *Predicting weather uncertainty with deep convnets*, 2019.

W. Gross, D. Tuia, U. Soergel, and W. Middelmann. Nonlinear feature normalization for hyperspectral domain adaptation and mitigation of nonlinear effects. *IEEE Transactions on Geoscience and Remote Sensing*, 57 (8):5975–5990, 2019.

J. Guo, B. Lei, C. Ding, and Y. Zhang. Synthetic aperture radar image synthesis by using generative adversarial nets. *IEEE Geoscience and Remote Sensing Letters*, 14(7):1111–1115, July 2017.

R. Guo, W. Wang, and H. Qi. Hyperspectral image unmixing using autoencoder cascade. In *2015 7th Workshop on Hyperspectral Image and Signal Processing: Evolution in Remote Sensing (WHISPERS)*, pages 1–4. IEEE, 2015.

X. Guo, Y. Chen, X. Liu, and Y. Zhao. Extraction of snow cover from high-resolution remote sensing imagery using deep learning on a small dataset. *Remote Sensing Letters*, 11(1):66–75, 2020. doi: 10.1080/2150704X.2019.1686548.

H.V. Gupta, K. Hsu, and S. Sorooshian. Effective and efficient modeling for streamflow forecasting. In R.S. Govindaraju and A. Ramachandra Rao, editors, *Artificial Neural Networks in Hydrology*, pages 7–22. Springer Netherlands, Dordrecht, 2000. ISBN 978-94-015-9341-0.

R. Gupta, B. Goodman, N. Patel, R. Hosfelt, S. Sajeev, E. Heim, J. Doshi, K. Lucas, H. Choset, and M. Gaston. Creating XBD: A dataset for assessing building damage from satellite imagery. In *The IEEE Conference on Computer Vision and Pattern Recognition (CVPR) Workshops*, June 2019.

M.U. Gutmann, R. Dutta, S. Kaski, and J. Corander. Likelihood-free inference via classification. *Statistics and Computing*, 28(2): 411–425, March 2018. ISSN 0960-3174. doi: 10.1007/s11222-017-9738-6. URL https://doi.org/10.1007/s11222-017-9738-6.

Y. Gwon, M. Cha, and H.T. Kung. Deep sparse-coded network (dsn). In *ICPR*, 2016.

H.M. Finn and R.S. Johnson. Adaptive detection mode with threshold control as a function of spatially sampled clutter-level estimates. In *RCA Review*, pages 1–8, 1968.

R.J. Haarsma, M.J. Roberts, P.L. Vidale, C.A. Senior, A. Bellucci, Q. Bao, P. Chang, S. Corti, N.S. Fučkar, V. Guemas, J. von Hardenberg, W. Hazeleger, C. Kodama, T. Koenigk, L.R. Leung, J. Lu, J.-J. Luo, J. Mao, M.S. Mizielinski, R. Mizuta, P. Nobre, M. Satoh, E. Scoccimarro, T. Semmler, J. Small, and J.-S. von Storch. High resolution model intercomparison project (highresmip v1.0) for cmip6. *Geoscientific Model Development*, 9(11):4185–4208, 2016. doi: 10.5194/gmd-9-4185-2016. URL https://www.geosci-model-dev.net/9/4185/2016/.

T. Hackel, N. Savinov, L. Ladicky, J.D. Wegner, K. Schindler, and M. Pollefeys. Semantic3d.net: A new large-scale point cloud classification benchmark. *ISPRS Annals of Photogrammetry, Remote Sensing and Spatial Information Science*, IV-1/W1, 2017.

D.M. Hall, J. Stewart, C. Tierney, and M. Govett. Adversarial networks for satellite to forecast model translation. In *8th International Workshop on Climate Informatics*, 2018.

R. Hallberg. Using a resolution function to regulate parameterizations of oceanic mesoscale eddy effects. *Ocean Modelling*, 72:92–103, 2013.

J.H. Ham, D.D. Lee, S. Mika, and B. Schölkopf. A kernel view of the dimensionality reduction of manifolds. *Departmental Papers (ESE)*, page 93, 2004.

X. Han, Th. Leung, Y. Jia, R. Sukthankar, and A.C. Berg. Matchnet: Unifying feature and metric learning for patch-based matching. In *The IEEE Conference on Computer Vision and Pattern Recognition (CVPR)*, June 2015.

X. Han, B. Shi, and Y. Zheng. Ssf-CNN: Spatial and spectral fusion with CNN for hyperspectral image super-resolution. In *Proceedings of the 25th IEEE Conference on Image Processing (ICIP)*, pages 2506–2510, 2018.

Y. Han, Y. Gao, Y. Zhang, J. Wang, and S. Yang. Hyperspectral sea ice image classification based on the spectral-spatial-joint feature with deep learning. *Remote Sensing*, 11(18), 2019. doi: 10.3390/rs11182170.

B. Hariharan, P. Arbeláez, R. Girshick, and J. Malik. Hypercolumns for object segmentation and fine-grained localization. In *Computer Vision and Pattern Recognition (CVPR)*, 2015.

S.A. Harris, H.M. French, J.A. Heginbottom, G.H. Johnston, B. Ladanyi, D.C. Sego, and R.O. van Everdingen. Glossary of permafrost and related ground-ice terms. *Technical Memorandum of The National Research Council of Canada*, Ottawa, 1988.

S. Hatfield, M. Chantry, P. Dueben, and T. Palmer. Accelerating high-resolution weather models with deep-learning hardware. In *Proceedings of the Platform for Advanced Scientific Computing Conference*, PASC'2019, pages 1:1–1:11, New York, NY, USA, 2019. ACM. ISBN 978-1-4503-6770-7. doi: 10.1145/3324989.3325711. URL http://doi.acm.org/10.1145/3324989.3325711.

J.B. Haurum, C.H. Bahnsen, and T.B. Moeslund. Is it raining outside? Detection of rainfall using general-purpose surveillance cameras, 2019. URL https://vbn.aau.dk/en/publications/is-it-raining-outside-detection-of-rainfall-using-general-purpose.

H. He, M. Chen, T. Chen, and D. Li. Matching of remote sensing images with complex background variations via Siamese convolutional neural network. *Remote Sensing*, 10(2), 2018. ISSN 2072-4292. doi: 10.3390/rs10020355.

H. He, M. Chen, T. Chen, D. Li, and P. Cheng. Learning to match multitemporal optical satellite images using multi-support-patches Siamese networks. *Remote Sensing Letters*, 10(6):516–525, 2019a. doi: 10.1080/2150704X.2019.1577572.

K. He, X. Zhang, S. Ren, and J. Sun. Deep residual learning for image recognition. In *2016 IEEE Conference on Computer Vision and Pattern Recognition (CVPR)*, pages 770–778, June 2016a.

K. He, X. Zhang, S. Ren, and J. Sun. Deep residual learning for image recognition. In *Computer Vision and Pattern Recognition (CVPR)*, pages 770–778, 2016b.

K. He, G. Gkioxari, P. Dollár, and R. Girshick. Mask R-CNN. In *ICCV*, pages 2980–2988. IEEE, 2017.

Q. He, D. Barajas-Solano, G. Tartakovsky, and A.M. Tartakovsky. Physics-informed neural networks for multiphysics data assimilation with application to subsurface transport. *Advances in Water Resources*, 141:103610, Jul 2020. ISSN 0309-1708. doi: 10.1016/J.ADVWATRES.2020.103610. URL https://www.sciencedirect.com/science/article/pii/S0309170819311649.

T.-L. He, D.B.A. Jones, B. Huang, Y. Liu, K. Miyazaki, Z. Jiang, E.C. White, H.M. Worden, and J.R. Worden. Recurrent U-net: Deep learning to predict daily summertime ozone in the United States. Aug 2019b. URL http://arxiv.org/abs/1908.05841.

Y. He, K. Kavukcuoglu, Y. Wang, A. Szlam, and Y. Qi. Unsupervised feature learning by deep sparse coding. In *Proceedings of the SIAM International Conference on Data Mining*, pages 902–910, 2014.

E. Hernández, V. Sanchez-Anguix, V. Julian, J. Palanca, and N. Duque. Rainfall prediction: A deep learning approach. In *International Conference on Hybrid Artificial Intelligence Systems*, pages 151–162. Springer, 2016.

B.C. Hewitson and R.G. Crane. Self-organizing maps: applications to synoptic climatology. *Climate Research*, 22(1):13–26, 2002.

T.D. Hewson. Objective fronts. *Meteorological Applications*, 5(1):37–65, 1998. ISSN 1469-8080. doi: 10.1017/S1350482798000553. URL http://dx.doi.org/10.1017/S1350482798000553.

A. Heye, K. Venkatesan, and J. Cain. Precipitation nowcasting: Leveraging deep recurrent convolutional neural networks. Technical report, 2017.

S. Hidetoshi. Improving predictive inference under covariate shift by weighting the log-likelihood function. *Journal of Statistical Planning and Inference*, 90 (2):227–244, 2000.

G.E. Hinton and R.R. Salakhutdinov. Reducing the dimensionality of data with neural networks. *Science*, 313(5786):504–507, July 2006a.

G.E. Hinton and R.S. Zemel. Autoencoders, minimum description length and helmholtz free energy. In *Neural Information Processing Systems*, 1994.

G.E. Hinton, S. Osindero, and Y.-W. Teh. A fast learning algorithm for deep belief nets. *Neural Computation*, 18(7):1527–1554, July 2006.

G.F. Hinton. A parallel computation that assigns canonical object-based frames of reference. In *Proceedings of the 7th International Joint Conference on Artificial Intelligence – Volume 2*, IJCAI'81, p-age 683–685, San Francisco, CA, USA, 1981. Morgan Kaufmann Publishers Inc.

T. Hoberg, F. Rottensteiner, R. Queiroz Feitosa, and C. Heipke. Conditional random fields for multitemporal and multiscale classification of optical satellite imagery. *IEEE Transactions on Geoscience and Remote Sensing (TGRS)*, 53(2):659–673, 2015. doi: 10.1109/TGRS.2014.2326886.

J. Hochreiter. Untersuchungen zu dynamischen neuronalen Netzen. Diploma thesis, Technical University of Munich, 1991.

S. Hochreiter and J. Schmidhuber. Long short-term memory. *Neural Computation*, 9(8):1735–1780, 1997. doi: 10.1162/neco.1997.9.8.1735.

S. Hochreiter, Y. Bengio, P. Frasconi, and J. Schmidhuber. Gradient flow in recurrent nets: the difficulty of learning long-term dependencies. In S.C. Kremer and J.F. Kolen, editors, *A Field Guide to Dynamical Recurrent Neural Networks*, pages 237–244. IEEE Press, Piscataway, NJ, USA, 2001.

J. Hoffman, E. Tzeng, T. Park, J.-Y. Zhu, P. Isola, K. Saenko, A. Efros, and T. Darrell. CyCADA: Cycle-consistent adversarial domain adaptation. In *International Conference on Machine Learning 2018*, pages 1989–1998, July 2018. URL http://proceedings.mlr.press/v80/hoffman18a.html.

E.J. Hoffmann, Y. Wang, M. Werner, J. Kang, and X.X. Zhu. Model fusion for building type classification from aerial and street view images. *Remote Sensing*, 11(11):1259, 2019a.

E.J. Hoffmann, M. Werner, and X.X. Zhu. Building instance classification using social media images. In *2019 Joint Urban Remote Sensing Event (JURSE)*, pages 1–4. IEEE, 2019b.

R.J. Hogan, C.A.T. Ferro, I.T. Jolliffe, and D.B. Stephenson. Equitability revisited: Why the "equitable threat score" is not equitable. *Weather and Forecasting*, 25(2):710–726, 2010.

D. Hong, N. Yokoya, J. Chanussot, and X. Zhu. An augmented linear mixing model to address spectral variability for hyperspectral unmixing. *IEEE Transactions on Image Processing*, 28(4):1923–1938, 2019.

C. Hope. The $10 trillion value of better information about the transient climate response. *Philosophical Transactions of the Royal Society A*, 373: 20140429, 2015. URL http://dx.doi.org/10.1098/rsta.2014.0429.

M. Horn, K. Walsh, M. Zhao, S.J. Camargo, E. Scoccimarro, H. Murakami, H. Wang, A. Ballinger, A. Kumar, D.A. Shaevitz, J.A. Jonas, and K. Oouchi. Tracking scheme dependence of simulated tropical cyclone response to idealized climate simulations. *Journal of Climate*, 27(24):9197–9213, 2014a. doi: 10.1175/JCLI-D-14-00200.1. URL https://doi.org/10.1175/JCLI-D-14-00200.1.

H. Hotelling. Analysis of a complex of statistical variables into principal components. *Journal of Educational Psychology*, 24(6):417, 1933.

B. Hou, Q. Liu, H.Wang, and Y.Wang. From W-Net to CDGAN: Bitemporal change detection via deep learning techniques. *IEEE Transactions on Geoscience and Remote Sensing*, pages 1–13, 2019.

R.A. Houze Jr,. Mesoscale convective systems. *Reviews of Geophysics*, 42 (4), 2004.

F. Hu, X. Tong, G. Xia, and L. Zhang. Delving into deep representations for remote sensing image retrieval. In *IEEE International Conference on Signal Processing*, pages 198–203, November 2016.

Y. Hua, L. Mou, and X.X. Zhu. Relation network for multilabel aerial image classification. *IEEE Transactions on Geoscience and Remote Sensing*, 2020.

B. Huang, K. Zhang, Y. Lin, B. Schölkopf, and C. Glymour. Generalized score functions for causal discovery. volume 2018, pages 1551–1560, 07 2018a.

C. Huang, H. Ai, Y. Li, and S. Lao. High-performance rotation invariant multiview face detection. *IEEE TPAMI*, 29(4):671–686, 2007.

K. Huang, J. Xia, Y. Wang, A. Ahlström, J. Chen, R.B. Cook, E. Cui, Y. Fang, J.B. Fisher, D.N. Huntzinger, Z. Li, A.M. Michalak, Y. Qiao, K. Schaefer, C. Schwalm, J. Wang, Y. Wei, X. Xu, L. Yan, C. Bian, and Y. Luo. Enhanced peak growth of global vegetation and its key mechanisms. *Nature Ecology & Evolution*, 2(12):1897–1905, December 2018b. ISSN 2397-334X. doi: 10.1038/s41559-018-0714-0.

L. Huang, J. Luo, Z. Lin, F. Niu, and L. Liu. Using deep learning to map retrogressive thaw slumps in the Beiluhe region (Tibetan Plateau) from CubeSat images. *Remote Sensing of Environment*, 237:111534, 2020. doi: https://doi.org/10.1016/j.rse.2019.111534.

L. Huang, L. Liu, L. Jiang, and T. Zhang. Automatic mapping of thermokarst landforms from remote sensing images using deep learning: A case study in the northeastern Tibetan plateau. *Remote Sensing*, 10 (12), 2018. doi: 10.3390/rs10122067.

R. Huang, H. Taubenböck, L. Mou, and X.X. Zhu. Classification of settlement types from tweets using LDA and LSTM. In *IGARSS 2018-2018 IEEE International Geoscience and Remote Sensing Symposium*, pages 6408–6411. IEEE, 2018c.

X. Huang and S. Belongie. Arbitrary style transfer in real-time with adaptive instance normalization. In *IEEE International Conference on Computer Vision*, pages 1501–1510, 2017.

X. Huang, M.-Y. Liu, S. Belongie, and L. Kautz. Multimodal unsupervised image-to-image translation. In *European Conference on Computer Vision*, pages 172–189, 2018d.

L.H. Hughes, M. Schmitt, L. Mou, Y.Wang, and X.X. Zhu. Identifying corresponding patches in SAR and optical images with a pseudo-Siamese CNN. *IEEE Geoscience and Remote Sensing Letters*, 15(5):784–788, May 2018. ISSN 1558-0571. doi: 10.1109/LGRS.2018.2799232.

H. Li, B.S. Manjunath, and S.K. Mitra. A contour-based approach to multisensor image registration. *IEEE Transactions on Image Processing*, 4(3):320–334, March 1995. ISSN 1941-0042. doi: 10.1109/83.366480.

V. Humphrey, L. Gudmundsson, and S.I. Seneviratne. A global reconstruction of climate-driven subdecadal water storage variability. *Geophysical Research Letters*, 44(5):2300–2309, 2017. doi: 10.1002/2017GL072564.

E. Ilg, N. Mayer, T. Saikia, M. Keuper, A. Dosovitskiy, and T. Brox. Flownet 2.0: Evolution of optical flow estimation with deep networks. In *CVPR*, 2017.

E. Ilg, O. Cicek, S. Galesso, A. Klein, O. Makansi, F. Hutter, and T. Brox. Uncertainty estimates and multi-hypotheses networks for optical flow. In *Proceedings of the European Conference on Computer Vision (ECCV)*, pages 652–667, 2018.

IMBIE. Mass balance of the Antarctic Ice Sheet from 1992 to 2017. *Nature*, 558:219–222, 2018.

IMBIE. Mass balance of the Greenland Ice Sheet from 1992 to 2018. *Nature*, 2019. doi: 10.1038/s41586-019-1855-2.

R. Imbriaco, C. Sebastian, E. Bondarev, and P.H.N. de With. Aggregated deep local features for remote sensing image retrieval. *Remote Sensing*, 11:493, February 2019.

D. Inamdar, G. Leblanc, R.J. Soffer, and M. Kalacska. The correlation coefficient as a simple tool for the localization of errors in spectroscopic imaging data. *Remote Sensing*, 10(2):231, 2018.

S. Ioffe and C. Szegedy. Batch normalization: Accelerating deep network training by reducing internal covariate shift. In *Proceedings of The 32nd International Conference on Machine Learning*, pages 448–456, 2015.

IPCC. *The Physical Science Basis. Contribution of Working Group I to the Fifth Assessment Report of the Intergovernmental Panel on Climate Change*. Cambridge University Press, 2013.

IPCC. *Special Report on Global Warming of* 1.5°C. Cambridge, United Kingdom and New York, NY, USA, 2018.

IPCC. Intergovernmental Panel on Climate Change (IPCC) Special Report on the Ocean and Cryosphere in a Changing Climate, 2019.

R.G. Isaacs, R.N. Hoffman, and L.D. Kaplan. Satellite remote sensing of meteorological parameters for global numerical weather prediction. *Reviews of Geophysics*, 24(4):701–743, 1986.

P. Isola, J.-Y. Zhu, T. Zhou, and A.A. Efros. Image-to-image translation with conditional adversarial networks. *arxiv*, 2016.

M. Jaderberg, K. Simonyan, A. Zisserman, and K. Kavukcuoglu. Spatial transformer networks. In C. Cortes, N.D. Lawrence, D.D. Lee, M. Sugiyama, and R. Garnett, editors, *Advances in Neural Information Processing Systems 28*. Curran Associates, Inc., 2015.

S. Jégou, M. Drozdzal, D. Vazquez, A. Romero, and Y. Bengio. The one hundred layers tiramisu: Fully convolutional densenets for semantic segmentation. In *Computer Vision and Pattern Recognition Workshops (CVPRW), 2017 IEEE Conference on*, pages 1175–1183. IEEE, 2017.

X. Ji, L. Lesack, J.M. Melack, S. Wang, W.J. Riley, and C. Shen. Seasonal and inter-annual patterns and controls of hydrological fluxes in an Amazon floodplain lake with a surface-subsurface processes model. *Water Resources Research*, 55(4):3056–3075, 2019. doi: 10.1029/2018WR023897.

Q. Jia, X. Wan, B. Hei, and S. Li. Dispnet based stereo matching for planetary scene depth estimation using remote sensing images. In *2018 10th IAPR Workshop on Pattern Recognition in Remote Sensing (PRRS)*, Aug 2018. doi: 10.1109/PRRS.2018.8486195.

X. Jia, B. De Brabandere, T. Tuytelaars, and L.V. Gool. Dynamic filter networks. In *Advances in Neural Information Processing Systems*, pages 667–675, 2016.

J. Jiang, J. Liu, C.-Z. Qin, and D. Wang. Extraction of urban waterlogging depth from video images using transfer learning. *Water*, 10(10), 2018a. ISSN 2073-4441. doi: 10.3390/w10101485. URL https://www.mdpi.com/2073-4441/10/10/1485.

K. Jiang, Z. Wang, P. Yi, G. Wang, T. Lu, and J. Jiang. Edge-enhanced gan for remote sensing image super-resolution. *IEEE Transactions on Geoscience and Remote Sensing*, 57(8):5799–5812, Aug 2019.

M. Jiang, Y. Wu, T. Zhao, Z. Zhao, and C. Lu. PointSIFT: A SIFT-like Network Module for 3d Point Cloud Semantic Segmentation, July 2018b.

S. Jiang, V. Babovic, Y. Zheng, and J. Xiong. Advancing opportunistic sensing in hydrology: A novel approach to measuring rainfall with ordinary surveillance cameras. *Water Resources Research*, 55(4): 3004–3027, 2019a. doi: 10.1029/2018WR024480. URL https://agupubs.onlinelibrary.wiley.com/doi/abs/10.1029/2018WR024480.

Y. Jiang, X. Zhu, X. Wang, S. Yang, W. Li, H. Wang, P. Fu, and Z. Luo. R2cnn: rotational region CNN for orientation robust scene text detection. *arXiv*:1706.09579, 2017.

Z. Jiang, K. Von Ness, J. Loisel, and Z. Wang. Arcticnet: A deep learning solution to classify arctic wetlands, 2019.

D. Jimenez Rezende and S. Mohamed. Variational inference with normalizing flows. 05 2015.

Y.-H. Jo, D.-W. Kim, and H. Kim. Chlorophyll concentration derived from microwave remote sensing measurements using artificial neural network algorithm. *Journal of the Academy of Marketing Science*, 26:102–110, 2018.

J. Emmanuel Johnson, V. Laparra, M. Piles, and G. Camps-Valls. Gaussianizing the earth: Multidimensional information measures for earth data analysis. Submitted.

I.T. Jolliffe. Principal components in regression analysis. In *Principal Component Analysis*, pages 129–155. Springer, 1986.

M.T. Jorgenson. Thermokarst terrains. *Treatise on Geomorphology*, 8:313–324, 2013.

R. Jozefowicz, W. Zaremba, and I. Sutskever. An empirical exploration of recurrent network architectures. In F. Bach and D. Blei, editors, *International Conference on Machine Learning (ICML), volume 37 of Proceedings of Machine Learning Research (PRML)*, pages 2342–2350, 2015.

M. Jung, C. Schwalm, M. Migliavacca, S. Walther, G. Camps-Valls, S. Koirala, P. Anthoni, S. Besnard, P. Bodesheim, N. Carvalhais, F. Chevallier, F. Gans, D.S. Groll, V. Haverd, K. Ichii, A.K. Jain, J. Liu, D. Lombardozzi, J.E.M.S. Nabel, J.A. Nelson, M. Pallandt, D. Papale, W. Peters, J. Pongratz, C. Rödenbeck, S. Sitch, G. Tramontana, U. Weber, M. Reichstein, P. Koehler, M. O'Sullivan, and A. Walker. Scaling carbon fluxes from eddy covariance sites to globe: Synthesis and evaluation of the FLUXCOM approach. *Biogeosciences Discussions*, pages 1–40, 2019. doi: https://doi.org/10.5194/bg-2019-368.

H.F. Kaiser. The varimax criterion for analytic rotation in factor analysis. *Psychometrika*, 23(3):187–200, 1958.

P. Kaiser, J.D.Wegner, A. Lucchi, M. Jaggi, T. Hofmann, and K. Schindler. Learning aerial image segmentation from online maps. *IEEE Transactions in Geoscience and Remote Sensing*, 55(11):6054–6068, 2017.

N. Kalchbrenner, I. Danihelka, and A. Graves. Grid long short-term memory. In *International Conference on Learning Representations (ICLR)*, 2016.

R.E. Kalman. A new approach to linear filtering and prediction problems. *Journal of Basic Engineering*, 82(1):35–45, 1960.

A. Kamilaris and F.X. Prenafeta-Boldú. Deep learning in agriculture: A survey. *Computers and Electronics in Agriculture*, 147:70–90, 2018.

M. Kampffmeyer, A.B. Salberg, and R. Jenssen. Semantic segmentation of small objects and modeling of uncertainty in urban remote sensing images using deep convolutional neural networks. In *Computer Vision and Pattern Recognition Workshops (CVPRw)*, 2016.

C.I. Kanatsoulis, X. Fu, N.D. Sidiropoulos, and W.-K. Ma. Hyperspectral super-resolution: A coupled tensor factorization approach. *IEEE Transactions on Signal Processing*, 66(24):6503–6517, 2018.

A. Kanazawa, A. Sharma, and D. Jacobs. Locally scale-invariant convolutional neural networks. In *Advances in Neural Information Processing Systems*, 2014.

G. Kang, L. Jiang, Y. Yang, and A.G. Hauptmann. Contrastive adaptation network for unsupervised domain adaptation. In *IEEE Conference on Computer Vision and Pattern Recognition*, pages 4893–4902, 2019.

J. Kang, M. Körner, Y. Wang, H. Taubenböck, and X.X. Zhu. Building instance classification using street view images. *ISPRS Journal of Photogrammetry and Remote Sensing*, 145:44–59, 2018.

K. Karantzalos, A. Sotiras, and N. Paragios. Efficient and automated multimodal satellite data registration through mrfs and linear programming. In *2014 IEEE Conference on Computer Vision and Pattern Recognition Workshops*, 2014.

H.S. Karimi, B. Natarajan, C.L. Ramsey, J. Henson, J.L. Tedder, and E. Kemper. Comparison of learning-based wastewater flow prediction methodologies for smart sewer management. *Journal of Hydrology*, 577:123977, 2019. ISSN 0022-1694. doi: https://doi.org/10.1016/j.jhydrol.2019.123977. URL http://www.sciencedirect.com/science/article/pii/S0022169419306973.

A. Karpathy and L. Fei-Fei. Deep visual-semantic alignments for generating image descriptions. In *CVPR*, 2015.

A. Karpatne, G. Atluri, J.H. Faghmous, M. Steinbach, A. Banerjee, A. Ganguly, S. Shekhar, N. Samatova, and Vipin Kumar. Theory-guided data science: A new paradigm for scientific discovery from data. *IEEE Transactions on Knowledge Data Engineering*, 29(10):2318–2331, 2017a. ISSN 1041-4347. doi: 10.1109/tkde.2017.2720168.

A. Karpatne, W. Watkins, J. Read, and V. Kumar. How can physics inform deep learning methods in scientific problems?: Recent progress and future prospects. Technical report, 2017c.

A. Karpatne, W. Watkins, J. Read, and V. Kumar. Physics-guided Neural Networks (PGNN): An application in lake temperature modeling. *ArXiv e-prints*, page 1710.11431, 2017d. URL http://arxiv.org/abs/1710.11431.

J. Karvonen. Baltic sea ice concentration estimation using Sentinel-1 SAR and AMSR2 microwave radiometer data. *IEEE Transactions on Geoscience and Remote Sensing*, 55(5):2871–2883, 2017.

K. Kashinath, A. Albert, R. Wang, M. Mustafa, and R. Yu. Physics-informed spatio-temporal deep learning models. *Bulletin of the American Physical Society*, 64, 2019.

K. Kavukcuoglu, P. Sermanet, Y.-L. Boureau, K. Gregor, M. Mathieu, and Y. LeCun. Learning convolutional feature hierachies for visual recognition. In *NIPS*, 2010.

Y. Kawachi, Y. Koizumi, and N. Harada. Complementary set variational autoencoder for supervised anomaly detection. In *2018 IEEE International Conference on Acoustics, Speech and Signal Processing (ICASSP)*, pages 2366–2370. IEEE, 2018.

M. Kaya and H.Ş. Bilge. Deep metric learning: A survey. *Symmetry*, 11(9), 2019. ISSN 2073-8994. doi: 10.3390/sym11091066. URL https://www.mdpi.com/2073-8994/11/9/1066.

B. Kellenberger, D. Marcos, S. Lobry, and D. Tuia. Half a percent of labels is enough: Efficient animal detection in uav imagery using deep CNNs and active learning. *IEEE Transactions on Geoscience and Remote Sensing*, 2019.

A. Kembhavi, D. Harwood, and L.S. Davis. Vehicle detection using partial least squares. *IEEE Transactions on Pattern Analysis and Machine Intelligence*, 33(6):1250–1265, 2010.

R. Kemker, C. Salvaggio, and C. Kanan. Algorithms for semantic segmentation of multispectral remote sensing imagery using deep learning. *ISPRS Journal of the International Society of Photogrammetry and Remote Sensing*, 145:60–77, 2018.

R. Kemker and C. Kanan. Self-taught feature learning for hyperspectral image classification. *IEEE TGRS*, 55(5):2693–2705, 2017.

A. Kendall and Y. Gal. What uncertainties do we need in Bayesian deep learning for computer vision? In *Advances in Neural Information Processing Systems*, pages 5574–5584, 2017.

M.G. Kendall. A new measure of rank correlation. *Biometrika*, 30(1/2): 81–93, 1938.

E. Khalid, P. McGuire, D. Power, and C. Moloney. Target detection in synthetic aperture radar imagery: a state-of-the-art survey. *Journal of Applied Remote Sensing*, 7(7):1–10, 2018.

S.H. Khan, X. He, F. Porikli, and M. Bennamoun. Forest change detection in incomplete satellite images with deep neural networks. *IEEE Transactions on Geoscience and Remote Sensing*, 55(9):5407–5423, 2017.

J. Kim, J.K. Lee, and K.M. Lee. Accurate image super-resolution using very deep convolutional networks. In *Computer Vision and Pattern Recognition (CVPR), 2016 IEEE Conference on*, pages 1646–1654, 2016.

J. Kim, K. Kim, J. Cho, Y.Q. Kang, H.-J. Yoon, and Y.-W. Lee. Satellite-based prediction of arctic sea ice concentration using a deep neural network with multi-model ensemble. *Remote Sensing*, 11(1), 2018. doi: 10.3390/rs11010019.

S.K. Kim, S. Ames, J. Lee, C. Zhang, A.C. Wilson, and D. Williams. Massive scale deep learning for detecting extreme climate events. *Climate Informatics*, 2017a.

T. Kim, M. Cha, H. Kim, J.K. Lee, and J. Kim. Learning to discover cross-domain relations with generative adversarial networks. *CoRR*, abs/1703.05192, 2017b. URL http://arxiv.org/abs/1703.05192.

Y.J. Kim, H.-C. Kim, D. Han, S. Lee, and J. Im. Prediction of monthly Arctic sea ice concentrations using satellite and reanalysis data based on convolutional neural networks. *The Cryosphere*, 14(3):1083–1104, 2020. doi: 10.5194/tc-14-1083-2020.

D.P. Kingma and M. Welling. Auto-encoding variational Bayes. *arXiv preprint arXiv:1312.6114*, 2013.

D.P. Kingma and J. Ba. Adam: A method for stochastic optimization. *arXiv preprint arXiv:1412.6980*, 2014.

D.P. Kingma and M. Welling. Auto-encoding variational Bayes. In *International Conference on Learning Representations (ICLR)*, 2014.

J. Kirkpatrick, R. Pascanu, N. Rabinowitz, J. Veness, G. Desjardins, A.A. Rusu, K. Milan, J. Quan, T. Ramalho, A. Grabska-Barwinska, D. Hassabis, C. Clopath, D. Kumaran, and R. Hadsell. Overcoming catastrophic forgetting in neural networks. *Proceedings of the National Academy of Sciences*, 114(13): 3521–3526, 2017.

I.A. Klampanos, A. Davvetas, S. Andronopoulos, C. Pappas, A. Ikonomopoulos, and V. Karkaletsis. Autoencoder-driven weather clustering for source estimation during nuclear events. *Environmental Modelling & Software*, 102:84–93, 2018.

B. Klein, L. Wolf, and Y. Afek. A dynamic convolutional layer for short range weather prediction. In *Proceedings of the IEEE Conference on Computer Vision and Pattern Recognition*, pages 4840–4848, 2015.

T.R. Knutson, J.J. Sirutis, S.T. Garner, I.M. Held, and R.E. Tuleya. Simulation of the recent multidecadal increase of atlantic hurricane activity using an 18-km-grid regional model. *Bulletin of the American Meteorological Society*, 88(10):1549–1565, 2007.

S. Kohl, B. Romera-Paredes, C. Meyer, J. De Fauw, J.R. Ledsam, K. Maier-Hein, S.M. Ali Eslami, D. Jimenez Rezende, and O. Ronneberger. A probabilistic u-net for segmentation of ambiguous images. In *Advances in Neural Information Processing Systems*, pages 6965–6975, 2018.

S. Koirala, P.J.-F. Yeh, Y. Hirabayashi, S. Kanae, and T. Oki. Global-scale land surface hydrologic modeling with the representation of water table dynamics. *Journal of Geophysical Research: Atmospheres*, 119: 75–89, 2014. doi: 10.1002/2013JD020398.

S.J. Kollet and R.M. Maxwell. Integrated surface–groundwater flow modeling: A free-surface overland flow boundary condition in a parallel groundwater flow model. *Advances inWater Resources*, 29(7):945–958, Jul 2006. ISSN 03091708. doi: 10.1016/j.advwatres.2005.08.006. URL http://dx.doi.org/10.1016/j.advwatres.2005.08.006.

G.J. Kooperman, M.S. Pritchard, M.A. Burt, M.D. Branson, and D.A. Randall. Robust effects of cloud superparameterization on simulated daily rainfall intensity statistics across multiple versions of the c ommunity e arth s ystem m odel. *Journal of Advances in Modeling Earth Systems*, 8(1):140–165, 2016.

E.N. Kornaropoulos, E.I. Zacharaki, P. Zerbib, C. Lin, A. Rahmouni, and N. Paragios. Deformable group-wise registration using a physiological model: Application to diffusion-weighted MRI. In *2016 IEEE International Conference on Image Processing (ICIP)*, pages 2345–2349, Sep. 2016. doi: 10.1109/ICIP.2016.7532778.

M. Kosmala, K. Hufkens, and A.D. Richardson. Integrating camera imagery, crowdsourcing, and deep learning to improve high-frequency automated monitoring of snow at continental-to-global scales. *PLoS ONE*, 13 (12):1–19, 2018. doi: 10/ggh5m7. URL https://doi.org/10.1371/journal.pone.0209649.

B. Kraft, M. Jung, M. Körner, C. Requena Mesa, J. Cortés, and M. Reichstein. Identifying dynamic memory effects of vegetation state using recurrent neural networks. *Frontiers in Big Data*, 2, 2019. doi: 10.3389/fdata.2019.00031.

B. Kraft, M. Jung, M. Körner, and M. Reichstein. Hybrid modeling: Fusion of a deep approach and physics-based model for global hydrological modeling. *The International Archives of Photogrammetry, Remote Sensing and Spatial Information Sciences*, 43:1537–1544, 2020.

M.A. Kramer. Nonlinear principal component analysis using autoassociative neural networks. *AIChE Journal*, 37(2):233–243, 1991.

C. Krapu, M. Borsuk, and M. Kumar. Gradient-based inverse estimation for a rainfall-runoff model. *Water Resources Research*, 55 (8):6625–6639, 2019. doi: 10.1029/2018WR024461. URL https://agupubs.onlinelibrary.wiley.com/doi/abs/10.1029/2018WR024461.

V.M. Krasnopolsky and Y. Lin. A neural network nonlinear multimodel ensemble to improve precipitation forecasts over continental us. *Advances in Meteorology*, 2012, 2012. doi: https://doi.org/10.1155/2012/649450.

V.M. Krasnopolsky and H. Schiller. Some neural network applications in environmental sciences. part i: forward and inverse problems in geophysical remote measurements. *Neural Networks*, (16):321–334, 2003.

V.M. Krasnopolsky, M.S. Fox-Rabinovitz, Y.T. Hou, S.J. Lord, and A. A. Belochitski. Accurate and fast neural network emulations of model radiation for the ncep coupled climate forecast system: Climate simulations and seasonal predictions. *Monthly Weather Review*, 138(5):1822–1842, 2010. doi: 10.1175/2009MWR3149.1.

V.M. Krasnopolsky, S. Nadiga, A. Mehra, E. Bayler, and D. Behringer. Neural networks technique for filling gaps in satellite measurements: Application to ocean color observations. *Computational Intelligence and Neuroscience*, 2016, 2016.

V.M. Krasnopolsky. The application of neural networks in the earth system sciences. *Neural Networks Emulations for Complex Multidimensional Mappings*, 46, 2013.

V.M. Krasnopolsky, M.S. Fox-Rabinovitz, H.L. Tolman, and A.A. Belochitski. Neural network approach for robust and fast calculation of physical processes in numerical environmental models: Compound parameterization with a quality control of larger errors. *Neural Networks*, 21 (2–3):535–543, 2008.

F. Kratzert, D. Klotz, G. Shalev, G. Klambauer, S. Hochreiter, and G. Nearing. Towards learning universal, regional, and local hydrological behaviors via machine learning applied to large-sample datasets. *Hydrology and Earth System Sciences*, 23(12):5089–5110, 2019. doi: 10.5194/hess-23-5089-2019. URL https://www.hydrol-earth-syst-sci.net/23/5089/2019/.

F. Kratzert, D. Klotz, C. Brenner, K. Schulz, and M. Herrnegger. Rainfall–runoff modelling using Long Short-Term Memory (LSTM) networks. *Hydrology and Earth System Sciences*, 22(11):6005–6022, Nov 2018. ISSN 1607-7938. doi: 10.5194/hess-22-6005-2018. URL https://www.hydrol-earth-syst-sci.net/22/6005/2018/.

K. Kreutz-Delgado, J.F. Murray, B.D. Rao, K. Engan, T.-W. Lee, and T.J. Sejnowski. Dictionary learning algorithms for sparse representation. *Neural computation*, 15(2):349–396, 2003.

M.A. Krinitskiy, Yulia A. Zyulyaeva, and S.K. Gulev. Clustering of polar vortex states using convolutional autoencoders. 2019.

A. Krizhevsky. Imagenet classification with deep convolutional neural networks. In *Neural Information Processing Systems*, 1992.

A. Krizhevsky, I. Sutskever, and G.E. Hinton. Imagenet classification with deep convolutional neural networks. In F. Pereira, C.J.C. Burges, L. Bottou, and K.Q. Weinberger, editors, *Advances in Neural Information Processing Systems 25*, pages 1097–1105. 2012.

K.E. Kunkel, D.R. Easterling, D.A.R. Kristovich, B. Gleason, L. Stoecker, and R. Smith. Meteorological causes of the secular variations in observed extreme precipitation events for the conterminous united states. *Journal of Hydrometeorology*, 13(3):1131–1141, 2012. doi: 10.1175/JHM-D-11-0108.1. URL https://doi.org/10.1175/JHM-D-11-0108.1.

K. Kuppala, S. Banda, and Th. R. Barige. An overview of deep learning methods for image registration with focus on feature-based approaches. *International Journal of Image and Data Fusion*, pages 1–23, 2020. doi: 10.1080/19479832.2019.1707720. URL https://doi.org/10.1080/19479832.2019.1707720.

T. Kurth, J. Yang, N. Satish, M. Patwary, E. Racah, N. Mitliagkas, I. Sundaram, M. Patwary, Prabhat, and Pradeep Dubey. Deep learning at 15pf: supervised and semi-supervised learning for scientific data. *to appear in Supercomputing*, 2017.

T. Kurth, S. Treichler, J. Romero, M. Mudigonda, N. Luehr, E. Phillips, A. Mahesh, M. Matheson, J. Deslippe, M. Fatica, Prabhat, and M. Houston. Exascale deep learning for climate analytics. In *Proceedings of the International Conference for High Performance Computing, Networking, Storage, and Analysis*, SC '18, pages 51:1–51:12, Piscataway, NJ, USA, 2018. IEEE Press. doi: 10.1109/SC.2018.00054. URL https://doi.org/10.1109/SC.2018.00054.

K. Kuwata and R. Shibasaki. Estimating corn yield in the United States with modis evi and machine learning methods. *ISPRS Annals of Photogrammetry, Remote Sensing & Spatial Information Sciences*, 3(8), 2016.

A. Lagrange, B.L. Saux, A. Beaupere, A. Boulch, A. Chan-Hon-Tong, S. Herbin, H. Randrianarivo, and M. Ferecatu. Benchmarking classification of earth-observation data: From learning explicit features to convolutional networks. In *IEEE International Geoscience and Remote Sensing Symposium (IGARSS)*, Milan, Italy, 2015.

B. Lakshminarayanan, A. Pritzel, and C. Blundell. Simple and scalable predictive uncertainty estimation using deep ensembles. *Advances in Neural Information Processing Systems*, 2017-Decem(Nips):6403–6414, 2017. ISSN 10495258.

R. LaLonde, D. Zhang, and M. Shah. Clusternet: Detecting small objects in large scenes by exploiting spatio-temporal information. In *CVPR*, June 2018.

E. Laloy, R. Hérault, D. Jacques, and N. Linde. Training-image based geostatistical inversion using a spatial generative adversarial neural network. *Water Resources Research*, 54(1):381–406, 2018. doi: 10.1002/2017WR022148. URL https://agupubs.onlinelibrary.wiley.com/doi/abs/10.1002/2017WR022148.

D. Lam, R. Kuzma, K. McGee, S. Dooley, M. Laielli, M. Klaric, Yaroslav Bulatov, and B. McCord. xview: Objects in context in overhead imagery, 2018. URL https://aps.arxiv.org/abs/1802.07856.

A.M. Lamb, A. Goyal alias Parth Goyal, Y. Zhang, S. Zhang, A.C. Courville, and Y. Bengio. Professor forcing: A new algorithm for training recurrent networks. In *NIPS*. 2016.

C. Lanaras, E. Baltsavias, and K. Schindler. Hyperspectral super-resolution with spectral unmixing constraints. *Remote Sensing*, 9(11): 1196, 2017.

L. Landrieu and M. Simonovsky. Large-scale point cloud semantic segmentation with superpoint graphs. In *Computer Vision and Pattern Recognition (CVPR)*, pages 4558–4567, Salt Lake City, UT, 2018.

Z.L. Langford, J. Kumar, F.M. Hoffman, A.L. Breen, and C.M. Iversen. Arctic vegetation mapping using unsupervised training datasets and convolutional neural networks. *Remote Sensing*, 11(1), 2019. doi: 10.3390/rs11010069.

V. Laparra and R. Santos-Rodríguez. Spatial/spectral information trade-off in hyperspectral images. In *2015 IEEE International Geoscience and Remote Sensing Symposium (IGARSS)*, pages 1124–1127, 2015.

V. Laparra, G. Camps, and J. Malo. Iterative gaussianization: from ICA to random rotations. *IEEE Transactions on Neural Networks*, 22(4):537–549, 2011.

S. Lapuschkin, S. Wäldchen, A. Binder, G. Montavon, W. Samek, and K.-R. Müller. Unmasking clever hans predictors and assessing what machines really learn. *Nature Communications*, 10:1096, 2019. doi: 10.1038/s41467-019-08987-4. URL http://dx.doi.org/10.1038/s41467-019-08987-4.

W. Larcher. *Physiological Plant Ecology: Ecophysiology and Stress Physiology of Functional Groups*. Springer-Verlag, Berlin Heidelberg, 2003. ISBN 978-3-540-43516-7.

H. Larochelle, Y. Bengio, J. Louradour, and P. Lamblin. Exploring strategies for training deep neural networks. *The Journal of Machine Learning Research*, 10:1–40, 2009.

D.A. Lavers, G. Villarini, R.P. Allan, E.F. Wood, and A.J. Wade. The detection of atmospheric rivers in atmospheric reanalyses and their links to british winter floods and the large-scale climatic circulation. *Journal of Geophysical Research: Atmospheres*, 117(D20), 2012.

B.N. Lawrence, M. Rezny, R. Budich, P. Bauer, J. Behrens, M. Carter, W. Deconinck, R. Ford, C. Maynard, S. Mullerworth, C. Osuna, A. Porter, K. Serradell, S. Valcke, N.Wedi, and S. Wilson. Crossing the chasm: how to develop weather and climate models for next generation computers? *Geoscientific Model Development*, 11(5):1799–1821, 2018. doi: 10.5194/gmd-11-1799-2018. URL https://www.geosci-model-dev.net/11/1799/2018/.

D.M. Lawrence, R.A. Fisher, C.D. Koven, K.W. Oleson, S.C. Swenson, G. Bonan, N. Collier, B. Ghimire, L. van Kampenhout, D. Kennedy, E. Kluzek, P.J. Lawrence, F. Li, H. Li, D. Lombardozzi, W.J. Riley, W.J. Sacks, Mingjie Shi, M. Vertenstein, W.R. Wieder, C. Xu, A.A. Ali, A.M. Badger, G. Bisht, M. van den Broeke, M.A. Brunke, S.P. Burns, J. Buzan, M. Clark, A. Craig, K. Dahlin, B. Drewniak, J.B. Fisher, M. Flanner, A.M. Fox, P. Gentine, F. Hoffman, G. Keppel-Aleks, R. Knox, S. Kumar, J. Lenaerts, L.R. Leung, W.H. Lipscomb, Y. Lu, A. Pandey, J.D. Pelletier, J. Perket, J.T. Randerson, D.M. Ricciuto, B.M. Sanderson, A. Slater, Z.M. Subin, J. Tang, R.Q. Thomas, M.V. Martin, and X. Zeng. The community land model version 5: Description of new features, benchmarking, and impact of forcing uncertainty. *Journal of Advances in Modeling Earth Systems*, n/a(n/a). doi: 10.1029/2018MS001583. URL https://agupubs.onlinelibrary.wiley.com/doi/abs/10.1029/2018MS001583.

Q.V. Le, N. Jaitly, and G.E. Hinton. A simple way to initialize recurrent networks of rectified linear units. 2015. URL https://aps.arxiv.org/abs/1504.00941.

J. Le Moigne, N. Netanyahu, and R. Eastman. *Image Registration for Remote Sensing*. Cambridge University Press, 2011. doi: 10.1017/CBO9780511777684.

B. Le Saux, N. Yokoya, R. Hansch, M. Brown, and G. Hager. 2019 data fusion contest [technical committees]. *IEEE Geoscience and Remote Sensing Magazine*, 7(1):103–105, 2019.

T.P. Leahy, F.P. Llopis, M.D. Palmer, and N.H. Robinson. Using neural networks to correct historical climate observations. *Journal of Atmospheric and Oceanic Technology*, 35(10):2053–2059, Oct 2018. ISSN 15200426. doi: 10.1175/JTECH-D-18-0012.1.

V. Lebedev, V. Ivashkin, I. Rudenko, A. Ganshin, A. Molchanov, S. Ovcharenko, R. Grokhovetskiy, I. Bushmarinov, and D. Solomentsev. Precipitation nowcasting with satellite imagery. In *Proceedings of the 25th ACM SIGKDD International Conference on Knowledge Discovery & Data Mining*, pages 2680–2688, 2019b.

Y. LeCun, B. Boser, D. Denker, J. S. Henderson, R.E. Howard, W. Hubbard, and L.D. Jackel. Backpropagation applied to handwritten zip code recognition. *Neural Computation*, 1(4):541–551, 1989.

Y. LeCun, L. Bottou, Y. Bengio, and P. Haffner. Gradient-based learning applied to document recognition. *Proceedings of the IEEE*, 86(11):2278–2324, 1998a.

Y. LeCun, L. Bottou, G. Orr, and K. Müller. Efficient backprop. In *Neural Networks: Tricks of the Trade*, pages 9–50. Springer Berlin, 1998b.

Y. LeCun, Y. Bengio, and G. Hinton. Deep learning. *Nature*, 521: 436, 2015. doi: 10.1038/nature14539.

J.S. Lee, L. Jurkevich, P. Dewaele, P. Wambacq, and A. Oosterlinck. Speckle filtering of synthetic aperture radar images: A review. In *Remote Sensing Reviews*, pages 313–340, 1994.

A.X. Lee, R. Zhang, F. Ebert, P. Abbeel, C. Finn, and S. Levine. Stochastic adversarial video prediction. *arXiv preprint arXiv:1804.01523*, 2018a.

H. Lee, C. Ekanadham, and A.Y. Ng. Sparse deep belief net model for visual area v2. In *Advances in Neural Information Processing Systems*, pages 873–880, 2008.

H. Lee and S. Kim. Ensemble classification for anomalous propagation echo detection with clustering-based subset-selection method. *Atmosphere*, 8(1):11, 2017.

H. Lee, R. Grosse, R. Ranganath, and A.Y. Ng. Convolutional deep belief networks for scalable unsupervised learning of hierarchical representations. In *Proceedings of the 26th Annual International Conference on Machine Learning*, ICML '09, pages 609–616, New York, NY, USA, 2009. ACM. ISBN 978-1-60558-516-1. doi: 10.1145/1553374.1553453. URL http://doi.acm.org/10.1145/1553374.1553453.

H.Y. Lee, H.Y. Tseng, J.B. Huang, M. Singh, and M.H. Yang. Diverse image-to-image translation via disentangled representations. In *European Conference on Computer Vision*, pages 35–51, 2018b.

J.A. Lee and M. Verleysen. *Nonlinear Dimensionality Reduction*. Springer, 2007.

S. Lefèvre, D. Tuia, J.D. Wegner, T. Produit, and A.S. Nassar. Toward seamless multiview scene analysis from satellite to street level. *Proceedings of the IEEE*, 105(10):1884–1899, 2017.

C.E. Leith. Diffusion approximation for two-dimensional turbulence. *The Physics of Fluids*, 11(3):671–672, 1968.

C. Leng, H. Zhang, B. Li, G. Cai, Z. Pei, and L. He. Local feature descriptor for image matching: A survey. *IEEE Access*, 12 2018. doi: 10.1109/ACCESS.2018.2888856.

W.J. Leong and H.J. Horgan. Deepbedmap: Using a deep neural network to better resolve the bed topography of antarctica. *The Cryosphere Discussions*, 2020:1–27, 2020. doi: 10.5194/tc-2020-74.

V. Levizzani, P. Bauer, and F.J. Turk. *Measuring Precipitation from Space: EURAINSAT and the Future*, volume 28. Springer Science & Business Media, 2007.

C. Li, C. Xu, Z. Cui, D. Wang, T. Zhang, and J. Yang. Feature-attentioned object detection in remote sensing imagery. pages 3886–3890, 2019.

H. Li and S. Misra. Prediction of subsurface NMR T2 distributions in a shale petroleum system using variational autoencoder-based neural networks. *IEEE Geoscience and Remote Sensing Letters*, 14(12):2395–2397, 2017.

J. Li, C. Qu, and J. Shao. Ship detection in SAR images based on an improved faster R-CNN. In *2017 SAR in Big Data Era: Models, Methods and Applications (BIGSARDATA)*, pages 1–6, 2017.

J. Li, R. Cui, B. Li, R. Song, Y. Li, Y. Dai, and Q. Du. Hyperspectral image super-resolution by band attention through adversarial learning. *IEEE Transactions on Geoscience and Remote Sensing*, pages 1–15, 2020.

K. Li, G. Wan, G. Cheng, L. Meng, and J. Han. Object detection in optical remote sensing images: A survey and a new benchmark. *ISPRS Journal of Photogrammetry and Remote Sensing*, 159:296–307, 2020.

P. Li and P. Ren. Partial randomness hashing for large-scale remote sensing image retrieval. *IEEE Geoscience and Remote Sensing Letters*, 14(3):464–468, March 2017.

P.W. Li, W.K. Wong, K.Y. Chan, and E.S.T. Lai. SWIRLS-an evolving nowcasting system. *Special Administrative Region Government*, 2000.

Q. Li, L. Mou, Q. Xu, Y. Zhang, and X.X. Zhu. R^3-net: A deep network for multi-oriented vehicle detection in aerial images and videos. *IEEE Transactions on Geoscience and Remote Sensing*, 57(7): 5028–5042, 2019.

S. Li, H. Yin, and L. Fang. Remote sensing image fusion via sparse representations over learned dictionaries. *IEEE Transactions on Geoscience and Remote Sensing*, 51(9):4779–4789, Sept 2013. ISSN 0196-2892. doi: 10.1109/TGRS.2012.2230332.

S. Li, R. Dian, L. Fang, and J.M. Bioucas-Dias. Fusing hyperspectral and multispectral images via coupled sparse tensor factorization. *IEEE Transactions on Image Processing*, 27(8):4118–4130, 2018a.

W. Li, H. Fu, L. Yu, P. Gong, D. Feng, C. Li, and N. Clinton. Stacked autoencoder-based deep learning for remote-sensing image classification: a case study of African land-cover mapping. *International Journal of Remote Sensing*, 37(23):5632–5646, 2016a.

Y. Li and T.R. Bretschneider. Semantic-sensitive satellite image retrieval. *IEEE Transactions on Geoscience and Remote Sensing*, 45(4):853–860, April 2007.

Y. Li, Y. Zhang, C. Tao, and H. Zhu. Content-based high-resolution remote sensing image retrieval via unsupervised feature learning and collaborative affinity metric fusion. *Remote Sensing*, 8(9):709, August 2016b.

Y. Li, R. Bu, M. Sun, W. Wu, X. Di, and B. Chen. PointCNN: Convolution on X-transformed points. In *Advances in Neural Information Processing Systems (NeurIPS)*, pages 820–830, 2018b. NIPS.

Y. Li, Y. Zhang, X. Huang, and J. Ma. Learning source-invariant deep hashing convolutional neural networks for cross-source remote sensing image retrieval. *IEEE Transactions on Geoscience and Remote Sensing*, 56(11):6521–6536, June 2018a.

Y. Li, Y. Zhang, X. Huang, H. Zhu, and J. Ma. Large-scale remote sensing image retrieval by deep hashing neural networks. *IEEE Transactions on Geoscience and Remote Sensing*, 56(2):950–965, February 2018b.

Z. Li and H. Leung. Contour-based multisensor image registration with rigid transformation. In *2007 10th International Conference on Information Fusion*, 2007.

Z. Li, C. Peng, G. Yu, X. Zhang, Y. Deng, and J. Sun. Light-head R-CNN: In defense of two-stage object detector. *arXiv:1711.07264*, 2017.

C. Liang, H. Li, M. Lei, and Q. Du. Dongting lake water level forecast and its relationship with the Three Gorges Dam based on a long short-term memory network. *Water*, 10(10):1389, Oct 2018. ISSN 2073-4441. doi: 10.3390/w10101389. URL http://dx.doi.org/10.3390/w10101389.

X. Liang, X. Shen, J. Feng, L. Lin, and S. Yan. Semantic object parsing with graph LSTM. In *European Conference on Computer Vision (ECCV)*, pages 125–143, 2016. doi: 10.1007/978-3-319-46448-0_8.

M. Liao, B. Shi, and X. Bai. Textboxes++: A single-shot oriented scene text detector. *IEEE TIP*, 27(8):3676–3690, 2018a.

M. Liao, Z. Zhu, B. Shi, G. Xia, and X. Bai. Rotation-sensitive regression for oriented scene text detection. In *CVPR*, pages 5909–5918, 2018b.

R. Liao, S. Miao, P. Tournemire, S. Grbic, A. Kamen, T. Mansi, and D. Comaniciu. An artificial agent for robust image registration. 2016. URL https://aps.arxiv.org/abs/1611.10336.

J.-L. Lin and C.W.J. Granger. Forecasting from non-linear models in practice. *Journal of Forecasting*, 13(1):1–9, 1994.

T.-Y. Lin, M. Maire, S. Belongie, J. Hays, P. Perona, D. Ramanan, P. Dollár, and C.L. Zitnick. Microsoft COCO: Common objects in context. In *ECCV*, pages 740–755, 2014.

T.-Y. Lin, P. Dollar, R. Girshick, K. He, B. Hariharan, and S. Belongie. Feature pyramid networks for object detection. In *IEEE Conference on Computer Vision and Pattern Recognition*, pages 2117–2125, 2017.

T. Lin, W.G. Horne, P. Tiňo, and C.L. Giles. Learning long-term dependencies in NARX recurrent neural networks. *IEEE Transactions on Neural Networks*, 7(6):1329–1338, 1996. doi: 10.1109/72.548162.

Y. Lin, T. Zhang, S. Zhu, and K. Yu. Deep coding network. In *Proceedings of NIPS*, pages 1405–1413, 2010.

Y. Lin, H. He, Z. Yin, and F. Chen. Rotation-invariant object detection in remote sensing images based on radial-gradient angle. *IEEE Geoscience and Remote Sensing Letters*, 12(4):746–750, 2015.

J. Ling, R. Jones, and J. Templeton. Machine learning strategies for systems with invariance properties. *Journal of Computational Physics*, 318:22–35, 2016a.

J. Ling, A. Kurzawski, and J. Templeton. Reynolds averaged turbulence modelling using deep neural networks with embedded invariance. *Journal of Fluid Mechanics*, 807:155–166, 2016b.

Z.C. Lipton, J. Berkowitz, and C. Elkan. A critical review of recurrent neural networks for sequence learning. *arXiv:1506.00019 [cs]*, 2015.

C. Liu, J. Ma, X. Tang, X. Zhang, and L. Jiao. Adversarial hash-code learning for remote sensing image retrieval. In *IEEE International Geoscience and Remote Sensing Symposium*, pages 4324–4327, July 2019.

H. Liu and K.C. Jezek. A complete high-resolution coastline of Antarctica extracted from orthorectified Radarsat SAR imagery. *Photogrammetric Engineering & Remote Sensing*, 70(5):605–616, 2004.

J. Liu, S. Ji, C. Zhang, and Z. Qin. Evaluation of deep learning based stereo matching methods: from ground to aerial images. *ISPRS – International Archives of the Photogrammetry, Remote Sensing and Spatial Information Sciences*, 422: 593–597, May 2018a. doi: 10.5194/isprs-archives-XLII-2-593-2018.

K. Liu and G. Máttyus. Fast multiclass vehicle detection on aerial images. *IEEE Geoscience and Remote Sensing Letters*, 12(9):1938–1942, 2015.

L. Liu and B. Lei. Can SAR images and optical images transfer with each other? In *IGARSS 2018 – 2018 IEEE International Geoscience and Remote Sensing Symposium*, pages 7019–7022, July 2018.

L. Liu, Z. Pan, X. Qiu, and L. Peng. Sar target classification with cyclegan transferred simulated samples. In *IGARSS 2018 – 2018 IEEE International Geoscience and Remote Sensing Symposium*, pages 4411–4414, July 2018b.

L. Liu, Z. Pan, and B. Lei. Learning a rotation invariant detector with rotatable bounding box. *CoRR*, abs/1711. 09405, 2017a.

M.Y. Liu and O. Tuzel. Coupled generative adversarial networks. In *Advances in Neural Information Processing Systems*, pages 469–477, 2016.

M.Y. Liu, T. Breuel, and J. Kautz. Unsupervised image-to-image translation networks. In *Advances in Neural Information Processing Systems*, pages 700–708, 2017b.

P. Liu, J. Wang, A.K. Sangaiah, Y. Xie, and X. Yin. Analysis and prediction of water quality using LSTM deep neural networks in iot environment. *Sustainability*, 11(7), 2019. ISSN 2071-1050. doi: 10.3390/su11072058. URL https://www.mdpi.com/2071-1050/11/7/2058.

W. Liu, F. Su, and X. Huang. Unsupervised adversarial domain adaptation network for semantic segmentation. *IEEE Geoscience and Remote Sensing Letters*, pages 1–5, 2019.

W. Liu, D. Anguelov, D. Erhan, C. Szegedy, S. Reed, C.-Y. Fu, and A.C. Berg. Ssd: Single shot multibox detector. *arXiv preprint arXiv:1512.02325*, 2015.

X. Liu, Y. Wang, and Q. Liu. Psgan: A generative adversarial network for remote sensing image pan-sharpening. In *2018 25th IEEE International Conference on Image Processing (ICIP)*, pages 873–877, 2018c.

Y. Liu, B. Fan, L. Wang, J. Bai, S. Xiang, and C. Pan. Semantic labeling in very high resolution images via a self-cascaded convolutional neural network. *ISPRS J. Int. Soc. Photo. Remote Sens.*, 145:78–95, 2018.

Y. Liu, C.R. Schwalm, K.E. Samuels-Crow, and K. Ogle. Ecological memory of daily carbon exchange across the globe and its importance in drylands. *Ecology Letters*, 22:1806–1816, 2019. doi: 10.1111/ele.13363.

Y. Liu, Prabhat, E. Racah, J. Correa, A. Khosrowshahi, D. Lavers, K. Kunkel, M. Wehner, and W. Collins. Application of deep convolutional neural networks for detecting extreme weather in climate datasets. In *Advances in Big Data Analytics*, pages 81–88, 2016b.

Y. Liu, Prabhat, E. Racah, J. Correa, A. Khosrowshahi, D. Lavers, K. Kunkel, M. Wehner, and William Collins. Extreme weather pattern detection using deep convolutional neural network. In *Proceedings of the 6th International Workshop on Climate Informatics*, pages 109–112, 2016c.

Z. Liu, H. Wang, L. Weng, and Y. Yang. Ship rotated bounding box space for ship extraction from high-resolution optical satellite images with complex backgrounds. *IEEE Geosci. Remote Sensing Letters*, 13 (8):1074–1078, 2016d.

Z. Liu, J. Hu, L. Weng, and Y. Yang. Rotated region based CNN for ship detection. In *ICIP*, pages 900–904. IEEE, 2017c.

D.B. Lobell, A. Sibley, and J.I. Ortiz-Monasterio. Extreme heat effects on wheat senescence in India. *Nature Climate Change*, 2:186–189, 2012. doi: 10.1038/nclimate1356.

S. Lobry, J. Murray, D. Marcos, and D. Tuia. Visual question answering from remote sensing images. In *IGARSS 2019-2019 IEEE International Geoscience and Remote Sensing Symposium*, pages 4951–4954. IEEE, 2019.

P.C. Loikith, B.R. Lintner, and A. Sweeney. Characterizing large-scale meteorological patterns and associated temperature and precipitation extremes over the northwestern united states using self-organizing maps. *Journal of Climate*, 30(8):2829–2847, 2017.

L. Loncan, L.B. De Almeida, J.M. Bioucas-Dias, X. Briottet, J. Chanussot, N. Dobigeon, S. Fabre, W. Liao, G.A. Licciardi, M. Simoes, et al. Hyperspectral pansharpening: A review. *IEEE Geoscience and remote sensing magazine*, 3(3):27–46, 2015.

J. Long, E. Shelhamer, and T. Darrell. Fully convolutional networks for semantic segmentation. In *Proceedings of the IEEE Conference on Computer Vision and Pattern Recognition*, pages 3431–3440, 2015.

E.N. Lorenz. Empirical orthogonal functions and statistical weather prediction. *Scientific Reports 1, Statistical Forecasting Project*, 1956.

E.N. Lorenz. Deterministic Nonperiodic Flow. *Journal of the Atmospheric Sciences*, 20(2):130–141, March 1963. doi: 10.1175/1520-0469(1963)020⟨0130:dnf⟩2.0.co;2.

E.N. Lorenz. The predictability of a flow which possesses many scales of motion. *Tellus*, 21(3):289–307, 1969. doi: 10.1111/j.2153-3490.1969.tb00444.x. URL https://onlinelibrary.wiley.com/doi/abs/10.1111/j.2153-3490.1969.tb00444.x .

E.N. Lorenz. Predictability: A problem partly solved. In *Proceedings of the Seminar on Predictability*, volume 1, 1996.

D.G. Lowe. Object recognition from local scale-invariant features. In *Proceedings of the Seventh IEEE International Conference on Computer Vision*, volume 2, Sep. 1999. doi: 10.1109/ICCV.1999.790410.

D.G. Lowe. Object recognition from local scale-invariant features. In *Proceedings of the Seventh IEEE International Conference on Computer Vision*, volume 2, pages 1150–1157. Ieee, 1999.

X. Lu, Y. Yuan, and X. Zheng. Joint dictionary learning for multispectral change detection. *IEEE Transactions on Cybernetics*, 47(4): 884–897, 2016.

T. Luo, K. Kramer, D.B. Goldgof, L.O. Hall, S. Samson, A. Remsen, and T. Hopkins. Active learning to recognize multiple types of plankton. *Journal of Machine Learning Research*, 6:589–613, 2005.

B. Lusch, J.N. Kutz, and S.L. Brunton. Deep learning for universal linear embeddings of nonlinear dynamics. *Nature Communications*, 9(1):4950, 2018.

N. Lv, C. Chen, T. Qiu, and A.K. Sangaiah. Deep learning and superpixel feature extraction based on contractive autoencoder for change detection in SAR images. *IEEE transactions on industrial informatics*, 14(12): 5530–5538, 2018.

M. Kang, K. Ji, X. Leng and Z. Lin. Contextual region-based convolutional neural network with multilayer fusion for SAR ship detection. *Remote Sensing*, 9(8): 860–, 2017.

H.-Y. Ma, S. Xie, S.A. Klein, K.D. Williams, J.S. Boyle, S. Bony, H. Douville, S. Fermepin, B. Medeiros, S. Tyteca, M. Watanabe, and D. Williamson. On the correspondence between mean forecast errors and climate errors in CMIP5 Models. *Journal of Climate*, 27(4):1781–1798, Nov 2013. ISSN 0894-8755. doi: 10.1175/JCLI-D-13-00474.1. URL https://doi.org/10.1175/JCLI-D-13-00474.1.

J. Ma, W. Shao, H. Ye, L. Wang, H. Wang, Y. Zheng, and X. Xue. Arbitrary-oriented scene text detection via rotation proposals. *IEEE Transactions on Multimedia*, 20(11):3111–3122, 2018.

K. Ma, D. Feng, K. Lawson, W.-P. Tsai, C. Liang, X. Huang, A. Sharma, and C. Shen. Transferring hydrologic data across continents – leveraging data-rich regions to improve hydrologic prediction in data-sparse regions. *Water Resources Research*, 57, e2020WR028600, 2021. https://doi.org/10.1029/2020wr028600.

L. Ma, Y. Liu, X. Zhang, Y. Ye, G. Yin, and B.A. Johnson. Deep learning in remote sensing applications: A meta-analysis and review. *ISPRS Journal of Photogrammetry and Remote Sensing*, 152:166–177, 2019.

W. Ma, J. Zhang, Y. Wu, L. Jiao, H. Zhu, and W. Zhao. A novel two-step registration method for remote sensing images based on deep and local features. *IEEE Transactions on Geoscience and Remote Sensing*, 57(7): 4834–4843, July 2019. ISSN 1558-0644. doi: 10.1109/TGRS.2019.2893310.

E. Maggiori, Y. Tarabalka, G. Charpiat, and P. Alliez. Convolutional neural networks for large-scale remote-sensing image classification. *IEEE Transactions in Geoscience and Remote Sensing*, 55(2):645–657, 2017a.

E. Maggiori, Y. Tarabalka, G. Charpiat, and P. Alliez. High-resolution aerial image labeling with convolutional neural networks. *IEEE Transactions in Geoscience and Remote Sensing*, 55(12):7092–7103, 2017b.

E. Maggiori, Y. Tarabalka, G. Charpiat, and P. Alliez. Can semantic labeling methods generalize to any city? The inria aerial image labeling benchmark. In *IEEE International Geoscience and Remote Sensing Symposium (IGARSS)*. IEEE, 2017c.

L. Magnusson and E. Källén. Factors influencing skill improvements in the ECMWF forecasting system. *Monthly Weather Review*, 141(9): 3142–3153, 2013. doi: 10.1175/MWR-D-12-00318.1.

D. Mahapatra and Z. Ge. Combining transfer learning and segmentation information with gans for training data independent image registration. *CoRR*, abs/1903.10139, 2019. URL http://arxiv.org/abs/1903.10139.

J. Mairal, M. Elad, and G. Sapiro. Sparse representation for color image restoration. *IEEE Transactions on Image Processing*, 17(1):53–69, 2008.

J. Mairal, F. Bach, and J. Ponce. Task-driven dictionary learning. *IEEE Transactions on Pattern Analysis and Machine Intelligence*, 34(4): 791–804, 2011.

D. Malmgren-Hansen, V. Laparra, A.A. Nielsen, and G. Camps-Valls. Statistical retrieval of atmospheric profiles with deep convolutional neural networks. *ISPRS Journal of Photogrammetry and Remote Sensing*, 158:231–240, 2019.

D. Malmgren-Hansen, V. Laparra, G. Camps-Valls, X. Calbet. IASI dataset v1. Technical University of Denmark. Dataset. 2020. https://doi.org/10.11583/DTU.12999642.v1

D. Malmgren-Hansen, L.T. Pedersen, A.A. Nielsen, H. Skriver, R. Saldo, M.B. Kreiner, and J. Buus-Hinkler. ASIP Sea Ice Dataset – version 1. 3 2020. doi: 10.11583/DTU.11920416.v1.

P.P. Mana and L. Zanna. Toward a stochastic parameterization of ocean mesoscale eddies. *Ocean Modelling*, 79: 1–20, 2014.

D. Marcos, R. Hamid, and D. Tuia. Geospatial correspondence for multimodal registration. In *IEEE International Conference on Computer Vision and Pattern Recognition (CVPR)*, 2016.

D. Marcos, M. Volpi, B. Kellenberger, and D. Tuia. Land cover mapping at very high resolution with rotation equivariant CNNs: Towards small yet accurate models. *ISPRS Journal of the International Society of Photogrammetry and Remote Sensing*, 145:96–107, 2018a.

D. Marcos, S. Lobry, and D. Tuia. Semantically interpretable activation maps: what-where-how explanations within CNNs. In *International Conference on Computer Vision Workshops (ICCVW)*, Workshop "Interpreting and explaining visual artificial intelligence models", Seoul, Korea, 2019.

D. Marcos, D. Tuia, B. Kellenberger, L. Zhang, M. Bai, R. Liao, and R. Urtasun. Learning deep structured active contours end-to-end. In *Proceedings of the IEEE Conference on Computer Vision and Pattern Recognition*, pages 8877–8885, 2018b.

G.P. Marino, D.P. Kaiser, L. Gu, and D.M. Ricciuto. Reconstruction of false spring occurrences over the southeastern United States, 1901–2007: an increasing risk of spring freeze damage? *Environmental Research Letters*, 6:024015, 2011. doi: 10.1088/1748-9326/6/2/024015.

D. Marmanis, M. Datcu, T. Esch, and U. Stilla. Deep learning Earth observation classification using ImageNet pretrained networks. *IEEE Geoscience and Remote Sensing Letters*, 13(1):105–109, 2015.

D. Marmanis, K. Schindler, J.D. Wegner, S. Galliani, M. Datcu, and U. Stilla. Classification with an edge: Improving semantic image segmentation with boundary detection. *ISPRS Journal of the International Society of Photogrammetry and Remote Sensing*, 135:158–172, 2018.

D. Marmanis, W. Yao, F. Adam, M. Datcu, P. Reinartz, K. Schindler, J.D. Wegner, and U. Stilla. Artificial generation of big data for improving image classification: A generative adversarial network approach on SAR data, 2017.

P. Márquez-Neila, M. Salzmann, and P. Fua. Imposing hard constraints on deep networks: Promises and limitations. *arXiv preprint arXiv:1706.02025*, 2017.

S.J. Marshall. *The Cryosphere*, volume 2. Princeton University Press, 2011.

Z.C. Marton, R.B. Rusu, and M. Beetz. On Fast Surface Reconstruction Methods for Large and Noisy Datasets. In *International Conference on Robotics and Automation (ICRA)*, May 12–17 2009.

M. Martone, P. Rizzoli, C. Wecklich, C. González, J.-L. Bueso-Bello, P. Valdo, D. Schulze, M. Zink, G. Krieger, and A. Moreira. The global forest/non-forest map from tandem-x interferometric SAR data. *Remote Sensing of Environment*, 205: 352–373, 2018.

J. Masci, U. Meier, D. Cireşan, and J. Schmidhuber. Stacked convolutional auto-encoders for hierarchical feature extraction. In *Proceedings of the 21th International Conference on Artificial Neural Networks – Volume Part I*, ICANN'11, pages 52–59, Berlin, Heidelberg, 2011. Springer-Verlag. ISBN 978-3-642-21734-0. URL http://dl.acm.org/citation.cfm?id=2029556.2029563.

G. Masi, D. Cozzolino, L. Verdoliva, and G. Scarpa. Pansharpening by convolutional neural networks. *Remote Sensing*, 8(7):594, 2016.

G. Mateo-García, V. Laparra, and L. Gómez-Chova. Domain adaptation of Landsat-8 and Proba-V data using generative adversarial networks for cloud detection. In *IGARSS 2019 – 2019 IEEE International Geoscience and Remote Sensing Symposium*, pages 712–715, July 2019. doi: 10.1109/IGARSS.2019.8899193.

G. Mateo-García, L. Gómez-Chova, J. Amorós-López, J. Muñoz-Marí, and G. Camps-Valls. Multitemporal cloud masking in the Google Earth engine. *Remote Sensing*, 10(7):1079, 2018.

G. Mateo-García, V. Laparra, D. López-Puigdollers, and L. Gómez-Chova. Transferring deep learning models for cloud detection between Landsat-8 and Proba-V. *ISPRS Journal of Photogrammetry and Remote Sensing*, 160:1–17, 2020.

G. Mateo-García, V. Laparra, D. López-Puigdollers, and L. Gómez-Chova. Transferring deep learning models for cloud detection between Landsat-8 and Proba-V. *ISPRS Journal of Photogrammetry and Remote Sensing*, 160:1–17, February 2020. ISSN 0924-2716. doi: http://dx.doi.org/10.1016/j.isprsjprs.2019.11.024 . URL https://authors.elsevier.com/a/1aCri3I9x1YsMH.

B.V. Matheussen, O.-C. Granmo, and J. Sharma. Hydropower optimization using deep learning. In F. Wotawa, G. Friedrich, I. Pill, R. Koitz-Hristov, and M. Ali, editors, *Advances and Trends in Artificial Intelligence. From Theory to Practice*, pages 110–122, Cham, 2019. Springer International Publishing. ISBN 978-3-030-22999-3.

M. Matsueda, A.Weisheimer, and T.N. Palmer. Calibrating climate change time-slice projections with estimates of seasonal forecast reliability. *Journal of Climate*, 29(10): 3831–3840, Mar 2016. ISSN 0894-8755. doi: 10.1175/JCLI-D-15-0087.1. URL https://doi.org/10.1175/JCLI-D-15-0087.1.

D. Maturana and S. Scherer. Voxnet: A 3d convolutional neural network for real-time object recognition. In *International Conference on Intelligent Robots and Systems (IROS)*, pages 922–928, 2015.

R. Maulik and O. San. A neural network approach for the blind deconvolution of turbulent flows. *Journal of Fluid Mechanics*, 831:151–181, 2017.

R. Maulik, O. San, A. Rasheed, and P. Vedula. Subgrid modelling for two-dimensional turbulence using neural networks. *Journal of Fluid Mechanics*, 858:122–144, 2019.

R.M. Maxwell, M. Putti, S. Meyerhoff, J.-O. Delfs, I.M. Ferguson, V. Ivanov, J. Kim, O. Kolditz, S.J. Kollet, M. Kumar, S. Lopez, J. Niu, C. Paniconi, Y.-J. Park, M.S. Phanikumar, C. Shen, E.A. Sudicky, and M. Sulis. Surface-subsurface model intercomparison: A first set of benchmark results to diagnose integrated hydrology and feedbacks. *Water Resources Research*, 50 (2):1531–1549, Feb 2014. ISSN 00431397. doi: 10.1002/2013WR013725. URL http://doi.wiley.com/10.1002/2013WR013725.

R. McAllister and J. Sheppard. Deep learning for wind vector determination. In *2017 IEEE Symposium Series on Computational Intelligence* (eds. Bonissone, P. & Fogel, D.), pages 1–8. IEEE, 2017.

R. McAllister and J. Sheppard. Evaluating spatial generalization of stacked autoencoders in wind vector determination. In *Proceedings of the Thirty-First International Florida Artificial Intelligence Research Society Conference, FLAIRS 2018, Melbourne, Florida, USA. May 21-23 2018* (eds. Brawner, K. & Rus, V.), pages 68–73, 2018.

A. McGovern, K.L. Elmore, D.J. Gagne, S.E. Haupt, C.D. Karstens, R. Lagerquist, T. Smith, and J.K. Williams. Using artificial intelligence to improve real-time decision-making for high-impact weather. *Bulletin of the American Meteorological Society*, 2017. ISSN 00030007. doi: 10.1175/BAMS-D-16-0123.1.

G.A. Meehl, C.A. Senior, V. Eyring, G. Flato, J.-F. Lamarque, R.J. Stouffer, K.E. Taylor, and M. Schlund. Context for interpreting equilibrium climate sensitivity and transient climate response from the CMIP6 earth system models. *Science Advances*, in review, 2020.

M.J. Mei, T. Maksym, B.Weissling, and H. Singh. Estimating early-winter antarctic sea ice thickness from deformed ice morphology. *The Cryosphere*, 13(11):2915–2934, 2019. doi: 10.5194/tc-13-2915-2019.

Q. Meng, D. Catchpoole, D. Skillicom, and P.J. Kennedy. Relational autoencoder for feature extraction. In *2017 International Joint Conference on Neural Networks (IJCNN)*, pages 364–371. IEEE, 2017.

N. Merkle, P. Fischer, S. Auer, and R. Müller. On the possibility of conditional adversarial networks for multi-sensor image matching. In *2017 IEEE International Geoscience and Remote Sensing Symposium (IGARSS)*, pages 2633–2636, July 2017. doi: 10.1109/IGARSS.2017.8127535.

N. Merkle,W. Luo, S. Auer, R. Müller, and R. Urtasun. Exploiting deep matching and SAR data for the geo-localization accuracy improvement of optical satellite images. *Remote Sensing*, 9(6), 2017. ISSN 2072-4292. doi: 10.3390/rs9060586.

N. Merkle, S. Auer, R. Müller, and P. Reinartz. Exploring the potential of conditional adversarial networks for optical and SAR image matching. *IEEE Journal of Selected Topics in Applied Earth Observations and Remote Sensing*, 11(6):1811–1820, June 2018. ISSN 2151-1535.

K. Miao, et al. Contextual region-based convolutional neural network with multilayer fusion for SAR ship detection. *Remote Sensing 9.8*, 860, 2017.

S. Minaee, Y. Boykov, F. Porikli, A. Plaza, N. Kehtarnavaz, and D. Terzopoulos. Image segmentation using deep learning: A survey. *arXiv preprint arXiv:2001.05566*, 2020.

L. Mingsheng, C. Yue, W. Jianmin, and L. Michael. Learning transferable features with deep adaptation networks. In *International Conference on Machine Learning*, volume 37, pages 97–105, 2015.

M. Mirza and S. Osindero. Conditional generative adversarial nets, 2014. URL http://arxiv.org/abs/1411.1784 . cite arxiv:1411.1784.

V. Mnih and G.E. Hinton. Learning to label aerial images from noisy data. In *Proceedings of the 29th International Conference on Machine Learning (ICML-12)*, pages 567–574, 2012.

S. Mo, N. Zabaras, X. Shi, and J. Wu. Deep autoregressive neural networks for high-dimensional inverse problems in groundwater contaminant source identification. *Water Resources Research*, 55 (5):3856–3881, 2019a. doi: 10.1029/2018WR024638. URL https://agupubs.onlinelibrary.wiley.com/doi/abs/10.1029/2018WR024638.

S. Mo, Y. Zhu, N. Zabaras, X. Shi, and J. Wu. Deep convolutional encoder-decoder networks for uncertainty quantification of dynamic multiphase flow in heterogeneous media. *Water Resources Research*, 55(1):703–728, 2019b. doi: 10.1029/2018WR023528. URL https://agupubs.onlinelibrary.wiley.com/doi/abs/10.1029/2018WR023528.

S. Mohajerani and P. Saeedi. Cloud-net: An end-to-end cloud detection algorithm for Landsat 8 imagery. In *IGARSS 2019 - 2019 IEEE International Geoscience and Remote Sensing Symposium*, pages 1029–1032, July 2019. doi: 10.1109/IGARSS.2019.8898776.

S. Mohajerani and P. Saeedi. Cloud-net$^+$: A cloud segmentation CNN for Landsat 8 remote sensing imagery optimized with filtered jaccard loss function, 2020.

Y. Mohajerani, M. Wood, I. Velicogna, and E. Rignot. Detection of glacier calving margins with convolutional neural networks: A case study. *Remote Sensing*, 11(1), 2019. doi: 10.3390/rs11010074.

M.R. Mohammadi. Deep multiple instance learning for airplane detection in high resolution imagery. *CoRR*, abs/1808.06178, 2018. URL http://arxiv.org/abs/1808.06178.

S. Molins, D. Trebotich, C.I. Steefel, and C. Shen. An investigation of the effect of pore scale flow on average geochemical reaction rates using direct numerical simulation. *Water Resources Research*, 48(3): n/a–n/a, Mar 2012. ISSN 00431397. doi: 10.1029/2011WR011404.

C. Molnar. Interpretable Machine Learning. 2019. https://christophm.github.io/interpretable-ml-book/.

G. Montavon, W. Samek, and K.-R. Müller. Methods for interpreting and understanding deep neural networks. *Digital Signal Processing*, 73:1–15, 2018.

R. Montes and C. Ureña. An overview of brdf models. *University of Grenada, Technical Report LSI-2012*, 1, 2012.

A. Moosavi, Ahmed Attia, and Adrian Sandu. Tuning Covariance Localization Using Machine Learning, 2019.

T. Moranduzzo and F. Melgani. Detecting cars in UAV images with a catalog-based approach. *IEEE Transactions in Geoscience and Remote Sensing*, 52(10): 6356–6367, 2014.

Á. Moreno-Martínez, G. Camps-Valls, J. Kattge, N. Robinson, M. Reichstein, P. van Bodegom, K. Kramer, J.H.C. Cornelissen, P. Reich, M. Bahn, et al. A methodology to derive global maps of leaf traits using remote sensing and climate data. *Remote Sensing of Environment*, 218:69–88, 2018.

L. Mou, L. Bruzzone, and X.X. Zhu. Learning spectral-spatial-temporal features via a recurrent convolutional neural network for change detection in multispectral imagery. *IEEE Transactions on Geoscience and Remote Sensing*, 57(2):924–935, 2018.

L. Mou, Y. Hua, and X.X. Zhu. A relation-augmented fully convolutional network for semantic segmentation in aerial scenes. In *Proceedings of the IEEE Conference on Computer Vision and Pattern Recognition*, pages 12416–12425, 2019.

S. Mouatadid, J.F. Adamowski, M.K. Tiwari, and J.M. Quilty. Coupling the maximum overlap discrete wavelet transform and long short-term memory networks for irrigation flow forecasting. *Agricultural Water Management*, 219:72–85, 2019. ISSN 0378-3774. doi: https://doi.org/10.1016/j.agwat.2019.03.045. URL http://www.sciencedirect.com/science/article/pii/S0378377418311831.

M. Mudigonda, S. Kim, A. Mahesh, S. Kahou, K. Kashinath, D. Williams, V. Michalski, T. O'Brien, and Prabhat. Segmenting and Tracking Extreme Climate Events using Neural Networks. Technical report, 2017.

M. Mudigonda, K. Kashinath, Prabhat, S. Kim, L. Kapp-Schoerer, E. Karaismailoglu, A. Graubner, L. von Kleist, K. Yang, C. Lewis, J. Chen, A. Greiner, T. Kurth, T. O'Brien, W. Chapman, C. Shields, K. Dagon, A. Albert, M. Wehner, and W. Collins. Climatenet: Bringing the power of deeplearning to weather and climate sciences via open datasets and architectures. In *SC18: International Conference for High Performance Computing, Networking, Storage and Analysis*, pages 649–660. IEEE, 2018.

M. Munir, S.A. Siddiqui, A. Dengel, and S. Ahmed. Deepant: A deep learning approach for unsupervised anomaly detection in time series. *IEEE Access*, 7:1991–2005, 2018.

H. Murakami. Tropical cyclones in reanalysis data sets. *Geophysical Research Letters*, 41(6):2133–2141, 2014. ISSN 1944-8007. doi: 10.1002/2014GL059519. URL http://dx.doi.org/10.1002/2014GL059519. 2014GL059519.

H. Murakami, Y. Wang, H. Yoshimura, R. Mizuta, M. Sugi, E. Shindo, Y. Adachi, S. Yukimoto, M. Hosaka, S. Kusunoki, T. Ose, and A. Kitoh. Future changes in tropical cyclone activity projected by the new high-resolution mri-agcm. *Journal of Climate*, 25(9):3237–3260, 2012. doi: 10.1175/JCLI-D-11-00415.1. URL https://doi.org/10.1175/JCLI-D-11-00415.1.

J.M. Murphy, D. M.H. Sexton, D.N. Barnett, G.S. Jones, M.J. Webb, M. Collins, and D.A. Stainforth. Quantification of modelling uncertainties in a large ensemble of climate change simulations. *Nature*, 430(7001):768, 2004.

V. Nair and G.E. Hinton. Rectified linear units improve restricted Boltzmann machines. In *Proceedings of the 27th International Conference on Machine Learning (ICML-10)*, pages 807–814, 2010.

J.E. Nash and J.V. Sutcliffe. River flow forecasting through conceptual models part i—a discussion of principles. *Journal of Hydrology*, 10(3):282–290, 1970. doi: 10.1016/0022-1694(70)90255-6.

National Academies of Sciences Engineering and Medicine. *Attribution of Extreme Weather Events in the Context of Climate Change*. The National Academies Press, Washington, DC, 2016. ISBN 978-0-309-38094-2. doi: 10.17226/21852. URL http://www.nap.edu/catalog/21852.

P.J. Neiman, F.M. Ralph, G.A. Wick, J.D. Lundquist, and M.D. Dettinger. Meteorological characteristics and overland precipitation impacts of atmospheric rivers affecting the West coast of North America based on eight years of SSM/I satellite observations. *Journal of Hydrometeorology*, 9(1):22–47, 2008.

U. Neu, M.G. Akperov, N. Bellenbaum, R. Benestad, R. Blender, R. Caballero, A. Cocozza, H.F. Dacre, Y. Feng, K. Fraedrich, et al. Imilast: A community effort to intercompare extratropical cyclone detection and tracking algorithms. *Bulletin of the American Meteorological Society*, 94(4):529–547, 2013.

J. Ngiam, P.W. Koh, Z. Chen, S. Bhaskar, and A.Y. Ng. Sparse filtering. In *NIPS*, pages 1125–1133, 2011.

J.-M. Nicolas and J. Inglada. *Image Geometry and Registration*, chapter 2, pages 33–52. John Wiley & Sons, Ltd, 2014. ISBN 9781118899106. doi: 10.1002/9781118899106.ch2. URL https://onlinelibrary.wiley.com/doi/abs/10.1002/9781118899106.ch2.

A.A. Nielsen. An extension to a filter implementation of a local quadratic surface for image noise estimation. In *Proceedings 10th International Conference on Image Analysis and Processing*, pages 119–124. IEEE, 1999.

M.A. Nielsen. Neural Networks and Deep Learning. 2015. URL http://neuralnetworksanddeeplearning.com.

R. Nijhawan, J. Das, and B. Raman. A hybrid of deep learning and hand-crafted features based approach for snow cover mapping. *International Journal of Remote Sensing*, 40(2):759–773, 01 2019. doi: 10.1080/01431161.2018.1519277.

S. Niu, Y. Luo, D. Li, S. Cao, J. Xia, J. Li, and M.D. Smith. Plant growth and mortality under climatic extremes: An overview. *Environmental and Experimental Botany*, 98:13–19, 2014. doi: 10.1016/j.envexpbot.2013.10.004.

X. Niu, M. Gong, T. Zhan, and Y. Yang. A conditional adversarial network for change detection in heterogeneous images. *IEEE Geoscience and Remote Sensing Letters*, 16(1):45–49, Jan 2019.

J. Nocedal and S. Wright. *Numerical Optimization*. Springer Science & Business Media, 2006.

K. Nock, E. Gilmour, P. Elmore, E. Leadbetter, N. Sweeney, and F. Petry. Deep learning on hyperspectral data to obtain water properties and bottom depths. In *Signal Processing,*

Sensor/Information Fusion, and Target Recognition XXVIII, volume 11018, page 110180Y. International Society for Optics and Photonics, 2019.

H. Noh, S. Hong, and B. Han. Learning deconvolution network for semantic segmentation. In *International Conference on Computer Vision (ICCV)*, pages 1520–1528, 2015.

D.S. Nolan and M.G. McGauley. Tropical cyclogenesis in wind shear: Climatological relationships and physical processes. In *Cyclones: Formation, Triggers, and Control*, pages 1–36. 2012.

P.D. Nooteboom, Q.Y. Feng, C. López, E. Hernández-García, and H. A. Dijkstra. Using network theory and machine learning to predict el niño. *Earth System Dynamics*, 9(3):969–983, 2018. doi: 10.5194/esd-9-969-2018. URL https://www.earth-syst-dynam.net/9/969/2018/.

D.O. North. An analysis of the factors which determine signal/noise discrimination in pulsed-carrier systems. *Proceedings of the IEEE*, 51(7): 1016–1027, 1963.

L.M. Novak and M.C. Burl. Optimal speckle reduction in polarimetric SAR imagery. *IEEE Transactions on Aerospace and Electronic Systems*, 26(2): 293–305, 1988.

K. Ogle, J.J. Barber, G.A. Barron-Gafford, L. Patrick Bentley, J.M. Young, T.E. Huxman, M.E. Loik, and D.T. Tissue. Quantifying ecological memory in plant and ecosystem processes. *Ecology Letters*, 18:221–235, 2015. doi: 10.1111/ele.12399.

P.A. O'Gorman and J.G. Dwyer. Using machine learning to parameterize moist convection: Potential for modeling of climate, climate change, and extreme events. *Journal of Advances in Modeling Earth Systems*, 10(10): 2548–2563, 2018.

A. Özgün Ok, Ç. Senaras, and B. Yüksel. Automated detection of arbitrarily shaped buildings in complex environments from monocular VHR optical satellite imagery. *IEEE Transactions on Geoscience and Remote Sensing*, 51(3-2): 1701–1717, 2013.

D.A.B. Oliveira, R.S. Ferreira, R. Silva, and E.V. Brazil. Improving seismic data resolution with deep generative networks. *IEEE Geoscience and Remote Sensing Letters*, 16(12):1929–1933, Dec 2019.

B. Olshausen and D.J. Field. Sparse coding with an overcomplete basis set: a strategy employed by V1? *Vision Research*, 37(23):3311–3325, 1997.

M. Omlin and P. Reichert. A comparison of techniques for the estimation of model prediction uncertainty. *Ecological Modelling*, 115:45–59, 1999.

I.H. Onarheim, T. Eldevik, L.H. Smedsrud, and J.C. Stroeve. Seasonal and regional manifestation of Arctic sea ice loss. *Journal of Climate*, 31(12):4917–4932, 2018.

A. van den Oord, S. Dieleman, H. Zen, K. Simonyan, O. Vinyals, A. Graves, N. Kalchbrenner, A. Senior, and K. Kavukcuoglu. Wavenet: A generative model for raw audio. *arXiv preprint arXiv:1609.03499*, 2016.

K. Oouchi, J. Yoshimura, H. Yoshimura, R. Mizuta, S. Kusonoki, and A. Noda. Tropical cyclone climatology in a global-warming climate as simulated in a 20 km-mesh global atmospheric model: Frequency and wind intensity analyses. *Journal of the Meteorological Society of Japan. Ser. II*, 84(2):259–276, 2006. doi: 10.2151/jmsj.84.259.

E. Othman, Y. Bazi, N. Alajlan, H. Alhichri, and F. Melgani. Using convolutional features and a sparse autoencoder for land-use scene classification. *International Journal of Remote Sensing*, 37(10):2149–2167, 2016.

B. Oueslati and G. Bellon. The double ITCZ bias in CMIP5 models: interaction between SST, large-scale circulation and precipitation. *Climate Dynamics*, 44(3-4):585–607, 2015.

W. Ouyang, K. Lawson, D. Feng, L. Ye, C. Zhang, C. Shen, Continental-scale streamflow modeling of basins with reservoirs: A demonstration of effectiveness and a delineation of challenges, 2020. https://arxiv.org/abs/2101.04423.

F. Pacifici, M. Chini, and W.J. Emery. A neural network approach using multi-scale textural metrics from very high-resolution panchromatic imagery for urban land-use classification. *Remote Sensing of Environment*, 113(6):1276–1292, 2009.

S. Paisitkriangkrai, J. Sherrah, P. Janney, and A. Van Den Hengel. Semantic labeling of aerial and satellite imagery. *IEEE Journal of Selected Topics in Applied Earth Observations and Remote Sensing*, 9(7):2868–2881, 2016.

A. Pal, S. Mahajan, and M.R. Norman. Using deep neural networks as cost-effective surrogate models for super-parameterized e3sm radiative transfer. *Geophysical Research Letters*, 46(11):6069–6079, 2019. doi: 10.1029/2018GL081646.

T.N. Palmer, F.J. Doblas-Reyes, A. Weisheimer, and M.J. Rodwell. Toward seamless prediction: Calibration of climate change projections using seasonal forecasts. *Bulletin of the American Meteorological Society*, 89:459–470, 2008. ISSN 0003-0007. doi: 10.1175/BAMS-89-4-459.

T.N. Palmer, A. Döring, and G. Seregin. The real butterfly effect. *Nonlinearity*, 27(9):R123–R141, Aug 2014.

F. Palsson, J.R. Sveinsson, and M.O. Ulfarsson. Multispectral and hyperspectral image fusion using a 3-D-convolutional neural network. *IEEE Geoscience and Remote Sensing Letters*, 14(5):639–643, 2017.

B. Pan, K. Hsu, A. AghaKouchak, and S. Sorooshian. Improving precipitation estimation using convolutional neural network. *Water Resources Research*, 55(3):2301–2321, 2019. doi: 10.1029/2018WR024090. URL https://agupubs.onlinelibrary.wiley.com/doi/abs/10.1029/2018WR024090.

J. Pan, Y. Yin, J. Xiong, W. Luo, G. Gui, and H. Sari. Deep learning-based unmanned surveillance systems for observing water levels. *IEEE Access*, 6: 73561–73571, 2018. ISSN 2169-3536. doi: 10.1109/ACCESS.2018.2883702.

C. Papagiannopoulou, D.G. Miralles, W.A. Dorigo, N.E.C. Verhoest, M. Depoorter, and W. Waegeman. Vegetation anomalies caused by antecedent precipitation in most of the world. *Environmental Research Letters*, 12:074016, 2017. doi: 10.1088/1748-9326/aa7145.

R. Pascanu, T. Mikolov, and Y. Bengio. On the difficulty of training recurrent neural networks. In Sanjoy Dasgupta and David McAllester, editors, *Proceedings of the 30th International Conference on Machine Learning, volume 28 of Proceedings of Machine Learning Research*, pages 1310–1318, 2013.

J. Pathak, Z. Lu, B.R. Hunt, M. Girvan, and E. Ott. Using machine learning to replicate chaotic attractors and calculate Lyapunov exponents from data. *Chaos: An Interdisciplinary Journal of Nonlinear Science*, 27(12):121102, Dec 2017. ISSN 1054-1500. doi: 10.1063/1.5010300. URL http://www.ncbi.nlm.nih.gov/pubmed/29289043http://aip.scitation.org/doi/10.1063/1.5010300.

J. Pathak, A. Wikner, R. Fussell, S. Chandra, B.R. Hunt, M. Girvan, and E. Ott. Hybrid forecasting of chaotic processes: Using machine learning in conjunction with a knowledge-based model. *Chaos*, 28(4), 2018. ISSN 10541500. doi: 10.1063/1.5028373.

V. Pellet and F. Aires. Bottleneck channels algorithm for satellite data dimension reduction: A case study for IASI. *IEEE Transactions on Geoscience and Remote Sensing*, 56(10):8360516, 6069–6081, 2018. ISSN 15580644, 01962892. doi: 10.1109/tgrs.2018.2830123.

G.D. Peterson. Contagious disturbance, ecological memory, and the emergence of landscape pattern. *Ecosystems*, 5:329–338, 2002. doi: 10.1007/s10021-001-0077-1.

O.L. Phillips, L.E.O.C. Aragão, S.L. Lewis, J.B. Fisher, J. Lloyd, G. López-González, Y. Malhi, A. Monteagudo, J. Peacock, C.A. Quesada, G. van der Heijden, S. Almeida, I. Amaral, L. Arroyo, G. Aymard, T.R. Baker, O. Bánki, L. Blanc, D. Bonal, P. Brando, J. Chave, Á.C. Alves de Oliveira, N.D. Cardozo, C.I. Czimczik, T.R. Feldpausch, M. Aparecida Freitas, E. Gloor, N. Higuchi, E. Jiménez, G. Lloyd, P. Meir, C. Mendoza, A. Morel, D.A. Neill, D. Nepstad, S. Patiño, M.C. Peñuela, A. Prieto, F. Ramírez, M. Schwarz, J. Silva, M. Silveira, S. Sota Thomas, H. ter Steege, J. Stropp, R. Vásquez, P. Zelazowski, E. Alvarez Dávila, S. Andelman, A. Andrade, K.-J. Chao, T. Erwin, A. Di Fiore, E. Honorio C, H. Keeling, T.J. Killeen, W.F. Laurance, A. Peña Cruz, N.C.A. Pitman, P. Núñez Vargas, H. Ramírez-Angulo, A. Rudas, R. Salamão, N. Silva, John Terborgh, and A. Torres-Lezama. Drought sensitivity of the Amazon rainforest. *Science*, 323:1344–1347, 2009. doi: 10.1126/science.1164033.

P.O. Pinheiro, T.-Y. Lin, R. Collobert, and P. Dollár. Learning to refine object segments. In *European Conference on Computer Vision (ECCV)*, pages 75–91. Springer, 2016.

R.S. Plant and G.C. Craig. A stochastic parameterization for deep convection based on equilibrium statistics. *Journal of the Atmospheric Sciences*, 65(1):87–105, 2008.

S.B. Pope. A more general effective-viscosity hypothesis. *Journal of Fluid Mechanics*, 72(2):331–340, 1975.

J. Porway, Q. Wang, and S.C. Zhu. A hierarchical and contextual model for aerial image parsing. *International Journal of Computer Vision*, 88(2):254–283, 2010.

J. Poterjoy, R.A. Sobash, and J.L. Anderson. Convective-scale data assimilation for the weather research and forecasting model using the local particle filter. *Monthly Weather Review*, 145(5):1897–1918, Mar 2017. ISSN 0027-0644. doi: 10.1175/mwr-d-16-0298.1.

J. Poterjoy, L. Wicker, and M. Buehner. Progress toward the application of a localized particle filter for numerical weather prediction. *Monthly Weather Review*, 147(4):1107–1126, Apr 2019. ISSN 15200493. doi: 10.1175/MWR-D-17-0344.1.

R. Prabha, M. Tom, M. Rothermel, E. Baltsavias, L. Leal-Taixé, and K. Schindler. Lake ice monitoring with webcams and crowd-sourced images. In *ISPRS Annals of Photogrammetry, Remote Sensing and Spatial Information Sciences*, 2020. (to appear).

Prabhat, O. Rübel, S. Byna, K. Wu, F. Li, M. Wehner, W. Bethel, et al. Teca: A parallel toolkit for extreme climate analysis. In *Third Worskhop on Data Mining in Earth System Science (DMESS) at the International Conference on Computational Science (ICCS)*, 2012.

Prabhat, S. Byna, V. Vishwanath, E. Dart, M. Wehner, and W.D. Collins. Teca: Petascale pattern recognition for climate science. In *Computer Analysis of Images and Patterns*, pages 426–436. Springer, 2015a.

Prabhat, K. Kashinath, T. Kurth, M. Mudigonda, A. Mahesh, B.A. Toms, J. Biard, S.K. Kim, S. Kahou, B. Loring, et al. Climatenet: Bringing the power of deep learning to the climate community via open datasets and architectures. In *AGU Fall Meeting Abstracts*, 2018.

W.K. Pratt. *Digital Image Processing*. John Wiley & Sons, Inc., USA, 1978. ISBN 0471018880.

N. Proia and V. Pagé. Characterization of a Bayesian ship detection method in optical satellite images. *IEEE Geoscience and Remote Sensing Letters*, 7(2):226–230, 2009.

R. Prudden, S. Adams, D. Kangin, N. Robinson, S. Ravuri, S. Mohamed, and A. Arribas. A review of radar-based nowcasting of precipitation and applicable machine learning techniques. 2020. URL https://arxiv.org/abs/2005.04988.

C.R. Qi, H. Su, K. Mo, and L.J. Guibas. Pointnet: Deep learning on point sets for 3d classification and segmentation. In *Computer Vision and Pattern Recognition (CVPR)*, pages 652–660, 2017a.

C.R. Qi, L. Yi, H. Su, and L.J. Guibas. PointNet^{++}: deep hierarchical feature learning on point sets in a metric space. In *Advances on Neural Information Processing Systems (NeurIPS)*, pages 5105–5114, 2017b.

C. Qiu, L. Mou, M. Schmitt, and X.X. Zhu. Local climate zone-based urban land cover classification from multi-seasonal Sentinel-2 images with a recurrent residual network. *ISPRS Journal of Photogrammetry and Remote Sensing*, 154:151–162, 2019.

C. Qiu, M. Schmitt, C. Geiß, T.-H.K. Chen, and X.X. Zhu. A framework for large-scale mapping of human settlement extent from Sentinel-2 images via fully convolutional neural networks. *ISPRS Journal of Photogrammetry and Remote Sensing*, 163:152–170, 2020.

M. Qiu, P. Zhao, K. Zhang, J. Huang, X. Shi, X. Wang, and W. Chu. A short-term rainfall prediction model using multi-task convolutional neural networks. In *2017 IEEE International Conference on Data Mining (ICDM)*, pages 395–404. IEEE, 2017.

Y. Qu, H. Qi, and C. Kwan. Unsupervised sparse Dirichlet-net for hyperspectral image super-resolution. In *Proceedings of the IEEE Conference on CVPR*, pages 2511–2520, 2018.

D. Quan, S. Wang, M. Ning, T. Xiong, and L. Jiao. Using deep neural networks for synthetic aperture radar image registration. In *2016 IEEE International Geoscience and Remote Sensing Symposium (IGARSS)*, pages 2799–2802, July 2016. doi: 10.1109/IGARSS.2016.7729723.

D. Quan, S. Wang, X. Liang, R. Wang, S. Fang, B. Hou, and L. Jiao. Deep generative matching network for optical and SAR image registration. In *IGARSS 2018 – 2018 IEEE International Geoscience and Remote Sensing Symposium*, pages 6215–6218, July 2018.

J. Quiñonero-Candela, M. Sugiyama, A. Schwaighofer, and N.D. Lawrence. *Dataset Shift in Machine Learning*. MIT Press, 2009.

L.R. Rabiner and B.-H. Juang. An introduction to hidden Markov models. *IEEE ASSP Magazine*, 3(1):4–16, 1986. doi: 10.1109/MASSP.1986.1165342.

E. Racah, C. Beckham, T. Maharaj, Prabhat, and C.J. Pal. Semi-supervised detection of extreme weather events in large climate datasets. *CoRR*, abs/1612.02095, 2016. URL http://arxiv.org/abs/1612.02095.

E. Racah, C. Beckham, T. Maharaj, S.E. Kahou, Prabhat, and C. Pal. Extremeweather: A large-scale climate dataset for semi-supervised detection, localization, and understanding of extreme weather events. In *Advances in Neural Information Processing Systems*, pages 3402–3413, 2017.

A. Radford, L. Metz, and S. Chintala. Unsupervised representation learning with deep convolutional generative adversarial networks. In *ICLR*, 2016.

F. Rahmani, K. Lawson, W. Ouyang, A. Appling, S. Oliver and C. Shen, Exploring the exceptional performance of a deep learning stream temperature model and the value of streamflow data. *Environmental Research Letters*, 2021. doi: 10.1088/1748-9326/abd501

R. Raina, A. Battle, H. Lee, B. Packer, and A.Y. Ng. Self-taught learning: transfer learning from unlabeled data. In *International Conference on Machine Learning*, pages 759–766. ACM, 2007.

M. Raissi, P. Perdikaris, and G.E. Karniadakis. Physics-informed neural networks: A deep learning framework for solving forward and inverse problems involving nonlinear partial differential equations. *Journal of Computational Physics*, 378:686–707, 2019. ISSN 0021-9991. doi: 10/gfzbvx. *URL* http://www.sciencedirect.com/science/article/pii/S0021999118307125.

D. Randall, M. Khairoutdinov, A. Arakawa, and W. Grabowski. Breaking the cloud parameterization deadlock. *Bulletin of the American Meteorological Society*, 84(11):1547–1564, 2003.

M.A. Ranzato, C. Poultney, S. Chopra, and Y. Lecun. Efficient learning of sparse representations with an energy-based model. In *NIPS*, pages 1137–1144, 2006.

M. Ranzato, A. Szlam, J. Bruna, M. Mathieu, R. Collobert, and S. Chopra. Video (language) modeling: A baseline for generative models of natural videos. *arXiv preprint arXiv:1412.6604*, 2014.

C.E. Rasmussen. *Evaluation of Gaussian Processes and Other Methods for Non-linear Regression*. University of Toronto, 1999.

S. Rasp and S. Lerch. Neural networks for postprocessing ensemble weather forecasts. *MonthlyWeather Review*, 146(11):3885–3900, 2018. doi: 10.1175/MWR-D-18-0187.1.

S. Rasp, M.S. Pritchard, and P. Gentine. Deep learning to represent subgrid processes in climate models. *Proceedings of the National Academy of Sciences*, 115(39):9684–9689, 2018.

S. Rasp, H. Schulz, S. Bony, and B. Stevens. Combining crowd-sourcing and deep learning to understand meso-scale organization of shallow convection. 2019. URL https://aps.arxiv.org/abs/1906.01906.

S. Rasp, P.D. Dueben, S. Scher, J.A. Weyn, S. Mouatadid, and N. Thuerey. WeatherBench: A benchmark dataset for data-driven weather forecasting, 2020. URL https://aps.arxiv.org/abs/2002.00469.

B.H. Raup, A. Racoviteanu, S.J.S. Khalsa, C. Helm, R. Armstrong, and Y. Arnaud. The GLIMS geospatial glacier database: a new tool for studying glacier change. *Global and Planetary Change*, 56(1-2):101–110, 2007.

B.H. Raup, L.M. Andreassen, T. Bolch, and S. Bevan. *Remote Sensing of Glaciers*, chapter 7, pages 123–156. John Wiley & Sons, 2014. doi: 10.1002/9781118368909.ch7.

J.S. Read, X. Jia, J. Willard, A.P. Appling, J.A. Zwart, S.K. Oliver, A. Karpatne, G.J.A. Hansen, P.C. Hanson, W. Watkins, M. Steinbach, and V. Kumar. Process-guided deep learning predictions of lake water temperature. *Water Resources Research*, 55 (11):9173–9190, 2019. doi: 10.1029/2019WR024922. URL https://agupubs.onlinelibrary.wiley.com/doi/abs/10.1029/2019WR024922.

T. Reato, B. Demir, and L. Bruzzone. An unsupervised multicode hashing method for accurate and scalable remote sensing image retrieval. *IEEE Geoscience and Remote Sensing Letters*, 16(2):276–280, October 2019.

I. Redko, E. Morvant, A. Habrard, M. Sebban, and Y. Bennani. *Advances in Domain Adaptation Theory*. Elsevier, 2019.

J. Redmon, S. Divvala, R. Girshick, and A. Farhadi. You only look once: Unified, real-time object detection. In *Proceedings of the IEEE Conference on Computer Vision and Pattern Recognition*, pages 779–788, 2016.

M. Reichstein, S. Besnard, N. Carvalhais, F. Gans, M. Jung, B. Kraft, and M. Mahecha. Modelling Landsurface Time-Series with Recurrent Neural Nets. In *IGARSS 2018 – 2018 IEEE International Geoscience and Remote Sensing Symposium*, pages 7640–7643, 2018. doi: 10.1109/IGARSS.2018.8518007.

M. Reichstein, G. Camps-Valls, N. Stevens, M. Jung, J. Denzler, N. Carvalhais, et al. Deep learning and process understanding for data-driven earth system science. *Nature*, 566(7743):195–204, 2019. doi: 10.1038/s41586-019-0912-1. URL https://doi.org/10.1038/s41586-019-0912-1.

S. Ren, K. He, R. Girshick, and J. Sun. Faster R-CNN: Towards real-time object detection with region proposal networks. In *Advances in Neural Information Processing Systems*, pages 91–99, 2015.

S. Ren, K. He, R.B. Girshick, and J. Sun. Faster R-CNN: Towards real-time object detection with region proposal networks. *IEEE TPAMI*, 39(6):1137–1149, 2017.

C. Requena-Mesa, M. Reichstein, M. Mahecha, B. Kraft, and J. Denzler. Predicting landscapes from environmental conditions using generative networks. In *German Conference on Pattern Recognition*, pages 203–217. Springer, 2019.

J. Revaud, Ph. Weinzaepfel, Z. Harchaoui, and C. Schmid. Deepmatching: Hierarchical deformable dense matching. *International Journal of Computer Vision*, 120, 04 2016. doi: 10.1007/s11263-016-0908-3.

M. Reyniers. *Quantitative Precipitation Forecasts Based on Radar Observations: Principles, Algorithms and Operational Systems*. Institut Royal Météorologique de Belgique Brussel, Belgium, 2008.

A.D. Richardson, K. Hufkens, T. Milliman, D.M. Aubrecht, M. Chen, J.M. Gray, M.R. Johnston, T.F. Keenan, S.T. Klosterman, M. Kosmala, E.K. Melaas, M.A. Friedl, and S. Frolking. Tracking vegetation phenology across diverse North American biomes using PhenoCam imagery. *Scientific Data*, 5(1):1–24, March 2018. ISSN 2052-4463. doi: 10/gc6crk. URL https://www.nature.com/articles/sdata201828.

M.B. Richman. Rotation of principal components. *Journal of Climatology*, 6(3): 293–335, 1986.

G. Riegler, A.O. Ulusoy, and A. Geiger. OctNet: Learning deep 3D representations at high resolutions. In *Computer Vision and Pattern Recognition (CVPR)*, pages 6620–6629, Honolulu, HI, 2017. IEEE.

C. Rieke et al. Awesome satellite imagery datasets. https://github.com/chrieke/awesome-satellite-imagery-datasets, 2020.

S. Rifai, P. Vincent, X. Muller, X. Glorot, and Y. Bengio. Contractive auto-encoders: Explicit invariance during feature extraction. In *Proceedings of the 28th International Conference on Machine Learning (ICML-11)*, pages 833–840, 2011.

I. Rigas, G. Economou, and S. Fotopoulos. Low-level visual saliency with application on aerial imagery. *IEEE Geoscience and Remote Sensing Letters*, 10(6): 1389–1393, Nov 2013. ISSN 1545-598X. doi: 10.1109/LGRS.2013.2243402.

E. Rocha Rodrigues, I. Oliveira, R. Cunha, and M. Netto. Deepdownscale: A deep learning strategy for high-resolution weather forecast. In *2018 IEEE 14th International Conference on e-Science (e-Science)*, pages 415–422, Oct 2018. doi: 10.1109/eScience.2018.00130.

J.L. Rojo-Álvarez, M. Martínez-Ramón, J. Muñoz-Marí, and G. Camps-Valls. *Digital Signal Processing with Kernel Methods*. Wiley & Sons, UK, Apr 2018. ISBN 978-1118611791. URL https://www.wiley.com/en-es/Digital+Signal+Processing+with+Kernel+Methods-p-9781118611791.

C. Römer, M.Wahabzada, A. Ballvora, F. Pinto, M. Rossini, C. Panigada, J. Behmann, J. Léon, C. Thurau, and C. Bauckhage. Early drought stress detection in cereals: simplex volume maximisation for hyperspectral image analysis. *Functional Plant Biology*, 39(11):878–890, 2012.

A. Romero, P. Radeva, and C. Gatta. Meta-parameter free unsupervised sparse feature learning. *IEEE Transactions on Pattern Analysis and Machine Intelligence*, 2015 Aug;37(8):1716–22. doi: 10.1109/TPAMI.2014.2366129.

O. Ronneberger, P. Fischer, and T. Brox. U-Net: Convolutional networks for biomedical image segmentation. In *International Conference on Medical Image Computing and Computer-assisted Intervention*, pages 234–241. Springer, 2015.

R. Roscher, C. Römer, B. Waske, and L. Plümer. Landcover classification with self-taught learning on archetypal dictionaries. In *IEEE International Geoscience and Remote Sensing Symposium*, pages 2358–2361, 2015. Symposium Prize Paper Award.

R. Roscher, B. Bohn, M.F. Duarte, and J. Garcke. Explainable machine learning for scientific insights and discoveries. *IEEE Access*, 8:42200–42216, 2020.

F. Rosenblatt. *Perceptions and the Theory of Brain Mechanisms*. Spartan Books, 1962.

S.T. Roweis and L.K. Saul. Nonlinear dimensionality reduction by locally linear embedding. *Science*, 290(5500):2323–2326, 2000.

S. Roy, E. Sangineto, B. Demir, and N. Sebe. Metric-learning-based deep hashing network for content-based retrieval of remote sensing images. *IEEE Geoscience and Remote Sensing Letters*, February 2020. doi: 10.1109/LGRS.2020.2974629.

D.E. Rumelhart, G.E. Hinton, and R.J. Williams. Learning representations by back-propagating errors. *Nature*, 323(6088): 533–536, 1986. doi: 10.1038/323533a0.

D.E. Rumelhart, G.E. Hinton, and R.J. Williams. Learning representations by error propagation. *Parallel Distributed Processing*, 1, 1985.

J. Runge, S. Bathiany, E. Bollt, G. Camps-Valls, D. Coumou, E. Deyle, C. Glymour, M. Kretschmer, M.D. Mahecha, J. Muñoz-Marí, et al. Inferring causation from time series in earth system sciences. *Nature Communications*, 10(1):2553, 2019.

O. Russakovsky, J. Deng, H. Su, J. Krause, S. Satheesh, S. Ma, Zhiheng Huang, Andrej Karpathy, A. Khosla, M. Bernstein, et al. Imagenet large scale visual recognition challenge. *International Journal of Computer Vision*, 115(3):211–252, 2015.

M. Rußwurm, M. Ali, Zhu X., Gal Y., and Körner M. Model and data uncertainty for satellite time series forecasting with deep recurrent models. In *IGARSS 2020 IEEE International Geoscience and Remote Sensing Symposium*, page pp. IEEE, 2020.

M. Rußwurm and M. Körner. Temporal vegetation modelling using long short-term memory networks for crop identification from medium-resolution multi-spectral satellite images. In *2017 IEEE Conference on Computer Vision and Pattern Recognition Workshops (CVPRW)*, pages 1496–1504, 2017a. doi: 10.1109/CVPRW.2017.193.

M. Rußwurm and M. Körner. Multi-temporal land cover classification with sequential recurrent encoders. *ISPRS International Journal of Geo-Information*, 7(4):129, 2018a. ISSN 2220-9964. doi: 10.3390/ijgi7040129. Feature Paper.

M. Rußwurm and M. Körner. Convolutional LSTMs for cloud-robust segmentation of remote sensing imagery. *arXiv preprint arXiv:1811.02471*, 2018b.

E.M. Ryan, K. Ogle, T.J. Zelikova, D.R. LeCain, D.G. Williams, J.A. Morgan, and E. Pendall. Antecedent moisture and temperature conditions modulate the response of ecosystem respiration to elevated CO_2 and warming. *Global Change Biology*, 21:2588–2602, 2015. doi: 10.1111/gcb.12910.

S. Saha, F. Bovolo, and L. Bruzzone. Unsupervised multiple-change detection in vhr multisensor images via deep-learning based adaptation. In *IGARSS 2019 – 2019 IEEE International Geoscience and Remote Sensing Symposium*, pages 5033–5036, July 2019a.

S. Saha, F. Bovolo, and L. Bruzzone. Unsupervised deep change vector analysis for multiplechange detection in vhr images. *IEEE Transactions on Geoscience and Remote Sensing*, 57(6):3677–3693, 2019b.

T.N. Sainath, O. Vinyals, A. Senior, and H. Sak. Convolutional, long short-term memory, fully connected deep neural networks. In *International Conference on Acoustics, Speech and Signal Processing (ICASSP)*, pages 4580–4584, 2015. doi: 10.1109/ICASSP.2015.7178838.

H. Salehipour and W.R. Peltier. Deep learning of mixing by two "atoms" of stratified turbulence. *arXiv preprint arXiv:1809.06499*, 2018.

W. Samek, T. Wiegand, and K.-R. Müller. Explainable artificial intelligence: Understanding, visualizing and interpreting deep learning models. *CoRR*, abs/1708.08296, 2017. URL http://arxiv.org/abs/1708.08296.

W. Samek, G.Montavon, A. Vedaldi, L. Kai Hansen, K.-R. Müller, editors. Explainable AI: Interpreting, Explaining and Visualizing Deep Learning, volume 11700. 2019. doi: 10.1007/978-3-030-28954-6. URL http://dx.doi.org/10.1007/978-3-030-28954-6.

O. San and R. Maulik. Extreme learning machine for reduced order modeling of turbulent geophysical flows. *Physical Review E*, 97(4):042322, 2018.

M. Sandler, A. Howard, A. Zhu, M. and Zhmoginov, and L.-C. Chen. MobileNetV2: Inverted residuals and linear bottlenecks. In *Computer Vision and Pattern Recognition (CVPR)*, 2018. URL https://aps.arxiv.org/abs/1801.04381.

A. Santoro, D. Raposo, D.G. Barrett, Mateusz Malinowski, Razvan Pascanu, Peter Battaglia, and Timothy Lillicrap. A simple neural network module for relational reasoning. In *Advances in Neural Information Processing Systems*, pages 4967–4976, 2017.

G. Scarpa, S. Vitale, and D. Cozzolino. Target-adaptive CNN-based pansharpening. *IEEE Transactions on Geoscience and Remote Sensing*, 56(9):5443–5457, Sept 2018.

C. Schär and G. Jendritzky. Hot news from Summer 2003. *Nature*, 432(7017):559–560, 2004.

T. Schaul, S. Zhang, and Y. LeCun. No more pesky learning rates. In *ICML*, 2013.

S. Scher. Toward data-driven weather and climate forecasting: Approximating a simple general circulation model with deep learning. *Geophysical Research Letters*, 45(22):12–616, 2018.

S. Scher and G. Messori. Predicting weather forecast uncertainty with machine learning. *Quarterly Journal of the Royal Meteorological Society*, 144(717):2830–2841, Oct 2018. ISSN 1477870X. doi: 10.1002/qj.3410.

S. Scher and G. Messori. Weather and climate forecasting with neural networks: using GCMs with different complexity as study-ground. *Geoscientific Model Development Discussions*, pages 1–15, Mar 2019. doi: 10.5194/gmd-2019-53.

M. Schmitt and X.X. Zhu. Data fusion and remote sensing: An ever-growing relationship. *IEEE Geoscience and Remote Sensing Magazine*, 4(4):6–23, 2016.

M. Schmitt, L.H. Hughes, C. Qiu, and X.X. Zhu. Sen12ms–a curated dataset of georeferenced multi-spectral Sentinel-1/2 imagery for deep learning and data fusion. *ISPRS Annals of Photogrammetry, Remote Sensing and Spatial Information Sciences*, IV-2/W7:153–160, Sep 2019. ISSN 2194-9050. doi: 10.5194/isprs-annals-iv-2-w7-153-2019. URL http://dx.doi.org/10.5194/isprs-annals-iv-2-w7-153-2019.

T. Schneider, S. Lan, A. Stuart, and J. Teixeira. Earth system modeling 2.0: A blueprint for models that learn from observations and targeted high-resolution simulations. *Geophysical Research Letters*, 44(24): 12–396, 2017a.

T. Schneider, J. Teixeira, C.S. Bretherton, F. Brient, K.G. Pressel, C. Schär, and A.P. Siebesma. Climate goals and computing the future of clouds. *Nature Climate Change*, 7(1):3–5, 2017b.

T. Schneider, C.M. Kaul, and K.G. Pressel. Possible climate transitions from breakup of stratocumulus decks under greenhouse warming. *Nature Geoscience*, 12(3):163, 2019.

B. Schölkopf, A. Smola, and K.-R. Müller. Nonlinear component analysis as a kernel eigenvalue problem. *Neural Computation*, 10 (5):1299–1319, 1998.

M. Schuster and K.K. Paliwal. Bidirectional recurrent neural networks. *IEEE Transactions on Signal Processing*, 45(11):2673–2681, 1997. doi: 10.1109/78.650093.

E.A.G. Schuur, A.D. McGuire, C. Schädel, G. Grosse, J.W. Harden, D.J. Hayes, G. Hugelius, C.D. Koven, P. Kuhry, D.M. Lawrence, et al. Climate change and the permafrost carbon feedback. *Nature*, 520(7546):171–179, 2015.

E. Scoccimarro. Modeling tropical cyclones in a changing climate, 2016. URL //naturalhazardscience.oxfordre.com/10.1093/acrefore/9780199389407.001.0001/acrefore-9780199389407-e-22.

A. Seale, P. Christoffersen, R.I. Mugford, and M. O'Leary. Ocean forcing of the Greenland Ice Sheet: Calving fronts and patterns of retreat identified by automatic satellite monitoring of eastern outlet glaciers. *Journal of Geophysical Research: Earth Surface*, 116:F03013, 2011.

A. Sedaghat and N. Mohammadi. Illumination-robust remote sensing image matching based on oriented self-similarity. *ISPRS Journal of Photogrammetry and Remote Sensing*, 153:21–35, 2019. ISSN 0924-2716. doi: https://doi.org/10.1016/j.isprsjprs.2019.04.018.

M. Segal-Rozenhaimer, A. Li, K. Das, and V. Chirayath. Cloud detection algorithm for multi-modal satellite imagery using convolutional neural-networks (CNN). *Remote Sensing of Environment*, 237:111446, 2020.

P. Sermanet, D. Eigen, X. Zhang, M. Mathieu, R. Fergus, and Y. LeCun. Overfeat: Integrated recognition, localization and detection using convolutional networks. In *International Conference on Learning Representations (ICLR2014)*. CBLS, April 2014. URL http://openreview.net/document/d332e77d-459a-4af8-b3ed-55ba.

B. Settles. Active learning. *Synthesis Lectures on Artificial Intelligence and Machine Learning*, 6(1):1–114, 2012.

D.M.H. Sexton, A.V. Karmalkar, J.M. Murphy, K.D. Williams, I.A. Boutle, C.J. Morcrette, A.J. Stirling, and S.B. Vosper. Finding plausible and diverse variants of a climate model. Part 1:

establishing the relationship between errors at weather and climate time scales. *Clim. Dyn.*, 53:989–1022, 2019. ISSN 1432-0894. doi: 10.1007/s00382-019-04625-3. URL https://doi.org/10.1007/s00382-019-04625-3.

B. Seyednasrollah, A.M. Young, K. Hufkens, T. Milliman, M.A. Friedl, S. Frolking, and A.D. Richardson. Publisher correction: Tracking vegetation phenology across diverse biomes using Version 2.0 of the PhenoCam Dataset. *Scientific Data*, 6(1):1–1, November 2019. ISSN 2052-4463. doi: 10/ggh5m4. URL https://www.nature.com/articles/ s41597-019-0270-8.

K. Sfikas, T. Theoharis, and I. Pratikakis. Exploiting the PANORAMA representation for convolutional neural network classification and retrieval. In *Eurographics Workshop on 3D Object Retrieval*, Lyon, France, 2017.

M. Shahzad, M. Maurer, F. Fraundorfer, Y.Wang, and X.X. Zhu. Buildings detection in vhr SAR images using fully convolution neural networks. *IEEE Transactions on Geoscience and Remote Sensing*, 57(2):1100–1116, 2019.

Z. Shao and J. Cai. Remote sensing image fusion with deep convolutional neural network. *IEEE Journal of Selected Topics in Applied Earth Observations and Remote Sensing*, 11(5):1656–1669, May 2018.

Z. Shao, Z. Lu, M. Ran, L. Fang, J. Zhou, and Y. Zhang. Residual encoder-decoder conditional generative adversarial network for pansharpening. *IEEE Geoscience and Remote Sensing Letters, pages* 1–5, 2019.

C. Shen. A trans-disciplinary review of deep learning research and its relevance for water resources scientists. *Water Resources Research*, 54(11): 8558–8593, Dec 2018a. doi: 10.1029/2018WR022643. URL https://doi.org/10.1029/2018WR022643.

C. Shen. Deep learning: A next-generation big-data approach for hydrology, April 2018b. URL https://eos.org/editors-vox/deep-learning-a-next-generation-big-data-approach-for-hydrology.

C. Shen, W.J. Riley, K.M. Smithgall, J.M. Melack, and K. Fang. The fan of influence of streams and channel feedbacks to simulated land surface water and carbon dynamics. *Water Resources Research*, 52(2): 880–902, 2016. doi: 10.1002/2015WR018086.

C. Shen, E. Laloy, A. Elshorbagy, A. Albert, J. Bales, F.-J. Chang, S. Ganguly, K.-L. Hsu, D. Kifer, Z. Fang, K. Fang, D. Li, X. Li, and W.-P. Tsai. HESS Opinions: Incubating deep-learning-powered hydrologic science advances as a community. *Hydrology and Earth System Sciences*, 22(11):5639–5656, Nov 2018. ISSN 1607-7938. doi: 10.5194/hess-22-5639-2018. URL https://www.hydrol-earth-syst-sci.net/22/5639/2018/.

S.C. Sheridan and C.C. Lee. The self-organizing map in synoptic climatological research. *Progress in Physical Geography*, 35(1):109–119, 2011.

J. Shermeyer, D. Hogan, J. Brown, A. Van Etten, N. Weir, F. Pacifici, R. Haensch, A. Bastidas, S. Soenen, T. Bacastow, and R. Lewis. Spacenet 6: Multi-sensor all weather mapping dataset. In *Computer Vision and Pattern Recognition Workshops (CVPRw)*, 2020.

J. Sherrah. Fully convolutional networks for dense semantic labelling of high-resolution aerial imagery. *arXiv preprint arXiv:1606.02585*, 2016.

S.C. Sherwood, S. Bony, and J.-L. Dufresne. Spread in model climate sensitivity traced to atmospheric convective mixing. *Nature*, 505(7481):37, 2014.

B. Shi, S. Bai, Z. Zhou, and X. Bai. DeepPano: Deep Panoramic Representation for 3-D Shape Recognition. *IEEE Signal Processing Letters*, 22(12): 2339–2343, December 2015a.

X. Shi, Z. Chen, H. Wang, D.-Y. Yeung, W.-K. Wong, W.-C. Woo, and Hong Kong Observatory. Convolutional LSTM network: A machine learning approach for precipitation nowcasting. *arxiv*, 2015b.

X. Shi, Z. Gao, L. Lausen, H. Wang, D.-Y. Yeung, W.-K. Wong, and W.-C. Woo. Deep learning for precipitation nowcasting: A benchmark and a new model. In *Advances in Neural Information Processing Systems*, pages 5617–5627, 2017.

Z. Shi, X. Yu, Z. Jiang, and B. Li. Ship detection in high-resolution optical imagery based on anomaly detector and local shape feature. *IEEE Transactions on Geoscience and Remote Sensing*, 52(8): 4511–4523, 2013.

C.A. Shields, J.J. Rutz, L.-Y. Leung, F.M. Ralph, M. Wehner, B. Kawzenuk, J.M. Lora, E. McClenny, T. Osborne, A.E. Payne, et al. Atmospheric river tracking method intercomparison project (ARTMIP): project goals and experimental design. *Geoscientific Model Development*, 11(6):2455–2474, 2018.

Z. Shu, M. Sahasrabudhe, A. Guler Riza, D. Samaras, N. Paragios, and I. Kokkinos. Deforming autoencoders: Unsupervised disentangling of shape and appearance. In *The European Conference on Computer Vision (ECCV)*, September 2018.

S. Siachalou, G. Mallinis, and M. Tsakiri-Strati. A hidden Markov models approach for crop classification: linking crop phenology to time series of multi-sensor remote sensing data. *Remote Sensing*, 7(4):3633–3650, 2015. doi: 10.3390/rs70403633.

H.T. Siegelmann and E.D. Sontag. Turing vomputability with neural nets. *Applied Mathematics Letters*, 4(6):77–80, 1991. doi: 10.1016/0893-9659(91)90080-f.

H.T. Siegelmann and E.D. Sontag. On the computational power of neural nets. *Journal of Computer and System Sciences*, 50(1):132–150, 1995. doi: 10.1006/jcss.1995.1013.

V.D. Silva and J.B. Tenenbaum. Global versus local methods in nonlinear dimensionality reduction. In *Advances in Neural Information Processing Systems*, pages 721–728, 2003.

D. Silver, A. Huang, C.J. Maddison, A. Guez, L. Sifre, G. van den Driessche, J. Schrittwieser, I. Antonoglou, V. Panneershelvam, M. Lanctot, S. Dieleman, D. Grewe, J. Nham, N. Kalchbrenner, I. Sutskever, T. Lillicrap, M. Leach, K. Kavukcuoglu, Thore Graepel, and D. Hassabis. Mastering the game of Go with deep neural networks and tree search. *Nature*, 529(7587): 484–489, Jan 2016. ISSN 0028-0836. doi: 10.1038/nature16961. URL http://www.nature.com/doifinder/10.1038/nature16961.

A.J. Simmons and A. Hollingsworth. Some aspects of the improvement in skill of numerical weather prediction. *Quarterly Journal of the Royal Meteorological Society: A Journal of the Atmospheric Sciences, Applied Meteorology and Physical Oceanography*, 128(580):647–677, 2002.

M. Simões, J. Bioucas-Dias, L.B. Almeida, and J. Chanussot. A convex formulation for hyperspectral image super-resolution via subspace-based regularization. *IEEE Transactions on Geoscience and Remote Sensing*, 53(6):3373–3388, 2015.

K. Simonyan and A. Zisserman. Very deep convolutional networks for large-scale image recognition. In *Internaltional Conference on Learning Representation (ICLR)*, 2015.

A. Singh, H. Kalke, M. Loewen, and N. Ray. River ice segmentation with deep learning, 2019. URL https://aps.arxiv.org/abs/1901.04412.

B. Singh, M. Najibi, and L.S. Davis. Sniper: Efficient multi-scale training. *arXiv:1805.09300*, 2018.

P. Singh and N. Komodakis. Cloud-gan: Cloud removal for Sentinel-2 imagery using a cyclic consistent generative adversarial networks. In *IGARSS 2018 – 2018 IEEE International Geoscience and Remote Sensing Symposium*, pages 1772–1775, July 2018.

N. Skific, J.A. Francis, and J.J. Cassano. Attribution of projected changes in atmospheric moisture transport in the arctic: A self-organizing map perspective. *Journal of Climate*, 22(15):4135–4153, 2009.

J. Smagorinsky. General circulation experiments with the primitive equations: I. The basic experiment. *Monthly Weather Review*, 91(3):99–164, 1963.

H.-G. Sohn and K.C. Jezek. Mapping ice sheet margins from ERS-1 SAR and SPOT imagery. *International Journal of Remote Sensing*, 20(15-16): 3201–3216, 1999.

W. Song, S. Li, and J. Benediktsson. Deep hashing learning for visual and semantic retrieval of remote sensing images. *arXiv preprint arXiv:1909.04614*, 2019.

A. Sotiras, C. Davatzikos, and N. Paragios. Deformable medical image registration: A survey. *IEEE Transactions on Medical Imaging*, 32(7), 2013.

J.T. Springenberg, A. Dosovitskiy, T. Brox, and M. Riedmiller. Striving for simplicity: The all convolutional net. *arXiv preprint arXiv:1412.6806*, 2014.

N. Srivastava, E. Mansimov, and R. Salakhutdinov. Unsupervised learning of video representations using LSTMs. In *ICML*, 2015.

S. Srivastava, J.E. Vargas, and D. Tuia. Understanding urban landuse from the above and ground perspectives: a deep learning, multimodal solution. *Remote Sensing of Environment*, 228:129–143, 2019.

B. Stevens and S. Bony. What are climate models missing? *Science*, 340(6136):1053–1054, 2013. ISSN 0036-8075. doi: 10.1126/science.1237554. URL https://science.sciencemag.org/content/340/6136/1053.

B. Stevens, S.C. Sherwood, S. Bony, and M.J. Webb. Prospects for narrowing bounds on Earth's equilibrium climate sensitivity. *Earth's Future*, 4(11):512–522, 2016.

B. Stevens, M. Satoh, L. Auger, J. Biercamp, C.S. Bretherton, X. Chen, P. Düben, F. Judt, M. Khairoutdinov, D. Klocke, et al. Dyamond: The dynamics of the atmospheric general circulation modeled on non-hydrostatic domains. *Progress in Earth and Planetary Science*, 6(1):61, 2019.

T.F. Stocker, D. Qin, G. Plattner, M. Tignor, S.K. Allen, J. Boschung, A. Nauels, Y. Xia, V. Bex, and P.M. Midgley. IPCC, 2013: summary for policymakers in climate change 2013: the physical science basis, contribution of Working Group I to the Fifth Assessment Report of the Intergovernmental Panel on Climate Change. *Camb. Univ. Press Camb. UKNY NY USA*, 2013.

J. Strachan and J. Camp. Tropical cyclones of 2012. *Weather*, 68 (5):122–125, 2013. ISSN 1477-8696. doi: 10.1002/wea.2096. URL http://dx.doi.org/10.1002/wea.2096.

R. Stull. *Practical Meteorology: An Algebra-based Survey of Atmospheric Science*. University of British Columbia, 2015. ISBN 9780888651761.

H. Su, S. Maji, E. Kalogerakis, and E. Learned-Miller. Multi-view convolutional neural networks for 3d shape recognition. In *Computer Vision and Pattern Recognition (CVPR)*, pages 945–953, 2015.

Y. Su, J. Li, A. Plaza, A. Marinoni, P. Gamba, and S. Chakravortty. Daen: Deep autoencoder networks for hyperspectral unmixing. *IEEE Transactions on Geoscience and Remote Sensing*, 57(7):4309–4321, 2019.

C.H. Sudre, W. Li, T. Vercauteren, S. Ourselin, and M.J. Cardoso. Generalised dice overlap as a deep learning loss function for highly unbalanced segmentations. In *Deep Learning in Medical Image Analysis and Multimodal Learning for Clinical Decision Support*, pages 240–248. Springer, 2017.

G. Sumbul, M. Charfuelan, B. Demir, and V. Markl. Bigearthnet: A large-scale benchmark archive for remote sensing image understanding. *IGARSS 2019 – 2019 IEEE International Geoscience and Remote Sensing Symposium*, Jul 2019. doi: 10.1109/igarss.2019.8900532. URL http://dx.doi.org/10.1109/IGARSS.2019.8900532.

A.Y. Sun. Discovering state-parameter mappings in subsurface models using generative adversarial networks. *Geophysical Research Letters*, 45(20):11,137–11,146, 2018. doi: 10.1029/2018GL080404. URL https://agupubs.onlinelibrary.wiley.com/doi/abs/10.1029/2018GL080404.

A.Y. Sun, B.R. Scanlon, Z. Zhang, D. Walling, S.N. Bhanja, A. Mukherjee, and Z. Zhong. Combining physically based modeling and deep learning for fusing grace satellite data: Can we learn from mismatch? *Water Resources Research*, 55(2):1179–1195, 2019. doi: 10.1029/2018WR023333. URL https://agupubs.onlinelibrary.wiley.com/doi/abs/10.1029/2018WR023333.

B. Sun and K. Saenko. *Deep CORAL: Correlation alignment for deep domain adaptation*. In *European Conference on Computer VisionWorkshops*, pages 443–450. Springer, 2016.

H. Sun, X. Sun, H.Wang, Y. Li, and X. Li. Automatic target detection in high-resolution remote sensing images using spatial sparse coding bag-of-words model. *IEEE Geoscience and Remote Sensing Letters*, 9(1):109–113, Jan 2012. ISSN 1545-598X. doi: 10.1109/LGRS.2011.2161569.

J. Sun, M. Xue, J.W. Wilson, I. Zawadzki, S.P. Ballard, J. Onvlee-Hooimeyer, P. Joe, D.M. Barker, P.-W. Li, B. Golding, et al. Use of nwp for nowcasting convective precipitation: Recent progress and challenges. *Bulletin of the American Meteorological Society*, 95 (3):409–426, 2014.

F. Sung, Y. Yang, L. Zhang, T. Xiang, Ph.H.S. Torr, and T.M. Hospedales. Learning to compare: Relation network for few-shot learning. In *The IEEE Conference on Computer Vision and Pattern Recognition (CVPR)*, June 2018.

I. Sutskever, O. Vinyals, and Q.V. Le. Sequence to sequence learning with neural networks. In *Advances in Neural Information Processing Systems*, pages 3104–3112, 2014.

S. Suzuki et al. Topological structural analysis of digitized binary images by border following. *Computer vision, graphics, and image processing*, 30(1): 32–46, 1985.

D.H. Svendsen, P. Morales-Álvarez, R. Molina, and G. Camps-Valls. Deep Gaussian processes for geophysical parameter retrieval. In *IGARSS 2018-2018 IEEE International Geoscience and Remote Sensing Symposium*, pages 6175–6178. IEEE, 2018.

D.H. Svendsen, L. Martino, and G. Camps-Valls. Active emulation of computer codes with gaussian processes – application to remote sensing. *Pattern Recognition*, 100(107103):1–12, 2020. doi: https://doi.org/10.1016/j.patcog.2019.107103.

C. Szegedy, W. Liu, Y. Jia, P. Sermanet, S. Reed, D. Anguelov, D. Erhan, V. Vanhoucke, and A. Rabinovich. Going deeper with convolutions. In *Proceedings of the IEEE Conference on Computer Vision and Pattern Recognition (CVPR)*, pages 1–9, 2015.

R. Szeliski. *Computer Vision: Algorithms and Applications*. Springer-Verlag, Berlin, Heidelberg, 1st edition, 2010. ISBN 1848829345.

K. Szura. The big data project: Enhancing public access to noaa's open data. In *AGU Fall Meeting Abstracts*, 2018.

K. Takata, S. Emori, and T. Watanabe. Development of the minimal advanced treatments of surface interaction and runoff. *Global and Planetary Change*, 38:209–222, 2003. doi: 10.1016/S0921-8181(03)00030-4.

S.S. Talathi and A. Vartak. Improving performance of recurrent neural network with ReLU nonlinearity. In *International Conference on Learning Representations (ICLR) Workshops*, 2015.

J. Tan, N. NourEldeen, K. Mao, J. Shi, Z. Li, T. Xu, and Z. Yuan. Deep learning convolutional neural network for the retrieval of land surface temperature from AMSR2 data in China. *Sensors*, 19(13), 2019. doi: 10.3390/s19132987.

G. Tang, D. Long, A. Behrangi, C. Wang, and Y. Hong. Exploring deep neural networks to retrieve rain and snow in high latitudes using multisensor and reanalysis data. *Water Resources Research*, 54(10): 8253–8278, 2018a. doi: 10.1029/2018WR023830. URL https://agupubs.onlinelibrary.wiley.com/doi/abs/10.1029/2018WR023830.

X. Tang, X. Zhang, F. Liu, and L. Jiao. Unsupervised deep feature learning for remote sensing image retrieval. *Remote Sensing*, 10(8):1243, August 2018b.

X. Tang, C. Liu, X. Zhang, J. Ma, C. Jiao, and L. Jiao. Remote sensing image retrieval based on semi-supervised deep hashing learning. In *IEEE International Geoscience and Remote Sensing Symposium*, pages 879–882, July 2019.

T. Tanikawa, W. Li, K. Kuchiki, T. Aoki, M. Hori, and K. Stamnes. Retrieval of snow physical parameters by neural networks and optimal estimation: case study for ground-based spectral radiometer system. *Optics Express*, 23(24):A1442–A1462, 2015.

Y. Tao, X. Gao, K. Hsu, S. Sorooshian, and A. Ihler. A deep neural network modeling framework to reduce bias in satellite precipitation products. *Journal of Hydrometeorology*, 17:931–945, 2016. doi: 10.1175/JHM-D-15-0075.1.

Y. Tao, X. Gao, A. Ihler, S. Sorooshian, K. Hsu, Y. Tao, X. Gao, A. Ihler, S. Sorooshian, and K. Hsu. Precipitation identification with bispectral satellite information using deep learning approaches. *Journal of Hydrometeorology*, 18(5):1271–1283, May 2017. ISSN 1525-755X. doi: 10.1175/JHM-D-16-0176.1. URL http://journals.ametsoc.org/doi/10.1175/JHM-D-16-0176.1.

Y. Tao, K. Hsu, A. Ihler, X. Gao, and S. Sorooshian. A two-stage deep neural network framework for precipitation estimation from bispectral satellite information. *Journal of Hydrometeorology*, 19(2):393–408, Feb 2018. doi: 10.1175/JHM-D-17-0077.1. URL http://journals.ametsoc.org/doi/10.1175/JHM-D-17-0077.1.

A.M. Tartakovsky, Carlos Ortiz Marrero, Paris Perdikaris, Guzel Tartakovsky, and David A. Barajas-Solano. Learning parameters and constitutive relationships with physics informed deep neural networks. 2018.

O. Tasar, Y. Tarabalka, and P. Alliez. Incremental learning for semantic segmentation of large-scale remote sensing data. *IEEE Journal of Selected Topics in Applied Earth Observations and Remote Sensing*, 12(9):3524–3537, 2019.

O. Tasar, S.L. Happy, Y. Tarabalka, and P. Alliez. ColorMapGAN: Unsupervised domain adaptation for semantic segmentation using color mapping generative adversarial networks. *IEEE Transactions on Geoscience and Remote Sensing*, 2020.

M. Tatarchenko, J. Park, V. Koltun, and Q.-Y. Zhou. Tangent convolutions for dense prediction in 3D. In *Computer Vision and Pattern Recognition (CVPR)*, pages 3887–3896, Salt Lake City, UT, USA, June 2018. ISBN 978-1-5386-6420-9.

K.E. Taylor, R.J. Stouffer, and G.A. Meehl. An overview of CMIP5 and the experiment design. *Bulletin of the American Meteorological Society*, 93(4):485–498, 2012.

L. Tchapmi, C. Choy, I. Armeni, J.Y. Gwak, and S. Savarese. SEGCloud: Semantic segmentation of 3D point clouds. In *2017 International Conference on 3D Vision (3DV)*, pages 537–547, Qingdao, 2017. 3DV.

C. Tebaldi and R. Knutti. The use of the multi-model ensemble in probabilistic climate projections. *Philosophical Transactions of the Royal Society A: Mathematical, Physical and Engineering Sciences*, 365(1857):2053–2075, Aug 2007. doi: 10.1098/rsta.2007.2076. URL https://doi.org/10.1098/rsta.2007.2076.

M. Tedesco. *Remote Sensing of the Cryosphere*. JohnWiley & Sons, 2014.

J. Teixeira and C.A. Reynolds. Stochastic nature of physical parameterizations in ensemble prediction: A stochastic convection approach. *Monthly Weather Review*, 136(2):483–496, 2008.

I. Tekeste and B. Demir. Advanced local binary patterns for remote sensing image retrieval. In *IEEE International Geoscience and Remote Sensing Symposium*, pages 6855–6858, July 2018.

J.B. Tenenbaum, V. De Silva, and J.C. Langford. A global geometric framework for nonlinear dimensionality reduction. *Science*, 290 (5500):2319–2323, 2000.

M. Tharani, N. Khurshid, and M. Taj. Unsupervised deep features for remote sensing image matching via discriminator network, 2018. URL https://aps.arxiv.org/abs/1810.06470.

The RGI Consortium. Randolph glacier inventory – a dataset of global glacier outlines: Version 6.0: technical report, Global Land Ice Measurements From Space, Colorado, USA, 2017.

J.J. Thiagarajan, K.N. Ramamurthy, and A. Spanias. Learning stable multilevel dictionaries for sparse representations. *IEEE Transactions on Neural Networks and Learning Systems*, 26(9):1913–1926, 2015.

H. Thomas, C.R. Qi, J.-E. Deschaud, B. Marcotegui, F. Goulette, and L. J. Guibas. KPConv: flexible and deformable convolution for point clouds, April 2019. URL https://arxiv.org/abs/1904.08889.

V. Thompson, N.J. Dunstone, A.A. Scaife, D.M. Smith, J.M. Slingo, S. Brown, and S.E. Belcher. High risk of unprecedented UK rainfall in the current climate. *Nature Communications*, 8(1):107, 2017. ISSN 2041-1723. doi: 10.1038/s41467-017-00275-3. URL https://doi.org/10.1038/s41467-017-00275-3.

X.-A. Tibau, C. Requena-Mesa, C. Reimers, J. Denzler, V. Eyring, M. Reichstein, and J. Runge. Supernovae: Vae based kernel-pca for analysis of spatio-temporal earth data. In *Proceedings of the 8th International Workshop on Climate Informatics: CI 2018*, pages 73–77. NCAR, 2018.

X.-A. Tibau, C. Reimers, V. Eyring, J. Denzler, M. Reichstein, and J. Runge. *Spatiotemporal model for benchmarking causal discovery algorithms*. EGU General Assembly Conference Abstracts, 2020.

R. Tipireddy, P. Perdikaris, P. Stinis, and A.M. Tartakovsky. A comparative study of physics-informed neural network models for learning unknown dynamics and constitutive relations. *ArXiv*, abs/1904.04058, 2019.

M.E. Tipping. Sparse kernel principal component analysis. In *Advances in Neural Information Processing Systems*, pages 633–639, 2001.

P. Tokarczyk, J.D. Wegner, S. Walk, and K. Schindler. Features, color spaces, and boosting: New insights on semantic classification of remote sensing images. *IEEE Transactions in Geoscience and Remote Sensing*, 53(1):280–295, 2015.

E. Tola, V. Lepetit, and P. Fua. Daisy: An efficient dense descriptor applied to wide-baseline stereo. *IEEE Transactions on Pattern Analysis and Machine Intelligence*, 32(5):815–830, May 2010. ISSN 1939-3539. doi: 10.1109/TPAMI.2009.77.

M. Tom, U. Kälin, M. Sütterlin, E. Baltsavias, and K. Schindler. Lake ice detection in low-resolution optical satellite images. *ISPRS Annals of Photogrammetry, Remote Sensing and Spatial Information Sciences*, IV-2, 2018.

M. Tom, R. Aguilar, S. Leinss, E. Baltsavias, and K. Schindler. Lake ice detection from Sentinel-1 SAR with deep learning. In *ISPRS Annals of Photogrammetry, Remote Sensing and Spatial Information Sciences*, 2020. (to appear).

B.A. Toms, E.A. Barnes, and I. Ebert-Uphoff. Physically interpretable neural networks for the geosciences: Applications to earth system variability. *arXiv preprint arXiv:1912.01752*, 2019a.

B.A. Toms, K. Kashinath, D. Yang, et al. Deep learning for scientific inference from geophysical data: The madden-julian oscillation as a test case. *arXiv preprint arXiv:1902.04621*, 2019b.

B.A. Toms, E.A. Barnes, and I. Ebert-Uphoff. Physically interpretable neural networks for the geosciences: Applications to earth system variability. *Journal of Advances in Modeling Earth Systems*, 12 (9), Aug 2020. ISSN 1942-2466. doi: 10.1029/2019ms002002. URL http://dx.doi.org/10.1029/2019ms002002.

J. Tonttila, Z. Maalick, T. Raatikainen, H. Kokkola, T. Kuhn, and S. Romakkaniemi. UCLALES-SALSA v1.0: a large-eddy model with interactive sectional microphysics for aerosol, clouds and precipitation. *Geoscientific Model Development*, 10(1):169–188, 2017. doi: 10.5194/gmd-10-169-2017.

B.D. Tracey, K. Duraisamy, and J.J. Alonso. A machine learning strategy to assist turbulence model development. In *53rd AIAA Aerospace Sciences Meeting*, page 1287, 2015.

G. Tramontana, M. Jung, C.R. Schwalm, K. Ichii, G. Camps-Valls, B. Ráduly, M. Reichstein, M.A. Arain, A. Cescatti, G. Kiely, L. Merbold, P. Serrano-Ortiz, S. Sickert, S. Wolf, and D. Papale. Predicting carbon dioxide and energy fluxes across global FLUXNET sites with regression algorithms. *Biogeosciences*, 13:4291–4313, 2016. doi: 10.5194/bg-13-4291-2016.

D. Tran, L. Bourdev, R. Fergus, L. Torresani, and M. Paluri. Learning spatiotemporal features with 3d convolutional networks. In *ICCV*, 2015.

Q.-K. Tran and S. Song. Computer vision in precipitation nowcasting: Applying image quality assessment metrics for training deep neural networks. *Atmosphere*, 10(5):244, 2019.

G. Tsagkatakis, A. Aidini, K. Fotiadou, M. Giannopoulos, A. Pentari, and P. Tsakalides. Survey of deep-learning approaches for remote sensing observation enhancement. *Sensors*, 19(18):3929, 2019.

Y.-L.S. Tsai, A. Dietz, N. Oppelt, and C. Kuenzer. Remote sensing of snow cover using spaceborne SAR: A review. *Remote Sensing*, 11 (12):1456, 2019.

W.-P. Tsai, D. Feng, M. Pan, H. Beck, K. Lawson, Y. Yang, J. Liu and C. Shen. From parameter calibration to parameter learning: Revolutionizing large-scale geoscientific modeling with big data, 2020. https://arxiv.org/abs/2007.15751

D. Tuia, M. Volpi, L. Copa, M. Kanevski, and J. Munoz-Mari. A survey of active learning algorithms for supervised remote sensing image classification. *IEEE Journal of Selected Topics in Signal Processing*, 5(3):606–617, 2011.

D. Tuia, M. Volpi, M. dalla Mura, A. Rakotomamonjy, and R. Flamary. Automatic feature learning for spatio-spectral image classification with sparse SVM. *IEEE Transactions in Geoscience and Remote Sensing*, 52(10):6062–6074, 2014.

D. Tuia, N. Courty, and R. Flamary. Multiclass feature learning for hyperspectral image classification: sparse and hierarchical solutions. *ISPRS Journal of the International Society for Photogrammetry and Remote Sensing*, 105:272–285, 2015.

D. Tuia, D. Marcos, and G. Camps-Valls. Multi-temporal and multi-source remote sensing image classification by nonlinear relative normalization. *ISPRS Journal of Photogrammetry and Remote Sensing*, 120:1–12, 2016a.

D. Tuia, C. Persello, and L. Bruzzone. Recent advances in domain adaptation for the classification of remote sensing data. *IEEE Geoscience and Remote Sensing Magazine*, 4(2):41–57, 2016b.

T. Tuytelaars and K. Mikolajczyk. Local invariant feature detectors: A survey. *Found. Trends. Comput. Graph. Vis.*, 3(3):177–280, July 2008. ISSN 1572-2740. doi: 10.1561/0600000017. URL https://doi.org/10.1561/0600000017.

J. Tyndall. I. The Bakerian Lecture. On the absorption and radiation of heat by gases and vapours, and on the physical connexion of radiation, absorption, and conduction. *Philosophical Transactions of the Royal Society of London*, 151, Jan 1861. doi: 10.1098/rstl.1861.0001. URL https://doi.org/10.1098/rstl.1861.0001.

J.R.R. Uijlings, K.E.A. Van De Sande, T. Gevers, and A.W.M. Smeulders. Selective search for object recognition. *International Journal of Computer Vision*, 104(2):154–171, 2013.

P. Ukkonen and A. Mäkelä. Evaluation of machine learning classifiers for predicting deep convection. *Journal of Advances in Modeling Earth Systems*, 11(6):1784–1802, 2019. doi: 10.1029/2018MS001561. URL https://agupubs.onlinelibrary.wiley.com/doi/abs/10.1029/2018MS001561.

P.A. Ullrich and C.M. Zarzycki. Tempestextremes: a framework for scale-insensitive pointwise feature tracking on unstructured grids. *Geoscientific Model Development*, 10(3):1069–1090, 2017. doi: 10.5194/gmd-10-1069-2017. URL https://www.geosci-model-dev.net/10/1069/2017/.

UNFCCC. *Report of the Conference of the Parties on its twenty-first session, held in Paris from 30 November to 13 December 2015.* 2015. URL http://unfccc.int/resource/docs/2015/cop21/eng/10.pdf.

B. Uzkent, C. Yeh, and S. Ermon. Efficient object detection in large images using deep reinforcement learning. 2019. URL http://arxiv.org/abs/1912.03966.

M. Vakalopoulou and K. Karantzalos. Automatic descriptor-based co-registration of frame hyperspectral data. *Remote Sensing*, 6(4),2014.

M. Vakalopoulou, K. Karantzalos, N. Komodakis, and N. Paragios. Graph-based registration, change detection, and classification in very high resolution multitemporal remote sensing data. *IEEE Journal of Selected Topics in Applied Earth Observations and Remote Sensing*, 9(7), 2016.

M. Vakalopoulou, S. Christodoulidis, M. Sahasrabudhe, S. Mougiakakou, and N. Paragios. Image registration of satellite imagery with deep convolutional neural networks. In *IGARSS 2019 – 2019 IEEE International Geoscience and Remote Sensing Symposium*, pages 4939–4942, July 2019. doi: 10.1109/IGARSS.2019.8898220.

A. van den Oord, Y. Li, and O. Vinyals. Representation learning with contrastive predictive coding. *CoRR*, abs/1807.03748, 2018. URL http://arxiv.org/abs/1807.03748.

L. van der Maaten and G. Hinton. Visualizing data using t-SNE. *Journal of Machine Learning Research*, 9:2579–2605, 2008.

J.E. Vargas, S. Lobry, A.X. Falcão, and D. Tuia. Correcting rural building annotations in OpenStreetMap using convolutional neural networks. *ISPRS Journal of the International Society for Photogrammetry and Remote Sensing*, 147:283–293, 2019.

N. Varney, V.K. Asari, and Q. Graehling. Dales: A large-scale aerial LiDAR data set for semantic segmentation. In *Proceedings of the IEEE/CVF Conference on Computer Vision and Pattern RecognitionWorkshops*, pages 186–187, 2020.

J. Verrelst, G. Camps-Valls, J. Muñoz-Marí, J. Pablo Rivera, F. Veroustraete, J.G.P.W. Clevers, and J. Moreno. Optical remote sensing and the retrieval of terrestrial vegetation bio-geophysical properties – a review. *ISPRS Journal of Photogrammetry and Remote Sensing*, 108:273–290, 2015.

J. Vial, J.-L. Dufresne, and S. Bony. On the interpretation of intermodel spread in cmip5 climate sensitivity estimates. *Climate Dynamics*, 41(11-12): 3339–3362, 2013.

P. Vincent, H. Larochelle, I. Lajoie, Y. Bengio, and P.-A. Manzagol. Stacked denoising autoencoders: Learning useful representations in a deep network with a local denoising criterion. *Journal of Machine Learning Research*, 11 (Dec):3371–3408, 2010.

P. Vincent and H. Larochelle. Stacked denoising autoencoders: Learning useful representations in a deep network with a local denoising criterion. *Journal of Machine Learning Research*, 11:3371–3408, 2010.

P. Vincent, H. Larochelle, Y. Bengio, and P.-A. Manzagol. Extracting and composing robust features with denoising autoencoders. In *Proceedings of the 25th International Conference on Machine Learning*, pages 1096–1103. ACM, 2008.

P. Viola and W. Wells. Alignment by maximization of mutual information. *Proceedings of IEEE International Conference on Computer Vision. volume* 24, pages 16–23, 1995. ISBN 0-8186-7042-8. doi: 10.1109/ICCV.1995.466930.

G. Vivone, L. Alparone, J. Chanussot, M. Dalla Mura, A. Garzelli, G. Licciardi, R. Restaino, and L. Wald. A critical comparison among pansharpening algorithms. *IEEE Transactions on Geoscience and Remote Sensing*, 53(5): 2565–2586, 2014.

G. Vivone, L. Alparone, J. Chanussot, M. Dalla Mura, A. Garzelli, G.A. Licciardi, R. Restaino, and L. Wald. A critical comparison among pansharpening algorithms. *IEEE Transactions on Geoscience and Remote Sensing*, 53(5): 2565–2586, May 2015.

M. Volpi and V. Ferrari. Semantic segmentation of urban scenes by learning local class interactions. In *Computer Vision and Pattern RecognitionWorkshops (CVPRw)*, 2015.

M. Volpi and D. Tuia. Dense semantic labeling of subdecimeter resolution images with convolutional neural networks. *IEEE Transactions in Geoscience and Remote Sensing*, 55 (2):881–893, 2017.

M. Volpi and D. Tuia. Deep multi-task learning for a geographically-regularized semantic segmentation of aerial images. *ISPRS Journal of the International Society for Photogrammetry and Remote Sensing*, 144:48–60, 2018.

M. Wahabzada, A.-K. Mahlein, C. Bauckhage, U. Steiner, E.-C. Oerke, and K. Kersting. Plant phenotyping using probabilistic topic models: Uncovering the hyperspectral language of plants. *Scientific Reports*, 6, 2016.

L. Wald, T. Ranchin, and M. Mangolini. Fusion of satellite images of different spatial resoltuions: assessing the quality of resulting images. *Photogrammetric Engineering and Remote Sensing*, 63(6):691–699, 1997.

L. Wald. Quality of high resolution synthesised images: Is there a simple criterion? In *Third Conference: "Fusion of Earth data: merging point measurements, raster maps and remotely sensed images"*, pages 99–103. SEE/URISCA, 2000.

K. Walsh, S. Lavender, E. Scoccimarro, and H. Murakami. Resolution dependence of tropical cyclone formation in cmip3 and finer resolution models. *Climate Dynamics*, 40(3):585–599, Feb 2013. ISSN 1432-0894. doi: 10.1007/s00382-012-1298-z. URL https://doi.org/10.1007/s00382-012-1298-z.

L. Wan, L. Zheng, H. Huo, and T. Fang. Affine invariant description and large-margin dimensionality reduction for target detection in optical remote sensing images. IEEE Geosci. Remote Sensing Lett., 2017.

B. Wang, J. Lu, Z. Yan, H. Luo, T. Li, Y. Zheng, and G. Zhang. Deep uncertainty quantification. In *Proceedings of the 25th ACM SIGKDD International Conference on Knowledge Discovery & Data Mining – KDD '19*, pages 2087–2095, New York, New York, USA, 2019a. ACM Press. ISBN 9781450362016. doi: 10.1145/3292500.3330704. URL http://dl.acm.org/citation.cfm?doid=3292500.3330704.

G. Wang, X. Wang, B. Fan, and C. Pan. Feature extraction by rotation-invariant matrix representation for object detection in aerial image. *IEEE Geoscience and Remote Sensing Letters*, 2017a.

H. Wang and D.-Y. Yeung. Towards Bayesian deep learning: A framework and some existing methods. *IEEE Transactions on Knowledge and Data Engineering*, 28(12):3395–3408, 2016.

H. Wang, F. Nie, and H. Huang. Robust and discriminative self-taught learning. In Sanjoy Dasgupta and David McAllester, editors, *Proceedings of the 30th International Conference on Machine Learning, volume 28 of Proceedings of Machine Learning Research*, pages 298–306, Atlanta, Georgia, USA, 17–19 Jun 2013a. PMLR. URL http://proceedings.mlr.press/v28/wang13g.html.

J. Wang, J. Ding, H. Guo, W. Cheng, T. Pan, and W. Yang. Mask obb: A semantic attention-based mask oriented bounding box representation for multi-category object detection in aerial images. *Remote Sensing*, 11(24):2930, 2019b.

L. Wang, K.A. Scott, L. Xu, and D.A. Clausi. Sea ice concentration estimation during melt from dual-pol SAR scenes using deep convolutional neural networks: A case study. *IEEE Transactions on Geoscience and Remote Sensing*, 54(8):4524–4533, Aug 2016. doi: 10.1109/TGRS.2016.2543660.

L. Wang, X. Xu, Y. Yu, R. Yang, R. Gui, Z. Xu, and F. Pu. Sar-to-optical image translation using supervised cycle-consistent adversarial networks. *IEEE Access*, 7:129136–129149, 2019.

L. Wang, K.A. Scott, and D.A. Clausi. Improved sea ice concentration estimation through fusing classified SAR imagery and AMSR-E data. *Canadian Journal of Remote Sensing*, 42(1):41–52, 2016.

L. Wang, K.A. Scott, and D.A. Clausi. Sea ice concentration estimation during freeze-up from SAR imagery using a convolutional neural network. *Remote Sensing*, 9(5), 2017. doi: 10.3390/rs9050408.

L. Wang, H. Geng, P. Liu, K. Lu, J. Kolodziej, R. Ranjan, and A.Y. Zomaya. Particle swarm optimization based dictionary learning for remote sensing big data. *Knowledge-Based Systems*, 79:43–50, 2015.

M. Wang and W. Deng. Deep visual domain adaptation: A survey. *Neurocomputing*, 312:135–153, 2018.

S. Wang, D. Quan, X. Liang, M. Ning, Y. Guo, and L. Jiao. A deep learning framework for remote sensing image registration. *ISPRS Journal of Photogrammetry and Remote Sensing*, 145:148–164, 2018. ISSN 0924-2716. doi: https://doi.org/10.1016/j.isprsjprs.2017.12.012. Deep Learning RS Data.

W. Wang, Y. Huang, Y. Wang, and L. Wang. Generalized autoencoder: A neural network framework for dimensionality reduction. In *Proceedings of the IEEE Conference on Computer Vision and Pattern Recognition Workshops*, pages 490–497, 2014.

Y. Wang, L. Zhang, X. Tong, L. Zhang, Z. Zhang, H. Liu, X. Xing, and P.T. Mathiopoulos. A three-layered graph-based learning approach for remote sensing image retrieval. *IEEE Transactions on Geoscience and Remote Sensing*, 54(10):6020–6034, October 2016.

Y. Wang, L. Wang, H. Lu, and Y. He. Segmentation based rotated bounding boxes prediction and image synthesizing for object detection of high resolution aerial images. *Neurocomputing*, 2020.

Y. Wang, C. Wang, H. Zhang, Y. Dong, and S. Wei. A SAR dataset of ship detection for deep learning under complex backgrounds. *Remote Sensing*, 11(7):765–, 2019a.

Y. Wang, M. Long, J. Wang, Z. Gao, and P.S. Yu. Predrnn: Recurrent neural networks for predictive learning using spatiotemporal LSTMs. In *Advances in Neural Information Processing Systems*, pages 879–888, 2017d.

Y. Wang, J. Zhang, H. Zhu, M. Long, J. Wang, and P.S. Yu. Memory in memory: A predictive neural network for learning higher-order non-stationarity from spatiotemporal dynamics. In *Proceedings of the IEEE Conference on Computer Vision and Pattern Recognition*, pages 9154–9162, 2019b.

Z. Wang and A.C. Bovik. A universal image quality index. *IEEE Signal Processing Letters*, 9(3):81–84, 2002.

Z. Wang, N.M. Nasrabadi, and T.S. Huang. Spatial-spectral classification of hyperspectral images using discriminative dictionary designed by learning vector quantization. *IEEE Transactions on Geoscience and Remote Sensing*, PP(99):1–15, 2013b. ISSN 0196-2892. doi: 10.1109/TGRS.2013.2285049.

Z. Wang and A.C. Bovik. Mean squared error: Love it or leave it? a new look at signal fidelity measures. *IEEE signal processing magazine*, 26(1): 98–117, 2009.

Z. Wang, E.P. Simoncelli, and A.C. Bovik. Multiscale structural similarity for image quality assessment. In *The Thrity-Seventh Asilomar Conference on Signals, Systems & Computers, 2003*, volume 2, pages 1398–1402. Ieee, 2003.

Z. Wang, A.C. Bovik, H.R. Sheikh, E.P. Simoncelli, et al. Image quality assessment: From error visibility to structural similarity. *IEEE Transactions on Image Processing*, 13(4):600–612, 2004.

P.A.G. Watson. Applying machine learning to improve simulations of a chaotic dynamical system using empirical error correction. Apr 2019. URL http://arxiv.org/abs/1904.10904.

P.A.G. Watson, J. Berner, S. Corti, P. Davini, J. von Hardenberg, C. Sanchez, A. Weisheimer, and T.N. Palmer. The impact of stochastic physics on tropical rainfall variability in global climate models on daily to weekly timescales. *Journal of Geophysics Research: Atmospheres*, 122: 5738–5762, 2017. ISSN 2169897X. doi: 10.1002/2016JD026386. URL http://doi.wiley.com/10.1002/2016JD026386.

M.F. Wehner, K.A. Reed, F. Li, Prabhat, J. Bacmeister, C.-T. Chen, C. Paciorek, P.J. Gleckler, K.R. Sperber, W.D. Collins, A. Gettelman, and C. Jablonowski. The effect of horizontal resolution on simulation quality in the community atmospheric model, CAM5.1. *Journal of Advances in Modeling Earth Systems*, 6(4):980–997, 2014. ISSN 1942-2466. doi: 10.1002/2013MS000276. URL http://dx.doi.org/10.1002/2013MS000276.

M.F. Wehner, K.A. Reed, and C.M. Zarzycki. *High-Resolution Multi-decadal Simulation of Tropical Cyclones*, pages 187–211. Springer International Publishing, Cham, 2017. ISBN 978-3-319-47594-3. doi: 10.1007/978-3-319-47594-3_8. URL https://doi.org/10.1007/978-3-319-47594-3_8.

Q. Wei, N. Dobigeon, J. Tourneret, J. Bioucas-Dias, and S. Godsill. R-fuse: Robust fast fusion of multiband images based on solving a sylvester equation. *IEEE Signal Processing Letters*, 23(11):1632–1636, Nov 2016.

Y. Wei, Q. Yuan, H. Shen, and L. Zhang. Boosting the accuracy of multispectral image pansharpening by learning a deep residual network. *IEEE Geoscience and Remote Sensing Letters*, 14(10):1795–1799, Oct 2017.

N. Weir, D. Lindenbaum, A. Bastidas, A. Van Etten, S. McPherson, J. Shermeyer, V. Kumar, and H. Tang. SpaceNet MVOI: A multi-view overhead imagery dataset, 2019. URL https://aps.arxiv.org/abs/1903.12239.

A. Weisheimer and T.N. Palmer. On the reliability of seasonal climate forecasts. *Journal of The Royal Society Interface*, 11(96):20131162, 2014. doi: 10.1098/rsif.2013.1162. URL https://royalsocietypublishing.org/doi/abs/10.1098/rsif.2013.1162.

M.L. Weisman, C. Davis, W. Wang, K.W. Manning, and J.B. Klemp. Experiences with 0–36-h explicit convective forecasts with the WRF-ARW model. *Weather and Forecasting*, 23(3):407–437, 2008.

G. Weiss, Y. Goldberg, and E. Yahav. On the practical computational power of finite precision RNNs for language recognition. In *Annual Meeting of the Association for Computational Linguistics*, volume 2, pages 740–745, 2018. doi: 10.18653/v1/P18-2117.

Q. Weng and Y. He. *High Spatial Resolution Remote Sensing: Data, Analysis, and Applications*. 2018. ISBN 9781498767682.

P.J. Werbos. Backpropagation through time: what it does and how to do it. *Proceedings of the IEEE*, 78(10):1550–1560, 1990a. ISSN 0018-9219 VO – 78. doi: 10.1109/5.58337.

P.J. Werbos. Backpropagation through time: what it does and how to do it. *Proceedings of the IEEE*, 78(10):1550–1560, 1990b. doi: 10.1109/5.58337.

S. Westra, H.J. Fowler, J.P. Evans, L.V. Alexander, P. Berg, F. Johnson, E.J. Kendon, G. Lenderink, and N.M. Roberts. Future changes to the intensity and frequency of short-duration extreme rainfall. *Review of Geophysics*, 52:522–555, 2014. ISSN 0096-3941. doi: 10.1002/2014RG000464.

J.A. Weyn, D.R. Durran, and R. Caruana. Can machines learn to predict weather? using deep learning to predict gridded 500-hpa geopotential height from historical weather data. *Journal* of Advances in Modeling Earth Systems, 11(8):2680–2693, 2019. doi: 10.1029/2019MS001705. URL https://agupubs.onlinelibrary.wiley.com/doi/abs/10.1029/2019MS001705.

D.S. Wilks. Effects of stochastic parametrizations in the Lorenz '96 system. *Quarterly Journal of the Royal Meteorological Society*, 131(606):389–407, 2005. ISSN 00359009. doi: 10.1256/qj.04.03. URL http://doi.wiley.com/10.1256/qj.04.03.

R.J. Williams and J. Peng. An efficient gradient-based algorithm for on-line training of recurrent network trajectories. *Neural Computation*, 2(4): 490–501, 1990. doi: 10.1162/neco.1990.2.4.490.

R.J. Williams and D. Zipser. A learning algorithm for continually running fully recurrent neural networks. *Neural Computation*, 1(2):270–280, 1989. doi: 10.1162/neco.1989.1.2.270.

R.J. Williams and D. Zipser. Gradient-based learning algorithms for recurrent networks and their computational complexity. In *Backpropagation: Theory, Architectures, and Applications*, pages 433–486. L. Erlbaum Associates Inc., 1995.

R.J. Williams, G.E. Hinton, and D.E. Rumelhart. Learning representations by back-propagating errors. *Nature*, 323(6088):533–536, 1986.

B. Willmore and D.J. Tolhurst. Characterizing the sparseness of neural codes. *Network*, 12(12):255–270, 2001.

B.J. Winer. Statistical principles in experimental design. 1962.

A. Wolanin, G. Mateo-García, G. Camps-Valls, L. Gómez-Chova, M. Meroni, G. Duveiller, L. You, and L. Guanter. Estimating and understanding crop yields with explainable deep learning in the indian wheat belt. *Environmental Research Letters*, 2020.

W. Woo and W. Wong. Operational application of optical flow techniques to radar-based rainfall nowcasting. *Atmosphere*, 8(3):48, 2017.

J. Wu, C. Chang, H.-Y. Tsai, and M.-C. Liu. C–registration between multisource remote-sensing images. *ISPRS – International Archives of the Photogrammetry, Remote Sensing and Spatial Information Sciences*, XXXIX–B3, 2012.

J. Wu, K. Kashinath, A. Albert, *Prabhat, and H.* Xiao. Physics-informed generative learning to emulate unresolved physics in climate models. In *AGU Fall Meeting Abstracts*, 2018.

Y. Wu and K. He. Group normalization. In *Proceedings of the European Conference on Computer Vision (ECCV)*, pages 3–19, 2018.

M. Wurm, T. Stark, X.X. Zhu, M. Weigand, and H. Taubenböck. Semantic segmentation of slums in satellite images using transfer learning on fully convolutional neural networks. *ISPRS Journal of the International Society for Photogrammetry and Remote Sensing*, 150: 59–69, 2019.

G.-S. Xia, X. Bai, J. Ding, Z. Zhu, S. Belongie, J. Luo, M. Datcu, M. Pelillo, and L. Zhang. Dota: A large-scale dataset for object detection in aerial images. In *The IEEE Conference on Computer Vision and Pattern Recognition (CVPR)*, June 2018.

M. Xia, W. Liu, B. Shi, L. Weng, and J. Liu. Cloud/snow recognition for multispectral satellite imagery based on a multidimensional deep residual network. *International Journal of Remote Sensing*, 40(1): 156–170, 2019. doi: 10.1080/01431161.2018.1508917.

S. Xian, W. Zhirui, S. Yuanrui, D. Wenhui, Z. Yue, and F. Kun. AIR-SARShip-1.0: High resolution SAR ship detection dataset. *Journal of Radars*, 8(6):852–862, 2019.

Z. Xiang and I. Demir. Distributed long-term hourly streamflow predictions using deep learning – A case study for State of Iowa. *Environmental Modelling & Software*, 131:104761,

Sep 2020. ISSN 1364-8152. doi: 10.1016/J.ENVSOFT.2020.104761. URL https://www.sciencedirect.com/science/article/pii/S1364815220301900.

Z. Xiang, J. Yan, and I. Demir. A rainfall-runoff model with LSTM-Based Sequence-to-sequence learning. *Water Resources Research*, 56 (1), Jan 2020. ISSN 0043-1397. doi: 10.1029/2019WR025326. URL https://onlinelibrary.wiley.com/doi/abs/10.1029/2019WR025326.

Z. Xiao, Y. Gong, Y. Long, D. Li, X. Wang, and H. Liu. Airport detection based on a multiscale fusion feature for optical remote sensing images. *IEEE Geoscience and Remote Sensing Letters*, 14(9):1469–1473, 2017.

M. Xie, N. Jean, M. Burke, D. Lobell, and S. Ermon. Transfer learning from deep features for remote sensing and poverty mapping. In *Thirtieth AAAI Conference on Artificial Intelligence*, 2016.

Q. Xie, M. Zhou, Q. Zhao, D. Meng, W. Zuo, and Z. Xu. Multispectral and hyperspectral image fusion by MS/HS fusion net. In *The IEEE Conference on Computer Vision and Pattern Recognition (CVPR)*, June 2019.

W. Xie, B. Liu, Y. Li, J. Lei, C. Chang, and G. He. Spectral adversarial feature learning for anomaly detection in hyperspectral imagery. *IEEE Transactions on Geoscience and Remote Sensing*, pages 1–14, 2019.

Y. Xie, J. Tian, and X.X. Zhu. A review of point cloud semantic segmentation. page 52, August 2019. URL https://aps.arxiv.org/abs/1908.08854.

Z. Xie, U.K. Haritashya, V.K. Asari, B.W. Young, M.P. Bishop, and J.S. Kargel. Glaciernet: A deep-learning approach for debris-covered glacier mapping. *IEEE Access*, 8:83495–83510, 2020.

S. Xingjian, Z. Chen, H. Wang, D.-Y. Yeung, W.-K. Wong, and W.-C. Woo. Convolutional LSTM network: A machine learning approach for precipitation nowcasting. In *Advances in Neural Information Processing Systems*, pages 802–810, 2015.

K. Xu, J. Ba, R. Kiros, A. Courville, R. Salakhutdinov, R. Zemel, and Y. Bengio. Show, attend and tell: Neural image caption generation with visual attention. In *ICML*, 2015.

M. Xu, X. Jia, M. Pickering, and A.J. Plaza. Cloud removal based on sparse representation via multitemporal dictionary learning. *IEEE Transactions on Geoscience and Remote Sensing*, 54(5):2998–3006, 2016.

T. Xu and A.J. Valocchi. Data-driven methods to improve baseflow prediction of a regional groundwater model. *Computers and Geosciences*, 85:124–136, 2015. ISSN 00983004. doi: 10.1016/j.cageo.2015.05.016. URL http://dx.doi.org/10.1016/j.cageo.2015.05.016.

Y. Xu and K.A. Scott. Impact of intermediate ice concentration training data on sea ice concentration estimates from a convolutional neural network. *International Journal of Remote Sensing*, 40(15):5799–5811, 08 2019. doi: 10.1080/01431161.2019.1582113.

T. Xue, J. Wu, K. Bouman, and W. Freeman. Visual dynamics: Probabilistic future frame synthesis via cross convolutional networks. In *Advances in neural information processing systems*, pages 91–99, 2016.

L. Yan, B. Fan, H. Liu, C. Huo, S. Xiang, and C. Pan. Triplet adversarial domain adaptation for pixel-level classification of vhr remote sensing images. *IEEE Transactions on Geoscience and Remote Sensing, pages* 1–16, 2019.

Q. Yan and W. Huang. Sea ice sensing from GNSS-R data using convolutional neural networks. *IEEE Geoscience and Remote Sensing Letters*, 15(10): 1510–1514, Oct 2018. doi: 10.1109/LGRS.2018.2852143.

Q. Yan, W. Huang, and C. Moloney. Neural networks based sea ice detection and concentration retrieval from GNSS-R delay-Doppler maps. *IEEE Journal of Selected Topics in Applied Earth Observations and Remote Sensing*, 10(8):3789–3798, 2017.

F. Yang, H. Fan, P. Chu, E. Blasch, and H. Ling. Clustered object detection in aerial images. In *CVPR*, pages 8311–8320, 2019.

G.-Z. Yang, D.J. Hawkes, D. Rueckert, A. Noble, and C. Taylor. *Medical Image Computing and Computer-Assisted Intervention–MICCAI 2009: 12th International Conference, London, UK, September 20–24, 2009, Proceedings*, volume 5761. Springer, 2009.

J. Yang, X. Fu, Y. Hu, Y. Huang, X. Ding, and J. Paisley. Pannet: A deep network architecture for pan-sharpening. In *2017 IEEE International Conference on Computer Vision (ICCV)*, pages 1753–1761, Oct 2017.

J. Yang, Z. Wang, Z. Lin, S. Cohen, and T. Huang. Coupled dictionary training for image super-resolution. *IEEE Transactions on Image Processing*, 21(8):3467–3478, 2012.

L. Yang, S. Treichler, T. Kurth, K. Fischer, D. Barajas-Solano, J. Romero, V. Churavy, A. Tartakovsky, M. Houston, M. Prabhat, and G. Karniadakis. Highly-ccalable, physics-informed gans for learning solutions of stochastic pdes. In *2019 IEEE/ACM Third Workshop on Deep Learning on Supercomputers (DLS)*, pages 1–11, Nov 2019. doi: 10.1109/DLS49591.2019.00006.

S. Yang, H. Jin, M. Wang, Y. Ren, and L. Jiao. Data-driven compressive sampling and learning sparse coding for hyperspectral image classification. *IEEE Geoscience and Remote Sensing Letters*, 11(2):479–483, Feb 2014. ISSN 1545-598X. doi: 10.1109/LGRS.2013.2268847.

S. Yang, D. Yang, J. Chen, and B. Zhao. Real-time reservoir operation using recurrent neural networks and inflow forecast from a distributed hydrological model. *Journal of Hydrology*, 579:124229, Dec 2019a. ISSN 0022-1694. doi: 10.1016/J.JHYDROL.2019.124229. URL https://www.sciencedirect.com/science/article/pii/S0022169419309643{#}!

T. Yang, F. Sun, P. Gentine, W. Liu, H. Wang, J. Yin, M. Du, and C. Liu. Evaluation and machine learning improvement of global hydrological model-based flood simulations. *Environmental Research Letters*, 14(11):114027, Nov 2019b.

X. Yang. Understanding the variational lower bound. 2017. URL http://legacydirs.umiacs.umd.edu/~xyang35/files/understanding-variational-lower.pdf

X. Yang, H. Sun, K. Fu, J. Yang, X. Sun, M. Yan, and Z. Guo. Automatic ship detection in remote sensing images from Google Earth of complex scenes based on multiscale rotation dense feature pyramid networks. *Remote Sensing*, 10(1):132, 2018a. doi: 10.3390/rs10010132. URL https://doi.org/10.3390/rs10010132.

X. Yang, H. Sun, X. Sun, M. Yan, Z. Guo, and K. Fu. Position detection and direction prediction for arbitrary-oriented ships via multiscale rotation region convolutional neural network. *arXiv:1806.04828*, 2018b.

X. Yang, J. Yang, J. Yan, Y. Zhang, T. Zhang, Z. Guo, X. Sun, and K. Fu. SCRDet: Towards more robust detection for small, cluttered and rotated objects. In *ICCV*, pages 8232–8241, 2019c.

Y. Yang and S. Newsam. Geographic image retrieval using local invariant features. *IEEE Transactions on Geoscience and Remote Sensing*, 51(2): 818–832, February 2013.

Z. Yang and S. Newsam. Bag-of-visual-words and spatial extensions for land-use classification. In *Proceedings of the 18th SIGSPATIAL International Conference on Advances in Geographic Information Systems*, pages 270–279, 2010.

Z. Yang, T. Dan, and Y. Yang. Multi-temporal remote sensing image registration using deep convolutional features. *IEEE Access*, 6:38544–38555, 2018. ISSN 2169-3536.

W. Yao, Z. Zeng, C. Lian, and H. Tang. Pixel-wise regression using u-net and its application on pansharpening. *Neurocomputing*, 312:364–371, 2018.

F. Ye, H. Xiao, X. Zhao, M. Dong, W. Luo, and W. Min. Remote sensing image retrieval using convolutional neural network features and weighted distance. *IEEE Geoscience and Remote Sensing Letters*, 15(10):1535–1539, Oct 2018.

F. Ye, W. Luo, M. Dong, H. He, and W. Min. SAR image retrieval based on unsupervised domain adaptation and clustering. *IEEE Geoscience and Remote Sensing Letters*, 16(9):1482–1486, Sep 2019.

L. Ye, L. Gao, R. Marcos-Martinez, D. Mallants, and B.A. Bryan. Projecting Australia's forest cover dynamics and exploring influential factors using deep learning. *Environmental Modelling & Software*, 119:407–417, 2019.

M.-H. Yen, D.-W. Liu, Y.-C. Hsin, C.-E. Lin, and C.-C. Chen. Application of the deep learning for the prediction of rainfall in Southern Taiwan. *Scientific Reports*, 9(1):1–9, September 2019. ISSN 2045-2322. doi: 10/ggcfxm. URL https://www.nature.com/articles/s41598-019-49242-6.

Z. Yi, H. Zhang, P. Tan, and M. Gong. Dualgan: Unsupervised dual learning for image-to-image translation. *2017 IEEE International Conference on Computer Vision (ICCV)*, pages 2868–2876, 2017.

N. Yokoya, P. Ghamisi, J. Xia, S. Sukhanov, R. Heremans, C. Debes, B. Bechtel, B. Le Saux, G. Moser, and D. Tuia. Open data for global multimodal land use classification: Outcome of the 2017 IEEE GRSS Data Fusion Contest. *IEEE Journal of Selected Topics in Applied Earth Observations and Remote Sensing*, 11(5):1363–1377, 2018.

N. Yokoya, T. Yairi, and A. Iwasaki. Coupled nonnegative matrix factorization unmixing for hyperspectral and multispectral data fusion. *IEEE Transactions on Geoscience and Remote Sensing*, 50(2):528–537, 2012.

N. Yokoya, C. Grohnfeldt, and J. Chanussot. Hyperspectral and multispectral data fusion: A comparative review of the recent literature. *IEEE Geoscience and Remote Sensing Magazine*, 5(2):29–56, 2017.

F. Yu and V. Koltun. Multi-scale context aggregation by dilated convolutions. In *ICLR*, 2016.

Y. Yu, X. Li, and F. Liu. Attention gans: Unsupervised deep feature learning for aerial scene classification. *IEEE Transactions on Geoscience and Remote Sensing*, 58(1):519–531, Jan 2020.

J. Yuan. Automatic building extraction in aerial scenes using convolutional networks. *arXiv preprint arXiv:1602.06564*, 2016.

J. Yuan, Z. Chi, X. Cheng, T. Zhang, T. Li, and Z. Chen. Automatic extraction of supraglacial lakes in southwest greenland during the 2014–2018 melt seasons based on convolutional neural network. *Water*, 12(3), 2020. ISSN 2073-4441. doi: 10.3390/w12030891.

Q. Yuan, Y.Wei, X. Meng, H. Shen, and L. Zhang. A multiscale and multidepth convolutional neural network for remote sensing imagery pan-sharpening. *IEEE Journal of Selected Topics in Applied Earth Observations and Remote Sensing*, 11(3):978–989, March 2018.

Q. Yuan, H. Shen, T. Li, Z. Li, S. Li, Y. Jiang, H. Xu, W. Tan, Q. Yang, J. Wang, et al. Deep learning in environmental remote sensing: Achievements and challenges. *Remote Sensing of Environment*, 241:111716, 2020b.

J. Yuval and P.A. O'Gorman. Use of machine learning to improve simulations of climate. Jan 2020. URL http://arxiv.org/abs/2001.03151.

J. Yuval and P.A. O'Gorman. Stable machine-learning parameterization of subgrid processes for climate modeling at a range of resolutions. *Nature Communications*, 11(1):1–10, 2020.

J. Zabalza, J. Ren, J. Zheng, H. Zhao, C. Qing, Z. Yang, P. Du, and S. Marshall. Novel segmented stacked autoencoder for effective dimensionality reduction and feature extraction in hyperspectral imaging. *Neurocomputing*, 185:1–10, 2016.

S. Zagoruyko and N. Komodakis. Learning to compare image patches via convolutional neural networks. In *2015 IEEE Conference on Computer Vision and Pattern Recognition (CVPR)*, pages 4353–4361. doi: 10.1109/CVPR.2015.7299064.

A. Zampieri, G. Charpiat, N. Girard, and Y. Tarabalka. Multimodal image alignment through a multiscale chain of neural networks with application to remote sensing. In *Computer Vision – ECCV 2018 – 15th European Conference, Munich, Germany*, September 8-14, 2018, Proceedings, *Part XVI*, pages 679–696, 2018. doi: 10.1007/978-3-030-01270-0_40.

L. Zanna, J.M. Brankart, M. Huber, S. Leroux, T. Penduff, and P.D. Williams. Uncertainty and scale interactions in ocean ensembles: From seasonal forecasts to multidecadal climate predictions. *Quarterly Journal of the Royal Meteorological Society*, 2018.

L. Zanna and T. Bolton. Data-driven discovery of mesoscale eddy closures. *Geophysical Research Letters*, 2020. doi: 10.1029/2020GL088376.

L. Zanna, P. Porta Mana, J. Anstey, T. David, and T. Bolton. Scale-aware deterministic and stochastic parametrizations of eddy-mean flow interaction. *Ocean Modelling*, 111:66–80, 2017.

M.D. Zeiler and R. Fergus. Visualizing and understanding convolutional networks. *arXiv preprint arXiv:1311.2901*, 2013.

D. Zhang, J. Lin, Q. Peng, D. Wang, T. Yang, S. Sorooshian, X. Liu, and J. Zhuang. Modeling and simulating of reservoir operation using the artificial neural network, support vector regression, deep learning algorithm. *Journal of Hydrology*, 565:720–736, 2018a. ISSN 0022-1694. doi: https://doi.org/10.1016/j.jhydrol.2018.08.050. URL http://www.sciencedirect .com/science/article/pii/S0022169418306462.

E. Zhang, L. Liu, and L. Huang. Automatically delineating the calving front of Jakobshavn Isbræ from multitemporal TerraSAR-X images: a deep learning approach. *The Cryosphere*, 13(6):1729–1741, 2019. doi: 10.5194/tc-13-1729-2019.

G. Zhang, P. Ghamisi, and X.X. Zhu. Fusion of heterogeneous earth observation data for the classification of local climate zones. *IEEE Transactions on Geoscience and Remote Sensing*, 57(10):7623–7642, 2019b.

H. Zhang, W. Ni, W. Yan, D. Xiang, J. Wu, X. Yang, and H. Bian. Registration of multimodal remote sensing image based on deep fully convolutional neural network. *IEEE Journal of Selected Topics in Applied Earth Observations and Remote Sensing*, 12(8):3028–3042, Aug 2019. ISSN 2151-1535. doi: 10.1109/JSTARS.2019.2916560.

J. Zhang, K. Howard, C. Langston, B. Kaney, Y. Qi, L. Tang, H. Grams, Y. Wang, S. Cocks, S. Martinaitis, et al. Multi-radar multi-sensor (mrms) quantitative precipitation estimation:

Initial operating capabilities. *Bulletin of the American Meteorological Society*, 97(4):621–638, 2016a.

J. Zhang, Y. Zhu, X. Zhang, M. Ye, and J. Yang. Developing a Long Short-Term Memory (LSTM) based model for predicting water table depth in agricultural areas. *Journal of Hydrology*, 561:918–929, Jun 2018b. ISSN 0022-1694. doi: 10.1016/J.JHYDROL.2018.04.065. URL https://www.sciencedirect.com/science/article/pii/S0022169418303184.

K. Zhang, B. Schölkopf, K. Muandet, and Z. Wang. Domain adaptation under target and conditional shift. In *International Conference on Machine Learning*, pages 819–827, 2013.

L. Zhang, D. Chen, J. Ma, and J. Zhang. Remote-sensing image super-resolution based on visual saliency analysis and unequal reconstruction networks. *IEEE Transactions on Geoscience and Remote Sensing*, pages 1–17, 2020a.

L. Zhang, L. Zhang, and B. Du. Deep learning for remote sensing data: A technical tutorial on the state of the art. *IEEE Geoscience and Remote Sensing Magazine*, 4(2):22–40, 2016b.

L. Zhang, Z. Shao, J. Liu, and Q. Cheng. Deep learning based retrieval of forest aboveground biomass from combined LiDAR and Landsat 8 data. *Remote Sensing*, 11(12):1459, 2019.

M. Zhang, M. Gong, Y. Mao, J. Li, and Y. Wu. Unsupervised feature extraction in hyperspectral images based on wasserstein generative adversarial network. *IEEE Transactions on Geoscience and Remote Sensing*, 57(5):2669–2688, May 2019a.

Q. Zhang, R.P. Phillips, S. Manzoni, R.L. Scott, A.C. Oishi, A. Finzi, E. Daly, R. Vargas, and K.A. Novick. Changes in photosynthesis and soil moisture drive the seasonal soil respiration-temperature hysteresis relationship. *Agricultural and Forest Meteorology*, 259:184–195, September 2018c. ISSN 0168-1923. doi: 10.1016/j.agrformet.2018.05.005.

R. Zhang, P. Isola, and A.A. Efros. Colorful image colorization. In *European Conference on Computer Vision (ECCV)*, 2016c.

W. Zhang, L. Han, J. Sun, H. Guo, and J. Dai. Application of multi-channel 3D-cube successive convolution network for convective storm nowcasting. 2017. URL https://aps.arxiv.org/abs/1702.04517.

W. Zhang, C. Witharana, A.K. Liljedahl, and M. Kanevskiy. Deep convolutional neural networks for automated characterization of arctic ice-wedge polygons in very high spatial resolution aerial imagery. *Remote Sensing*, 10(9), 2018. doi: 10.3390/rs10091487.

Y. Zhang, C. Liu, M. Sun, and Y. Ou. Pan-sharpening using an efficient bidirectional pyramid network. *IEEE Transactions on Geoscience and Remote Sensing*, 57(8):5549–5563, Aug 2019b.

Y. Zhang, Z. Liu, T. Liu, B. Peng, X. Li, and Q. Zhang. Large-scale point cloud contour extraction via 3-d-guided multiconditional residual generative adversarial network. *IEEE Geoscience and Remote Sensing Letters*, 17(1): 142–146, Jan 2020b.

Z. Zhang, W. Guo, S. Zhu, and W. Yu. Toward arbitrary-oriented ship detection with rotated region proposal and discrimination networks. *IEEE Geoscience and Remote Sensing Letters*, (99):1–5, 2018d.

Z. Zhang, Y. Xu, J. Yang, X. Li, and D. Zhang. A survey of sparse representation: algorithms and applications. *IEEE Access*, 3:490–530, 2015.

Z. Zhang and H. Zha. Principal manifolds and nonlinear dimensionality reduction via tangent space alignment. *SIAM Journal on Scientific Computing*, 26(1):313–338, 2004.

H. Zhao, J. Shi, X. Qi, X. Wang, and J. Jia. Pyramid scene parsing network. In *Proceedings of the IEEE Conference on Computer Vision and Pattern Recognition*, pages 2881–2890, 2017.

J. Zhao, W. Guo, Z. Zhang, and Y. Wenxian. A coupled convolutional neural network for small and densely clustered ship detection in SAR images. In *SCIENCE CHINA Information Sciences*, pages 1–16, 2019.

W. Zhao, L. Mou, J. Chen, Y. Bo, and W.J. Emery. Incorporating metric learning and adversarial network for seasonal invariant change detection. *IEEE Transactions on Geoscience and Remote Sensing*, pages 1–12, 2019.

W.L. Zhao, P. Gentine, M. Reichstein, Y. Zhang, S. Zhou, Y. Wen, C. Lin, X. Li, and G.Y. Qiu. Physics-constrained machine learning of evapotranspiration. *Geophysical Research Letters*, page 2019GL085291, Dec 2019. ISSN 0094-8276. doi: 10.1029/2019GL085291. URL https://onlinelibrary.wiley.com/doi/abs/10.1029/2019GL085291.

W. Zhi, D. Feng, W.-P. Tsai, G. Sterle, A. Harpold, C. Shen and L. Li. From hydrometeorology to river water quality: Can a deep learning model predict dissolved oxygen at the continental scale? *Environmental Science & Technology*, 2021. doi: 10.1021/acs.est.0c06783.

L. Zhong, L. Hu, and H. Zhou. Deep learning based multi-temporal crop classification. *Remote Sensing of Environment*, 221:430–443, 2019.

B. Zhou, A. Andonian, A. Oliva, and A. Torralba. Temporal relational reasoning in videos. In *Proceedings of the European Conference on Computer Vision (ECCV)*, pages 803–818, 2018.

G.-B. Zhou, J. Wu, C.-L. Zhang, and Z.-H. Zhou. Minimal gated unit for recurrent neural networks. *International Journal of Automation and Computing*, 13:226–234, 2016. doi: 10.1007/s11633-016-1006-2.

J. Zhou, D. Civco, and J. Silander. A wavelet transform method to merge Landsat TM and SPOT panchromatic data. *International Journal of Remote Sensing*, 19(4):743–757, 1998.

W. Zhou, Z. Shao, C. Diao, and Q. Cheng. High-resolution remote-sensing imagery retrieval using sparse features by auto-encoder. *Remote Sensing Letters*, 6(10):775–783, October 2015.

W. Zhou, S. Newsam, C. Li, and Z. Shao. Learning low dimensional convolutional neural networks for high-resolution remote sensing image retrieval. *Remote Sensing*, 9(5):489, May 2017a.

Y. Zhou, Q. Ye, Q. Qiu, and J. Jiao. Oriented response networks. In *CVPR, pages 4961–4970.* IEEE, 2017b.

Z.-H. Zhou. A brief introduction to weakly supervised learning. *National Science Review*, 5:44–53, 2018.

Z. Zhou, G. He, S. Wang, and G. Jin. Subgrid-scale model for large-eddy simulation of isotropic turbulent flows using an artificial neural network. *Computers & Fluids, page* 104319, 2019.

H. Zhu, L. Jiao, W. Ma, F. Liu, and W. Zhao. A novel neural network for remote sensing image matching. *IEEE Transactions on Neural Networks and Learning Systems*, 30(9):2853–2865, Sep 2019. ISSN 2162-2388.

J.-Y. Zhu, T. Park, P. Isola, and A.A. Efros. Unpaired image-to-image translation using cycle-consistent adversarial networks. In *IEEE International Conference on Computer Vision*, pages 2223–2232, 2017.

P. Zhu, L. Wen, X. Bian, H. Ling, and Q. Hu. Vision meets drones: A challenge. *arXiv:1804.07437*, 2018.

R. Zhu, D. Yu, Sh. Ji, and M. Lu. Matching rgb and infrared remote sensing images with densely-connected convolutional neural networks. *Remote Sensing*, 11(23), 2019a. ISSN 2072-4292. doi: 10.3390/rs11232836.

R. Zhu, L. Yan, N. Mo, and Y. Liu. Semi-supervised center-based discriminative adversarial learning for cross-domain scene-level land-cover classification of aerial images. *ISPRS Journal of Photogrammetry and Remote Sensing*, 155:72–89, 2019b.

X. Zhu, D. Tuia, L. Mou, G. Xia, L. Zhang, F. Xu, and F. Fraundorfer. Deep learning in remote sensing: A comprehensive review and list of resources. *IEEE Geoscience and Remote Sensing Magazine*, 5(4):8–36, 2017c.

X. Zhu, J. Hu, C. Qiu, Y. Shi, J. Kang, L. Mou, H. Bagheri, M. Haberle, Y. Hua, R. Huang, L.H. Hughes, H. Li, Y. Sun, G. Zhang, S. Han, M. Schmitt, and Y. Wang. So2Sat LCZ42: A benchmark dataset for global local climate zones classification. *IEEE Geoscience and Remote Sensing Magazine*, in press.

X.X. Zhu, D. Tuia, L. Mou, G.-S. Xia, L. Zhang, F. Xu, and F. Fraundorfer. Deep learning in remote sensing: A comprehensive review and list of resources. *IEEE Geoscience and Remote Sensing Magazine*, 5(4):8–36, Oct 2017d. doi: 10.1109/MGRS.2017.2762307.

X.X. Zhu, R. Huang, L.H. Hughes, H. Li, Y. Sun, G. Zhang, S. Han, M. Schmitt, Y. Wang, J. Hu, et al. So2sat lcz42: A benchmark dataset for global local climate zones classification. *IEEE Geoscience and Remote Sensing Magazine*, in press, 2020. ISSN 2373-7468. doi: 10.1109/mgrs.2020.2964708. URL http://dx.doi.org/10.1109/mgrs.2020.2964708.

Y. Zhu and N. Zabaras. Bayesian deep convolutional encoder–decoder networks for surrogate modeling and uncertainty quantification. *Journal of Computational Physics*, 366:415–447, 2018. ISSN 0021-9991. doi: 10/gdt3jp. *URL* http://www.sciencedirect.com/science/article/pii/S0021999118302341.

Y. Zhu, N. Zabaras, P.-S. Koutsourelakis, and P. Perdikaris. Physics-constrained deep learning for high-dimensional surrogate modeling and uncertainty quantification without labeled data. *Journal of Computational Physics*, 394:56–81, Oct 2019c. ISSN 0021-9991. doi: 10.1016/J.JCP.2019.05.024. URL https://www.sciencedirect.com/science/article/pii/S0021999119303559.

B. Zitova and J. Flusser. Image registration methods: a survey. *Image and Vision Computing*, 21(11), 2003.

Index

Deep Learning for the Earth Sciences: A Comprehensive Approach to Remote Sensing, Climate Science, and Geosciences, First Edition. Edited by Gustau Camps-Valls, Devis Tuia, Xiao Xiang Zhu, and Markus Reichstein.

t

u

v

w

weather forecasting/prediction 7, 10, 210, 216, 222–239, 324

Printed and bound by CPI Group (UK) Ltd, Croydon, CR0 4YY
17/09/2026
14964267-0005